Anatomy and Physiology of Animals

Anatomy and Physiology of Animals

Edited by **Kenneth Hayes**

SYRAWOOD
PUBLISHING HOUSE
New York

Published by Syrawood Publishing House,
750 Third Avenue, 9th Floor,
New York, NY 10017, USA
www.syrawoodpublishinghouse.com

Anatomy and Physiology of Animals
Edited by Kenneth Hayes

International Standard Book Number: 978-1-68286-073-1 (Hardback)

Contents

Permissions

List of Contributors

Preface

I am honored to present to you this unique book which encompasses the most up-to-date data in the field. I was extremely pleased to get this opportunity of editing the work of experts from across the globe. I have also written papers in this field and researched the various aspects revolving around the progress of the discipline. I have tried to unify my knowledge along with that of stalwarts from every corner of the world, to produce a text which not only benefits the readers but also facilitates the growth of the field.

Anatomy and Physiology are complementary fields of study especially for disciplines associated with biology. This book exclusively covers the topics related to anatomy and physiology of animals. It aims to shed light on the multidisciplinary facets of zoology by focusing on the structural, physiological and evolutionary advancements in animals which have been extensively covered in this book. Students, researchers, experts and all associated with zoology, veterinary sciences and related fields will benefit alike from this book.

Finally, I would like to thank all the contributing authors for their valuable time and contributions. This book would not have been possible without their efforts. I would also like to thank my friends and family for their constant support.

Editor

Copulation or sensory cues from the female augment Fos expression in arginine vasopressin neurons of the posterodorsal medial amygdala of male rats

Shantala Arundathi Hari Dass and Ajai Vyas[*]

Abstract

Background: The posterodorsal part of the medial amygdala is essential for processing reproductively salient sensory information in rodents. This is the initial brain structure where information from olfactory system and male hormones intersect. The neurochemical identity of the neurons participating in the sensory processing in medial amygdala remains presently undetermined. Many neurons in this brain structure express arginine vasopressin in a testosterone-dependent manner, suggesting that this neuropeptide is maintained by the androgenic milieu.

Method: Here we use Fos, a protein expressed by recently active neurons, to quantify activation of arginine vasopressin neurons after exposure to odor from physically inaccessible female. We compare it to mating with accessible female and to reproductively innocuous odor.

Results: We show that inaccessible female activate arginine vasopressin neurons in the male posterodorsal medial amygdala. The magnitude of activation is not further enhanced when physical access with resultant mating is granted, even though it remains undetermined if same population of AVP neurons is activated by both inaccessible female and copulation. We also show that arginine vasopressin activation cannot be fully accounted for by mere increase in the number of Fos and AVP neurons.

Conclusion: These observations posit a role for the medial amygdala arginine vasopressin in reproductive behaviors, suggesting that these neurons serve as integrative node between the hormonal status of the animal and the availability of reproductive opportunities.

Keywords: Affiliation, Mating, Neuropeptide, Nonapeptide, Pheromone, Sexual behavior, Social behavior, Testosterone, Vasotocin

Introduction

The medial amygdala (MeA) plays an important role during male reproductive behavior in the rodents [1]. Lesions of the MeA reduce reproductive behavior in hamsters [2], rats [3] and gerbils [4]. In hamsters, the anterior MeA is involved in discrimination of conspecific odor from same-sex versus opposite-sex donors, while the posterodorsal MeA is selectively activated by opposite-sex conspecifics [5]. Fiber-sparing lesions of anterior MeA in this species reduce number of Fos immunoreactive cells in the posterodorsal MeA and efferent forebrain regions [6], suggesting a unidirectional flow of chemosensory information through anterior MeA to its downstream targets. In male rats, MeA lesions drastically reduce penile erections in response to an inaccessible estrous female [7]. Interestingly, such lesions do not affect reflexive erections in response to penile sheath retraction. These observations suggest that MeA involvement in male reproductive behavior is restricted to motivation and not to the downstream initiation of mating in this species.

MeA is a sexually dimorphic structure (reviewed in [8]), characterized by more neurons and larger neuronal soma in males compared to females [9,10]. Important

* Correspondence: avyas@ntu.edu.sg
School of Biological Sciences, Nanyang Technological University, 60 Nanyang Drive, Nanyang 637551, Republic of Singapore

from the perspective of this report, male MeA also contains sunstantial number of extra-hypothalamic parvocellular population of arginine vasopressin (AVP) neurons [11]. Testosterone is required for sexual dimorphism of the MeA neurons [8] and also for AVP expression in the MeA [12,13]. The essential nature of testosterone is further supported by the observations that antagonism of androgen receptors in the MeA inhibits penile erection in male rats in presence of estrus females [14]; an effect reversed by testosterone implants within the MeA of castrates [15,16].

Further evidence suggests that the contribution of the MeA to reproductive behavior is anchored in its ability to integrate pheromonal information with hormonal milieu. Soiled bedding from females increases MeA-Fos in male mice with or without aromatase [17] and in testosterone-primed gonadectomized rats [18,19]. Since Fos is regarded as a proxy for recent neuronal activity [20], this observation suggests that pheromones enhance the activity of MeA neurons. Similarly, female vaginal fluid increases Fos expression in the MeA of mandarin voles [21]. These observations provide correlational support for activation of the MeA in pheromonal processing. Experiments in Syrian hamster further strengthen this. Mating in this species requires intact ability to smell female pheromones and presence of testicular testosterone; absence of either countermands copulation. Interestingly, implantation of testosterone in the MeA of castrates can reinitiate copulatory behavior [22]. Yet, the ability of the testosterone to rescue effects of the castration is dependent on olfactory inputs, such that surgical removal of olfactory bulbs renders testosterone implants ineffective [22,23]. Since both olfactory inputs afferent to MeA and testosterone within MeA are required for the rescue, it suggests that MeA integrates information from sensory environment and internal androgenic milieu.

Despite the role of the MeA in processing of reproductively salient sensory cues, the neurochemical identity of the pertinent cell groups is yet undetermined. As mentioned before, many MeA neurons also express AVP in a testosterone dependent manner. This is important because the AVP mediates several social and sexual behaviors, e.g. monogamy in voles [24] and social recognition of juveniles in male rats [25]. Moreover, AVP neurons are activated during copulation in bed nucleus of stria terminalis [26], a brain region with significant neuro-architectural similarity to the MeA.

In view of androgen-dependent expression of the MeA-AVP and the role of the testosterone in the pheromonal processing, we hypothesized that sensory cues from females selectively activate AVP producing neurons in the MeA. We tested this hypothesis by quantifying co-labeling of AVP and Fos, an immediate early gene product that marks recently activated neurons, post-exposure to rabbit odor or inaccessible estrus female or copulation. Posteroventral and posterodorsal sub-nuclei of MeA were quantified separately in view of their disparate neurochemistry and differential involvement in processing of olfactory signals [27].

Results and discussion

Figure 1 depicts a representative image acquired after histological staining for arginine vasopressin immunoreactive (AVP-ir) and Fos immunoreactive (Fos-ir) neurons in the MePD. Total number of DAPI cells imaged did not significantly differ between the brain regions or the experimental treatments (ANOVA: $p > 0.29$).

Reproductive stimuli suppressed number of Fos-ir neurons in the MePV, but increased it in the MePD

We utilized repeated measure ANOVA to compare number of Fos-ir neurons in MePV and MePD across experimental treatments. Main effect of the experimental treatments reached statistical significance ($F_{(2,14)} = 4.44$; $p = 0.032$). ANOVA revealed significant differences for the main effect of the sub-nuclei (MePD > MePV, 59.2% change in marginal means; $F_{(1,14)} = 35.46$; $p < 0.001$) and the interaction ($F_{(2,14)} = 29.21$; $p < 0.001$).

Specifically in the MePV, animals exposed to an estrus female or copulation contained a significantly reduced number of Fos-ir neurons compared to a reproductively neutral stimulus (Figure 2A, > 65% reduction; post-hoc

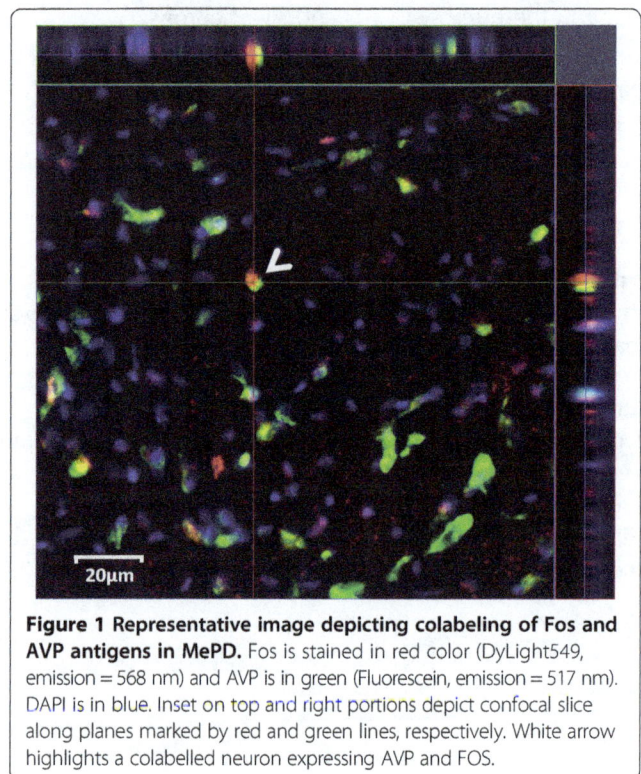

Figure 1 Representative image depicting colabeling of Fos and AVP antigens in MePD. Fos is stained in red color (DyLight549, emission = 568 nm) and AVP is in green (Fluorescein, emission = 517 nm). DAPI is in blue. Inset on top and right portions depict confocal slice along planes marked by red and green lines, respectively. White arrow highlights a colabelled neuron expressing AVP and FOS.

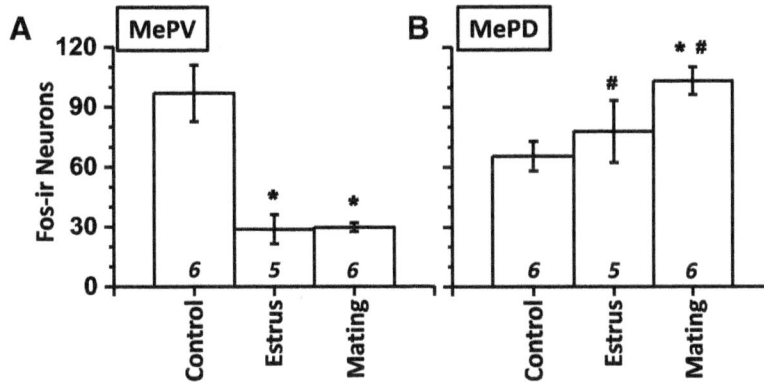

Figure 2 Fos expression in MePV and MePD. Ordinate depicts call counts of Fos-ir neurons () in MePV **(A)** and MePD **(B)**. N is indicated above abscissa (*italics*). *, $p < 0.05$, post-hoc comparison between experimental treatments within a particular sub-nuclei. #, $p < 0.05$, post-hoc paired comparison between sub-nuclei for a particular experimental treatment. Mean ± SEM.

LSD test: $p \leq 0.001$). It is not obvious whether this represents a reduction in Fos-ir due to reproductive salient stimuli or if rabbit urine increased Fos-ir above the unquantified quiescent baseline. No statistical difference was evident between the males exposed to estrus females with or without the opportunity of mating (Figure 2A; $p = 0.181$).

In contrast to the MePV, the number of Fos-ir neurons in the MePD exhibited significant increase after copulation, compared to rabbit odor (Figure 2B, 57% increase; $p < 0.05$). Exposure to an estrus female did not result in a significant increase ($p = 0.259$). The number of Fos-ir neurons was greater in the MePD compared to the MePV when animals were treated with reproductive stimuli (post-hoc paired t-test: $p < 0.05$) and not statistically different during control stimulus ($|t_5| = 2.35, p = 0.066$).

MePD contained greater number of AVP-ir neurons
ANOVA revealed a significant main effect of the brain regions ($F_{(1,14)} = 177.3$; $p < 0.001$). Main effect of the experimental treatment did not reach statistical significance ($F_{(2,14)} = 2.13$; $p < 0.156$). Interaction between treatment and sub-nuclei revealed significant differences ($F_{(2,14)} = 4.58$; $p < 0.030$). Marginal mean of AVP-ir number for MePD was substantially greater than that for MePV (181% more, MePD relative to MePV; $p < 0.001$). Similarly for all experimental treatments, AVP-ir in MePD surpassed observations in MePV (Figure 3; post-hoc paired Student's t-test: $p \leq 0.001$). In view of significant interaction, we further investigated AVP-ir values between experimental groups in MePV and MePD (post-hoc LSD test). Significant inter-group differences were not observed in MePV (Figure 3A; $p = 0.16$). In MePD, no significant differences were observed between control and estrous group (Figure 3B; $p = 0.482$). In contrast, males exposed to estrus females with opportunity

to mate exhibited significantly greater number of AVP-ir compared to control stimuli (27% increase; $p < 0.05$).

Reproductive stimuli increased number of colabeled neurons in MePD, but not in MePV
ANOVA revealed significant main effects of experimental treatments ($F_{(2,14)} = 27.28$; $p < 0.001$) and of sub-nuclei ($F_{(1,14)} = 373.72$; $p < 0.0001$). Interaction between treatments and sub-nuclei was also highly significant ($F_{(2,14)} = 73.10$; $p < 0.001$).

Consistent with lesser number of AVP-ir neurons in the MePV, this sub-nuclei also contained lower number of colabeled neurons (AVP-ir and Fos-ir; marginal mean: MePV = 8.33 ± 1.04, MePD = 38.88 ± 5.56). Across all experimental groups, the MePD contained greater number of colabeled neurons than MePV (Figure 4; post-hoc paired Student's t-test: $p < 0.01$). Within MePV, experimental treatments did not significantly change the number of colabeled neurons (Figure 4A; $p > 0.3$).

In the MePD, exposure to estrus females robustly increased number of colabeled neurons compared to the control stimuli (Figure 4B, 233% increase; post-hoc LSD test: $p < 0.00001$). Similarly, mating with females also increased number of colabeled neurons in the male MePD (300% increase; $p < 0.00001$).

Reproductively salient stimuli specifically activated AVP-ir neurons in the MePD
In the MePD, >45% of Fos-ir neurons activated by either estrus female or mating expressed AVP (Table 1; ≈ 3-fold increase compared to rabbit odor). Similarly, >58% of all imaged MePD AVP-ir neurons also expressed Fos (Table 1) after exposure to female or copulation. Data described above demonstrates an increase in MePD colabeled neurons. To analyze if the number of colabeled neurons were reflective of more Fos-ir and AVP-ir, we compared

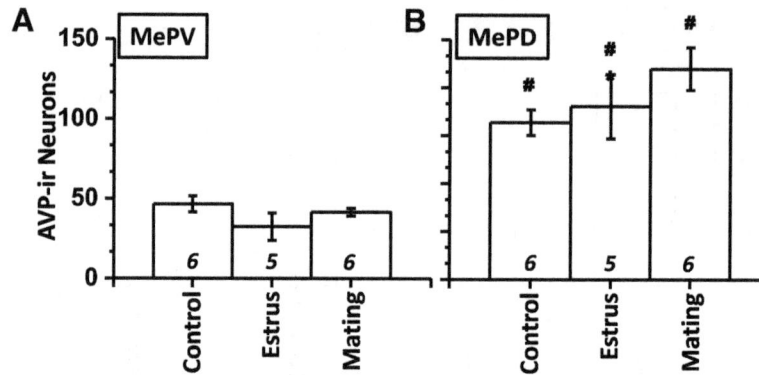

Figure 3 AVP expression in MePV and MePD. Cell counts of AVP-ir neurons () in MePV **(A)** and MePD **(B)**. N is indicated above abscissa (*italics*). *, $p < 0.05$, post-hoc comparison between experimental treatments within a particular sub-nuclei. #, $p < 0.05$, post-hoc paired comparison between sub-nuclei for a particular experimental treatment.

observed and expected (based on mathematical product of AVP-ir and Fos-ir frequencies) values using a repeated measure ANOVA.

In case of the MePV, ANOVA revealed that the interaction between treatment and the observed/expected values were not significantly different ($F_{(2,14)} = 1.94$; $p < 0.181$). On the other hand, observed values for the MePD were substantially divergent from the expectations (main effect: $F_{1,14)} = 348.93$, $p < 0.0001$; 119% difference, expected < observed), suggesting a selective Fos activation of AVP-ir neurons. Amongst experimental groups, animals exposed to reproductively salient stimuli exhibited greater departure of the observed values from the expectations (Figure 5A; chance is depicted by diagonal gray line). We further recapitulated this departure by calculating a divergence scale that was indifferent to the distance of the expected/observed Cartesian points from the origin. For each point in Figure 5A, we calculated divergence by dividing $(x-y)^2$ with $(x + y)^2$, expressed in percentage (Figure 5B). A one-way ANOVA revealed that exposure

to reproductively salient stimuli significantly enhanced divergence between expected and observed values ($F_{(2,14)} = 18.45$, $p < 0.001$; Figure 5B; > 350% increase; $p < 0.001$).

Amongst the various nuclei of the extended amygdala, the MeA is especially important for the appetitive aspects of the reproduction because it is the initial site during pheromonal processing where main and accessory olfactory information intersects [5,28,29]. It robustly expresses androgen receptors and is sexually dimorphic in the structure [30-32]. The androgen responsiveness suggests that the MEA has access to and can plausibly integrate both the internal hormonal milieu and external pheromonal environment. Lesions of this structure ablate penile erections in male rats in response to an inaccessible female [23], while leaving reflexive erections intact [7]. Importance of MeA efferents for mating can be ascertained by the observation that simultaneous unilateral MeA lesion coupled with contralateral medial preoptic lesion ablates mating behavior [33]. The MeA also shares bidirectional innervation with bed nucleus of stria terminalis (BNST).

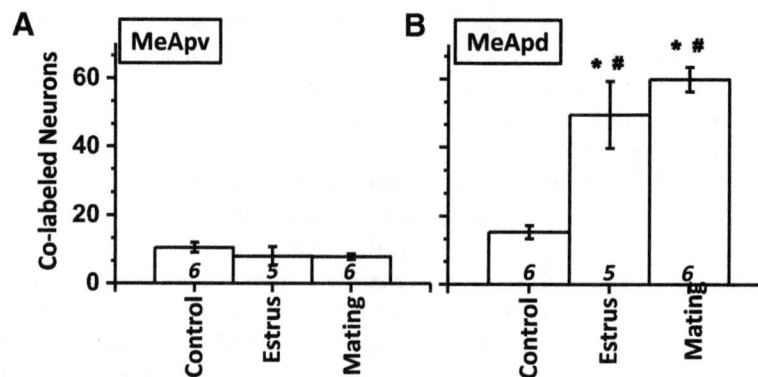

Figure 4 Colabeling in MePV and MePD. Cell counts of colabeled neurons () in MePV **(A)** and MePD **(B)**. *, $p < 0.05$, post-hoc comparison between experimental treatments within a particular sub-nuclei. #, $p < 0.05$, post-hoc paired comparison between sub-nuclei for a particular experimental treatment.

Table 1 Mean ± SeM values pertaining to co-labeled AVP-ir and Fos-ir cells

	Mean ± SeM		
	Rabbit (n = 6)	Estrous (n = 5)	Copulation (n = 6)
Percentage of Fos-ir cells that were also AVP-ir			
MePV	23.6 ± 3.9	23.4 ± 6.3	19.2 ± 2.3
MePD	5.5 ± 1.0	45.6 ± 2.8*#	47.0 ± 3.6*#
Percentage of AVP-ir cells that were also Fos-ir			
MePV	11.8 ± 1.8	28.2 ± 5.7*	27.2 ± 4.1*
MePD	24.3 ± 3.9 #	64.5 ± 4.3*#	58.4 ± 1.6*#

Main effect of intra-animal brain regions (MePV versus MePD): $F_{(1,14)} > 8.2$; $p < 0.001$.
Main effect of inter-animals treatments: $F_{(2,14)} > 3.5$; $p < 0.001$.
Interaction: $F_{(2,14)} > 23.7$; $p < 0.001$.
*$p < 0.05$, post-hoc comparison between experimental treatments within a particular sub-nuclei. #, $p < 0.001$, post-hoc paired comparison between sub-nuclei for a particular experimental treatment.

For example, in male hamsters after exposure to a female odor, greater number of MePD neurons colabeled with Fos and a retrograde tracer injected in the BNST [34]. This demonstrates that female odors increases activity in BNST-projecting neurons of the MePD, compared to an odor from male conspecifics. The greater communication between MePD and the BNST is consistent with the ontological and hodological similarities between these brain regions [35].

Despite the importance of neurons in the MeA in the processing of reproductively salient cues, neurochemical identity of the activated neurons is still undetermined. Data presented in this report shows that the exposure to a physically inaccessible female rat can selectively activate AVP neurons in the MePD, in contrast to the MePV or in contrast to reproductively innocuous rabbit urine. Moreover, the extent of AVP neurons being activated cannot be explained by arithmetic changes in AVP or Fos neurons, thus suggesting that AVP neurons are activated

in a non-random manner. While the present report examined two sub-nuclei of the MeA, it is possible that other MeA sub-nuclei or other parts of the social brain network are also activated during exposure to the reproductively salient stimuli.

We demonstrate a rapid increase in the density of Fos neurons in the MePD after exposure to an inaccessible female or post-copulation. This suggests that Fos in the MePD is dynamically regulated by the presence of sensory cues. Yet, this increase in Fos density is not sufficient to explain the full extent of MePD-AVP activation. This is substantiated by the fact that a greater number of active neurons express AVP and a greater fraction of the AVP neurons become active. Thus, AVP neurons are activated during exposure to sexually salient environmental signals. This is consistent with the prior observations in desert finches expressing vasotocin, a neuropeptide homologous to the AVP. This bird is an opportunistic breeder similar to the rat. In this species co-housing with females enhances activation of the pre-existing vasotocin neurons in the bed nucleus of stria terminalis of the males, compared to the unisexual group of males [36].

The extent of MePD-AVP activation does not seem to be affected by presence or absence of the physical contact or mating itself. In other words, sensory signals of a physically inaccessible female and the actual mating induce equivalent amount of AVP activation. It should be however noted that experiments described in this report cannot differentiate if same or different population of MePD-AVP neurons are activated by inaccessible female and copulation. Localized manipulations in MeA does increase reproductive motivation of the males even when females are inaccessible or even when soiled bedding or vaginal fluids are used instead of the females [30]. Testosterone implants in the MeA rescue the effects of castration on appetitive aspects of the reproductive behavior. It

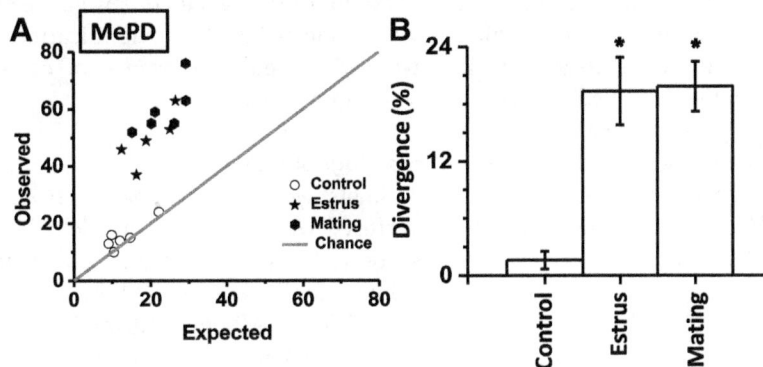

Figure 5 Departure of observed colabeling from theoretical prediction. Expected and observed values of colabeled neurons in the MePD **(A)**. Expected probability (abscissa) was calculated as product of individual probabilities for FOS and AVP neurons. The diagonal gray line from the origin depicts chance level (expected probability equals observed probability). Divergence of observed values from the chance **(B)**. Divergence was calculated for each Cartesian point in panel A by dividing $(x-y)^2$ with $(x+y)^2$. Divergence is expressed as percentage on the ordinate.
*$p < 0.05$, post-hoc comparison between experimental treatments.

is not clear if testosterone in these cases directly activates androgen receptor and/or is aromatized to estradiol and subsequently binds to estrogen receptors. Mice lacking aromatase gene exhibit MeA-Fos activation comparable to the wild-type individuals when exposed to female odors [17], suggesting direct role for the androgen receptors. These observations and data in present report suggest, but do not prove, that MePD is involved in processing of information with no additional role in mating. It should be noted that, contrary to the suggestion of non-additional role of the MeA testosterone in consummatory reproductive behavior, androgen receptor occupancy in MeA does facilitate intromission and ejaculation if simultaneous estrogen receptor binding is available throughout the brain and the periphery [37].

AVP neurons mediate social and sexual behaviors in a variety of species and paradigms. Relevant examples include social recognition in rats [38-40] and development of pair bonding in monogamous voles [24]. Homologous neuropeptides are involved in flocking and territoriality in birds [41]; and in mate recognition and mating in nematode *Caenorhabditis elegans* [42]. AVP neurons in medial bed nucleus of stria terminalis are also activated during copulation in mice [26,43]. While these strands of evidences suggest a pervasive role for the AVP in a variety of social and sexual behaviors, it is currently unknown if the sexual sensory cues themselves activate AVP neurons. In this backdrop, we show that exposure to inaccessible females is able to activate AVP neurons in MePD, the actual act of copulation not being obligatory. This is consistent with the prior observations that testosterone acting within the MeA promotes sexual arousal to the odor of female rats, without any apparent effect on mating behavior itself.

Reproductive investment and testosterone levels in animals are frequently calibrated to incipient metabolic conditions and mating opportunities [44,45]. It is plausible that AVP within the MePD serves as integrative node between reproductive status of the animal (signaled by testosterone mediated expression of AVP) and availability of reproductive opportunities (signaled by activation of AVP neurons by sexual pheromones). Related to this, social isolation after weaning reduces the volume of MeA and blunts the sexual behavior in rats [46]. Similarly, ageing reduces MeA-AVP and AVP innervation to brain regions efferent to the MeA [47].

Conclusion
In conclusion, data presented here show that AVP neurons in MePD are activated during the processing of reproductive cues. It is likely that these neurons play critical role in the mediation of pheromone-directed reproductive behaviors. In addition this report provides support to the role of extra-hypothalamic AVP neurons in reproductive and affiliative behaviors.

Methods
Animals
Wistar rats (47–50 days old) were obtained from the vivarium at National University of Singapore. Animals (housed 2/cage) were maintained on 12 hour light–dark cycle with *ad libitum* food and water (lights on at 0700 hours). The Nanyang Technological University institutional animal care and use committee reviewed and approved all procedures. These procedures are compliant with the NIH guidelines. All the male and female subjects used in this paper were sexually naïve at the start of the experiment.

Exposure to reproductive stimuli
Males were habituated for ten successive days ((ten minutes each day, between 1100 and 1400 hours) in a rectangular arena, in which they would eventually receive the stimuli (46 × 9 cm; 15 cm high). On the day of exposure males were shifted into the procedure room just before the beginning of the light phase (7 AM). After a four hour rest period they were exposed to either a physically inaccessible estrus female behind translucent perforated plastic partition (N = 5) or allowed to mate with an accessible receptive female (N = 6). Females were allowed to explore the entire arena for two hours before the start of the trial, during which time they placed urine marks in the arena. To control for novelty of the odor, a third group of males was exposed to rabbit urine on an inaccessible towel (N = 6). All animals were sacrificed two hour after the onset of stimulus exposure.

Naturally cycling females were used as the stimulus. Estrus phase was determined using examination of vaginal lavage, obtained by gentle flushing of cells from vaginal lining using 20 µl buffered saline (between 1030 and 1100 hours). Unstained lavages were examined on a glass slide using 20× magnification. Females in estrus were identified by presence of cornified cells and absence of nucleated cells.

Histological staining
Animals were deeply anaesthetized and transcardially perfused with 4% paraformaldehyde. Free floating brain sections (40 µm thick) were incubated in a cocktail of primary antibodies for 72 hours at 4°C (guinea pig anti-AVP, 1:500, Bachem; and, rabbit anti-Fos, 1:100, Santa Cruz Biotenchology). This was followed by incubation with secondary antibodies at room temperature for 2 hours (biotinylated anti guinea pig; 1: 200 + anti rabbit-DyLight 549;1:200; obtained from Vector Laboratories). The biotinylated antibody signal was developed using Vectastain elite ABC kit (Vector Laboratories) and tyramide signal

amplification system (Perkin Elmer). Sections were counter stained with DAPI for 1 minute.

Brain sections between Bregma levels −2.76 mm and −3.24 mm (Interaural 7.28 to 7.08) were selected for analysis. Sections were imaged at 40× magnification and 1.2× digital zoom using a confocal microscope (optically sliced at 4 μm, three set of stacks per animal, Carl Zeiss LSM 710). Neurons positive for DAPI, Fos and AVP were counted. Scores were cumulated per animal.

Calculation of observed and expected frequencies

We calculated the expected probability of encountering colabeled neurons by multiplying individual probabilities of AVP-ir and Fos-ir neurons. Individual probabilities for AVP-ir were calculated by division of number of AVP-ir neurons with total number of DAPI positive neurons counted (i.e. probability that a particular DAPI positive neuron will be also be AVP-ir). Individual probabilities for Fos-ir were also counted in the similar manner. A product of these probabilities defines the baseline expectation of colabeling by mere chance and assuming biological independence between Fos and AVP activation. The observed numbers of the colabeled cells were compared to the expected baseline, with null hypothesis of colabeling being a mere mathematical coincidence (adapted from [48]).

Statistics

Repeated measures analysis of variance (ANOVA) was used to quantify statistical significance for main effects and interactions. In case of within-subject comparisons, paired Student's t-test was employed for post-hoc significance testing. In case of between-subject comparisons, LSD test was used. Values reported are mean ± SEM.

Competing interests
The authors declare that they have no competing interests.

Authors' contribution
SAHD performed all the experiments. AV and SAHD designed the experiments. AV and SAHD analysed the data. AV wrote the paper. Both authors read and approved the final manuscript.

Acknowledgements
Funded by Nanyang Technological University and Ministry of Education, Singapore.

References

1. Greco B, Edwards DA, Michael RP, Clancy AN: Androgen receptors and estrogen receptors are colocalized in male rat hypothalamic and limbic neurons that express Fos immunoreactivity induced by mating. Neuroendocrinology 1998, 67:18–28.

2. Lehman MN, Winans SS, Powers JB: Medial nucleus of the amygdala mediates chemosensory control of male hamster sexual behavior. Science 1980, 210:557–60.

3. Kondo Y: Lesions of the medial amygdala produce severe impairment of copulatory behavior in sexually inexperienced male rats. Physiol Behav 1992, 51:939–43.

4. Heeb MM, Yahr P: Cell-body lesions of the posterodorsal preoptic nucleus or posterodorsal medial amygdala, but not the parvicellular subparafascicular thalamus, disrupt mating in male gerbils. Physiol Behav 2000, 68:317–31.

5. Maras PM, Petrulis A: Chemosensory and steroid-responsive regions of the medial amygdala regulate distinct aspects of opposite-sex odor preference in male Syrian hamsters. Eur J Neurosci 2006, 24:3541–52.

6. Maras PM, Petrulis A: The anterior medial amygdala transmits sexual odor information to the posterior medial amygdala and related forebrain nuclei. Eur J Neurosci 2010, 32:469–82.

7. Kondo Y, Sachs BD, Sakuma Y: Importance of the medial amygdala in rat penile erection evoked by remote stimuli from estrous females. Behav Brain Res 1998, 91:215–22.

8. Cooke BM: Steroid-dependent plasticity in the medial amygdala. Neuroscience 2006, 138:997–1005.

9. Mizukami S, Nishizuki M, Arai Y: Sexual difference in nuclear volume and its ontogeny in the rat amygdala. Exp Neurol 1983, 79:569–75.

10. Cooke BM, Tabibnia G, Breedlove SM: A brain sexual dimorphism controlled by adult circulating androgens. Proc Natl Acad Sci U S A 1999, 96:7538–40.

11. Wang Z, De Vries GJ: Androgen and estrogen effects on vasopressin messenger RNA expression in the medial amygdaloid nucleus in male and female rats. J Neuroendocrinol 1995, 7:827–31.

12. DeVries GJ, Buijs RM, Van Leeuwen FW, Caffe AR, Swaab DF: The vasopressinergic innervation of the brain in normal and castrated rats. J Comp Neurol 1985, 233:236–54.

13. Auger CJ, Coss D, Auger AP, Forbes-Lorman RM: Epigenetic control of vasopressin expression is maintained by steroid hormones in the adult male rat brain. Proc Natl Acad Sci 2011, 108:4242–7.

14. Bialy M, Nikolaev-Diak A, Kalata U, Nikolaev E: Blockade of androgen receptor in the medial amygdala inhibits noncontact erections in male rats. Physiol Behav 2011, 103:295–301.

15. Bialy M, Sachs BD: Androgen implants in medial amygdala briefly maintain noncontact erection in castrated male rats. Horm Behav 2002, 42:345–55.

16. Cooke BM, Breedlove SM, Jordan CL: Both estrogen receptors and androgen receptors contribute to testosterone-induced changes in the morphology of the medial amygdala and sexual arousal in male rats. Horm Behav 2003, 43:336–46.

17. Aste N, Honda S, Harada N: Forebrain Fos responses to reproductively related chemosensory cues in aromatase knockout mice. Brain Res Bull 2003, 60:191–200.

18. Bressler SC, Baum MJ: Sex comparison of neuronal Fos immunoreactivity in the rat vomeronasal projection circuit after chemosensory stimulation. Neuroscience 1996, 71:1063–72.

19. Paredes RG, Lopez ME, Baum MJ: Testosterone augments neuronal Fos responses to estrous odors throughout the vomeronasal projection pathway of gonadectomized male and female rats. Horm Behav 1998, 33:48–57.

20. Hoffman GE, Smith MS, Verbalis JG: c-Fos and related immediate early gene products as markers of activity in neuroendocrine systems. Front Neuroendocrinol 1993, 14:173–213.

21. He F, Wu R, Yu P: Study of Fos, androgen receptor and testosterone expression in the sub-regions of medial amygdala, bed nucleus of stria terminalis and medial preoptic area in male mandarin voles in response to chemosensory stimulation. Behav Brain Res 2014, 258:65–74.

22. Wood RI, Coolen LM: Integration of chemosensory and hormonal cues is essential for sexual behaviour in the male Syrian hamster: role of the medial amygdaloid nucleus. Neuroscience 1997, 78:1027–35.

23. Wood RI, Newman SW: Integration of chemosensory and hormonal cues is essential for mating in the male Syrian hamster. J Neurosci 1995, 15:7261–9.

24. Lim MM, Hammock EA, Young LJ: The role of vasopressin in the genetic and neural regulation of monogamy. J Neuroendocrinol 2004, 16:325–32.

25. Bluthe RM, Schoenen J, Dantzer R: Androgen-dependent vasopressinergic neurons are involved in social recognition in rats. Brain Res 1990, 519:150–7.

26. Ho JM, Murray JH, Demas GE, Goodson JL: Vasopressin cell groups exhibit strongly divergent responses to copulation and male-male interactions in mice. Horm Behav 2010, 58:368–77.

27. Choi GB, Dong H-w, Murphy AJ, Valenzuela DM, Yancopoulos GD, Swanson LW, Anderson DJ: **Lhx6 delineates a pathway mediating innate reproductive behaviors from the amygdala to the hypothalamus.** *Neuron* 2005, **46**:647–60.

28. Samuelsen CL, Meredith M: **The vomeronasal organ is required for the male mouse medial amygdala response to chemical-communication signals, as assessed by immediate early gene expression.** *Neuroscience* 2009, **164**:1468–76.

29. Meredith M: **Vomeronasal, olfactory, hormonal convergence in the brain. cooperation or coincidence?** *Ann N Y Acad Sci* 1998, **855**:349–61.

30. Morris JA, Jordan CL, King ZA, Northcutt KV, Breedlove SM: **Sexual dimorphism and steroid responsiveness of the posterodorsal medial amygdala in adult mice.** *Brain Res* 2008, **1190**:115–21.

31. Blake CB, Meredith M: **Change in number and activation of androgen receptor-immunoreactive cells in the medial amygdala in response to chemosensory input.** *Neuroscience* 2011, **190**:228–38.

32. Zhou L, Blaustein JD, De Vries GJ: **Distribution of androgen receptor immunoreactivity in vasopressin- and oxytocin-immunoreactive neurons in the male rat brain.** *Endocrinology* 1994, **134**:2622–7.

33. Kondo Y, Arai Y: **Functional association between the medial amygdala and the medial preoptic area in regulation of mating behavior in the male rat.** *Physiol Behav* 1995, **57**:69–73.

34. Been LE, Petrulis A: **Chemosensory and hormone information are relayed directly between the medial amygdala, posterior bed nucleus of the stria terminalis, and medial preoptic area in male Syrian hamsters.** *Horm Behav* 2011, **59**:536–48.

35. Johnston JB: **Further contributions to the study of the evolution of the forebrain.** *J Comp Neurol* 1923, **35**:337–481.

36. Kabelik D, Morrison JA, Goodson JL: **Cryptic regulation of vasotocin neuronal activity but not anatomy by sex steroids and social stimuli in opportunistic desert finches.** *Brain Behav Evol* 2010, **75**:71–84.

37. Baum MJ, Tobet SA, Starr MS, Bradshaw WG: **Implantation of dihydrotestosterone propionate into the lateral septum or medial amygdala facilitates copulation in castrated male rats given estradiol systemically.** *Horm Behav* 1982, **16**:208–23.

38. Dluzen DE, Muraoka S, Engelmann M, Landgraf R: **The effects of infusion of arginine vasopressin, oxytocin, or their antagonists into the olfactory bulb upon social recognition responses in male rats.** *Peptides* 1998, **19**:999–1005.

39. Tobin VA, Hashimoto H, Wacker DW, Takayanagi Y, Langnaese K, Caquineau C, Noack J, Landgraf R, Onaka T, Leng G, Meddle SL, Engelmann M, Ludwig M: **An intrinsic vasopressin system in the olfactory bulb is involved in social recognition.** *Nature* 2010, **464**:413–7.

40. Wacker DW, Ludwig M: **Vasopressin, oxytocin, and social odor recognition.** *Horm Behav* 2012, **61**:259–65.

41. Goodson JL, Kelly AM, Kingsbury MA: **Evolving nonapeptide mechanisms of gregariousness and social diversity in birds.** *Horm Behav* 2012, **61**:239–50.

42. Garrison JL, Macosko EZ, Bernstein S, Pokala N, Albrecht DR, Bargmann CI: **Oxytocin/vasopressin-related peptides have an ancient role in reproductive behavior.** *Science* 2012, **338**:540–3.

43. Young LJ, Nilsen R, Waymire KG, MacGregor GR, Insel TR: **Increased affiliative response to vasopressin in mice expressing the V1a receptor from a monogamous vole.** *Nature* 1999, **400**:766–8.

44. Raab A, Haedenkamp G: **Impact of social conflict between mice on testosterone binding in the central nervous system.** *Neuroendocrinology* 1981, **32**:272–7.

45. Blanchard DC, Sakai RR, McEwen B, Weiss SM, Blanchard RJ: **Subordination stress: behavioral, brain, and neuroendocrine correlates.** *Behav Brain Res* 1993, **58**:113–21.

46. Cooke BM, Chowanadisai W, Breedlove SM: **Post-weaning social isolation of male rats reduces the volume of the medial amygdala and leads to deficits in adult sexual behavior.** *Behav Brain Res* 2000, **117**:107–13.

47. Van Zwieten EJ, Kos WT, Ravid R, Swaab DF: **Decreased number of vasopressin immunoreactive neurons in the medial amygdala and locus coeruleus of the aged rat.** *Neurobiol Aging* 1993, **14**:245–8.

48. Lin D, Boyle MP, Dollar P, Lee H, Lein ES, Perona P, Anderson DJ: **Functional identification of an aggression locus in the mouse hypothalamus.** *Nature* 2011, **470**:221–6.

2

Development and juvenile anatomy of the nemertodermatid *Meara stichopi* (Bock) Westblad 1949 (Acoelomorpha)

Aina Børve and Andreas Hejnol*

Abstract

Introduction: Nemertodermatida is the sister group of the Acoela, which together form the Acoelomorpha, a taxon that comprises bilaterally symmetric, small aquatic worms. While there are several descriptions of the embryology of acoel species, descriptions of nemertodermatid development are scarce. To be able to reconstruct the ground pattern of the Acoelomorpha it is crucial to gain more information about the development of several nemertodermatid species. Here we describe the development of the nemertodermatid *Meara stichopi* using light and fluorescent microscopic methods.

Results: We have collected *Meara stichopi* during several seasons and reconstruct the complex annual reproductive cycle dependent on the sea cucumber *Parastichopus tremulus*. Using common fluorescent markers for musculature (BODIPY FL-phallacidin) and neurons (antibodies against FMRFamide, serotonin, tyrosinated-tubulin) and live imaging techniques, we followed embryogenesis which takes approximately 9–10 weeks. The cleavage pattern is stereotypic up to the 16-cell stage. Ring- and longitudinal musculature start to develop during week 6, followed by the formation of the basiepidermal nervous system. The juvenile is hatching without mouth opening and has a basiepidermal nerve net with two dorsal neurite bundles and an anterior condensation.

Conclusions: The development of *Meara stichopi* differs from the development of Acoela in that it is less stereotypic and does not follow the typical acoel duet cleavage program. During late development *Meara stichopi* does not show a temporal anterior to posterior gradient during muscle and nervous system formation.

Keywords: Nemertodermatida, Acoelomorpha, Development, Muscle development, Neurogenesis, Cleavage, Cell lineage

Introduction

The clade Nemertodermatida comprises only nine described species of small, completely ciliated, exclusively marine, hermaphroditic worms that live mostly in interstitial habitats [1,2]. Nemertodermatids possess a medio-ventral mouth that is the sole opening to the epithelial, sack-like gut. The nervous system is located basiepidermally, and all nemertodermatid species possess a characteristic double-statocyst or gravitational sensory organ [3]. Nemertodermatida and Acoela (together forming the Acoelomorpha [4]) have recently gained attention because of their disputed phylogenetic

position, which greatly impacts our understanding of the evolution of animal body plans [5,6]. These rather simple worms have been placed as sister group to all remaining Bilateria [7-14] – in some studies as separate branches [15,16] - and thus helpful to understand the evolutionary transition of the cnidarian-bilaterian stem species into the bilaterian stem species [6]. Alternative hypotheses place acoelomorphs either as sister group to all remaining deuterostomes [10] or as sister group to the Ambulacraria (Echinodermata + Hemichordata) [10]. In both latter cases, the lack of some morphological features in acoelomorphs, such as nephridia and gill slits, would be interpreted as independent losses [17]. Nemertodermatids play a key role for determining the direction of character evolution in the Acoelomorpha [18]. Nemertodermatids share plesiomorphic characters such as a basiepidermal nervous

* Correspondence: andreas.hejnol@sars.uib.no
Sars International Centre for Marine Molecular Biology, University of Bergen, Thormøhlensgate 55, 5008 Bergen, Norway

system, monoflagellate sperm, and an epithelial gut [4,18,19] and lack acoel novelties, including a subepidermal brain and parenchymal tissues [18,19]. Nemertodermatids share these characters with members of the Xenoturbellida, a possible sister group of the Acoelomorpha [9,10,13]. A thorough comparison of the morphology and development of xenoturbellids, nemertodermatids and acoels is essential to gain a deeper insight into the ancestral character states of this taxon and the changes during cell type and organ system evolution.

Meara stichopi [20] and *Nemertoderma westbladi* [21] are the two most accessible nemertodermatid species, and both species can be collected relatively easily from the field. Embryos from both species can be obtained for developmental studies (present study and [22]), but detailed descriptions of the embryology are still missing. Here we describe the development of *Meara stichopi* and compare it with previous studies of acoel and nemertodermatid embryos.

Results

The annual reproductive cycle of *Meara stichopi* and presence in the host *Parastichopus tremulus* (Gunnerus, 1767)

Our sampling over four years revealed novel insights into the life cycle of *Meara stichopi* and its seasonal reproduction. As reported in the species description [20], *M. stichopi* is mainly found in the first 3 cm of the foregut of its host, the sea cucumber *Parastichopus tremulus* (Figure 1). We observed that *P. tremulus* collected on coarse sandy bottoms (e.g. Sognefjord, Hardangerfjord) did not contain any *M. stichopi*, possibly because the sand grains prevent *M. stichopi* from attaching to the foregut wall. We observed *M. stichopi* only inside sea cucumbers living on muddy bottoms, often in large numbers (up to 100–200 individuals) (Figure 1D), where they are mainly affiliated with the gut wall and largely absent from the gut content. We have observed that most individuals are oriented with the mouth directed toward the gut content.

We have detected an annual pattern of presence and size variation of *M. stichopi* in the gut of the host. With few exceptions *M. stichopi* was completely absent from the gut of the sea cucumbers between the months of November and February (Figure 2E). In samples from mid March onward, small individuals (150 μm long) are present in the foregut of the sea cucumber, initially in small numbers. The number of individuals in the foregut increased to 150–200 over the course of the following months. From April to October, individuals observed in the foregut are larger in size, measuring up to 5 mm in length (Figure 2A). From August on, we observed different staged oocytes in the gonads of the adults, with the matured oocytes located close to the gut tissue (Figure 2B). Nemertodermatids do not possess gonads that are surrounded by epithelia. The number of individuals slowly decreased from August until November, when *M. stichopi* is no longer observed in the sea cucumber. When searching for *M. stichopi* during the end of October and examining the entire gut of the sea

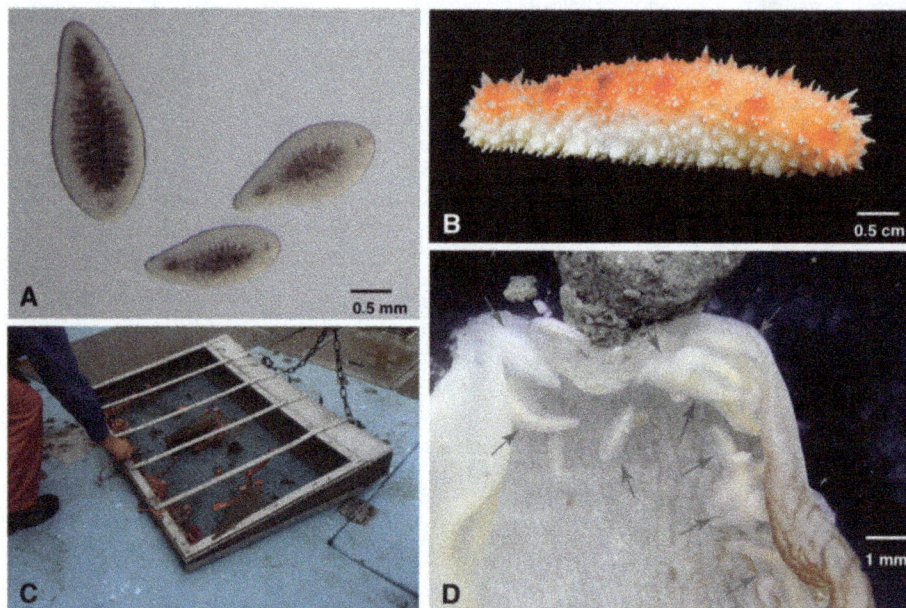

Figure 1 Collection of *Meara stichopi*. A) Three individuals of *Meara stichopi* from a collection in June. Individuals are not gravid and the size range is between 1–2 mm. **B)** Sea cucumber *Parastichopus tremulus*, the host of *M. stichopi* (photo courtesy of Mattias Ormestad, kahikai.org, anterior to the right). **C)** The "Schander sled", after dredging in 250 m depth in the Lysefjorden. Red *P. tremulus* sea cucumbers visible in the mesh. **D)** Opened foregut of *P. tremulus* with adult *M. stichopi* (arrows). Gut content visible on top.

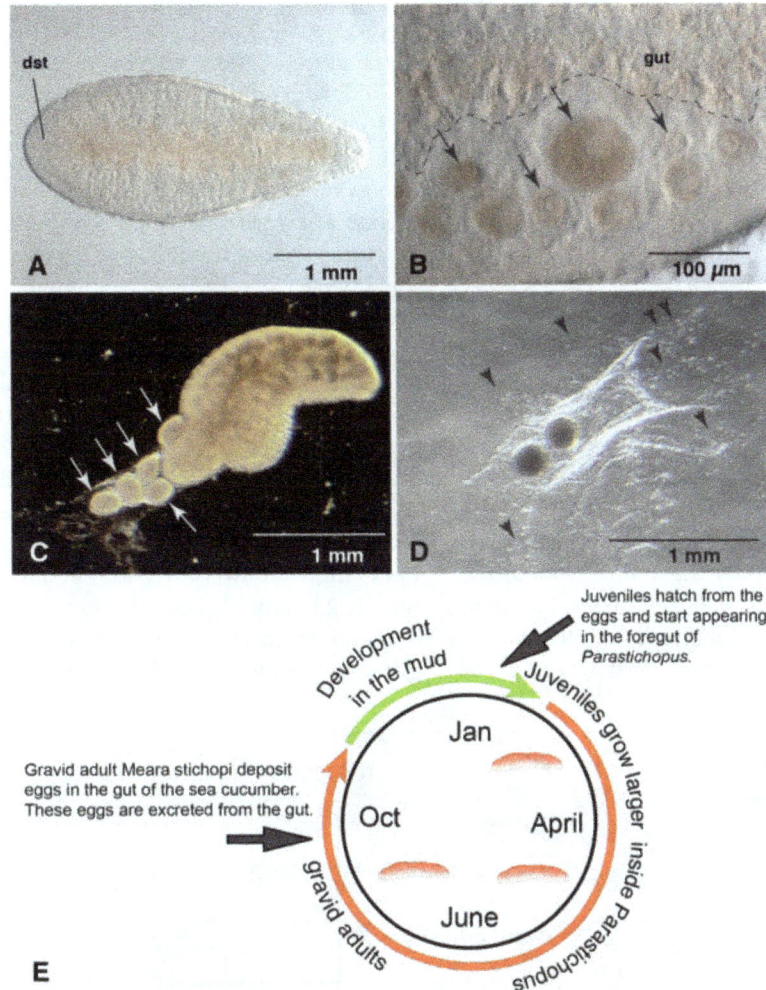

Figure 2 Egg deposition of gravid *M. stichopi* and model of annual life cycle. A) Gravid adult of *Meara stichopi* collected in September. The characteristic double statocyst (dst) at the anterior end is indicated. **B)** Close-up of oocytes in different stages of the adult (black arrows). Note that smaller oocytes are located more distally than the large oocytes. Anterior to the left, the gut is labeled with the dotted line. **C)** Five eggs (white arrows) deposited in a glass bowl by an adult *M. stichopi* oriented with the anterior end to the embryos. **D)** Two eggs in jelly deposited on the bottom of the glass bowl. Small dots surrounding the eggs are motile spermatozoa (arrowheads). **E)** Model for annual cycle of *M. stichopi*. According to the model, fertilized eggs exit the sea cucumber through the gut and develop between 9–12 weeks in the sediment. After hatching, the juveniles initially do not have a mouth opening and survive from the nutrients of the yolk. We presume the juveniles are ingested by the sea cucumbers, where they are able to adhere to the foregut of the sea cucumber, and live as commensals. The juveniles grow to adults over the next months and start to become gravid in August-October. Fertilization occurs in the foregut of the sea cucumber and eggs are deposited, probably exiting the gut of the sea cucumber through the anus. The adults disintegrate after egg deposition and are digested by the sea cucumber. The approximate variation of development covers a period of three months, which also includes the time window when gravid adults are observed to deposit eggs.

cucumber, we found partially digested large individuals in the midgut. We have never found living *M. stichopi* in this gut region nor did we find any embryos anywhere in the digestive tract of the sea cucumber. From these findings, we surmise that *M. stichopi* has an annual life cycle with embryonic and juvenile stages outside of the sea cucumber (see Figure 2E and Discussion).

Reproduction and fertilization

Gravid animals begin to deposit eggs following their transfer to small glass bowls (Figure 2C, D). Since the only body openings are the mouth and male gonopore, immediately following egg deposition, we fixed individuals (n = 10) and labeled them with BODIPY FL-phallacidin to examine possible ruptures of the musculature. However, we could not detect any ruptures in the muscle net and presume that eggs are deposited through the mouth, although we have never observed egg deposition directly. Gravid individuals deposit up to six oval, yolky eggs of ~100 µm length into a mucus-sheath on the bottom of the glass bowl (Figure 2C, D). We often observed motile spermatozoa around the oocytes (Figure 2D). Immediately after being deposited, oocytes are irregularly shaped and lack an eggshell. These eggs become spherical and develop a clear, oval-shaped eggshell, likely a result of fertilization (Figure 3). Against previous assumptions [20], we speculate that fertilization is external due to the observance of sperm near the oocytes, but we have not determined if the sperm originated from the same or different individuals.

Cleavage and gastrulation

The development of *M. stichopi* can be characterized as fairly slow. When cultured at 6-8°C, embryos developed for 9–10 weeks until the hatching of the juvenile. Our observations using light microscopy and 4D-microscopy show that the zygotes extrude two polar bodies after fertilization, with the first cleavage observed three days after egg deposition (Figure 3). The first polar body is observed approximately 24 hours after fertilization (Figure 3). The polar bodies mark the animal pole of the embryo, however they are not visible later in development, making it difficult to orientate embryos in later stages. The first cell division takes place about 24 hours after the

2nd polar body has been given off and is equal and meridional (Figures 3 and 4A). BODIPY FL-phallacidin labels the F-actin of the cell cortex of the blastomeres and propidium iodide stains nucleic acids of the nucleus and cytoplasm, as well as the centrosome (Figure 4). The second cleavage is equatorial and unequal, resulting in two smaller animal micromeres and two vegetal macromeres. The micromeres are not centered on top on the macromeres, but are instead slightly shifted in relation to the animal-vegetal axis of the embryo (Figures 3 and 4B). The interval between first, second and third cleavage is about 24 hours (Figure 3). At the 8-cell stage, the planes of cell division are all equatorial and equal, forming a tier of four blastomeres at the animal pole and four larger blastomeres at the vegetal pole (Figure 4C). The four animal blastomeres are situated directly on top of the vegetal blastomeres and not between the vegetal blastomeres as it is the case in spiralian embryos. The following cell divisions are equal and asynchronous (up to 24 hours apart, see Figure 3) and the cleavage planes vary in their angle between the blastomeres. In general, the cleavage planes are parallel to the surface

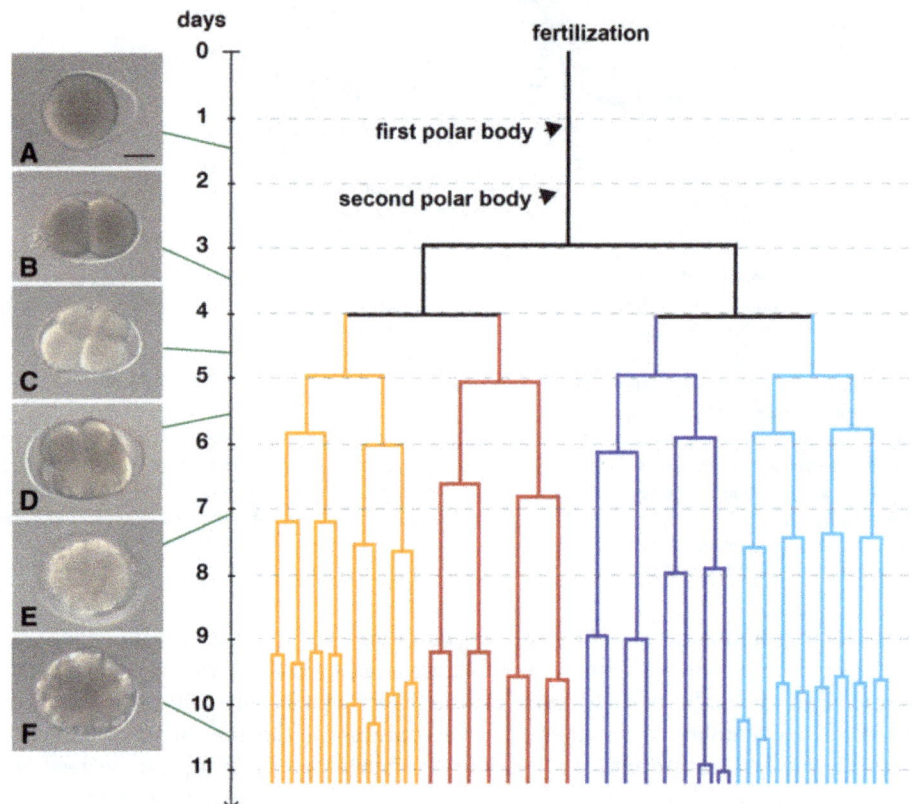

Figure 3 Timing of the cell divisions of an embryo of *M. stichopi* up to the 50-cell stage. Cell divisions of a single embryo recorded with time-lapse microscopy. The lineage of the vegetal blastomeres indicated with dark blue and dark red branches, the animal blastomeres in light blue and orange branches. The duration of the cell cycle increases during the course of development from 24 hours to 3 days. **A-E)** Fertilized egg and cleavage stages imaged with Nomarski optics. **A)** Fertilized egg with egg shell, **B)** 2-cell stage. **C)** 4-cell stage, **D)** 8-cell stage **E)** 16-cell stage, **A-E**, same embryo. **F)** different embryo in a 48-cell stage. Scale bar: 30 μm.

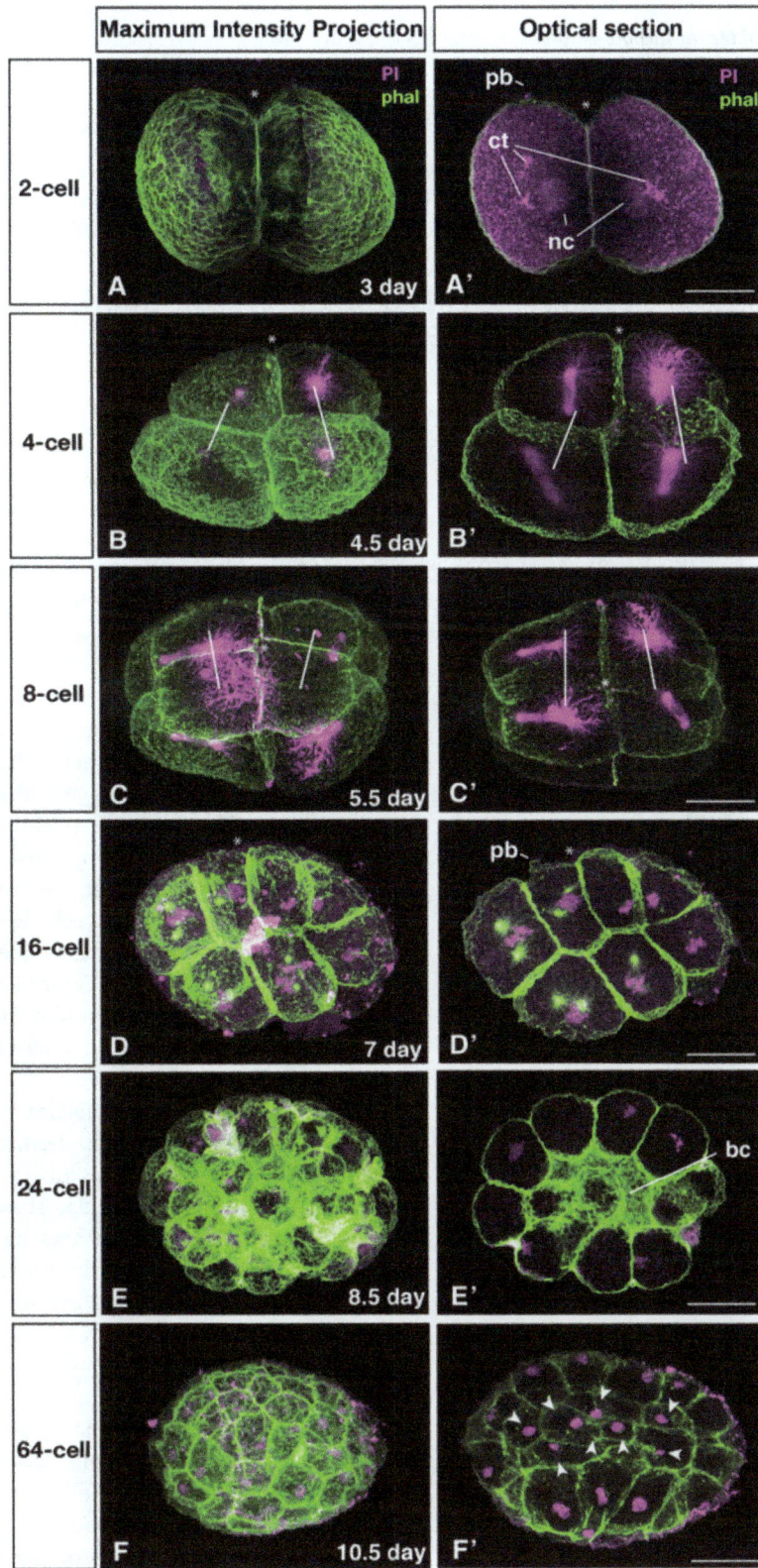

Figure 4 (See legend on next page.)

(See figure on previous page.)
Figure 4 Early cleavage pattern of *Meara stichopi* embryos. Nuclear labeling with Propidium Iodide (magenta), cell cortices and spindle with BODIPY FL-Phallacidin (green) Left row Maximum Intensity Projections, right row optical sections. **A)** 2-cell stage (3 days after fertilization). One of the polar bodies (pb) is visible at the animal pole. **A')** shows an optical section through the same embryo. Propidium iodide is labeling the chromatin in the nucleus (nc) as well as the centrosomes (ct). Both blastomeres are equal in size. **B)** After 4.5 days, the 4-cell stage has large blastomeres at the vegetal pole, and two smaller daughter blastomeres at the animal pole. **B')** shows a section of the embryo in B). The spindles are arranged for the future direction of cell division. **C)** After 5.5 days the 8-cell stage is composed out of four larger cells at the vegetal pole with four blastomeres at the animal pole. **C'** shows an optical section of the embryo of C), with spindles arranged to the future plane of division. **D)** 16-cell stage reached 7 days after fertilization. The size differences between the blastomeres are less prominent and the arrangement is variable. **D')** BODIPY FL-phallacidin labeled cell borders as well as the centrosomes, while the chromatin is labeled by propidium iodide. **E)** 24-cell stage after 8.5 days. **E')** shows a median section of the embryo shown in E). The blastocoel is bordered with the phallacidin labeled cell cortex of the outer blastomeres. **F)** 64-cell stage 10.5 days after fertilization. **F'** shows the cells that have been internalized (blastomeres labeled with arrowhead) during the transition from the 24 to the 64 cell stage. Sister blastomeres are connected by white bars, animal pole is indicated with an asterisk. Scale bar: 30 μm.

of the embryo producing equally sized blastomeres (Figure 4D, E). The durations of the cell cycles vary from less than 24 hours up to 43 hours (Figure 3). The live recording of embryos reveals that the cell cycles of the vegetal blastomeres are longer compared to the cell cycles of the animal blastomeres (Figure 3). At the 24-cell stage, individual cells can only be identified via cell tracing (Figure 4E) but not based on their size or shape. A small blastocoel is visible by the phallacidin labeling of the cell cortices (Figure 4E'). Nine days after fertilization, two or more cells are located internally, indicating the beginning of gastrulation (n = 6). After a round of cell divisions, several more cells are now located inside the embryo (Figure 4F). The internalized cells appear smaller than the outer cells (Figure 4F'). The internalized blastomeres are probably the endo-mesodermal precursor cells.

Further development and morphogenesis
Approximately two weeks after fertilization, the embryo is composed of approximately 180 cells, with an inner cell mass of larger blastomeres that are surrounded by an outer layer of smaller non-epithelial cells (Figure 5A). The internal cells are larger in size than the outer cells. This may indicate that the inner cells undergo fewer divisions than the outer cells (Figure 5A'). After three weeks, the embryo is composed out of approximately 500 cells (Figure 5B). Interestingly, the nuclei are located at the margin of each cell in an irregular pattern, suggesting planar cell polarity is not yet established (Figure 5B'). Four weeks after fertilization, the embryo is composed out of approximately 700 cells (Figure 5C). The optical section through the center of the embryo shows that some of the nuclei of the outer cell layer are located at the apical side of the cells (Figure 5C'). Muscle fibers become visible just below the outer cell layer and reveal the formation of actin bundles of the musculature, indicating the epithelial character of the outer cell layer (Figure 5C'). Five to six weeks after fertilization, more muscle fibers become visible and are arranged in an irregular network that extends along the anterior-posterior axis (Figure 5D, D'). The nuclei of the

outer layer of the embryo are in three different positions: I. In an apical position, indicating the development of the flat, multiciliary epidermal cells (Figure 5D', white arrows); II. In the center of cylindrical cells that form the main epidermal cell layer (Figure 5D', arrowheads); III. At the base of the epidermal layer forming the differentiating neurons of the future nerve net (Figure 5D', red arrows). Additional nuclei are located below the base of the outer cell layer and are affiliated with the muscle fibers. Phallacidin labeled fibers are also visible in the internal region of the embryo, indicating that cross-musculature begins to form (Figure 5D'). Six to seven weeks after fertilization, the network of muscle fibers is more dense, but still irregular (Figure 5E). The anti-tubulin staining indicates that the epidermal cells begin to form cilia (Figure 5E). Nerve fibers are also visible at the base of the epidermis (Figure 5E, insert). In the 7–8 week old embryo, the muscular fibers are arranged in a regular pattern of ring musculature and longitudinal muscle (Figure 5F). The epidermis of the embryo is now clearly organized into the outer cells of the integument, cylindrical epithelial cells, and basiepidermal neurons (Figure 5F'). Between the cylindrical cells, we observe smaller cells with extensions to the nerve net that are likely sensory cells of the epidermis (Figure 5F'). In the juvenile, the sub-epidermal muscular network is now more prominent and forms a muscular sheath surrounding the internal region. The juvenile also has a well-developed basiepidermal neural network, however we could not detect any nerve condensations or indications of the forming digestive system.

Anatomy of the hatchling
After 9–10 weeks of development, the hatchling emerges from the eggshell. The juvenile worm is slightly larger than the length of the eggshell (approximate 100 μm). The characteristic double-statocyst is clearly visible in the hatchling and the major parts of the nervous system are established (Figure 6). We could not detect any epithelia of the digestive system in the juvenile nor is a mouth opening present (Figure 6). The juvenile epidermis is composed of flat, multiciliary integument cells (Figure 6A, Additional

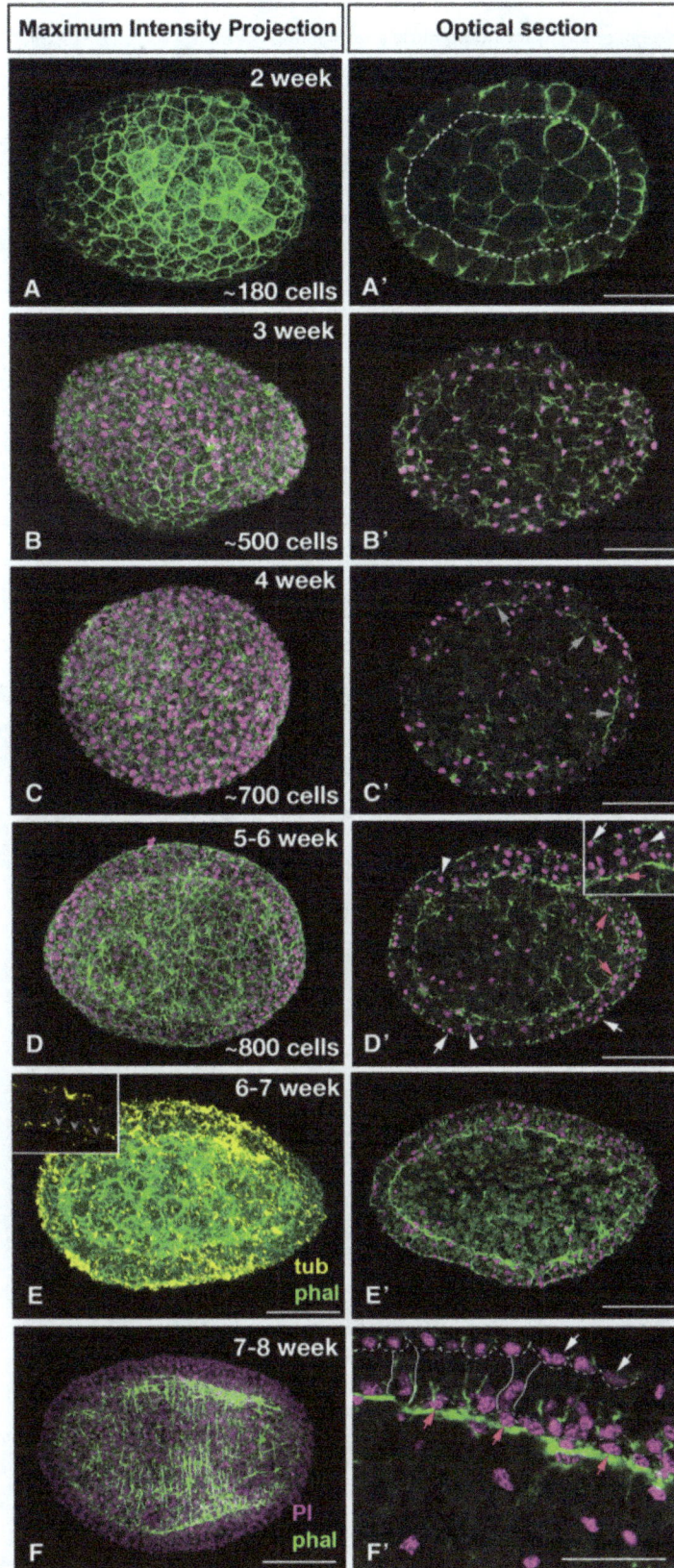

Figure 5 (See legend on next page.)

(See figure on previous page.)
Figure 5 Later development of *M. stichopi* embryos including muscle formation. Nuclear labeling with Propidium Iodide (magenta), muscle fibers with BODIPY FL-Phallacidin (green) and anti-tyrosinated tubulin (yellow). Left row Maximum Intensity Projections, right row optical sections. **A)** Embryo two weeks after fertilization with ~180 cells labeled with BODIPY FL-phallacidin. **A')** shows the inner cell mass (encircled by dotted line) in an optical section of the embryo shown in **A)**. **B)** Embryo with ~500 cells three weeks after fertilization. **B')** shows the nuclei close to the cell membrane of each cell. **C)** 4-week old embryo composed out of approximately 700 cells. **C')** Optical section of **C)**, with actin filaments visible that indicate the beginning of the formation of muscle fibers (arrows). **D)** Dorsal view on 5–6 week old embryo composed out of ~800 cells. The actin fibers of the myocytes are visible in all areas of the embryo. **D')** Optical section of **D)** with subepidermal signal of BODIPY FL-phallacidin visible in multiple areas of the embryo. **E)** The labeling of tyrosinated-tubulin in 6–7 week old embryo shows the cilia in the epidermis of the embryo (yellow), dorsal view. The phallacidin labeling of the musculature has become more prominent but is still irregular. **E')** Optical cross section through another embryo in the same age as **E)** The propidium iodide labeled nuclei and the musculature, dorsal view. **F)** The 7–8 week old embryo shows regularly arranged muscle fibers corresponding to the future pattern of the ring-musculature. **F')** most nuclei are located at the apical pole of the epidermal cells (white arrows). Other nuclei are located also at the base of the epidermis (red arrows), likely the nuclei of the neural precursors of the basiepidermal nerve net. Scale bar 30 μm in all images except F' 10 μm.

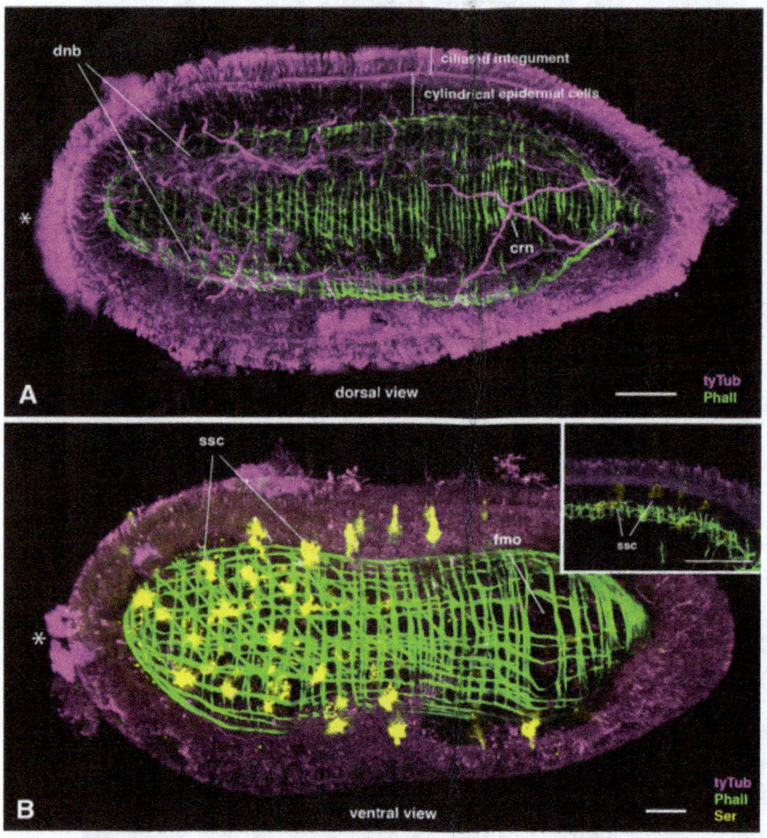

Figure 6 *Meara stichopi* hatchlings, general morphology and serotonergic cells. Optical stacks of different juveniles labeled with antibodies and BODIPY FL-Phallacidin. Anterior is indicated with an asterisk. **A)** Dorsal view of hatchling labeled with anti-tyrosinated tubulin antibody (magenta) and BODIPY-phallacidin (green). The basiepidermal nerve net is located just above the ring and longitudinal musculature of the juvenile. Two bilateral neurite bundles (dnb) are extending from anterior to the posterior along the body with a more anterior concentration of axon tracks. A prominent cross nerve (crn) is visible more posterior. The musculature is forming a spindle-shaped sheath around the body and is composed out of ring musculature and longitudinal muscles. **B)** Ventral view of hatchling of *Meara stichopi* labeled with anti-tyrosinated tubulin antibody (magenta), BODIPY FL-phallacidin (green) and anti-serotonin antibody (yellow). The location of the future mouth is indicated (fmo), but the mouth is not formed yet. The anti-serotonin antibody is labeling cells that are located in the epidermis on the ventral side of the animal. The shape of these cells is indicating a sensory function and a higher concentration of these cells is found anterior. Similar sensory cells are also found on the dorsal side of the hatchling (not shown). The inlet shows a close up of an optical section of the hatchling. The epidermal serotonergic sensory cells (ssc) are directly connected to the muscular system and possess extensions to the outer epidermis. Scale bar 15 μm

file 1A) that cover a thicker layer composed of cylindrical and sensory cells (Figure 6A). At the base of the epidermis, a dense network of axon tracts extends through different regions of the body (Figure 6A). On the dorsal side of the juvenile, multiple axon tracts are bundled into two bilateral condensations that extend from anterior to posterior (Figure 6A). These bundles are broader at the anterior end and are connected by a commissural bundle (Figure 6A). These bundles are anlage of the more prominent basiepidermal dorsal nerve condensations of the adults. At the posterior region, nerves cross in the median of the body (Figure 6A; Figure 7A), a feature observed in both juveniles and adults.

On the ventral side, no such condensations of axon tracts are observed (Figure 6B). Serotonin-positive sensory cells are located in the epidermis, and are connected to the basiepidermal nerve net and possess extensions through the layer of ciliated cells (Figure 6B inlet, Additional file 1H, I). There are more serotonergic cells detected in the anterior ventral region than in the posterior regions and the dorsal side (Figure 6B). The nervous system of the *M. stichopi* juvenile appears to have some specialized neurons, as there is a subset of FMRFamide positive neurons within the basiepidermal anterior bundles and commissure (Figure 7, Additional File 1B-F). Additionally, there are serotonin positive sensory cells, including axon tracts in the anterior region (Additional file 1 H, I). Since the statocyst is located internally, below the muscle sheet, axon tracts connect the cells of the double-statocyst to the basiepidermal nerve net (Additional file 1I). The statocyst is also connected to the muscle sheet (Additional file 1G). It is likely that these muscles help to keep the statocyst in place. In addition to the FMRFamide-positive cells of the dorsal neural bundles, we also detect positive cells that are more ventrally and internally located, whose function remains unknown (Figure 7A-C).

The muscle sheet of the hatchling is regular and composed of ring musculature and longitudinal musculature. No mouth opening is visible in the hatchlings (Figure 8A), which is similar to the juveniles of *Nemertoderma westbladi* [23]. All juveniles collected from the gut of the sea cucumber had a mouth opening (Figure 8B), so this could be either be due to progressing differentiation or an inductive effect by the sea cucumber. The optical cross sections indicate that the muscle fibers connect the dorsal and ventral musculature (Figure 8C). This dorso-ventrally arranged musculature follows a regular pattern along the anterior-posterior axis of the juvenile (Figure 8D).

Discussion

A reconstruction of the life cycle of *Meara stichopi*

Our samplings and observations show that *M. stichopi* has an annual life cycle that is strongly connected to the host sea cucumber (Figure 2E). Our collections allow us

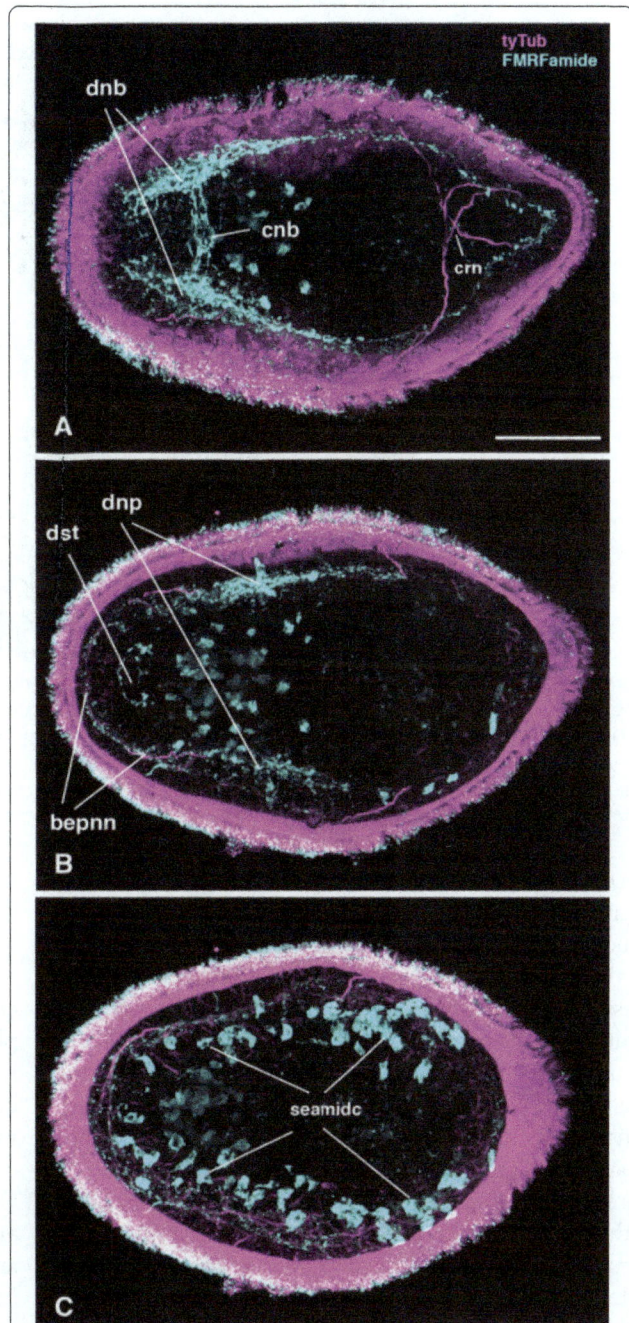

Figure 7 Morphology hatchlings of *Meara stichopi*: FMRFamide signal. Different optical sections through a hatchling of *Meara stichopi* labeled with anti-tyrosinated tubulin (magenta) and anti-FMRFamide (cyan) antibodies, anterior to the left. **A)** Dorsal section shows neurite bundles (dnb). A basiepidermal 'commissural' neurite bundle (cnb) is connecting the two bilateral longitudinal bundles. The longitudinal neurites extend to the posterior end, where the two strands are connected. The dorsal crossing nerves are visible (crn). **B)** More ventral optical section of the confocal stack. The basiepidermal nerve net (bepnn) is visible and FMRFamide-signal is detected internally around the double statocyst. **C)** Ventral optical section of the same hatchling as in **A)** and **B)**. Subepidermal cells that are labeled with the anti-FMRFamide antibody are visible (seamidc). The nature of these cells remains unclear. Scale bar 20 μm.

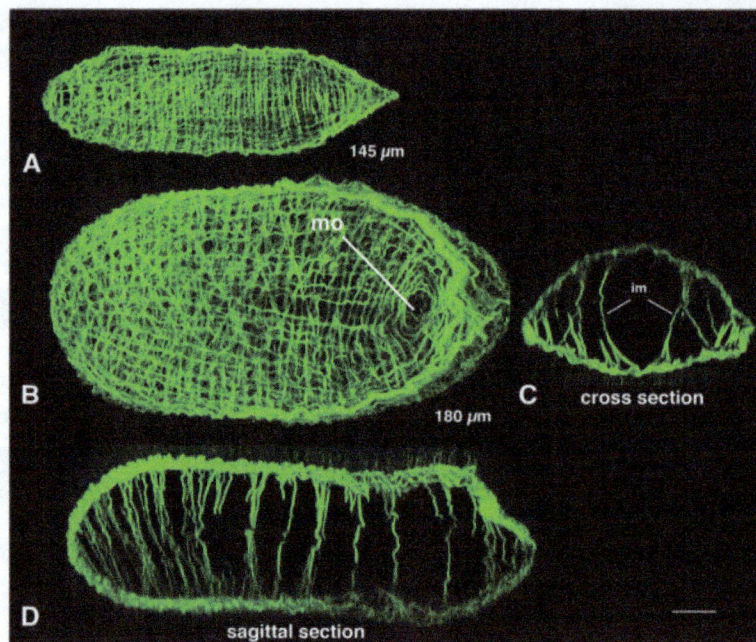

Figure 8 Musculature of hatchlings of *Meara stichopi*. Musculature of two different stages of *Meara stichopi* juveniles, anterior to the left. **A)** Ventral view on a juvenile that hatched in the laboratory. The muscle sheath is surrounding the whole body and no mouth opening is formed yet. **B)** A ventral view on a larger and older juvenile collected from the gut of the sea cucumber with the mouth opening (mo) present. **C)** Optical cross-section through the animal shown in **B)**. Internal muscle strands (im) extend from the dorsal to the ventral side. **D)** Longitudinal optical section through juvenile shown in **B)**. The dorso-ventral internal muscle is arranged along the anterior-posterior axis in a serial fashion. Scale bar 10 µm.

to reconstruct that upon entering the foregut of the sea cucumber, individuals grow inside the host until the reproductive phase in August-October. After depositing the eggs, the adults are then digested by the host. The embryos possess a tough eggshell that probably allows them to exit the gut of the sea cucumber unharmed. Embryogenesis and early postembryonic development takes up to three months and likely happens in the muddy sediment during winter. The hatchlings seem to survive on the remaining yolk until they are taken up by the sea cucumbers in January-March (Figure 2E). We also observed in November and December that the gut of the sea cucumbers is mostly empty of food, and gut parasites, such as the gastropod *Enteroxenos*, which infest the host. Although first described as a 'parasite', Westblad [20] considered *M. stichopi* to be commensal because if there is damage to the host, it is only minimal. Our findings that following the reproductive phase, adults even get digested by the host, suggesting that the impact of *M. stichopi* on the sea cucumber is even less than previously assumed. The possible loss of energy is confined to the homeostasis of the individual worms and to the yolk deposition into the eggs that leave the sea cucumber.

The development and architecture of the nervous system

The nemertodermatid nervous system has previously been investigated using histological [20,21] and immunocytochemical [24,25] methods and is described as entirely basiepidermal. Unlike acoels, nemertodermatids have no portions of the nervous system internalized in a way that they are located below the muscle sheath. The exception is the innervation of the statocyst, which is connected via nerve fibers to the outer basiepidermal plexus. There are no brain-like structures described for nemertodermatids – the anterior condensations are exclusively basiepidermal and ring-shaped (*Nemertoderma westbladi* [24,25]) or just connected by a commissure composed out of neurite bundles (*Meara stichopi* [25]). Our results confirm this structure for *Meara stichopi* and show that the dorsal neurite bundles persist from an anlage in the hatchling to the fully formed structure in the adult. The use of the tyrosinated-tubulin antibody reveals the presence of a larger net of neurons that extend axon tracts also to the internal of the body, while just a subset is stained by the anti-serotonin and anti-FMRFamide antibodies. The dorsal anlage of the two bilateral, longitudinal, thickenings of the nerve plexus are wider than previously described, with a more prominent anterior thickening. Interestingly, such dorsal longitudinal condensations are not found in *Nemertoderma westbladi*, which instead has a pair of ventral and lateral condensations [24]. A previous study by Raikova et al. [25] describes the presence of 'parenchymal fibre bundles' in *M. stichopi*. Our results using anti-tyrosinated, anti-FMRFamide and anti-serotonin antibodies, along with

BODIPY FL-phallacidin, shows that these 'fibre bundles' are basiepidermal, located above the muscle sheet and not internally. Contrary to previous observations [24,25], we have detected positive immunoreactivity around the statocyst using anti-serotonin and anti-FMRFamide antibodies (Additional file 1B-F). Axon tracts connect the statocyst to anterior epithelial cells and to the dorsal basiepidermal nerve condensations. In accordance with previous reports, we could not detect any stomatogastric nervous system in the juvenile of *M. stichopi*. The nervous system of *M. stichopi*, as well as that of other nemertodermatids, is devoid of any prominent internalized structures, such as brains or neurite bundles, which are present in some acoel groups. The nervous system of nemertodermatids is more similar to the nervous system of xenoturbellids, which lacks condensations and only consists of a basiepidermal nerve plexus [26]. Recent phylogenomic analyses [9,10,13] suggest that *Xenoturbella* is closely related to the Acoelomorpha (Xenacoelomorpha). Since Xenoturbella and nemertodermatids both lack subepidermal condensations, this condition has to be considered as plesiomorphic for the whole group and the internalized brain and neurite bundles ('cords') found in some acoel taxa have been secondarily evolved from a basiepidermal nerve net. This interpretation hinges on the phylogenetic position of the Xenacoelomorpha as a whole. In the case of Xenacoelomorpha within the Deuterostomia [10], multiple losses of brain-like and cord-like structures in the Xenacoelomorpha must be considered. However, it is difficult to explain why some lineages display only dorsal condensations (*M. stichopi*), and some lineages only ventral and lateral condensations (*Nemertoderma*) [24], as remnants of an ancestral ventrally condensed nervous system. Further molecular studies are necessary to place the Acoelomorpha in the animal tree of life and to clarify the homology of specific substructures found in this fascinating group of animals.

Comparison of the development with *Nemertoderma westbladi* and acoels

Studies of acoel development describe a characteristic 'duet-cleavage' for all investigated species so far [27-33] (Figure 9I-L). In the 'duet-cleavage' program, the blastomeres of the 2-cell stage give off two smaller micromeres to the animal pole (Figure 9J) The embryo thus has one 'duet' of micromeres at the animal pole and two macromeres at the vegetal pole. This arrangement of blastomeres in the 4-cell stage is similar between the acoel and the nemertodermatid embryos studied so far and can be interpreted as an apomorphy for the Acoelomorpha (Figure 9B, F, J). The following round of divisions differs between the nemertodermatid and the acoel embryo: in the acoel embryo, the vegetal macromeres divide again equatorially and unequally (Figure 9K), while in both nemertodermatid species the macromeres divide meridional and equally (Figure 9C, G). In the acoel embryo, the cleavage plane is shifted in an angle of about 45 degrees to the animal-vegetal axis of the embryo (Figure 9K), while in both nemertodermatid species the cleavage plane of the micromeres is strictly meridional (Figure C, G). The acoel cleavage program differs significantly from our present description of *M. stichopi* and the previous description of *N. westbladi* [22] (Figure 9). The only unequal division observed in both nemertodermatid species is the 2nd cell division (Figure 9B, F), while the last unequal division in the acoel is described in the 3rd division of the two vegetal macromeres (Figure 9L). Acoel embryos possess a more stereotypic arrangement of blastomeres up to the 32-cell stage (see for example Gardiner [30]). In acoels, two vegetal macromeres will gastrulate and form the entire endomesoderm of the embryo. These two cells gastrulate during the transition of the 12-cell to the 24-cell stage [29,30,34]. In *M. stichopi*, gastrulation happens one to two cell cycles later, between the 24-cell and 64-cell stage. The nemertodermatid pattern of 4 vegetal macromeres and 4 animal micromeres is reminiscent of the 8-cell stage of a spiralian embryo, although it is formed in a completely different way.

Although the general pattern of the first divisions of the *M. stichopi* embryo is similar to the cleavage of *N. westbladi* [22], the major differences are the more spherical shape of the *N. westbladi* embryo versus the oval shape of the *M. stichopi* embryo and the considerable size differences between the micromeres (Figure 9A-H). The later development of *M. stichopi* is characterized by an inner cell mass of large, equal-sized blastomeres, which are surrounded by a monolayer of smaller blastomeres. A similar pattern is also present in acoel embryos [6,29,30,33]. The first structure that emerges in acoelomorph embryos is the muscular grid that can be identified by fluorescently labeled phallotoxins [33] (Figure 5). In the acoel *Isodiametra pulchra*, the musculature starts to form at the animal pole (=anterior) of the embryo and progresses to the posterior end of the embryo [33]. In contrast, no such gradient is present in *M. stichopi*, as the musculature appears simultaneously along the entire body axis. Similar to *I. pulchra*, the ring musculature of *M. stichopi* is formed before the longitudinal musculature and both are formed before elements of the nervous system are detectable. The formation of the muscular sheath coincides with the differentiation of the outer epidermis and the formation of the cilia (Figure 5). Since the nervous system of *M. stichopi* is basiepidermal, one should not expect epidermal cells to immigrate internally below the muscle sheet. An exception might be the statocyst sensory complex at the anterior end, but its formation remains unclear. This is different from the nervous system

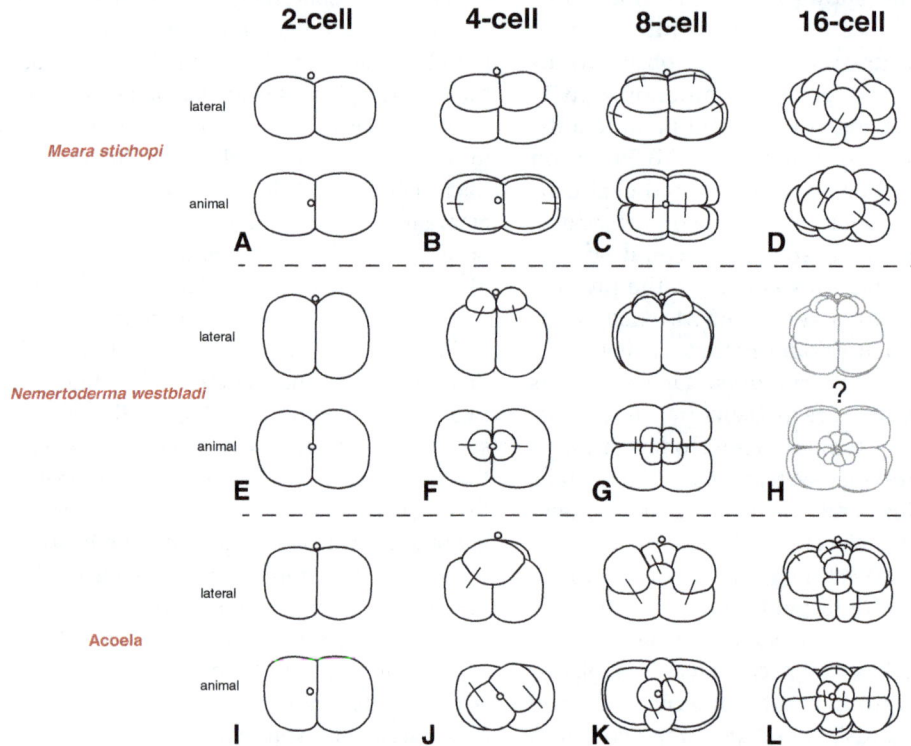

Figure 9 Schematic drawings of the comparisons of early acoelomorph embryos up to the 16-cell stage. Comparison between the development up to the 16-cell stage between the nemertodermatids *M. stichopi* **A)** 2-cell **B)** 4-cell, **C)** 8-cell **D)** 16-cell stage and previously described *N. westbladi* (**E-F**, same arrangement of the stages as for *M. stichopi*) and acoel embryos (**I-L**, same arrangement of the stages as for *M. stichopi*) (see discussion in the text). The 16-cell stage of *N. westbladi* is shaded and labeled with a question mark because it has not been documented in [22] with photographs and own observations could not confirm this blastomere arrangement. Bars connect sister blastomeres in all stages.

development in acoels where the nervous system is formed by all micromeres [32] and cells from the outer sheet migrate to form neural structures [35].

Conclusions

The nemertodermatid *Meara stichopi* has an annual life cycle with a main reproductive period inside the foregut of the holothurian *Parastichopus tremulus*. The development of the embryos undergoes an early stereotypic cleavage, and embryogenesis takes 9–10 weeks until the juvenile hatches. The cleavage program of *M. stichopi* shows significant differences to that of the sister group, Acoela. The musculature is formed before the nervous system, similar to what has been described in acoel embryos. Our study demonstrates the variability of the development in the Acoelomorpha and that further studies are needed to reconstruct the ground pattern of acoelomorph development.

Methods

Collection and maintenance of *Meara stichopi*

Sea cucumbers of the species *Parastichopus tremulus* (Gunnerus, 1767), the host of *Meara stichopi*, were collected throughout the year between Winter 2009/ 2010 – Winter 2013/2014 at collection sites around Bergen, Norway. Between 10 – 50 sea cucumbers were collected from 200 – 350 m depth using the "Schander Sled" (Figure 1) during each collection trip and brought to the lab at the Sars Centre for dissection. Sea cucumbers were opened and the digestive tracts were examined for the presence of *M. stichopi*. Approximately 3000 *M. stichopi* have been collected in total and about 800 individuals deposited a total of about 2500 embryos during the reproductive seasons. Juveniles and adults of *M. stichopi* were transferred to filtered seawater and kept at 5–8 degrees in glass bowls. Seawater was changed every 2–3 days and animals were kept for up to 2–3 months in the lab. Gravid adults deposited oocytes into bottoms of the glass bowls in a jelly-mesh and sperm was surrounding the eggs (Figure 1). Fertilized eggs were cultured in petri dishes containing seawater supplemented with Penicillin (100 Units/ml) and Streptomycin (100 µg/ml) and kept at 5–8 degrees.

Collection spots:

Hauglandsosen (60 24.533 N 5 06.566 E).

Lysefjorden (60 12.347 N 05 17.903 E).

Hjeltefjorden (60 24.366 N 05 06.111 E).

Raunefjorden (60 15.896 N 05 08.448 E).

Laboratory work on the species *M. stichopi* does not raise ethical issues. Therefore approval from a research ethics committee is not required.

Egg shell penetration and fixation of embryos and adults

Egg shells of embryos and pre-hatchlings were penetrated using 1% Thioglycolate/0.05% Pronase (Sigma Aldrichs 5147) in sea water (pH 8.0) for 4 hours at 4°C. During development, the eggshell extends slightly along the long axis and softens, such that late stage embryos are easier to penetrate with a needle than early cleavage stages. Before fixation, embryos, juveniles and adults were relaxed with 7.5% $MgCl_2$ in Millipore water, and fixed using 4% Paraformaldehyde in filtered sea water for 1 hour at 4°C. Fixed specimens were washed four times in PBS containing 0.1% Triton X (PTx) and stored at 4°C before subsequent staining.

Antibody and phallacidin staining

Before the antibody staining, holes were poked into the eggshell of the fixed embryos with insect pins to facilitate the penetration of the antibodies (total n = 500). Antibodies against tyrosinated tubulin (Sigma) and BODIPY®FL labeled phallacidin (Molecular Probes) were used to label the embryos and juveniles following a standard procedure [36]. The phallacidin was used to visualize the F-actin of the cell cortex and the muscle fibers of the embryo, however, it also labeled the centrosomes of some early embryos (e.g. 7 day old, Figure 4D). Anti-serotonin (Sigma) and anti-FMRFamide (Sigma) antibodies were used to label substructures of the nervous system. Specimens were blocked with two 15 min washes in PTx + 0.1% BSA (Bovine Serum Albumin) followed by a 30 min incubation in PTx + 5% normal goat serum. Specimens were incubated with the primary antibody (mouse anti-tyr-tub 1:500, rabbit anti-serotonin 1:200, rabbit anti-FMRFamide 1:200) in PTx + 5% goat serum overnight at 4°C on a shaker. Primary antibody was removed with three 5 min and four 30 min washes in PTx + BSA and an additional blocking step in PTx + normal goat serum for 30 min. Specimens were incubated with the secondary antibody (Cy3 labeled anti-mouse IgM and Cy5 labeled anti-rabbit IgM) diluted 1:200 in PTx + normal goat serum overnight. The secondary antibody was removed with three 5 min and four 30 min washes in PTx + BSA. BODIPY FL phallacidin was added to some samples by first washing the specimens in PBS and incubating in 3–10 Units/ml PTx for 2 hours. Specimens were then washed three times in PBS and prepared for mounting. Propidium Iodide was used to stain the nuclei in some of the samples following a standard protocol in which 0.01 mg/ml propidium iodide was added to the incubation with BODIPY FL-phallacidin.

Confocal microscopy

Specimens were mounted in 'Murray's Clear' (2:1 mixture of benzyl benzoate and benzyl alcohol). Prior to transfer to Murray's Clear, specimens were subjected to a series of isopropanol washes (70%, 85%, 95%, 100%). Specimens were imaged using a Leica SP5 confocal microscope. Image stacks were rendered using Imaris 7.6 (Bitplane).

4D-microscopy

Embryos were recorded using a 4D-microscopy system (modified system after Hejnol & Schnabel [37]). Zygote and 2-cell stages were mounted in seawater, covered with a coverslip and sealed with Vaseline. Recordings (n = 3) were conducted at 10°C and Z-stacks composed out of 50 images were taken every 10 minutes. Cells were traced using the software SIMI°BioCell.

Documentation

Images of juveniles and adults were taken using a Canon 5D Mark III mounted on a Leica 120 M dissecting scope or with a Zeiss AxioCam HRc mounted on a Zeiss Axio Skope.A1.

Additional file

Additional file 1: Details of *M. stichopi* hatchling. A) Epidermis cells of the integument of a *M. stichopi* juvenile labeled with anti-tyrosinated-tubulin. B) Optical section showing the FMRFamidergic innervation (arrowheads) of the statocyst. C) anti-FMRFamide (cyan) and anti-tyrosinated tubulin labeling (magenta) of a anterior part of a hatchling. The signal of the anti-FMRFamide antibody shows neurons of the anterior, dorsal basiepidermal nerve condensations. The bilateral neurite bundles are connected with a commissural neurite bundle (cnb) D) Sagittal optical section through a hatchling labeled with anti-tyrosinated tubulin (magenta) and anti-FMRFamide (cyan). Subepidermal cells (seamidc) of unknown function are labeled with the anti-FMRFamide antibody. E) Optical cross section through same juvenile as B, C and D showing FMRFamidergic neurons (arrowheads) connecting the statocyst with the basiepidermal dorsal nerve condensations (dnc). F) Optical cross section through juvenile showing ventral FMRFamide positive cells (magenta, arrows) that are below the epidermis (red label, anti-tyrosinated tubulin). G) Optical section of the anterior part of a hatchling showing the position of the subepidermal statocyst (dst). Four statocyst muscles (stm) are connected to the cells around the statocyst. H) Optical section showing the innervation of statocyst (dst) by serotonergic neurons (yellow). I) Hatchling labeled with anti-tyrosinated tubulin (magenta), BODIPY-FL phallacidin (green) and anti-serotonin (yellow). Optical section showing the innervation of statocyst (dst) by serotonergic neurons (yellow). The statocyst is internal from the muscle sheath (BODIPY FL-phallacidin, green), Scale bar 15 μm, anterior is indicated with an asterisk.

Competing interest
The authors declare that they have no competing interests.

Authors' contributions
AB carried out the confocal studies and edited the manuscript. AH designed the study and conducted the 4D-microscopic analysis, 3D-reconstruction of the confocal data and analysis of the data and wrote the manuscript. AB and AH collected the animals, cultured the embryos and conducted the labeling and documentation. Both authors read and approved the final manuscript.

Acknowledgements
We thank the crew of the "Hans Brattström" and Henrik Glenner, Christiane Todt, Glenn Bistrow, Kenneth Meland, Christopher Noever for the continuous help and supply of *Parastichopus tremulus*. Jonas Bengtsen and Sabrina Schiemann have been helpful with the 4D-microscopy. We thank all S9 team members for helping with the dissection of the sea cucumbers. Kevin Pang edited and improved the manuscript. The study received support by a Marie Curie International Re-Integration grant to AH (FP7-PEOPLE-2009-RG 256450).

References

1. Lundin K, Sterrer W: **The Nemertodermatida.** In *Interrelationships of the Platyhelminthes*. Edited by Littlewood DTJ, Bray RA. London: Taylor & Francis Ltd; 2001:24–27.
2. Sterrer W: **New and known Nemertodermatida (Platyhelminthes-Acoelomorpha) - A Revision -.** *Belg J Zool* 1998, **128**:55–92.
3. Ehlers U: **Comparative morphology of statocysts in the Plathelminthes and the Xenoturbellida.** *Hydrobiologia* 1991, **227**:263–271.
4. Ehlers U: *Das phylogenetische System der Plathelminthes.* Stuttgart: Gustav Fischer Verlag; 1985.
5. Baguñá J, Riutort M: **The dawn of bilaterian animals: the case of acoelomorph flatworms.** *Bioessays* 2004, **26**:1046–1057.
6. Hejnol A, Martindale MQ: **Acoel development supports a simple planula-like urbilaterian.** *Phil Trans Royal Soc Series B* 2008, **363**:1493–1501.
7. Carranza S, Baguñá J, Riutort M: **Are the Platyhelminthes a monophyletic primitive group? An assessment using 18S rDNA sequences.** *Mol Biol Evol* 1997, **14**:485–497.
8. Egger B, Steinke D, Tarui H, De Mulder K, Arendt D, Borgonie G, Funayama N, Gschwentner R, Hartenstein V, Hobmayer B, Hooge M, Hrouda M, Ishida S, Kobayashi C, Kuales G, Nishimura O, Pfister D, Rieger R, Salvenmoser W, Smith J, Technau U, Tyler S, Agata K, Salzburger W, Ladurner P: **To be or not to be a flatworm: the acoel controversy.** *PLoS One* 2009, **4**:e5502.
9. Hejnol A, Obst M, Stamatakis A, Ott M, Rouse GW, Edgecombe GD, Martinez P, Baguñá J, Bailly X, Jondelius U, Wiens M, Müller WEG, Seaver E, Wheeler WC, Martindale MQ, Giribet G, Dunn CW: **Assessing the root of bilaterian animals with scalable phylogenomic methods.** *Proc Royal Soc Series B* 2009, **276**:4261–4270.
10. Philippe H, Brinkmann H, Copley RR, Moroz LL, Nakano H, Poustka AJ, Wallberg A, Peterson KJ, Telford MJ: **Acoelomorph flatworms are deuterostomes related to *Xenoturbella*.** *Nature* 2011, **470**:255–258.
11. Ruiz-Trillo I, Paps J, Loukota M, Ribera C, Jondelius U, Baguñá J, Riutort M: **A phylogenetic analysis of myosin heavy chain type II sequences corroborates that Acoela and Nemertodermatida are basal bilaterians.** *Proc Nat Acad Sci USA* 2002, **99**:11246–11251.
12. Ruiz-Trillo I, Riutort M, Littlewood DT, Herniou EA, Baguna J: **Acoel flatworms: earliest extant bilaterian Metazoans, not members of Platyhelminthes.** *Science* 1999, **283**:1919–1923.
13. Srivastava M, Mazza-Curll KL, van Wolfswinkel JC, Reddien PW: **Whole-Body Acoel Regeneration Is Controlled by Wnt and Bmp-Admp Signaling.** *Curr Biol* 2014, **24**:1107–1113.
14. Telford MJ, Lockyer AE, Cartwright-Finch C, Littlewood DTJ: **Combined large and small subunit ribosomal RNA phylogenies support a basal position of the acoelomorph flatworms.** *Proc Royal Soc Series B* 2003, **270**:1077–1083.
15. Paps J, Baguña J, Riutort M: **Bilaterian phylogeny: a broad sampling of 13 nuclear genes provides a new Lophotrochozoa phylogeny and supports a paraphyletic basal Acoelomorpha.** *Mol Biol Evol* 2009, **26**:2397–2406.
16. Wallberg A, Curini-Galletti M, Ahmadzadeh A, Jondelius U: **Dismissal of Acoelomorpha: Acoela and Nemertodermatida are separate early bilaterian clades.** *Zool Scr* 2007, **36**:509–523.
17. Edgecombe GD, Giribet G, Dunn CW, Hejnol A, Kristensen RM, Neves RC, Rouse GW, Worsaae K, Sørensen MV: **Higher-level metazoan relationships: recent progress and remaining questions.** *Org Divers Evol* 2011, **11**:151–172.
18. Smith J, Tyler S: **The acoel turbellarians: kingpins of metazoan evolution or a specialized offshoot?** In *The origins and relationships of lower invertebrates*. Edited by Conway Morris S, George JD, Gibson R, Platt HM. Oxford: Calderon Press; 1985:123–142.
19. Rieger R, Tyler S, Smith JPS, Rieger GE: **Platyhelminthes: Turbellaria.** In *Microscopic anatomy of invertebrates. Volume 3*. Edited by Harrison FW, Bogitsch BJ. New York: John Wiley & Sons; 1991:7–140.
20. Westblad E: **On *Meara stichopi* (Bock) Westblad, a new representative of Turbellaria archoophora.** *Arkiv Zoologi* 1949, **1**:43–57.
21. Westblad E: **Die Turbellarien-Gattung Nemertoderma Steinböck.** *Acta Soc pro Fauna et Flora Fenn* 1937, **60**:45–89.
22. Jondelius U, Larsson K, Raikova OI: **Cleavage in *Nemertoderma westbladi* (Nemertodermatida) and its phylogenetic significance.** *Zoomorphology* 2004, **123**:221–225.
23. Meyer-Wachsmuth I, Raikova OI, Jondelius U: **The muscular system of *Nemertoderma westbladi* and *Meara stichopi* (Nemertoderma, Acoelomorpha).** *Zoomorphology* 2013, **132**:239–252.
24. Raikova OI, Reuter M, Gustafsson MK, Maule AG, Halton DW, Jondelius U: **Basiepidermal nervous system in *Nemertoderma westbladi* (Nemertodermatida): GYIRFamide immunoreactivity.** *Zoology* 2004, **107**:75–86.
25. Raikova OI, Reuter M, Jondelius U, Gustafsson MKS: **The brain of the Nemertodermatida (Platyhelminthes) as revealed by anti-5HT and anti-FMRFamide immunostainings.** *Tissue Cell* 2000, **32**:358–365.
26. Raikova OI, Reuter M, Jondelius U, Gustafsson MKS: **An immunocytochemical and ultrastructural study of the nervous and muscular systems of Xenoturbella westbladi (Bilateria inc. sed.).** *Zoomorphology* 2000, **120**:107–118.
27. Apelt G: **Fortpflanzungsbiologie, Entwicklungszyklen und vergleichende Frühentwicklung acoeler Turbellarien.** *Marine Biol* 1969, **4**:267–325.
28. Boyer BC: **Regulative development in a spiralian embryo as shown by cell deletion experiments on the Acoel, *Childia*.** *J Exp Zool* 1971, **176**:97–105.
29. Bresslau E: **Die Entwicklung der Acoelen.** *Verh Deutsch Zoologisch Gesell* 1909, **19**:314–323.
30. Gardiner EG: **Early development of *Polychoerus caudatus*, Mark.** *J Morph* 1895, **11**:155–176.
31. Georgévitch J: **Etude sur le développement de la *Convoluta roscoffensis* Graff.** *Arch Zool Expérim* 1899, **3**:343–361.
32. Henry JQ, Martindale MQ, Boyer BC: **The unique developmental program of the acoel flatworm, *Neochildia fusca*.** *Dev Biol* 2000, **220**:285–295.
33. Ladurner P, Rieger R: **Embryonic muscle development of *Convoluta pulchra* (Turbellaria-acoelomorpha, platyhelminthes).** *Dev Biol* 2000, **222**:359–375.
34. Hejnol A, Martindale MQ: **Acoel development indicates the independent evolution of the bilaterian mouth and anus.** *Nature* 2008, **456**:382–386.
35. Hejnol A, Martindale MQ: **Coordinated spatial and temporal expression of Hox genes during embryogenesis in the acoel *Convolutriloba longifissura*.** *BMC Biol* 2009, **7**:65.
36. *Some simple methods and tips for embryology.* [http://celldynamics.org/celldynamics/downloads/methods/methodsAndTips.doc]
37. Hejnol A, Schnabel R: **What a couple of dimensions can do for you: Comparative developmental studies using 4D-microscopy - examples from tardigrade development.** *Integ Comp Biol* 2006, **46**:151–161.

3

Branchial NH$_4^+$-dependent acid–base transport mechanisms and energy metabolism of squid (*Sepioteuthis lessoniana*) affected by seawater acidification

Marian Y Hu[1], Ying-Jey Guh[1], Meike Stumpp[1], Jay-Ron Lee[1], Ruo-Dong Chen[1], Po-Hsuan Sung[2], Yu-Chi Chen[2], Pung-Pung Hwang[1] and Yung-Che Tseng[2*]

Abstract

Background: Cephalopods have evolved strong acid–base regulatory abilities to cope with CO$_2$ induced pH fluctuations in their extracellular compartments to protect gas transport via highly pH sensitive hemocyanins. To date, the mechanistic basis of branchial acid–base regulation in cephalopods is still poorly understood, and associated energetic limitations may represent a critical factor in high power squids during prolonged exposure to seawater acidification.

Results: The present work used adult squid *Sepioteuthis lessoniana* to investigate the effects of short-term (few hours) to medium-term (up to 168 h) seawater acidification on pelagic squids. Routine metabolic rates, NH$_4^+$ excretion, extracellular acid–base balance were monitored during exposure to control (pH 8.1) and acidified conditions of pH 7.7 and 7.3 along a period of 168 h. Metabolic rates were significantly depressed by 40% after exposure to pH 7.3 conditions for 168 h. Animals fully restored extracellular pH accompanied by an increase in blood HCO$_3^-$ levels within 20 hours. This compensation reaction was accompanied by increased transcript abundance of branchial acid–base transporters including V-type H$^+$-ATPase (VHA), Rhesus protein (RhP), Na$^+$/HCO$_3^-$ cotransporter (NBC) and cytosolic carbonic anhydrase (CAc). Immunocytochemistry demonstrated the sub-cellular localization of Na$^+$/K$^+$-ATPase (NKA), VHA in basolateral and Na$^+$/H$^+$-exchanger 3 (NHE3) and RhP in apical membranes of the ion-transporting branchial epithelium. Branchial VHA and RhP responded with increased mRNA and protein levels in response to acidified conditions indicating the importance of active NH$_4^+$ transport to mediate acid–base balance in cephalopods.

Conclusion: The present work demonstrated that cephalopods have a well developed branchial acid–base regulatory machinery. However, pelagic squids that evolved a lifestyle at the edge of energetic limits are probably more sensitive to prolonged exposure to acidified conditions compared to their more sluggish relatives including cuttlefish and octopods.

Keywords: Acid–base regulation, Invertebrate, Metabolism, Ocean acidification, Rh proteins

* Correspondence: yct@ntnu.edu.tw
[2]Department of Life Science, National Taiwan Normal University, Taipei City, Taiwan
Full list of author information is available at the end of the article

Introduction

CO_2 induced acid–base disturbances are a unifying physiological phenomenon that all animals are confronted with. Depending on the degree of metabolic activity an organism generates metabolic CO_2 leading to intra- and extra-cellular pCO_2 fluctuations causing acid–base disturbances after the hydration of CO_2 in body fluids. Cephalopods have probably evolved the highest physiological complexity among all invertebrate taxa. It is believed that convergent evolutionary features including sensory and locomotory abilities derived from the competition with fish for similar resources in the marine environment. As a trade off to their less efficient swimming mode and active lifestyle they have the highest metabolic rates among marine animals [1]. As a consequence cephalopods are confronted with strong metabolic CO_2 induced temporal acid–base disturbances during jetting and fast swimming [2]. To accommodate strong temporal CO_2 induced acid–base challenges cephalopods have evolved moderate to strong acid–base regulatory abilities to stabilize blood pH during exercise and hypercapnic exposure by accumulating up to 7 mM HCO_3^- [2,3]. It is believed that well developed extracellular pH (pHe) regulatory abilities are an essential feature for cephalopods to protect their highly pH sensitive extracellular O_2 transporting hemocyanins [4,5]. Oxygen affinity of the extracellular hemocyanin is inversely related to both, pH and pCO_2 which is expressed by the Bohr effect [6]. The high pH sensitivity of squid hemocyanin is an important feature to efficiently load oxygen at the gill and unload in tissues. On one hand, high Bohr coefficients of cephalopods were proposed to be a critical physiological characteristic that would make cephalopods particularly sensitive to acid–base disturbances [7]. On the other hand well developed acid–base regulatory abilities were hypothesized to represent a unifying feature that makes ectothermic marine animals robust to seawater acidification as projected for the coming century [8]. Studies using adult cuttlefish could in fact demonstrate that these animals can tolerate exposure to pH 7.1 over several weeks without compromising growth rates and even increase calcification rates of the internal cuttlebone [9]. Moreover the same study demonstrated that metabolic rates of cuttlefish exposed to decreased seawater pH of 7.1 (0.6 kPa CO_2) remained unchanged along the experimental period of 24 h. In contrast the pelagic squid *Dosidicus gigas* responded with depressed metabolic rates (31%) and activity levels (45%) in response to acute exposure to 0.1 kPa CO_2 [1]. However, medium- to long-term (several days) acidification experiments using squids with a pelagic lifestyle are rare as these animals are extremely difficult to keep under laboratory conditions. Such experiments require large-scale experimental facilities with access to natural, high quality

seawater. Recent research conducted with squid and cuttlefish embryonic stages which are easier to handle under laboratory conditions demonstrated that acidified conditions evoke a developmental delay associated with an increase in proton secretion activity [10]. The upregulation of acid–base regulatory genes including Na^+/H^+ exchanger (NHE3), V-type H^+-ATPase (VHA), Rhesus protein (RhP) and Na^+/HCO_3^- cotransporter (NBC) expressed in epidermal ionocytes suggest that these transporters are key players of acid–base regulation in cephalopod early life stages. The current model for epidermal ionocytes of squid embryos denotes the presence of NHE3 in apical membranes whereas Na^+/K^+-ATPase (NKA) and VHA are localized in basolateral membranes. Proton secretion by epidermal ionocytes is sensitive to ethyl-isopropyl amiloride (EIPA) suggesting a central role of NHE proteins in proton secretion pathways. Less information is available for the branchial acid–base regulatory machinery of adult cephalopods. Earlier studies identified and localized acid–base transporters including NKA, VHA, NBC and carbonic anhydrase (CA) in specialized ion-transporting cells of the cephalopod gill [11-13]. In contrast to fish and crustacean gills, the gill of decapod cephalopods is a highly folded epithelium consisting of two epithelial layers that line a blood sinus (Figure 1). The two epithelial layers are linked by pilaster cells, that interdigitate deeply with muscle cells differentiated on the basal lamina of the inner and the outer epithelium. The thin outer epithelium

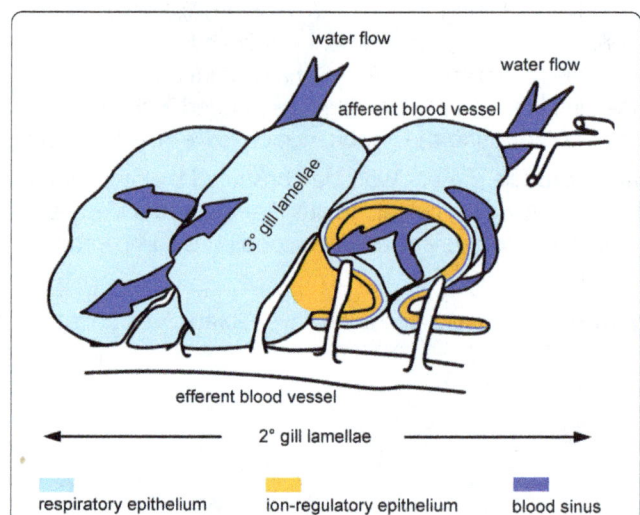

Figure 1 Graphical illustration of the squid gill morphology.
The 3° gill lamellae arise from the 2 order gill lamellae and create a highly folded epithelium with semi-tubular spaces. This epithelium is well perfused by blood vessels, and divided into an outer and inner epithelium lining a blood sinus re-drawn from [54]. The outer epithelium (turquois) is thin and mainly involved in gas exchange whereas the inner epithelium (orange) located in the concave areas of the third order lamellae is involved in ion regulatory processes indicated by high mitochondrial density and expression of ion transporters [11,12].

is mainly involved in respiratory processes, whereas the inner, mitochondria rich epithelium is responsible for ion regulation and, probably, nitrogen excretion [11,12,14].

In teleost fish acid–base regulating epithelia and organs have been extensively studied and the sub-cellular organization of ion transporters localized in mitochondria-rich cells is well described e.g. [15-17]. Besides primary active proton extrusion mechanisms via VHA these models suggest an import of HCO_3^- and export of protons by secondary active transporters such as NHEs and Na^+-dependent HCO_3^- transporters of the SLC4 solute transporter family energized by the NKA located in basolateral membranes. Although a large body of knowledge is available for teleosts, a comparatively small number of studies investigated acid–base and ion regulatory mechanisms in non-model invertebrates like mollusks, echinoderms and crustaceans. For the latter a number of studies exist with osmotic and acid–base regulatory mechanisms recently summarized by Henry et al. [18], demonstrating the complexity of ion-transport processes in gill epithelia of crustaceans. Although the picture seems more complete for osmoregulatory mechanisms in crustacean gill epithelia the mechanistic basis for acid–base regulation seems largely unexplored as well. Crustaceans were also characterized as strong acid–base regulators that are capable of accumulating high concentrations of HCO_3^- in their body fluids to buffer an excess of protons. It has been pharmacologically demonstrated that low pH conditions trigger the response of carbonic anhydrase (CA), apical VHA and basolateral Na^+/HCO_3^- exchanger in perfused gills of the euryhaline crab *Neohelice* (*Chasmagnathus*) *granulata* [19]. Furthermore, NHE-dependent acid–base regulation has been suggested in the blue crab that responded with an increase in Na^+ uptake when exposed to hypercapnic conditions of 1% CO_2 [20]. Pharmacological studies suggested the presence of NHEs and VHA in apical membranes of crustacean gills. Another potential model for proton equivalent secretion in crustacean gills has been proposed by Weihrauch and colleagues [21], suggesting trapping of NH_4^+ in VHA-rich vesicles and subsequent exocytosis across the apical membrane. The entrance of NH_4^+ across the basolateral membrane is achieved by the NKA, which may also accept NH_4^+ as a substrate or by K^+/NH_4^+ channels. In this context a Rhesus protein (RhP) cloned from Dungeness crab *Metacarcinus magister* is proposed to be substantially involved in branchial NH_3/NH_4^+ regulation during exposure to a high NH_4^+ environment [22].

Interestingly, a range of marine invertebrates including bivalves, echinoderms and crustaceans responded with increased NH_4^+ excretion rates when exposed to acidified conditions [23-26]. It has been suggested that this phenomenon may be associated with enhanced protein metabolism to fuel increased acid–base regulatory costs,

and/or may support NH_4^+ based proton equivalent secretion to mediate pH homeostasis. In vertebrates, the excretion of NH_4^+ mediated by Rhesus C glycoprotein (Rhcg) and Rhesus B glycoprotein (Rhbg) in combination with NHE3 or VHA located in apical membranes has been demonstrated to be connected to net export of protons [27-29]. Thus, it can be hypothesized that NH_4^+-based proton secretion also represents a fundamental pH regulatory pathway, probably connected to a reallocation of energy sources, in marine invertebrates.

The present work aims at identifying and characterizing the acid–base regulatory mechanisms by looking at H^+ extrusion and HCO_3^- import pathways in gill epithelia of adult squid. Additionally, metabolism and excretion are monitored during exposure to acidified conditions, which are important indices for altered energetic features potentially associated with acid–base regulatory efforts. Immunocytochemical techniques in combination with gene expression analyses were applied in order to study the branchial acid–base regulatory machinery. It can be hypothesized that similar to the situation in embryonic epidermal ionocytes, the branchial acid–base regulatory machinery of adult squid involves ion-transporters including NHE3, V-type H^+-ATPase, Rh-Protein as well as CAc and NBC, which allows the animal to cope with CO_2 induced acid–base disturbances. In this context special attention has been dedicated to the potential role of RhP in mediating pH homeostasis during environmental hypercapnia by supporting proton equivalent transport across membranes. A potential coupling of NH_3 and H^+ excretion/secretion is proposed, which may represent a fundamental pathway of pH regulation in marine ammonotelic organisms.

Results
Metabolic rates and NH_4^+ excretion
Metabolic rates of squid *Sepiteuthis lessoniana* kept under control conditions were 25.20 ± 4.97 µmol O_2 h^{-1} g_{FM}^{-1} (Figure 2A + B). During short-term exposure to hypercapnic conditions at the time point of 20 h no difference in metabolic rates was observed for animals from different treatment groups. However, after medium-term exposure to pH 7.3, at the time point of 168 h, metabolic rates were significantly decreased down to 16.3 ± 0.79 µmol O_2 h^{-1} g_{FM}^{-1} compared to animals from the pH 8.1 treatment (Figure 2B). Despite a positive correlation between acidified conditions and NH_4^+ excretion rates no significant differences (p = 0.28) were found between control and low pH treated animals after 20 h (Figure 2C). When exposed to acidified conditions for 168 h NH_4^+ excretion rates were decreased although statistical analysis could not demonstrate a significant effect (p = 0.155) in pH 7.3 treated animals (Figure 2D). No change in O:N ratio were observed in low pH treated animals during both, short- and medium-term exposure to acidified conditions (Figure 2E + F).

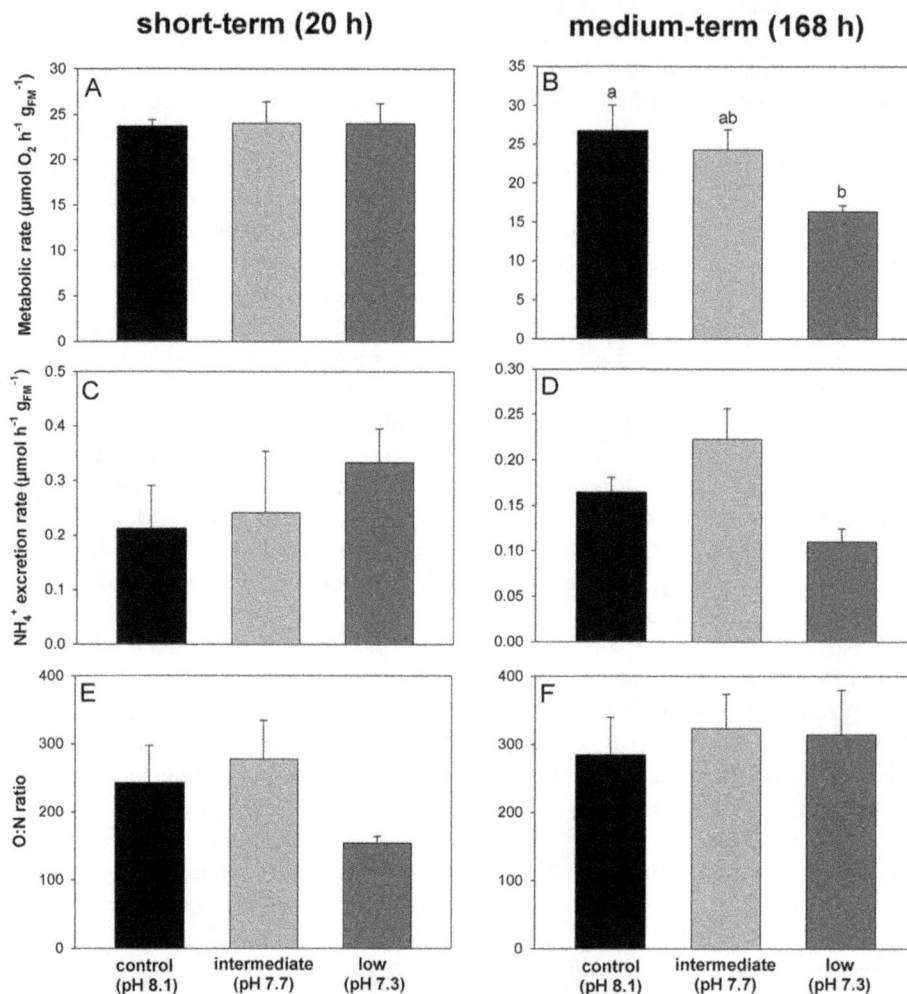

Figure 2 Effects of seawater acidification on metabolic rates and ammonia excretion. Routine metabolic rates were determined after 20 h **(A)** and 168 h **(B)** exposure to acidified conditions. NH_4^+ excretion rates determined after 20 h and 168 h exposure to different pH conditions are presented in **(C)** and **(D)**, respectively. O:N ratios calculated from oxygen consumption and NH_4^+ excretion are given for short-term **(E)** and medium-term **(F)** exposure to different pH conditions. Letters indicate significant differences between pH treatments (p < 0.05). Bars represent mean ± SE (n = 3; with the average of 2 animals from each replicate).

Extracellular acid–base status

Extracellular pH (pHe) measured in venous blood from control animals was 7.34 ± 0.07 (Figure 3A). In response to acidified conditions of pH 7.7 and 7.3 pHe dropped by approximately 0.05 to 0.08 pH units compared to control animals at the time point of 6 h. pHe was quickly restored after 20 h and maintained stable between pH 7.35 and pH 7.4 in both, control and low pH treated animals. Blood HCO_3^- levels were found to range from 2.24 to 2.68 mM in control animals along the incubation period of 168 h (Figure 3B). In response to acidified conditions of pH 7.3 blood HCO_3^- levels were significantly increased by 2 mM along the entire incubation period of 168 h. Only at the 168 h time point blood HCO_3^- levels were found to be increased by approximately 1 mM in pH 7.7 treated animals (Figure 3B).

Blood pCO_2 levels ranged from 0.32 to 0.39 kPa in control animals along the incubation period of 168 h (Figure 3C). In response to CO_2 induced seawater acidification, blood pCO_2 levels increased to peak values of 0.51 ± 0.08 kPa and 0.71 ± 0.29 kPa in pH 7.7 and pH 7.3 treated animals, respectively (Figure 3C). Acid–base compensatory characteristics are depicted in Figure 4 using a davenport diagram. For clarity reasons the two low pH treatments were separated into two graphs. Figure 4A compares control (pH 8.1) to intermediate (pH 7.7) treated animals whereas Figure 4B compares control to low (pH 7.3) pH treated animals. In response to 6 h exposure to acidified conditions a slight initial respiratory acidosis was observed in both treatments. However, pHe is fully restored at the time point of 20 h of low pH exposure and is maintained along the entire period of 168 h.

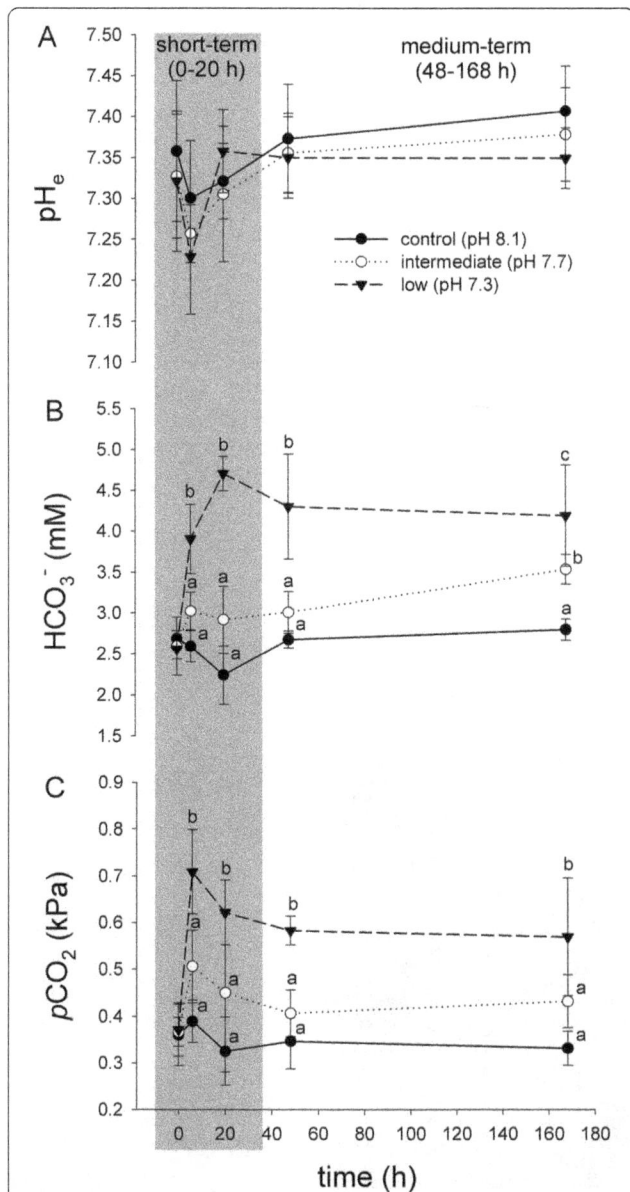

Figure 3 Extracellular acid–base status in *Sepioteuthis lessoniana* during acclimation to acidified conditions. Time course of *in vivo* changes in blood acid–base parameters including extracellular pH (pHe) **(A)**, HCO_3^- **(B)** and pCO_2 **(C)** during 168 h exposure to control (pH 8.1) and acidified (pH 7.7 and pH 7.3) conditions. The exposure period was separated into a short-term (grey) and medium-term (white) acclimation period. Letters indicate significant differences between pH treatments (p > 0.05). Bars represent mean ± SE (n = 3 replicate tanks, with six biological replicates per time point and treatment).

Localization of acid–base relevant transporters in gill epithelia

Using antibodies specifically designed for cephalopod VHA, NHE3 and RhP the sub-cellular localization of these transporters in gill epithelia has been clarified (Figure 5). Using double staining of NKA and NHE3 high concentrations of NKA in basolateral and NHE3

in apical membranes was demonstrated for cells belonging to the inner (concave) ion-transporting epithelium of the cephalopod gill (Figure 5A). The basolateral signal of NKA can be characterized by deep infoldings into the cell cytoplasm of ionocytes. In contrast, the apical signal of NHE3 can be characterized by a sharp lining along the apical membrane. Both signals were clearly co-localized in cells belonging to the ion-transporting inner part of the gill epithelium. Double staining of NHE3 and VHA demonstrated the localization of VHA in basolateral membranes of the ion-transporting inner epithelium. Moreover VHA immunoractivity was also observed in pilaster cells spanning between the outer and the inner epithelium (Figure 5B + D). Using an antibody specifically designed against the squid RhP that was cloned in a previous study [10] positive immunoreactivity has been demonstrated in apical membranes of cells belonging to the inner ion-transporting epithelium (Figure 5C). Negative controls by omitting the primary antibody demonstrated no unspecific fluorescence signal (Additional file 1: Figure S1). Western blot analyses of the four antibodies used in this study demonstrate specific immunoreactivity with proteins including NKA (115 kDa), NHE3 (90 kDa), VHA (65 kDa) and RhP (50 kDa) (Figure 5E).

NKA and VHA activity

Enzyme activities of NKA measured in gill homogenates of control animals at the time point of 0 h were 150 ± 29.17 µmol ATP h^{-1} g_{FM}^{-1}. After 168 h NKA maximum activities decreased down to 43.3 ± 12.3 µmol ATP h^{-1} g_{FM}^{-1} (Figure 6A). Along the entire experimental period no significant differences (p = 0.342) were observed between control and low pH treated animals. Enzyme activities of VHA measured in gill homogenates of control animals at the time point of 0 h were 172 ± 39.40 µmol ATP h^{-1} g_{FM}^{-1}, and control activities remained relatively stable over the entire incubation period of 168 h (Figure 6B). Despite a tendency (p = 0.061) of increased activities in gills of pH 7.3 treated animals at the time point of 20 h no significant differences were observed between control and low pH treated animals.

Protein concentrations

Determination of relative protein concentrations in gill homogenates at the time point of 20 h (short-term) demonstrated that NKA protein concentration normalized to β-actin had no statistical difference between pH treatments (Figure 7A). VHA protein concentrations in gill tissues were significantly increased by 54% (p = 0.009) and 47% (p = 0.033) in response to low pH treatments of pH 7.7 and 7.3, respectively (Figure 7B). No change in relative protein concentrations was observed for NHE3 in gill homogenates (Figure 7C). Finally, RhP protein

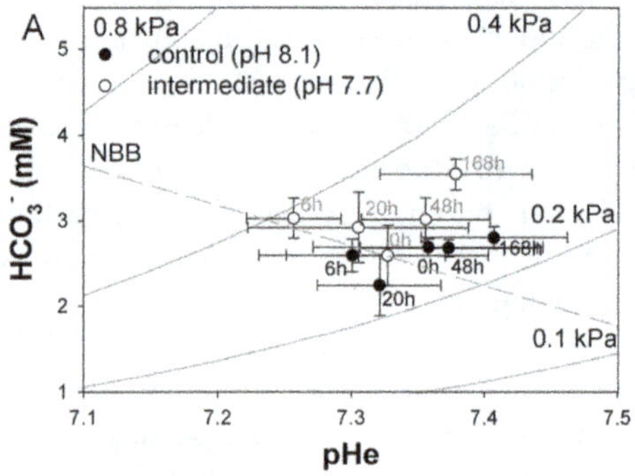

Figure 4 pH-bicarbonate (Davenport) diagram demonstrating the time course of acid–base compensation. Blood acid–base status was determined along the experimental period of 168 h for animals exposed to control (pH 8.1), intermediate (pH 7.7) and low (pH 7.3) conditions. For clarity reasons the intermediate pH **(A)** and low pH **(B)** treatments are presented in two separate graphs, including control acid–base conditions along the incubation period of 168 h. The non-bicarbonate buffer line for *Sepioteuthis lessoniana* is indicated by a grey dashed line and was adopted from Lykkeboe and Johansen [39]. The solid curved lines represent pCO_2 isopleths. Numbers in brackets indicate sampling time points. Bars represent mean ± SE (n = 3 replicate tanks, with six biological replicates per time point and treatment).

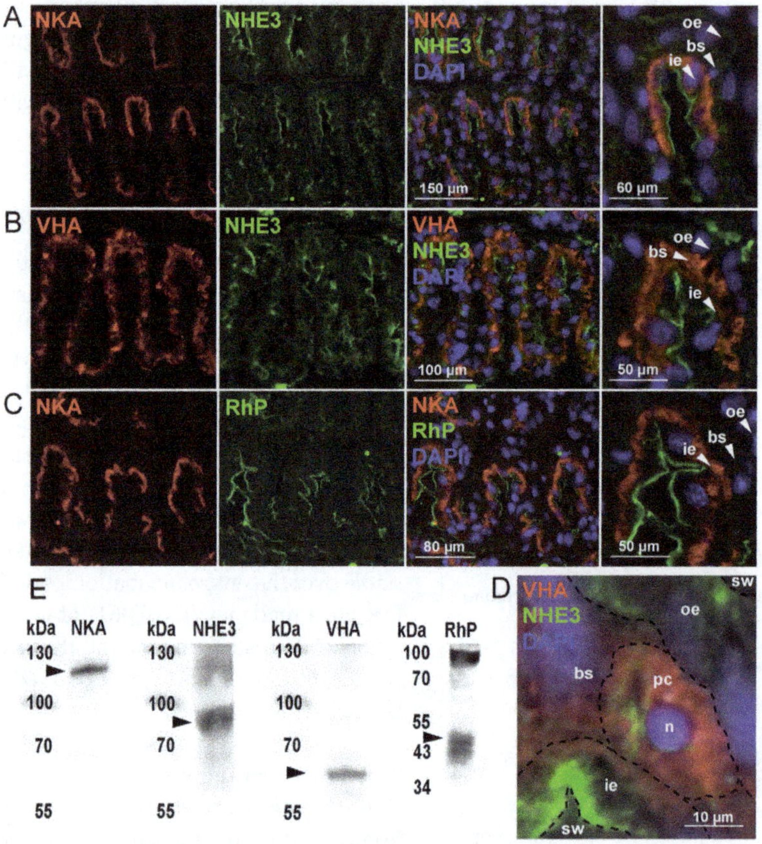

Figure 5 Localization of acid–base transporters in gill epithelia of squid *Sepioteuthis lessoniana*. Immunohistochemical analyses demonstrated co-localization of acid–base transporters including Na^+/K^+-ATPase (NKA), Na^+/H^+-exchanger 3 (NHE3), V-type H^+-ATPase (VHA) and Rhesus protein (RhP) in the ion-transporting epithelium of the squid gill. NKA is located in basolateral membranes whereas the NHE3 specific antibody shows positive immunoreactivity in apical membranes **(A)**. VHA is located in basolateral membranes as well as pillar-cells spanning through the blood sinus **(B)**. The RhP specific antibody shows positive immunoreactivity in apical membranes of the inner, ion-transporting branchial epithelium **(C)**. High magnification image of a pilaster (pillar) cell showing positive VHA immunoreactivity. Dashed lines indicate the contour of the cell and the epithelial cells lining the blood sinus **(D)**. Western blot analyses using gill homogenates, indicating specific immune-reactivity of the different antibodies with proteins in the predicted size range (indicated by arrows) **(E)**. blood sinus (bs); inner epithelium (ie); outer epithelium (oe).

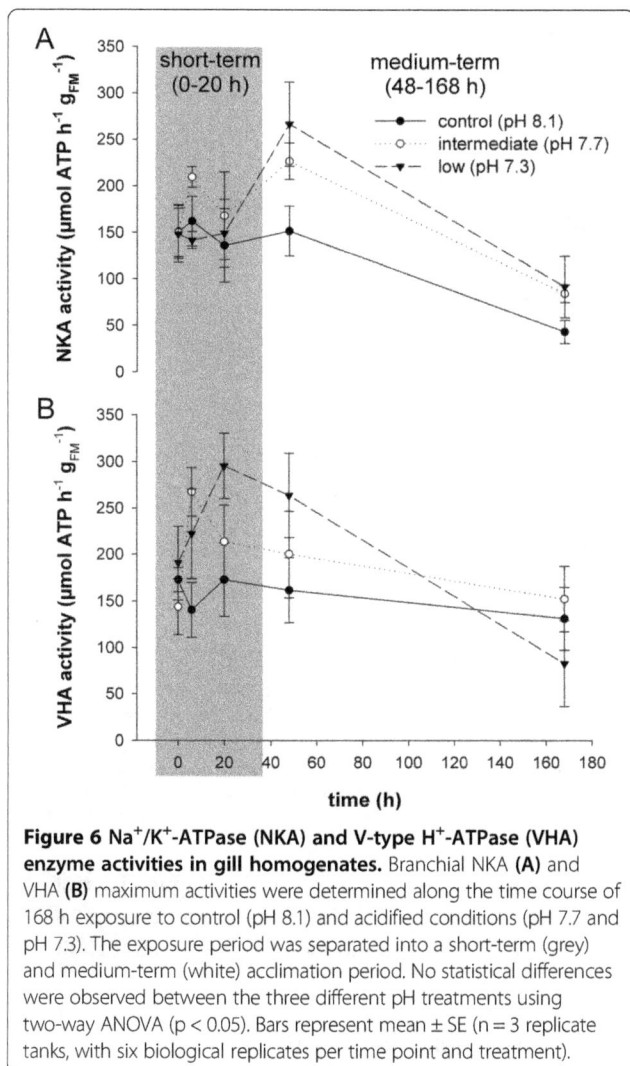

Figure 6 Na$^+$/K$^+$-ATPase (NKA) and V-type H$^+$-ATPase (VHA) enzyme activities in gill homogenates. Branchial NKA **(A)** and VHA **(B)** maximum activities were determined along the time course of 168 h exposure to control (pH 8.1) and acidified conditions (pH 7.7 and pH 7.3). The exposure period was separated into a short-term (grey) and medium-term (white) acclimation period. No statistical differences were observed between the three different pH treatments using two-way ANOVA (p < 0.05). Bars represent mean ± SE (n = 3 replicate tanks, with six biological replicates per time point and treatment).

concentrations were significantly increased by 40% in response to the pH 7.3 treatment (p = 0.045) (Figure 7).

Gene expression

Gene expression studies demonstrated differences along the incubation period of 168 h in control and low pH treated animals. Despite an up regulation pattern of NKA and NHE3 during short-term exposure to acidified conditions no significant differences (p = 0.1 for NKA and p = 0.15 for NHE3) were detected between control and low pH treatments (Figure 8). In contrast, significant up regulations were detected for the genes VHA, RhP and NBC between control and low pH treated animals. VHA expression increased rapidly at the time point of 6 h in pH 7.7 and pH 7.3 treated animals by 53% and 71%, respectively (Figure 8). At this time point statistically significant differences for VHA transcript levels were found between pH 8.1 and pH 7.3 (p < 0.001) and between pH 7.7 and pH 7.3

(p = 0.032) treatments. Although no significant differences were detected for the following time points VHA transcript levels of low pH treated animals were elevated by approximately 30% compared to control levels along the entire period of 168 h (Figure 8). In low pH treated animals RhP increased transcript levels during both short-term and medium-term exposure to acidified conditions. In response to 48 h exposure to acidified conditions RhP transcript levels were significantly increased by 50% (p < 0.001) and 55% (p = 0.02) in pH 7.7 and pH 7.3 treated animals, respectively when compared to pH 8.1 conditions (Figure 8). This significant increase of RhP mRNA levels in low pH treated animals compared to control animals was still evident after 168 h in pH 7.7 (p = 0.038) and pH 7.3 (p = 0.025), respectively. Compared to control animals NBC was significantly up regulated at 48 h in pH 7.7 and pH 7.3 treated animals by 55% (p < 0.001) and 33% (p = 0.002), respectively (Figure 8). No statistical differences for NBC mRNA levels were found between control and low pH treated animals during medium-term exposure of 168 h. During short-term (6 h) and medium-term (168 h) exposure to acidified conditions carbonic anhydrase transcript levels were significantly increased (p < 0.001) in gills of squid exposed to pH 7.7 conditions when compared to control (pH 8.1) animals (Figure 8). No significant differences were observed for branchial CAc transcript levels between control and pH 7.3 treated animals.

Discussion
Metabolism and excretion
The present work demonstrated that routine metabolic and NH$_4^+$ excretion rates in squid *Sepioteuthis lessoniana* are comparable to those determined for other squid and cuttlefish species [1,30-33]. While NH$_4^+$ excretion rates were not significantly affected by acidified conditions, metabolic rates were reduced upon prolonged exposure to pH 7.3. To date only a few studies investigated the effects of acidified conditions on metabolic responses in cephalopods. One study using the pelagic squid *Dosidicus gigas* demonstrated that in response to acute (several minutes) CO$_2$ induced acidified conditions of pH 7.6 (0.1 kPa pCO_2) animals respond with depressed metabolic rates accompanied by decreased activity levels [1]. In contrast, routine metabolic rates of the demersal cuttlefish *Sepia officinalis* remained unaffected by CO$_2$ induced seawater acidification down to pH 7.1 during acute (24 h) exposure [9]. Differential responses observed for the three cephalopod species can be attributed to their very different life styles and abilities to swim and maintain neutral to positive buoyancy in the water column. Migratory pelagic squids like *D. gigas* swim and maintain positive buoyancy by continuously jetting water through their funnel, whereas *S. officinalis* has a decoupled

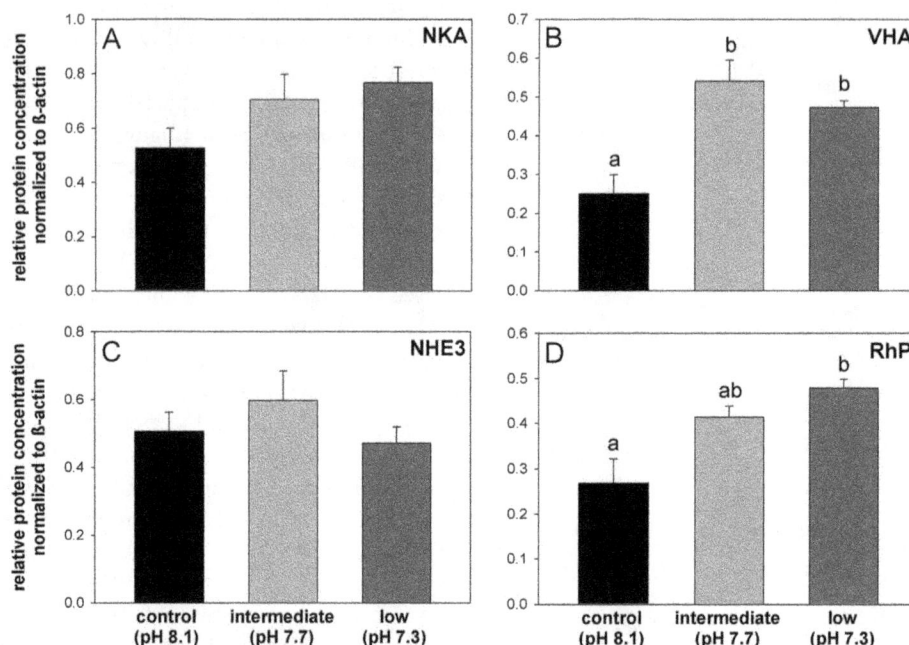

Figure 7 Relative protein concentrations of branchial acid–base transporters. Branchial protein concentrations of acid–base transporters including Na$^+$/K$^+$-ATPase (NKA) **(A)**, V-type H$^+$-ATPase (VHA) **(B)**, Na$^+$/H$^+$-exchanger 3 (NHE3) **(C)** and Rhesus protein (RhP) **(D)** determined after 20 h exposure to different pH conditions. Protein concentrations were normalized to ß-actin protein concentrations as an internal control. Letters indicate significant differences between pH treatments (p < 0.05). Bars represent mean ± SE (n = 3; with the average of 2 animals from each replicate).

swimming mode, by using jetting and fin undulation. Additionally, *Sepia* spp. have a gas filled cuttlebone to control buoyancy without continuous muscular activity that significantly increases locomotor efficiency [34]. *S. lessoniana* is a large-finned pelagic squid that has evolved a partially decoupled swimming mode by additionally maintaining buoyancy using their enlarged fins running the full length of its mantle. Powerful pelagic squids that have no decoupled swimming mode need to spend a larger fraction of their energy budget to swim and to maintain neutral buoyancy [34,35]. This higher fraction of energy that is spent for maintaining buoyancy could be a critical factor leading to higher sensitivities. In fact, the energy budget of marine invertebrates has been demonstrated to be compromised by seawater acidification by shifting a larger fraction of energy towards compensatory processes (e.g. acid–base regulation) leading to less energy available for growth and development [23]. The fact that in the present work two individuals died after exposure to pH 7.3 conditions for one week indicates a higher sensitivity towards acidified conditions than cuttlefish *S. officinalis* that survived with a five-fold increase in body mass during exposure to a seawater pH of 7.1 for 6 weeks [9]. It can be suggested that pelagic squids that evolved a life-style at the edge of energetic limitations, might react more sensitively to seawater acidification due to energetic limitations,

compared to less "tuned" cuttlefish and octopus. To test this hypothesis, studies addressing the energetic costs of acid–base regulation in cephalopods will be an important future task.

Acid–base regulation during seawater acidification

The present work demonstrated that squid *S. lessoniana* can fully compensate for an extracellular acidosis evoked by seawater acidification up to pH 7.3. Stabilization of pHe is accompanied by an increase in blood HCO_3^- levels, which is a conserved and efficient mechanism to counter a respiratory acidosis found in several taxa, including fish, crustaceans and cephalopods [2,3,18,36]. The hyperbolic increase in blood HCO_3^- levels in response to a respiratory acidosis described for other powerful acid–base regulators is in general accordance to the findings for *S. lessoniana* e.g. [37,38]. Under control conditions venous HCO_3^- levels of *S. lessoniana* (2.5 mM) were found to be in the range as described for other cephalopod species including the squid *Illex illecebrosus* (2.2 mM) and the cuttlefish *Sepia officinalis* (3.4 mM) [2,3]. An earlier study using the cuttlefish *Sepia officinalis* demonstrated control blood HCO_3^- levels of 3.4 mM and a partial compensation of pHe via HCO_3^- accumulation during exposure to environmental hypercapnia (0.6 kPa pCO$_2$; pH 7.1) [3]. In the same study it was suggested that a partial compensation of

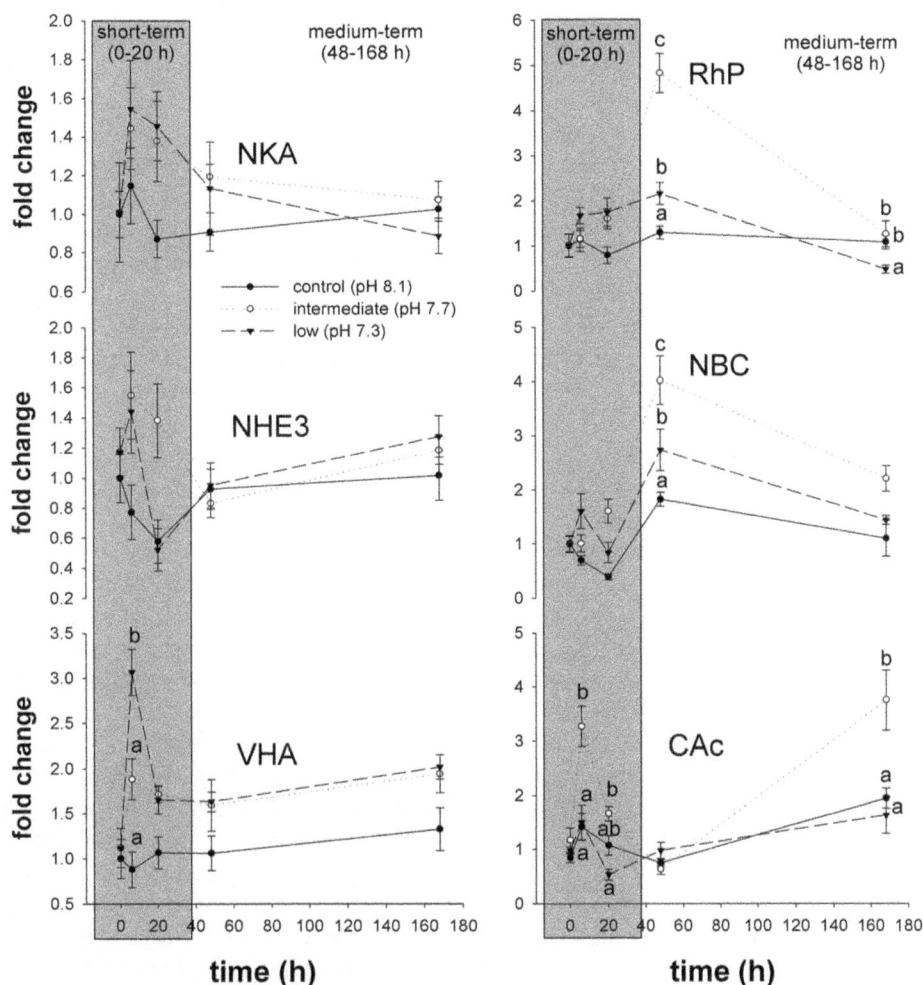

Figure 8 Effects of acidification on transcript abundance of gill acid–base transporters. Branchial mRNA expression levels of acid–base relevant candidates including Na+/K+-ATPase (NKA), Na+/H+-exchanger 3 (NHE3), V-type-H+-ATPase (VHA), Rhesus protein (RhP), Na+/HCO$_3^-$ co-transporter (NBC) and cytosolic carbonic anhydrase (CAc) during exposure to different pH conditions including pH 8.1 (control), pH 7.7 (intermediate) and pH 7.3 (low) along the time course of 168 h. Expression of the gene candidates are normalized to UBC and presented as relative change. The exposure period was separated into a short-term (grey) and medium-term (white) acclimation period. Letters indicate significant differences between pH treatments (p < 0.05). Bars represent mean ± SE (n = 3; with the average of 2 animals from each replicate).

0.2 pH units below control levels is sufficient to achieve sufficient gas transport via the blood pigment hemocyanin under acidified conditions in this less active cephalopod species. However, for the squid *S. lessoniana* a full compensation of extracellular pH was evident after 20 h during exposure to acidified conditions (pH 7.3). It has been hypothesized that more sluggish cephalopod species like cuttlefish and octopods may not rely on pH dependent oxygen transport to the same extent as more active pelagic squid species [3,4,6]. Interestingly, blood HCO$_3^-$ levels in *S. officinalis* increased by approximately 7.5 mM within 48 h in response to 0.6 kPa CO$_2$ exposure whereas in this study blood [HCO$_3^-$] was only increased by 2 mM when exposed to a similar acidification level. This indicates the presence of differential pH buffering/regulatory

mechanisms, including non-bicarbonate buffering and H+ extrusion mechanisms among cephalopods. Non-bicarbonate buffer values determined for squid species ranged between 5 mmol l^{-1} pH unit^{-1} (*Illex illecebrosus*), 5.8 mmol l^{-1} pH unit^{-1} (*Loligo pealei*) and 4.7 mmol l^{-1} pH unit^{-1} (*S. lessoniana*) whereas those determined for cuttlefish, *S. officinalis* were 10 mmol l^{-1} pH unit^{-1} [3,6,39] indicating an even lower HCO$_3^-$ independent buffering potential in squid species. According to these observations it can be suggested that control of extracellular pH in squids is likely to be attributed to efficient H+ extrusion mechanisms. Earlier studies using fish and crustaceans demonstrated that the compensation of acid–base disturbances elicited by hypercapnia is always associated with significant export of proton equivalents [36,40,41]. This feature is particularly important, as HCO_3^- formation through the hydration of CO_2 is always accompanied with the generation of H^+. Thus, on the long run organisms that stabilize blood pH via increased HCO_3^- accumulation require H^+ secretion mechanisms as well. These observations are in line with the results of the present work demonstrating that environmental acidification stimulates expression of branchial acid–base transporters involved in HCO_3^- (NBC, CA) and H^+ transport (VHA and RhP). Although an increase of VHA in response to acidified conditions on both the protein and mRNA level has been demonstrated, no significant (p = 0.061) increases in branchial VHA enzyme activities were found. It can be suggested that despite a trend of increased VHA activity during short-term low pH acclimation, statistical analyses failed to prove this effect due to a relatively low experimental "n" (three experimental replicates with six biological replicates) which is always the limitation when working with non-model organisms. Nonetheless, whole animal observations and molecular findings suggest that besides HCO_3^- buffering H^+ secretion pathways across gill epithelia represent probably an even more important mechanism to compensate for acid–base disturbances in active squids. Thus, a special focus of the present work has been dedicated to a better understanding of branchial proton equivalent secretion mechanisms in ammonotelic cephalopods.

Branchial acid–base regulatory machinery
In convergence to fish and crustaceans, cephalopods evolved branchial ion regulatory epithelia, which are equipped with ion transporters including NKA, VHA and NBCe beneficial for coping with acid–base disturbances [16,17,25,42,43]. The present work further demonstrates that gene transcripts coding for Na^+/H^+ exchanger 3 (NHE3) and Rh protein (RhP), which are essential for proton equivalent transport in vertebrates [27,28,44,45] are also expressed in the cephalopod gill. NKA-rich cells (NaRs) located in the ion-transporting inner epithelium of the 3 order lamellae of the cephalopod gill showed positive immunoreactivity for VHA (basolateral), NHE3 (apical) and RhP (apical) using antibodies specifically designed for this species. These polyclonal antibodies were designed against conserved regions of the respective protein, and western blot analyses of a previous study [10] and the present work demonstrated specific immunoreactivity with proteins in the predicted size range. Using in situ hybridization an earlier study demonstrated that an electrogenic Na^+/HCO_3^- cotransporter (NBC) is also highly expressed in the ion-transporting epithelium of the cuttlefish (Sepia officinalis) gill [12]. Together with the results of the present work it can be suggested that this transporter represents an important player in branchial epithelia that mediates extracellular accumulation of

HCO_3^- in cephalopods. Due to the lack of sequence information the existence and role of anion exchangers (e.g. AE1) which were demonstrated to contribute to acid–base homeostasis in teleosts [46] remains unexplored for cephalopods.

Interestingly, positive VHA immunoractivity was additionally found in pilaster (or pillar) cells spanning through the blood sinus between the inner and the outer epithelium. Little information exists regarding a potential function of pillar cells in ion-regulatory or respiratory processes. Pillar cells in the dogfish (Squalus acanthias) gill were demonstrated to represent an important cell type that may contribute to gas exchange. These pillar cells are characterized by high concentrations of extracellular membrane bound carbonic anhydrase (CA) IV summarized in [47]. This extracellular membrane bound CAIV has been suggested to facilitate the formation of CO_2 from HCO_3^- in concert with basolateral VHA contributing to CO_2 excretion across branchial epithelia in dogfish [48]. In Cephalopods carbonic anhydrase has been demonstrated to be associated with the inner ion-transporting epithelium [11], but information regarding the expression of CA by pilaster cells is not available at present. However, it can be hypothesized that analogous to the situation in dogfish high concentrations of VHA associated with pillar cells may also support gas exchange by providing protons for the formation of CO_2.

The subcellular localization of ion-transporters in squid gills is similar to that found in epidermal ionocytes of cephalopod embryonic stages [10,49]. Interestingly, ion regulatory epithelia in both, adults and embryonic stages seem to have a basolateral orientation of the VHA. This feature has been described for base-secreting type B intercalated cells in the mammalian kidney, and was mainly associated with base secretory processes in other vertebrate systems [50-52]. Although information is scarce for pH regulatory systems of invertebrates a recent study suggested the interplay of VHA, NKA, NHE and CA in the NH_3/NH_4^+ secretion mechanism of the freshwater planarian, Schmidtea mediterranea [53]. Similar to this early invertebrate the present work indicates that also in cephalopod molluscs acid–base regulation in branchial epithelia is associated with increased VHA and RhP protein and mRNA levels. As VHA can only pump H^+ out of the cell, with the catalytic (V1 complex) site located within the cytoplasm, it can be proposed that increased demands of VHA may be explained by two possibilities: i) extracellular NH_4^+ formation in the basolateral boundary layer to import ammonium ions via basolateral NKA and / or ii) local and timely control of blood pH homeostasis in the branchial blood sinus to maintain optimum conditions for gas exchange by hemocyanins. The latter, is particularly interesting, as the cephalopod gill has to simultaneously serve acid–base regulatory (excretory) and respiratory functions, which to some extend, must be connected due to morphological features of the cephalopod gill (see Figures 1 and 9). At the site of gas exchange and excretion, a substantial amount of free protons are removed from the blood due to CO_2 diffusion across the respiratory epithelium into the seawater and via NH_4^+ secretion across the ion-transporting epithelium. During transit of blood through the gills (between ctenidial artery and vein) up to 0.62 mM l^{-1} NH_4^+ is excreted indicating an equimolar loss in H^+ ions in *Octopus dolfleini* [54]. Furthermore, cannulation experiments using unrestrained squid demonstrated that during exercise an alkalosis occurs in venous blood due to an alkalizing effect of hemocyanin deoxygenation [2]. Although a fraction of protons of approximately $7*10^{-6}$ mM l^{-1} see [6] are bound to hemocyanin during deoxygenation and *vice versa* it is unlikely that the release of protons during oxygenation at the gills would compensate for the total loss of proton equivalents (e.g. NH_4^+) leading to a local alkalosis in branchial blood sinuses. Accordingly, it can be hypothesized that VHA located in basolateral membranes and pillar cells could potentially contribute to temporal and local pH homeostasis (during exercise/environmental hypercapnia), to optimize branchial O_2 uptake and especially CO_2 release via the highly pH sensitive hemocyanins (see Figure 9). This is very important as cephalopod hemocyanin appear to have an O_2-dependent CO_2 binding mechanisms that

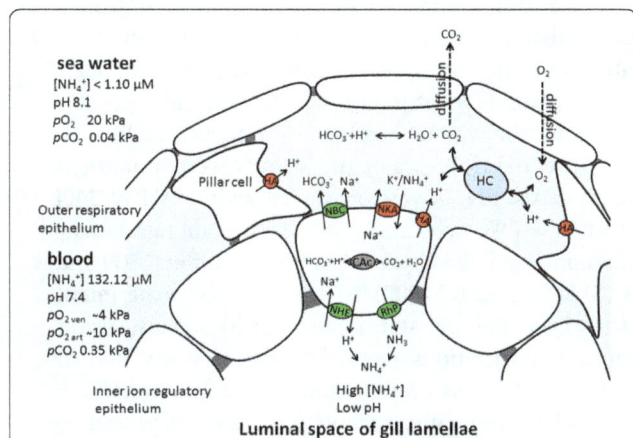

Figure 9 Hypothetical model for the coupling of acid–base regulation and gas exchange in the squid gill. Schematic illustration of the inner (ion-transporting) and the outer (respiratory) epithelium connected by pillar cells. Ionocytes of the ion-transporting epithelium express Na^+/K^+-ATPase (NKA), V-type H^+-ATPase (HA), Na^+/H^+ exchanger 3 (NHE3), Rhesus protein (RhP) and Na^+/HCO_3^- co-transporter (NBC) which are involved in HCO_3^- buffering and NH_4^+ excretion. NH_4^+ is trapped in the acidified tubular space of the 3° gill lamellae whereas the outer respiratory epithelium is spatially separated from the excretory epithelium. HA expressed in pillar cells may contribute to blood pH homeostasis to support gas transport by the highly pH sensitive hemocyanin (HC).

is particularly well developed in active squid species [55]. Unfortunately hemocyanin functioning in cephalopods mainly focused on O_2 release and CO_2 uptake/release in tissues, leaving the characteristics of O_2 uptake, and particularly CO_2 release mechanisms in branchial epithelia underrepresented [56]. Thus, future studies are needed to address the acid–base parameters at the site of gas exchange, (within gills) in combination with a more detailed characterization of potential pillar cell functions.

Branchial NH_3/NH_4^+ transport mechanisms

The dual function of the cephalopod gill in gas exchange and extracellular acid–base regulation is achieved by different epithelia in this organ (depicted in Figures 1 and 9). The thin outer epithelium is believed to be involved in gas exchange by diffusive processes, whereas the inner ion-transporting epithelium is responsible for active transport of acid–base equivalents [11-13]. Apical localization of RhP and NHE3 in the inner transporting epithelium supports the hypothesis that the cephalopod gill is a major site of NH_4^+ excretion. Earlier studies demonstrated that in various cephalopod species the largest fraction of ammonia produced through amino acid metabolism is excreted via branchial epithelia [14,54]. Blood NH_4^+/NH_3 concentrations determined for octopus [54], cuttlefish [14] and squid [57] range from 100 to 500 μmol l^{-1} and are comparable to those determined for *S. lessoniana* in the present work (132.11 ± 37.79 μmol l^{-1}; n = 4). Earlier studies hypothesized that ammonia is excreted as ammonia (NH_3) accompanied with an excretion of protons to form the ammonium ion (NH_4^+) [14,54]. This net export of proton equivalents further suggests that NH_4^+ excretion represents an important mechanism that contributes to acid–base balance in cephalopods. Interestingly, the cephalopod gill shows many morphological and functional similarities to the collecting duct of the mammalian kidney, where NH_4^+ is transported to the luminal space via Rh glycoproteins and V-type H^+-ATPase [45]. In the cephalopod gill, the semi-tubular structure of the 3° gill lamellae creates a luminal space into which NH_4^+ is secreted by the interplay of RhP and NHE3 (Figure 9). The involvement of NHE3 instead of VHA in this process is thermodynamically favored by the strong Na^+ gradient between cytosol (30 mM) and seawater (470 mM). In teleosts Rh proteins including Rhcg and Rhbg were identified as important players in branchial ammonia excretion pathways, as well [58]. The current model denotes the presence of Rhbg in basolateral membranes to facilitate the entry of NH_3 into the cell, whereas Rhcg in combination with VHA and NHE2/3 is located in apical membranes. This interplay of H^+ and NH_3 secretion provides an acid trapping mechanisms for apical NH_4^+ secretion [58]. Accordingly it can be hypothesized that similar to the situation in teleosts and the mammalian kidney, cephalopods excrete ammonia across gill epithelia by trapping NH_4^+ in the semi tubular space of the 3° lamellae (Figure 9).

Conclusion

The present work demonstrated that cephalopods have evolved an efficient pH regulatory machinery in branchial epithelia. Acid–base transporters potentially involved in both, HCO_3^- accumulation and H^+ equivalent secretion were identified and localized in gill epithelia suggesting that this represents the major site for acid–base regulatory in cephalopods. Although significant HCO_3^- buffering capacities to control extracellular pH were only described for few marine species (fish, crustaceans and cephalopods) proton or proton equivalent secretion mechanisms may represent a more direct and ubiquitious pH regulatory pathway. Particularly the coupling of H^+ and NH_3 secretion can be regarded a fundamental and evolutionary ancient pathway of excretion and acid–base regulation. The present work underlines the importance of NH_4^+ based proton secretion via RhP that may contribute to well developed acid–base regulatory abilities in cephalopod molluscs.

The rapid compensation of pHe during exposure to acidified conditions is accompanied by a stimulation of branchial acid–base transporters on the protein and mRNA level, suggesting that maintenance of pHe represents a critical and energy consuming process for cephalopods to maintain vital functions. The present work further demonstrated that squids can tolerate short-term exposure without compromising aerobic energy metabolism while medium-term (one week) exposure to acidified conditions evoked decreased metabolic rates and could even lead to mortality. These observations are in accordance to other studies using pelagic squids [1] but are contrasting to studies conducted on cephalopods that are able to switch to locomotory energy saving modes (e.g. cuttlefish) by burrowing in sediment or maintaining positive buoyancy by using their gas filled cuttlebones and fins [9]. Thus, this study indicates that energetic limitations may represent a critical feature that defines the degree of sensitivity towards seawater acidification. Pelagic squids that evolved a lifestyle at the edge of energetic limits due to high locomotory costs can be expected to be particularly sensitive to prolonged reallocations of energy towards compensatory processes despite their efficient proton equivalent secretion mechanisms. The

identification of physiological principles that may lead to differential sensitivities even within one taxa represent an important task for future directions to better predict species sensitivities in times of rapid environmental change.

Methods
Acidification experiments

Sepioteuthis lessoniana with mantle lengths ranging from 8 to 10.5 cm were obtained from a local dealer in Keelung, Taiwan (ROC) in June 2013 and reared in a flow through system (6000 l total volume, nitrification filter, salinity 29–30, temperature 29°C, constant 12 h dark: 12 h light cycle) at the Jiao-Shi marine station of the Institute of Cellular and Organismic Biology, Academia Sinica. The natural seawater was pumped directly from coastal waters off the east coast of Taiwan, which is the natural habitat of *S. lessoniana* to the culturing facilities of the marine station. Animals were fed twice per day with live *Palaemon* shrimps (approximately 20% of squid body mass). For the CO_2 perturbation experiment a total of 90 animals were used and distributed into nine 300 l tanks (10 animals per tank). The nine tanks, with three replicate tanks for each pH treatment were connected to a flow through system providing filtered, natural seawater. Flow rates were adjusted to approximately 3 l min^{-1} to guarantee high water quality inside the test aquaria. A light regime with a 12 h: 12 h light/dark-cycle was chosen. The aquaria were continuously equilibrated with the appropriate gas mixtures (pH 8.1, pH 7.7 and pH 7.3) using a continuous pH-stat system (pH controller, MACRO) that controlled the addition of CO_2 into the seawater, and aquaria were additionally continuously aerated with air (O_2 saturation > 90%). Specific seawater conditions for the various incubations are given in Table 1. Temperature, pH (NBS scale) and salinity were monitored on a daily basis. pH_{NBS} was measured with a WTW 340i meter and WTW SenTix 81 electrode calibrated daily with Radiometer IUPAC precision pH_{NBS} buffers 7.00 and 10.00 (S11M44, S11 M007) to monitor the experiment and to adjust the pH-stat system. Additionally, water ammonia concentrations were determined every two to three days and levels were maintained < 5.55 µmol l^{-1}. Total dissolved inorganic carbon (C_T) was measured in triplicate (100 µL each) using a Corning 965 carbon dioxide analyzer (Olympic Analytical Service (OAS),

Table 1 Seawater physiochemical parameters during the 168 h pH perturbation experiment including pH (NBS scale), CO_2 partial pressure (pCO_2), total CO_2 (TCO_2), total alkalinity (TA), salinity (Sal) and temperature (Temp)

Treatment	pH_{NBS}	pCO2 µatm	TCO2 mM	TA mM	SAL	Temp°C
pH 8.1	8.06 ± 0.002	624.71 ± 11.90	2.27 ± 0.04	2.51 ± 0.05	29.89 ± 0.19	28.63 ± 0.03
pH 7.7	7.72 ± 0.019	1585.59 ± 253.20	2.47 ± 0.15	2.56 ± 0.14	29.44 ± 0.19	28.71 ± 0.06
pH 7.3	7.34 ± 0.017	4134.16 ± 168.76	2.52 ± 0.12	2.47 ± 0.12	29.56 ± 0.19	28.75 ± 0.08

Malvern, U.K.). Seawater carbonate chemistry speciation was calculated from C_T and pH_{NBS} with the software CO2SYS [59] using the dissociation constants of Mehrbach et al. [60] as refitted by Dickson & Millero [61]. Along the incubation time of 168 h, extracellular acid–base parameters were determined, and tissue samples were taken and quickly shock frozen in liquid nitrogen for gene expression and protein analyses. Sampling and measurements were carried out at five different time points (0, 6, 20, 48, and 168 h). For each pH treatment (pH 8.1: control; pH 7.7: intermediate and pH 7.3: low) three independent replicate tanks were used leading to a total number of nine experimental tanks. At each sampling time point two animals from every experimental tank were sampled, and thus biological replication was n = 6. For statistical analyses the two animals from each replicate tank were averaged leading to a statistical n = 3. The experimental protocols were approved by the National Taiwan Normal University Institutional Animal Care and Utilization Committee (approval no.: 101005).

Metabolic rates and ammonia excretion

Determination of metabolic and ammonia (NH_4^+) excretion rates were determined at the 20 h and 168 h time points. *Sepiotheuthis lessoniana* from the pH experiments were starved overnight (12 h) and were gently transferred to glass respiration chambers with a volume of 4 L containing 0.2 µm filtered seawater equilibrated with the appropriate pCO_2 level. The digestion process in pelagic squids is finished within 2-6h [62] and the starvation time of 12 h is sufficient to prevent effects on metabolic rates due to digestion processes [63]. Respiration chambers were closed, and oxygen saturation was measured continuously (once every 30 s) for 20–30 min at 28–29°C using oxygen sensors (PreSens sensor spots, type PSt3) placed in the lid of respiration chambers, connected to an OXY-4 mini multichannel fiber optic oxygen transmitter (PreSens, Regensburg, Germany). The sensors were calibrated according to the manufacturer's instructions. Preliminary experiments demonstrated that the ventilatory current of the animal could sufficiently mix the water inside the respiration chamber and oxygen concentration decreased linearly. Animal fresh mass was determined on a precision scale after all water was removed from the mantle cavity. When oxygen concentration reached the 75% air saturation level, animals were removed from the respiration chamber. Additionally, a separate glass chamber was incubated without animals to determine background readings of filtered seawater for ammonium excretion and respiration of bacteria. Bacterial respiration was 3.35 ± 3.88 µmol O_2 h^{-1} compared to average oxygen consumption by squids (400.67 ± 161.16 µmol O_2 h^{-1}) leading to less than 1% of animal respiration. For calculation of oxygen consumption rates, the linear decrease in oxygen concentration during measuring

intervals between 10 min after start and the end of the measurement period was considered. Oxygen consumption rates (MO_2) are expressed as µmol O_2 g_{FM}^{-1} h^{-1}.

Ammonium excretion rates were determined from NH_4^+ concentration measurements prior to and following incubation of squids for respiration measurements. Before and after closing the respiration chambers 10 ml of seawater (stock) were sampled. For NH_4^+ determinations a 100 µL subsample was taken from the stock and 25 µL of reagent containing orthophthaldialdehyde, sodium sulphite and sodium borate was added [64]. Samples were then incubated for 2 h at room temperature in the dark until fluorescence was determined at an excitation and emission wavelength of 360 and 422 nm, respectively, using a microplate reader (Molecular Device, Spectra Max, M5). Ammonia (NH_3) was not measured as NH_3 concentrations are negligible at pH values of 8.0-7.1 (0.2–2% of total ammonium/ammonia, [65]. NH_4^+ concentrations were determined in triplicates and excretion rates were expressed as µmol NH_4^+ g_{FM}^{-1} h^{-1}. Blood NH_4^+ concentrations of four control animals were determined with the same method. This method is suitable for NH_4^+ determinations in blood samples as it is specific to NH_4^+ and insensitive to amino acids and proteins [64].

Extracellular acid–base status

Blood samples were collected from the vena cava via a gas-tight Hamilton syringe by dissecting the funnel and mantle from the ventral side. Determination of pH_e was performed in 500 µl samples inside a temperature controlled water bath (29°C) using a microelectrode (WTW Mic-D) and a WTW pHi 340 pH meter (precision ± 0.01 units) that was calibrated with Radiometer precision buffers 7 and 10 (S11M44, S11 M007). For blood cell disposal, withdrawn blood was centrifuged for 30 s (6000 rpm) using a minifuge (Spectrafuge, Labnet International INC.). The supernatant was transferred into a new sample tube for the determination of total dissolved inorganic carbon (C_T). (C_T) was determined in duplicates (100 µL each) via a Corning 965 carbon dioxide analyzer (precision ±0.1 mmol L^{-1}; Olympic Analytical Service, England) that was calibrated by generating a sodium bicarbonate standard curve with a fresh dilution series of 20, 10, 5, 2.5 and 1.25 mM bicarbonate in distilled water. Carbonate system speciation (i.e. pCO_2, [HCO_3^-]) within the coelomic fluid of S. lessoniana was calculated from extracellular pH (pH_e) and (C_T) measurements according to the Henderson–Hasselbalch equation

$$pCO_2 = CT\left(\alpha\left(10^{(pH-pk1')} + 1\right)\right)^{-1} \quad (1)$$

$$e[HCO_3^-] = CT - (\alpha pCO_2) \quad (2)$$

where α (0.039 µmol L^{-1} Pa) is the solubility coefficient of CO_2 in seawater and pK_1' (5.94) the dissociation constant of carbonic acid at a salinity of 30, and a temperature of 29°C [66].

Immunohistochemistry and western blot analyses

For immunohistochemistry tissues were fixed by direct immersion for 24 h in Bouin's fixative followed by rinses in 75% ethanol. Samples were fully dehydrated in a graded ethanol series and embedded in Paraplast (Paraplast Plus, Sigma, P3683). Sections of 4 μm were cut on a Leica RM2265 microtome, collected on poly-L-lysine-coated slides. The slides were deparaffinized in Histoclear II° for 10 min and passed through a descending alcohol series (100%, 95%, 90%, 70%, and 50% for 5 min each). Slides were washed in phosphate- buffered saline (PBS), pH 7.3. Subsequently, samples were transferred to a PBS solution containing 5% bovine serum albumin (BSA) for 30 min to block non-specific binding. The primary antibodies, a rabbit polyclonal antibody H-300, raised against the human α subunit of the Na$^+$/K$^+$-ATPase (NKA) (Santa Cruz Biotechnology, INC) and *Sepioteuthis lessoniana* specific polyclonal antibodies raised against part of the carboxyl-terminal region (IYRVRKVGYDEQ FIMSY) of Na$^+$/H$^+$-exchanger3 (NHE3), the subunit A region (SYSKYTRALDEFYDK) of the V-type-H$^+$-ATPase (VHA) for more detail see [10] and the Rhesus protein (RhP) (antibody designed against the synthetic peptideTRAGYQEFKW) were diluted in PBS (1:50–100) and placed in small droplets of 200 μl onto the sections, and incubated for 12 h at 4°C in a wet chamber. To remove unbound antibodies, the sections were then washed (3 × 5 min) in PBS and incubated for 1 h with small droplets (200 μl) of secondary antibody, anti-mouse Alexa Fluor 488 or anti- rabbit Alexa Fluor 568 (Invitrogen) (dilution 1:250). To allow double-color immunofluorescence staining, one of the polyclonal antibodies was directly labeled with Alexa Fluor dyes using the Zenon antibody labeling kit (Molecular Probes, Eugene, OR, USA). After rinses in PBS (3 × 5 min), sections were examined with a fluorescence microscope (Zeiss imager A1) equipped with an appropriate filter set.

For immunoblotting, 15 μL of crude extracts from gill tissues were used. Proteins were fractionated by SDS-PAGE on 10% polyacrylamide gels, according to Lämmli [67], and transferred to PVDF membranes (Millipore), using a tank blotting system (Bio-Rad). Blots were pre-incubated for 1 h at room temperature in TBS-Tween buffer (TBS-T, 50 mM Tris -HCl, pH 7.4, 0.9% (wt/vol) NaCl, 0.1% (vol/vol) Tween20) containing 5% (wt/vol) blocking reagent (Roche, Mannheim, Germany). Blots were incubated with the primary antibody (see previous section) diluted 1:250–500 at 4°C overnight. After washing with TBS-T, blots were incubated for 2 h with horseradish conjugated goat anti-rabbit IgG antibody (diluted 1:1,000-2,000, at room temperature;

Amersham Pharmacia Biotech). Protein signals were visualized by using the enhanced chemiluminescence system (ECL, Amersham Pharmacia Biotech) and recorded using Biospectrum 600 imaging system (UVP, Upland, CA, USA). Signal intensity was calculated using the free software "Image J" e.g. [68].

Enzyme activity

ATPase activity was measured in crude extracts in a coupled enzyme assay with pyruvate kinase (PK) and lactate dehydrogenase (LDH) by using the method of Schwartz et al. [69]. Crude extracts were obtained by quickly homogenizing the tissue samples using a tissue lyzer (Quiagen) in 10 volumes of ice-cold buffer containing 50 mM imidazole, pH 7.5, 250 mM sucrose, 1 mM EDTA, 5 mM β-mercaptoethanol, 0.1% (w/v) deoxycholate, proteinase inhibitor cocktail from Sigma-Aldrich (catalogue no. P8340). Cell debris was removed by centrifugation for 10 min at 1000 g, 4°C. The supernatant was used as a crude extract. The reaction was started by adding 2 μl of the sample homogenate to the reaction buffer containing 100 mM imidazole, pH 7.5, 80 mM NaCl, 20 mM KCl, 5 mMMgCl$_2$, 5 mM ATP, 0.24 mM Na-(NADH 2), 2 mM phosphoenolpyruvate, and about 12 U/ml PK and 17 U/ml LDH in a PK/LDH enzyme mix (Sigma-Aldrich). The oxidation of NADH coupled to the hydrolysis of ATP was followed photometrically at 29°C in a temperature controlled plate reader (Molecular Device, Spectra Max, M5), over a period of 15 min, with the decrease of extinction being measured at λ =339 nm. The fraction of Na$^+$/K$^+$ -ATPase or H$^+$-ATPase activity in total ATPase (TA) activity was determined by the addition of 2 μl ouabain (5 mM final concentration) or bafilomycin (Bafilomycin A1, Sigma-Aldrich) (10 μM final concentration) to the assay, respectively. The concentrations of inhibitors applied were demonstrated to be sufficient to fully inhibit the NKA [70] and VHA [71], respectively. Each sample was measured in six replicates (3 with inhibitor dissolved in DMSO and 3 with DMSO). Enzyme activity was calculated by using an extinction coefficient for NADH of ε =6.31 mM^{-1}·cm^{-1} and given as micromoles of ATP consumed per gram tissue fresh mass (g$_{FM}$) per hour.

Preparation of mRNA

Gill tissues (without branchial gland) were homogenized in Trizol reagent (Invitrogen, Carlsbad, CA, USA) using a Tissue lyser (Quiagen). Total RNA was extracted from the aqueous phase after addition of chloroform to Trizol homogenates and purified by addition of isopropanol. DNA contamination was removed with DNase I (Promega, Madison, WI, USA). The mRNA for the RT-PCR was obtained with a QuickPrep Micro mRNA Purification Kit (Amersham Pharmacia, Piscataway, NJ, USA) according

Table 2 Primers used for qRT-PCR

Gene name	Abbreviation	Primer sequence	Amplicon size (bp)	Accession numbers
Sodium–hydrogen exchanger 3	NHE3	F 5'- GGCTGTCTTCCAAGAAATGGGTGT -3'	168	KJ451615
		R 5'- AAGAACTTGGCAACACCAAGAGCG -3'		
Vacuolar-type H + –ATPase	VHA	F 5'- ACGTGAGGGCAGTGTCAGTATTGT -3'	161	ADM67602.1
		R 5'- TGATCAGCCAGTTGATGGAAGGGA -3'		
Na+, K + –ATPase	NKA	F 5'- CCGTGCTGAATTTAAGGCAGGTCA -3'	83	GQ153672.1
		R 5'- GCAAAGCTGATTCAGAAGCGTCAC -3'		
Rhesus protein	RhP	F 5'-GCACAAAGGAAAGCTGGACATGGT-3'	179	KJ451616
		R 5'-AATGATACCAGCCACCACTCCGA-3'		
Sodium-bicarbonate cotransporter	NBC	F 5'-AATTCGCTGCATGATTGTCCGTCC-3'	188	HM157263.1
		R 5'-TTCCGGAGAACTGACGACCGATTT-3'		
Cytosolic Carbonic anhydrase	CAc	F 5'-GTGAAGCCAACATGGAAGTC-3'	108	KJ451614
		R 5'-GCAGTTTGTAAGGAGTTGTCTC-3'		
Reference gene				
Ubiquitin-conjugated enzyme	UBC	F 5'- ATGCAGATGGCAGTATTTGCCTGG -3'	127	HM157280.1
		R 5'- TTATTGGCTGGGCTGTTTGGGTTC -3'		

F, forward primer; R, reverse primer.

to the supplier protocol. The amount of mRNA was determined by spectrophotometry (ND-2000, NanoDrop Technol, Wilmington, DE), and the mRNA quality was checked by running electrophoresis in RNA denatured gels. All mRNA pellets were stored at –80°C.

Real-time quantitative PCR (qPCR)

The mRNA expressions of target genes were measured by qPCR with the Roche LightCycler® 480 System (Roche Applied Science, Mannheim, Germany). Primers for all genes were designed using Primer Premier software (vers. 5.0; PREMIER Biosoft International, Palo Alto, CA). The sequences and primers were used in a previous study [10] and are depicted in Table 2. In addition to the previously cloned sequences of squid acid–base transporters a cytosolic carbonic anhydrase (CAc) sequence (KJ451614) was cloned from squid (*S. lessoniana*). PCRs contained 40 ng of cDNA, 50 nM of each primer, and the LightCycler® 480 SYBR Green I Master (Roche) in a final volume of 10 μl. All qPCR reactions were performed as follows: 1 cycle of 50°C for 2 min and 95°C for 10 min, followed by 45 cycles of 95°C for 15 sec and 60°C for 1 min (the standard annealing temperature of all primers). PCR products were subjected to a melting-curve analysis, and representative samples were electrophoresed to verify that only a single product was present. All primer pairs used in this PCR had efficiencies >96%. Control reactions were conducted with nuclease-free water to determine levels of background. Additionally, no PCR product was obtained by using DNAse I treated RNA samples as template demonstrating the success of the DNase I

treatment. The standard curve of each gene was confirmed to be in a linear range with ubiquitin conjugated protein (*UBC*) as reference genes. The expression of this reference gene has been demonstrated to be stable in cephalopods among ontogenetic stages and during CO_2 treatments [10,12].

Statistical analyses

Statistical analyses were performed using Sigma Stat 3.0 (Systat) software. Statistical differences between pH treatments were analyzed by two-way and one-way ANOVA followed by Tukey's post-hoc test. Data sets were normally distributed (Kolmogorov-Smirnov test). Equal variance was tested using the Levene median test. The significance level was set to $p < 0.05$.

Additional file

> **Additional file 1: Figure S1.** Negative controls for immunohistochemical analyses.

Competing interests
The authors declare that they have no competing interests.

Authors' contributions
MYH, PPH and YCT designed and conducted experiment, analyzed the data and compiled the manuscript. MYH conducted immunohistochemical experiments and analyzed the data. YJG and YCT carried out the molecular cloning studies. MS conducted metabolic rates and ammonia measurement and analyzed the data. JRL, RDC, PHS and YCC conducted CO_2 perturbation experiments and sample preparation. All authors read and approved the final manuscript.

Acknowledgements
This study was financially supported by the grants to Y. C. Tseng
from the National Science Council, Taiwan, Republic of China
(NSC 102-2321-B-003-002) and an Alexander von Humbold/National
Science Council (Taiwan) grant awarded to M. H (NSC 102-2911-I-001-002-2)
and M.S. (NSC 103-2911-I-001-506). We gratefully thank Mr. H. T. Lee
(assistant of Institute of Cellular and Organismic Biology marine station)
for his assistance to maintain experimental systems.

Author details
[1]Institute of Cellular and Organismic Biology, Academia Sinica, Taipei City,
Taiwan. [2]Department of Life Science, National Taiwan Normal University,
Taipei City, Taiwan.

References
1. Rosa R, Seibel BA: Synergistic effects of climate-related variables suggest future physiological impairment in a top oceanic predator. *Proc Natl Acad Sci* 2008, **105**:20776–20780.
2. Pörtner H-O, Webber DM, Boutilier RG, O'Dor RK: Acid–base regulation in exercising squid (*Illex illecebrosus, Loligo pealei*). *Am J Physiol Regul Integr Comp Physiol* 1991, **261**:R239–R246.
3. Gutowska MA, Melzner F, Langenbuch M, Bock C, Claireaux G, Pörtner H-O: Acid–base regulatory ability of the cephalopod (*Sepia officinalis*) in response to environmental hypercapnia. *J Comp Physiol B* 2010, **180**:323–335.
4. Pörtner HO: Coordination of metabolism, acid–base regulation and haemocyanin function in cephalopods. *Mar Fresh Behav Physiol* 1994, **25**:131–148.
5. Brix O, Lykkeboe G, Johansen K: The significance of the linkage between the Bohr and Haldane effects in cephalopod bloods. *Respir Physiol* 1981, **44**:177–186.
6. Pörtner H-O: An analysis of the effects of pH on oxygen binding by squid (*Illex illecebrosus, Loligo pealei*) haemocyanin. *J Exp Biol* 1990, **150**:407–424.
7. Pörtner HO, Langenbuch M, Reipschläger A: Biological impact of elevated ocean CO_2 concentrations: lessons from animal physiology and earth history. *J Oceanogr* 2004, **60**:705–718.
8. Melzner F, Gutowska MA, Langenbuch M, Dupont S, Lucassen M, Thorndyke MC, Bleich M, Pörtner H-O: Physiological basis for high CO_2 tolerance in marine ectothermic animals: pre-adaptation through lifestyle and ontogeny? *Biogeosciences* 2009, **6**:2313–2331.
9. Gutowska MA, Pörtner H-O, Melzner F: Growth and calcification in the cephalopod *Sepia officinalis* under elevated seawater pCO_2. *Mar Ecol Prog Ser* 2008, **373**:303–309.
10. Hu MY, Lee J-R, Lin L-Y, Shih T-H, Stumpp M, Lee M-F, Hwang P-P, Tseng Y-C: Development in a naturally acidified environment: Na^+/H^+-exchanger 3-based proton secretion leads to CO_2 tolerance in cephalopod embryos. *Front Zool* 2013, **10**:51–67.
11. Schipp R, Mollenhauer S, Boletzky S: Electron microscopical and histochemical studies of differentiation and function of the cephalopod gill (*Sepia officinalis* L.). *Zoomorph* 1979, **93**:193–207.
12. Hu MY, Tseng Y-C, Stumpp M, Gutowska MA, Kiko R, Lucassen M, Melzner F: Elevated seawater pCO_2 differentially affects branchial acid–base transporters over the course of development in the cephalopod *Sepia officinalis*. *Am J Physiol Regul Integr Comp Physiol* 2011, **300**:R1100–R1114.
13. Hu MY, Sucré E, Charmantier-Daures M, Charmantier G, Lucassen M, Melzner F: Localization of ion regulatory epithelia in embryos and hatchlings of two cephalopods. *Cell Tiss Res* 2010, **441**:571–583.
14. Donaubauer HH: Sodium- and potassium-activated adenosine triphosphatase in the excretory organs of *Sepia officinalis* (Cephalopoda). *Mar Biol* 1981, **63**:143–150.
15. Perry SF, Gilmour KM: Acid–base balance and CO_2 excretion in fish: Unanswered questions and emerging models. *Respir Physiol Neurobiol* 2006, **154**:199–215.
16. Hwang PP, Perry SF: Ionic And Acid–Base Regulation. In *Zebrafish*, vol. 29. Edited by Perry SF, Ekker M, Farrel AP, Brauner CJ. London: Elsevier; 2010:311–344.
17. Evans DH, Piermarini PM, Choe KP: The multifunctional fish gill: Dominant site of gas exchange, osmoregulation, acid–base regulation, and excretion of nitrogenous waste. *Physiol Rev* 2005, **85**:97–177.
18. Henry RP, Lucu C, Onken H, Weihrauch D: Multiple functions of the crustacean gill: osmotic/ionic reglation, acid–base balance, ammonia excretion, and bioaccumulation of toxic metals. *Front Physiol* 2012, **3**:431.
19. Tresguerres M, Parks S, Sabatini SE, Goss GG, Luquet CM: Regulation of ion transport by pH and [HCO3-] in isolated gills of the crab *Neohelice* (Chasmagnathus) *granulata*. *Am J Physiol Regul Integr Comp Physiol* 2008, **294**(3):R1033–R1043.
20. Cameron JN: Effects of hypercapnia on blood acid–base status, NaCl fluxes and trans-gill potential in freshwater blue crabs, *Callinectes sapidus*. *J Comp Physiol B* 1978, **123**:137–141.
21. Weihrauch D, Ziegler A, Siebers D, Towle DW: Active ammonia excretion across the gills of the green shore crab *Carcinus maenas*: participation of Na^+/k^+-ATPase, V-type H^+-ATPase and functional microtubules. *J Exp Biol* 2002, **205**:2765–2775.
22. Martin M, Fehsenfeld S, Sourial MM, Weihrauch D: Effects of high environmental ammonia on branchial ammonia excretion rates and tissue Rh-protein mRNA expression levels in seawater acclimated Dungeness crab *Metacarcinus magister*. *Comp Biochem Physiol A* 2011, **160**:267–277.
23. Stumpp M, Trübenbach K, Brennecke D, Hu MY, Melzner F: Resource allocation and extracellular acid–base status in the sea urchin Strongylocentrotus droebachiensis in response to CO_2 induced seawater acidification. *Aqua Toxicol* 2012, **110–111**:194–207.
24. Thomsen J, Melzner F: Moderate seawater acidification does not elicit long-term metabolic depression in the blue mussel *Mytilus edulis*. *Mar Biol* 2010, **157**:2667–2676.
25. Fehsenfeld S, Weihrauch D: Differential acid–base regulation in various gills of the green crab Carcinus maenas: effects of elevated environmental pCO_2. *Comp Biochem Physiol A* 2012, doi: 10.1016/J.cbpa.2012.09.016.
26. Hu MY, Casties I, Stumpp M, Ortega-Martinez O, Dupont S: Energy metabolism and regeneration impaired by seawater acidification in the infaunal brittlestar Amphiura filiformis. *J Exp Biol* doi:10.1242/jeb.100024.
27. Wu S-C, Horng J-L, Liu S-T, Hwang PP, Wen Z-H, Lin C-S, Lin LY: Ammonium-dependent sodium uptake in mitochondrion-rich cells of medaka (*Oryzias latipes*) larvae. *Am J Physiol Cell Physiol* 2010, **298**:C237–C250.
28. Nawata CM, Hirose S, Nakada T, Wood CM, Katoh A: Rh glycoprotein expression is modulated in pufferfish (*Takifugu rubripes*) during high environmental ammonia exposure. *J Exp Biol* 2010, **213**:3150–3160.
29. Wagner CA, Devuyst O, Belge H, Bourgeois S, Houillier P: The rhesus protein Rhcg: a new perspective in ammonium transport and distal urinary acidification. *Kidney Int* 2011, **79**:154–161.
30. Trübenbach K, Pegado MR, Seibel BA, Rosa R: Ventilation rates and activity levels of juvenile jumbo squid under metabolic suppression in the oxygen minimum zone. *J Exp Biol* 2013, **216**:359–368.
31. Webber DM, O'Dor RK: Monitoring the metabolic rate and activity of free-swimming squid with telemetered jet pressure. *J Exp Biol* 1986, **126**:205–224.
32. Boucher-Rodoni R, Mangold K: Comparative aspects of ammonia excretion in cephalopods. *Malacologica* 1988, **29**:145–151.
33. Boucher-Rodoni R, Mangold K: Respiration and nitrogen excretion by the squid *Loligo forbesi*. *Mar Biol* 1989, **103**:333–338.
34. O'Dor RK: Telemetered cephalopod energetics: swimming, soaring, and blimping. *Integr Comp Biol* 2002, **42**:1065–1070.
35. O'Dor RK, Webber DM: Invertebrate athletes: trade-offs between transport efficiency and power density in cephalopod evolution. *J Exp Biol* 1991, **160**:93–112.
36. Heisler N: *Acid–Base Regulation In Animals*. Amsterdam: Elsevier Biomedical Press; 1986.
37. Claiborne JB, Evans DH: Acid–base balance and ion transfers in the spiny dogfish (*Squalus acanthias*) during hypercapnia - a role for ammonia excretion. *J Exp Zool* 1992, **261**:9–17.
38. Toews DP, Holeton GF, Heisler N: Regulation of the acid–base status during environmental hypercapnia in the marine teleost fish *Conger conger*. *J Exp Biol* 1983, **107**:9–20.
39. Lykkeboe G, Johansen K: A cephalopod approach to rethinking about the importance of the Bohr and Haldane effects. *Pac Sci* 1982, **36**:305–313.
40. Heisler N: *Acid–Base Regulation In Fishes*. New York: Academic; 1984.
41. Cameron JN: Acid–Base Equilibria In Invertebrates. In *Acid–Base Regulation In Animals*. Edited by Heisler N. Amsterdam: Elsevier Biomedical Press; 1986.
42. Hwang PP, Lee TH, Lin LY: Ion regulation in fish gills: recent progress in the cellular and molecular mechanisms. *Am J Physiol Regul Integr Comp Physiol* 2011, **301**(1):R28–R47.

43. Charmantier G, Charmantier-Daures M: Ontogeny of osmoregulation in crustaceans: the embryonic phase. *Am Zool* 2001, **41**:1078–1089.

44. Watanabe S, Niida M, Maruyama T, Kaneko T: Na+/H + exchanger isoform 3 expressed in apical membrane of gill mitochondrion-rich cells in Mozambique tilapia Oreochromis mossambicus. *Fish Sci* 2008, **74**:813–821.

45. Bishop JM, Verlander JW, Lee H-W, Nelson RD, Weiner AJ, Handlogten ME, Weiner ID: Role of Rhesus glycoprotein, Rh B glycoprotein, in renal ammonia excretion. *Am J Physiol Renal Physiol* 2010, **299**:F1065–F1077.

46. Lee YC, Yan JJ, Cruz SA LHJ, Hwang PP: Anion exchanger 1b, but not sodium-bicarbonate cotransporter 1b, plays a role in transport functions of zebrafish H⁺-ATPase-rich cells. *Am J Physiol Regul Integr Comp Physiol* 2011, **300**:C295–C307.

47. Gilmour KM, Perry SF: Carbonic anhydrase and acid–base regulation in fish. *J Exp Biol* 2009, **212**:1647–1661.

48. Gilmour KM, Bayaa M, Kenny L, McNeill B, Perry SF: Type IV carbonic anhydrase is present in the gills of spiny dogfish (*Squalus acanthias*). *Am J Physiol Integr Comp Physiol* 2007, **292**:R556–R567.

49. Hu MY, Tseng Y-C, Lin L-Y, Chen P-Y, Charmantier-Daures M, Hwang PP, Melzner F: New insights into ion regulation of cephalopod molluscs: a role of epidermal ionocytes in acid–base regulation during embryogenesis. *Am J Physiol Regul Integr Comp Physiol* 2011, **301**:1700–1709.

50. Wagner CA, Finberg KE, Breton S, Marshanski V, Brown D, Geibel JP: Renal vacuolar H⁺-ATPase. *Physiol Rev* 2003, **84**:1263–1314.

51. Tresguerres M, Parks SK, Katoh F, Goss GG: Microtubule-dependent relocation of branchial V-H⁺-ATPase to the basolateral membrane in the Pacific spiny dogfish (*Squalus acanthias*): a role in base secretion. *J Exp Biol* 2006, **209**:599–609.

52. Piermarini PM, Evans DH: Immunochemical analysis of the vacuolar proton-ATPase B-subunit in the gills of a euryhaline stingray (*Dasyatis sabina*): effects of salinity and relation to Na⁺/K⁺-ATPase. *J Exp Biol* 2001, **204**:3251–3259.

53. Weihrauch D, Chan AC, Meyer H, Döring C, Sourial MM, O'Donnell MJ: Ammonia excetion in the freshwater planarian Schmidtea mediterranea. *J Exp Biol* 2012, doi:10.1242/jeb.067942. *J Exp Biol* 2012, **215**:3242–3253.

54. Potts WTW: Ammonia excretion in *Octopus dolfeini*. *Comp Biochem Physiol* 1965, **14**:339–355.

55. Lykkeboe G, Brix O, Johansen K: Oxygen-linked CO₂ binding independent of pH in cephalopod blood. *Nature* 1980, **287**:330–331.

56. Brix O, Bardgard A, Cau A, Colosimo SGC, Giardina B: Oxygen-binding properties of cephalopod blood with special reference to environmental temperatures and ecological distribution. *J Exp Zool* 1989, **252**:34–42.

57. Voight JR, Pörtner HO, O'Dor RK: A review of ammonia-mediated buoyancy in squids (Cephalopoda: Teuthoidea). *Mar Fresh Behav Physiol* 1994, **25**:193–203.

58. Wright PA, Wood CM: A new paradigm for ammonia excretion in aquatic animals: role of Rhesus (Rh) glycoproteins. *J Exp Biol* 2009, **212**:2303–2312.

59. Lewis E, Wallace DWR: *Program developed for CO2 system calculations*. Oak Ridge: Oak Ridge National Laboratory ORNL/CDIAC-105; 1998.

60. Mehrbach C, Culberso C, Hawley J, Pytkowic R: Measurement of apparent dissociation constants of carbonic acid in seawter at atmospheric pressure. *Limnol Oceanogr* 1973, **18**:897–907.

61. Dickson A, Millero F: A comparison of the equilibrium constants for the dissociation of carbonic acid in seawater media. *Deep Sea Res A* 1987, **34**:1733–1743.

62. Lipiński MR: Changes in pH in the caecum of *Loligo vulgaris reynaudii* during digestion. *S Afr J Mar Sci* 2010, **9**(1):43–51.

63. Katsanevakis S, Protopapas N, Miliou H, Verriopoulos G: Effect of temperature on specific dynamic action in the common octopus *Octopus vulgaris* (Cephalopoda). *Mar Biol* 2005, **146**:733–738.

64. Holmes RM, Aminot A, Kérouel R, Hooker BA, Peterson BJ: A simple and precise method for measuring ammonium in marine and freshwater ecosystems. *Can J Fish Aquat Sci* 1999, **56**(10):1801–1808.

65. Körner S, Das SK, Veenstra S, Vermaat JE: The effect of pH variation at the ammonium/ammonia equilibrium in wastewater and its toxicity to *Lemna gibba*. *Aquat Bot* 2001, **71**:71–78.

66. Weiss RF: Carbon dioxide in water and seawater: the solubility of a non-ideal gas. *Mar Chem* 1974, **2**:203–215.

67. Lämmli UK: Cleavage of structural proteins during the assembly of the head of Bacteriophage T4. *Nature* 1970, **227**:680–685.

68. Schneider CA, Rasband WS, Eliceiri KW: NIH Image to ImageJ: 25 years of image analysis. *Nat Methods* 2012, **9**:671–675.

69. Schwartz AA, Allen JC, Harigaya S: Possible involvement of cardiac Na⁺/K⁺-adenosine triphosphatase in the mechanism of action of cardiac glycosides. *J Pharmacol Exp Ther* 1969, **168**:31–41.

70. Morris JF, Ismail-Beigi F, Jr BVP, Gati I, Lichtstein D: Ouabain-sensitive Na+, K(+)-ATPase activity in toad brain. *Comp Biochem Physiol A* 1997, **118**:599–606.

71. Dröse S, Altendorf K: Bafilomycins and concanamycins as inhibitors of V-ATPases and P-ATPases. *J Exp Biol* 1997, **200**:1–8.

4

Are ticks venomous animals?

Alejandro Cabezas-Cruz[1,2] and James J Valdés[3*]

Abstract

Introduction: As an ecological adaptation venoms have evolved independently in several species of Metazoa. As haematophagous arthropods ticks are mainly considered as ectoparasites due to directly feeding on the skin of animal hosts. Ticks are of major importance since they serve as vectors for several diseases affecting humans and livestock animals. Ticks are rarely considered as venomous animals despite that tick saliva contains several protein families present in venomous taxa and that many Ixodida genera can induce paralysis and other types of toxicoses. Tick saliva was previously proposed as a special kind of venom since tick venom is used for blood feeding that counteracts host defense mechanisms. As a result, the present study provides evidence to reconsider the venomous properties of tick saliva.

Results: Based on our extensive literature mining and *in silico* research, we demonstrate that ticks share several similarities with other venomous taxa. Many tick salivary protein families and their previously described functions are homologous to proteins found in scorpion, spider, snake, platypus and bee venoms. This infers that there is a structural and functional convergence between several molecular components in tick saliva and the venoms from other recognized venomous taxa. We also highlight the fact that the immune response against tick saliva and venoms (from recognized venomous taxa) are both dominated by an allergic immunity background. Furthermore, by comparing the major molecular components of human saliva, as an example of a non-venomous animal, with that of ticks we find evidence that ticks resemble more venomous than non-venomous animals. Finally, we introduce our considerations regarding the evolution of venoms in Arachnida.

Conclusions: Taking into account the composition of tick saliva, the venomous functions that ticks have while interacting with their hosts, and the distinguishable differences between human (non-venomous) and tick salivary proteins, we consider that ticks should be referred to as venomous ectoparasites.

Keywords: Ticks, Venom, Secreted proteins, Toxicoses, Pathogens, Convergence

Introduction

As haematophagous (blood sucking) arthropods, ticks are mainly considered as ectoparasites that use their salivary constituents to successfully obtain a blood meal by targeting major physiological pathways involved in host defense mechanisms [1]. Ticks constitute an important pest affecting agricultural development, as well as domestic animal and human health since they transmit a variety of infectious agents. Tick saliva has been described as a complex mixture of pharmacologically active compounds with implications for pathogen transmission [1]. From a functional and evolutionary point of view, Fry and colleagues [2], considered the feeding secretions of some haematophagous invertebrates (such as ticks) as a specialized subtype of venom. Certainly, Ixodida, that includes hard and soft tick species, is proven to be a venomous taxonomic Order in Chelicerata [3]. In fact, the bite from a single tick can produce several types of toxicoses [4]; paralysis being the most common and recognized form of tick-induced toxicoses [3,5].

Tick paralysis is an ascending motor paralysis produced by an impairment of neurotransmission, possibly due to the blockade of ion channels involved in the depolarization of nervous tissue [6]. This form of polyneuropathy is mainly associated with the acquisition of a blood meal by female ticks and will spread to the upper limbs of the host, causing incoordination and, in some cases, ending with respiratory failure and death [4]. Nevertheless, evident signs of toxicoses (e.g., paralysis) are not a *sine qua non* effect from the tick bite as in the

* Correspondence: valdjj@gmail.com
[3]Institute of Parasitology, Biology Centre of the Academy of Sciences of the Czech Republic, České Budějovice 37005, Czech Republic
Full list of author information is available at the end of the article

case of other venomous taxa, such as snakes, spiders, scorpions or pseudoscorpions. This observational scarcity is perhaps the reason ticks are not considered venomous animals. Thus, tick saliva as venom has rarely been mentioned in parasitological literature, with the exception of a few examples (e.g., as in [7]).

Traditionally, venom was defined as a toxic fluid that inflicts an abrupt death or paralysis in the host and/or prey. This archaic concept, however, partially highlights the deleterious effects of venom on the host/prey and lacks ecological relevance. After investigating many venomous animals, Fry and colleagues [2,8] extended this limited definition of venom *"as secretions produced in specialized glands and delivered through a wound (regardless of the wound size), that interferes with normal physiological processes to facilitate feeding or defense by the animal that produces the venom"*. By interfering with normal host physiological processes infers that all toxins are venomous, but not all venomous proteins are toxic. This new paradigm allows us to consider a wider spectrum of envenomation produced by a myriad of macromolecules. In our study we hypothesize that due to their salivary composition ticks are venomous animals within the phylum Chelicerata. We base our hypothesis on the following points: (i) the various toxic effects induced by ticks (ii) the convergent protein families present in spiders, scorpions and ticks; (iii) the immunomodulatory properties found in ticks saliva is also found in other venomous taxa (iv) the pattern of immune response against toxins by the host/prey is similar in both ticks and other venomous taxa; (v) the structural similarities in members of major protein families between known venomous taxa and ticks; (vi) the bimodal structural dichotomy between human (non-venomous) and tick saliva; and, finally (vii), the phylogenetic position of parasitiformes (Ixodida, Holothyrida and Mesostigmata) as a sister clade of pseudoscorpiones based on [9].

Results and discussion
Toxicoses phenomena within ixodida
The Australian *Ixodes holocyclus* is perhaps the best example of a tick that induces paralysis on livestock [10], pet animals [11], and humans [12]. Tick-induced paralysis, however, is not limited to this tick species but has been reported for ~8% of all tick species from major tick genera, except Carios and Aponomma [3] (69 out of approximately 869 tick species; 55 hard tick species and 14 soft tick species). Some of these paralyses inducing tick species represented in Figure 1 are also endemic to and abundant in several geographic regions [4]. Examples in the distribution of such ticks species are the North American *Ixodes scapularis*, *Dermacentor variabilis* and *Amblyomma americanum* [13,14], the South American *Amblyomma cajannense* [15], the European

Ixodes ricinus [16], and the globally distributed *Rhipicephalus sanguineus* [17].

Additionally, several lethal and paralysis inducing toxins have been identified in ticks. For example, the 15.4 kDa acidic salivary toxin secreted by *Ornithodoros savignyi* is highly abundant and its purified form kills a mouse within 90 minutes at a concentration of 400 μg/10 g of mouse weight [20]. Another purified basic toxin from the same tick species was shown to kill a 20 g mouse within 30 minutes after administration of 34 μg of the toxin [21]. Verified via Western blot, a 20 kDa trimeric neurotoxin was identified in the salivary glands of *Rhipicephalus evertsi evertsi* that paralyzed muscle contractions in an *in vitro* assay [22,23]. Maritz and colleagues (2001) identified a 60 kDa toxin in *Argas walkerae* that reduces [3H]glycine release from crude rat brain synaptosomes, indicating a paralytic effect. Other toxins have also been identified in tick egg extract from *Amblyomma hebraeum*, *R. e. evertsi*, *R. microplus*, *R. decoloratus* and *Hyalomma truncatum*, (revised in [4]). The presence of these toxins in tick eggs may be related to the protection of the egg mass against predation in natural environments – adding a new function for venoms in ticks, i.e., defense.

Besides tick paralysis, other types of toxicoses can be induced by a particular tick species, including sand tampan toxicoses by *O. savignyi*, sweating sickness, Mhlosinga, Magudu, and necrotic stomatitis nephrosis syndrome by *H. truncatum*, spring lamp paralysis in South Africa by *R. e. evertsi*, and, finally, specific toxicoses induced by *R. microplus*, *D. marginatus*, *R. appendiculatus*, *I. rekicorzevi* and *O. gurneyi* (revised in [3]). Toxicoses by *R. microplus*, *H. truncatum* and *R. appendiculatus* induce an anorexigenic effect [3], as induced by the secreted toxin Bv8 from the skin of the fire-bellied toad, *Bombina variegata* [24]. Symptoms of general toxicoses were also reported after soft tick bites that include pain, blisters, local irritation, oedema, fever, pruritus, inflammation and systemic disturbances [25]. Recently, human and canine toxicoses induced by the argasid tick *O. brasiliensis*, known as "mouro" tick, were reported and the most frequent symptoms of toxicoses induced by this tick species were local pruritus, slow healing lesions, local edema and erythema, and local skin rash [26]. Different types of immune reactions can also be included in the general scope of tick toxicoses [3,27]. Immediate and delayed skin hypersensitivity was reported in cattle exposed to *R. microplus* and *R. decoloratus* antigens [28,29], and in dogs exposed to *A. cajennense* antigens [30].

There are important factors in considering the severity of tick-induced toxicoses. (i) As stated by Paracelsus, *the dose makes the poison*. For example, *I. rubicundus* induces Karoo paralysis in South African livestock only when critical infestation densities are reached during repletion [31]. (ii) The anatomical location where the tick

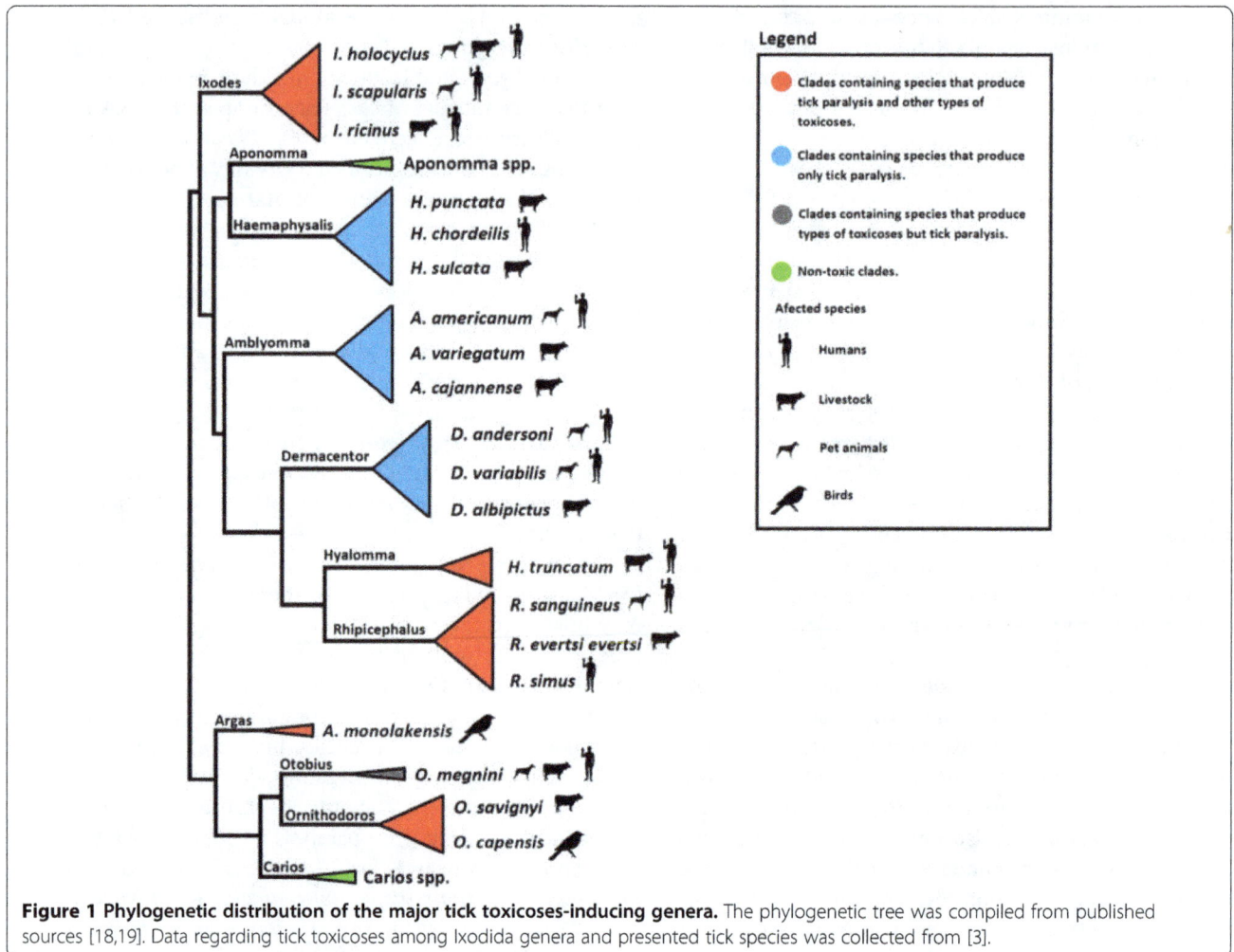

Figure 1 Phylogenetic distribution of the major tick toxicoses-inducing genera. The phylogenetic tree was compiled from published sources [18,19]. Data regarding tick toxicoses among Ixodida genera and presented tick species was collected from [3].

saliva is inoculated also seems to play a role in the toxic output. Although the tick species was not identified, a case report described a 3 year-old Indian boy with an acute onset of left-sided facial palsy secondary to tick infestation in the left ear [32]. Therefore, the proximity to a nerve (in this case the facial nerve) was important for the clinical toxic output (left-sided facial palsy). A similar case was also reported in a 3 year-old Turkish girl [33]. (iii) The duration of tick feeding is also an important factor of induced toxicoses [4]. Venzal and colleagues [34] showed that, after 3 days, laboratory mice infested with *Ornithodoros aff. puertoricensis* had initial signs of hyperaemia, followed by respiratory symptoms on day 4, and finally after 4 days the mice displayed nervous incoordination. A final factor (iv) to consider is the presence of common antigens between tick saliva and hosts. Recent episodes of human anaphylaxis after allergic sensitizations induced by bites of *A. americanum* have been reported. Patients with a history of *A. americanum* bites produced increased levels of pro-allergenic immunoglobulin E (IgE). The increased anti-tick IgE levels

in these patients were correlated to anaphylactic reactions to one anti-cancer monoclonal antibody (Cetuximab) and red meats [35]. Anaphylaxis induced by *A. americanum* is provoked by the presence of specific IgE to the carbohydrate galactose-alpha-1,3-galactose (alpha-gal) that is also present in Cetuximab and red meat [35]. Interestingly, alpha-gal was recently found in the gut of *I. ricinus*, a tick that also induces anaphylaxis [36].

The unified view of venom immune modulation and anti-venom immune responses

The haematotoxic and neurotoxic effects associated with venom exposure are widely recognized (revised in [2]). Nevertheless, all venomous animals are also constantly challenged by the host/prey or predator immune response. Studies have shown that the immune response of laboratory animals successfully counteracted venomous toxins [37,38]. In fact, natural resistance to snake venom was reported in both prey [39,40] and predator [41]. Thus, the immune system of the host/prey must constitute an important target of venoms in order to be effective. In fact, manipulating host defense mechanisms by venoms has been reported for some venomous animals like the parasitoid wasp, *Nasonia vitripennis* [42]. *N. vitripennis*, like ticks, are considered to be an ectoparasite since the *Nasonia* larvae feed on their hosts (invertebrates) without entering the host body [42]. The venom of *N. vitripennis* must suppress the immune response of the hosts in such a way that the host "allows" the parasitoid infection while simultaneously, the host will be able to control infections by other microorganism that otherwise would compete with the *N. vitripennis* larvae development [43]. Two major host defense cascades were suppressed by *N. vitripennis* venom: the phenoloxidase cascade and the coagulation cascade [43]. Several components of *N. vitripennis* venom have been suggested to modulate the host immune system, e.g., serine protease inhibitors, serine proteases, cystein-rich/Kunitz venom proteins and cysteine-rich/trypsin inhibitor-like venom proteins [43].

Manipulation of the host/prey immune system is not restricted to venomous Hymenoptera, e.g., *N. vitripennis*; for example, the haematophagous bat *Desmodus rotundus*, a venomous animal based on its salivary composition and feeding behavior [44], possesses two members of TNF-α-stimulated gene 6 (TSG-6) family that are highly expressed in its salivary glands. The TSG-6 family members have specific anti-inflammatory properties, such as the inhibition of neutrophil migration to interact with macrophage CD44 and modulation of NF-κB signaling [45]. This suggests that TSG-6 may play a feeding-facilitating role by suppressing the immune system. One well-studied example is the immune modulation induced by ticks in their hosts. The immune system manipulation by ticks is a complex process that has been recently revised [1].

Ticks are unique among hematophagous arthropods since they attach to host skin and feed for several days, while other blood-feeding arthropods (e.g. Triatomes or mosquitoes) feed little and often. Therefore, ticks need to counteract both the immediate innate immunity and the slower-developing adaptive immune responses in their vertebrate hosts. One first line of defense will be to counteract pain and itching responses of the host by targeting, for example, histamine, an immune-related mediator of pain and itch (revised in [1]). A few histamine-binding lipocalins was reported in the hard tick, *R. appendiculatus* [46]. In this regard, tick venom differs from canonical venoms since most venomous animals (e.g., wasps, bees, snakes, scorpions, spiders and jellyfish) will induce pain or an itch response. These venomous animals use their venom systems as a defensive or predatory function [47] with the desired effects of pain or itch to produce a deterrent effect. In contrast, similar to ticks, venomous haematophagous animals, like *D. rotundus* or triatomes bugs, should counteract prey/host awareness in order to feed until repletion.

After the skin is injured by a tick bite, the inflammatory response of the host will be activated. Ticks require a molecular arsenal to suppress both the cellular and molecular components of the host defenses. Tick salivary extract have been shown to reduce endothelial cell expression of the adhesion molecules ICAM-I and VCAM-I (*Dermacentor andersoni*) and P-selectin (*I. scapularis*). Reduction in adhesion molecules will reduce the extravasation of leukocytes at the site of tick attachment. The alternative pathway of complement activation is also one of the targets of the immunomodulation induced by tick saliva and thus complement inhibition activity has been reported in saliva of *D. andersoni*, *I. scapularis*, *I. ricinus*, *I. hexagonus*, *I. uriae* and *O. moubata* (revised in [1]). In addition, as a general trend, the saliva from haematophagous arthropods, including ticks, inhibit the proliferation of naïve T cell and the production of Th1 citokines [48]. One interesting example of modulating the adaptive immune response by hard ticks is Japanin. Japanin is a lipocalin that specifically reprograms human dendritic cells by hijacking the normal maturation process, even in the presence of "danger" signals like bacterial lipopolysaccharide [49]. Interestingly, Japanin promotes secretion of the anti-inflammatory cytokine IL-10 and increases expression of programmed death-ligand 1 (PD-L1), and both are involved in suppressing T cell immunity and induction of tolerogenic responses [49]. Such degree of molecular specialization has neither been described in other haematophagous arthropods nor in other venomous taxa. However, despite the immune suppression induced by tick saliva, some tick-host interactions result in immune-mediated acquired resistance to ticks after subsequent tick challenge.

Given the dynamics between induction of immune suppression by venoms and host/prey resistance development, an arms race between the host immune system and venomous components has been proposed [40]. The balance of this arms race will result in a susceptible or resistant host, prey or predators. In our revision of the topic, we found a convergence in the type of immune response that mammals display against both venom and tick saliva. Type 2 immune responses are mediated by lymphocytes T helper type 2 (Th2), IgE and IgG1 antibodies, but also by eosinophils, mast cells, basophils and, alternatively, by activated macrophages. This Th2 immune response encompasses a wider concept, namely allergies [50]. In mammals, venoms can induce allergic sensitization and development of specific IgE [37,38,51]; tick feeding also induces a Th2 polarization [1], specific IgE [52,53], and causes allergic sensitization [35]. The complex association between allergen IgE recognition with histamine secretion by mast cells and basophils that subsequently provoke uncomfortable reactions in animals has been highlighted [50]. This association goes beyond a specific neutralizing IgE antibodies response to a more complex detection of sensory stimuli by the olfactory, gustatory and visual systems that, surprisingly, may eventually result in developing aversive behaviors to specific locations or foods [50]. This suggests that the evolution of a differentiated pattern of immunity against venoms, including tick saliva, may have yet unexplored ecological implications. Another example of immune response convergence against venoms is that mast cells can be activated by the venom of scorpions without the concomitant presence of specific IgE [54], suggesting that the protective activities of mast cells is independent to the high affinity binding of IgE to the IgE receptor (FcεRI) present on mast cells. Wada and colleagues [55] recently showed that the protective role of mast cells in resistant mice to the tick *Haemaphysalis longicornis* was also independent of FcεRI. The above referenced studies show that (i) immune modulation may be a major function of venoms and (ii) the type of immune response elicited against the venom of ticks and other venomous taxa undergo similar immune pathways, thus tick saliva may possess venom-like molecules.

Tick saliva; or, the structural convergence of venomous proteins with venomous functions

The types of toxicoses induced by tick bites (ranging from lethal paralysis to local hypersensitivity) is not limited to the presence of lethal toxins but also to the presence of specific tick salivary protein families common among other venomous taxa. Recent advances in sequencing technologies have revealed an amazing body of information from the salivary glands of both hard and soft ticks [56-64]. From these high-throughput investigations, several protein families have been identified that are involved in tick-host interactions. Such protein families are found in the venoms of several other Metazoan species [2]. Examples of such venomous protein families found in tick saliva are defensins [65], lectins [66], cystatins [67], lipocalins [21,68-71], hyaluronidase [72], phospholipase A2 [73], Kunitz-like peptides [56,74,75], metalloproteases [76], AVIT [77], CAP proteins (Cysteine-Rich Secretion Proteins, Antigen 5, and Pathogenesis-Related) [2] and sphingomyelinase D [2].

Not only are these protein families present in tick saliva, but they also possess major functions described in conventional venomous systems. These functions include inhibition of thrombin, fXa, fVII/tissue factor system, platelet aggregation (i.e., collagen-induced, ADP-induced), act as a GPIIb/IIIa receptor antagonist, or affect fibrino (geno)lytic activity (revised in [2]). At the molecular level, venomous agents display common characteristics, despite their numerous biochemical activities and sequence variability, such as (i) possessing a signal peptide, (ii) displaying functional versatility within a protein family, (iii) targeting short-term physiological processes and,

(iv) stabilizing their tertiary structures via disulfide bonds. Finally, after being recruited as a functionally stable venomous agent, (v) duplication events occur to reinforce its adaptation (for a thorough description of said characteristics see [2]). An exception to this last property (gene duplication) is seen in platypus venom [78]. In the following sections we show the structural convergence between tick Kunitz peptides, cystatins, defensins, lipocalins, lectins and phospholipase A_2 and their conventional venomous counterparts.

Kunitz peptides

Kunitz peptides were named after Moses Kunitz who first discovered it in 1936 from bovine pancreas [79]. Since then, expression of the Kunitz protein family has been found in basically all kingdoms of life. Recent reports show that Kunitz peptides have undergone a massive gene expansion by gene duplication in the salivary glands of both *I. scapularis* [56] and *I. ricinus* [80], possibly due to specific selective pressures during the evolution of the tick-host interaction [81]. The Kunitz structure has been described with diverse functions in several venomous animals, including spiders and scorpions. Some Kunitz peptides from venomous animals possess dual activities by inhibiting both proteases and ion channels; examples of such toxins are LmKKT-1a from the scorpion *Lychas mucronatus* [82] and Huwentoxin-XI (HWTX-XI) from the spider *Ornithoctonus huwena* [83]. These venomous toxins have diversified their amino acid sequence causing a positive net charge on the all-atom Kunitz landscape (see Figure 2A). Reports have shown that toxins possessing a positive surface are most likely to target ion channels [84].

To date, only a few salivary secreted tick Kunitz peptides have been structurally resolved; however, these few reports reveal the venomous nature of these salivary peptides compared with other Kunitz structures from venomous animals. Figure 2A shows that the archetypal Kunitz fold is highly conserved for these tick salivary peptides and that they are structurally similar to HWTX-XI and LmKKT-1a. These structurally resolved tick salivary peptides show a structural conservation in their disulfide bridges (indicated by roman numerals), β-hairpin and the C-terminus α-helix. The only deviant from the archetypical Kunitz tertiary structure is Ra-KLP, since it is missing the second (II) disulfide bridge and possess a modified apex due to two atypical disulfide bridges (1 and 2; Figure 2A). Figure 2A also shows that the electrostatic potential of HWTX-XI and LmKKT-1a is strikingly similar to both TdPI and Ra-KLP, both from the salivary glands of *R. appendiculatus*. Ra-KLP has been reported as an ion channel modulator [85] like LmKKT-1a and HWTX-XI with no protease activity. It remains to be tested, however, if and how TdPI affects ion channels. Figure 2B shows a Cα backbone

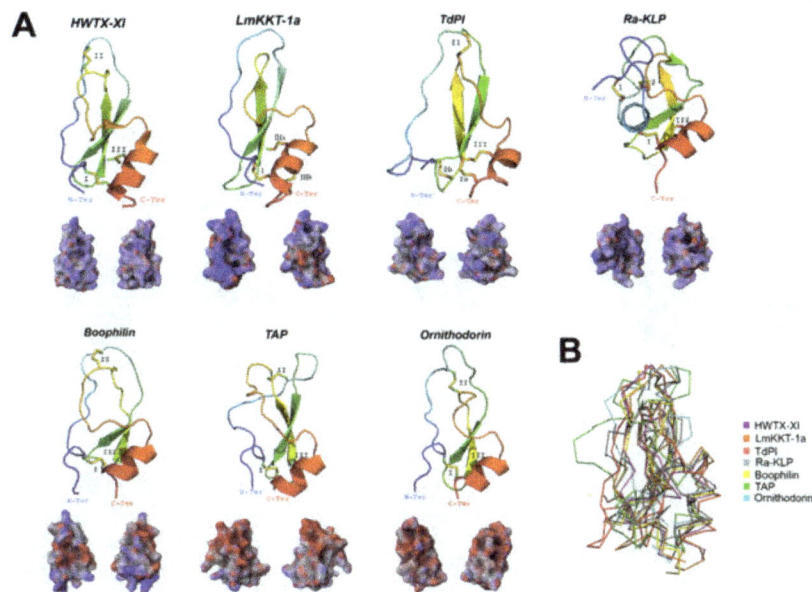

Figure 2 Tertiary structures of tick salivary Kunitz peptides. Panel **A** displays tertiary structure of toxins from spider (HWTX-XI; PDB: 2JOT) and scorpion (LmKKT-1a; PDB: 2 M01), and five tick salivary Kunitz-like peptides (PDBs: TAP-1D0D; ornithodorin-1TOC; boophilin-2ODY; TdPI-2UUX; Ra-KLP-2W8X). The tertiary structures depict the conserved disulfide bridges (indicated by roman numerals), loops, β-sheets that forms the β-hairpin, and α-helices. All structures are colored from the N-terminus (blue) to the C-terminus (red). Below each tertiary structure is the respective electrostatic potential in 180° turns (blue = positive; red = negative; white = neutral). A tertiary structural alignment in Panel **B** depicts the Cα protein backbone (color codes for each structure is presented on the right). (Note: For Panel A we used the C-terminus domain for both ornithodorin and boophilin since these possess two Kunitz-domains).

protein structural alignment of the represented Kunitz peptides. The root mean square deviation compared with HWTX-XI does not exceed 3 Å (TAP = 3 Å; ornithodorin = 2.4 Å; boophilin = 1.7 Å; TdPI = 2.8 Å; Ra-KLP = 2.8 Å); the structural difference with LmKKT-1a slightly varies from these deviations, but does not exceed 3.3 Å. Regardless of the conservative nature in the Kunitz fold, these tick salivary peptides display functional versatility and target different short-term physiological processes [85-89]. Therefore, as one of the most abundant tick salivary protein families [80], we consider Kunitz peptides as a typical example of a venomous agent that fit all five properties (i-v) referred above and described by Fry and colleagues [2].

Cystatins

Although cystatins have been identified from the venomous glands of spiders [90], snakes [91] and caterpillars [92], the venomous function of these cystatins remain elusive. Protease inhibition is the most common activity reported for these cystatins, as in one of the earliest studied cystatins isolated from the venom glands of the African puff adder (*Bitis arietans*) that inhibits papain, cathepsin B and dipeptidyl peptidase I [93]. The inhibitory sites of cystatins that bind during protein-protein interactions are the N-terminal loop and the two β-hairpin loop regions (indicated in Figure 3A as 1-3). A total of 95

cystine knot toxins have been identified in the venom glands of the tarantula *Chilobrachys jingzhao* and several of these toxins were reported to inhibit ion channels [90]. Two disulfide bonds form cystine knot toxins with their backbone connected by a third disulfide bond and the overall structure is invariably stabilized by β-sheets. Examples of these cystine knot toxins are Kunitz and defensin peptides. Although its toxic effects remain elusive, the cystatin JZTX-75 was among the 95 cystine knot toxins identified in the venom glands of the tarantula *C. jingzhao* [90]. The predicted tertiary structure of JZTX-75 (shown in Figure 3A) possesses a slightly positive electrostatic potential.

Over 80 cystatins have been reported in the salivary glands of hard and soft tick species [57,61,63,64,77,94]. For a full description on the physiological role of tick cystatins refer to [67]. In general, tick cystatins are potent inhibitors of papain-like cysteine proteases and play important roles during tick feeding. Tick salivary cystatins have been shown to serve as host immune modulators but their basic functions in tick saliva are unknown. A secreted cystatin has also been identified in the tick gut of *H. longicornis* that increases in expression during feeding on its host (Hlcyst-2; Figure 3A) [95]. Three crystal structures of cystatins secreted by tick salivary glands of *I. scapularis* (sialostatinL and sialostatinL2) and *O. moubata* (OmC2) have been resolved. Although

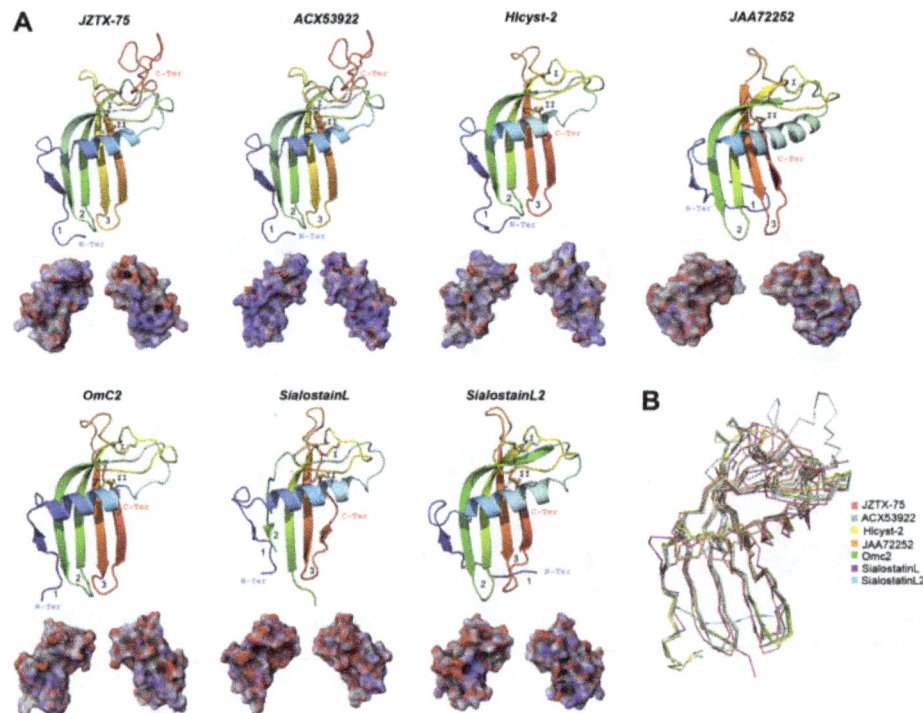

Figure 3 Tertiary structures of tick salivary cystatins. Panel **A** displays predicted tertiary structure of a cystatin from spider venom (JZTX75; GenBank: ABY71743) and, from tick salivary glands, we present three predicted cystatin structures (GenBank: ACX53922, JAA72252 and Hlcyst2-ABV71390) and three crystal structures (PDBs: OmC2-3L0R; sialostatinL-3LI7; sialostatinL2-3LH4). The tertiary structures depict the conserved disulfide bridges (indicated by roman numerals), β-sheets, the α-helix, and the inhibitory loop regions (1-3). All structures are colored from the N-terminus (blue) to the C-terminus (red). Below each tertiary structure is the respective electrostatic potential in 180° turns (blue = positive; red = negative; white = neutral). A tertiary structural alignment in Panel **B** depicts the Cα protein backbone (color codes for each structure is presented on the right). (Note: For Panel **A** we used the C-terminus domain for sialostatinL).

the binding of these tick cystatins remain elusive, an *in silico* study showed that these inhibitory loop regions for sialostatinL2 are conserved (Figure 3) [67]. A recent study showed that several tick cystatins were constantly expressed during a 5-day feeding period; among these was the cystatin ACX53922 [96]. Compared with the other five cystatins in Figure 3A, ACX53922 displays a more positive electrostatic potential throughout its all-atom landscape while still maintaining the archetypal tertiary backbone structure (Figure 3B; all structures have <3.0 Å root mean square deviation compared with JZTX-75).

Defensins

As in the Kunitz family, defensin peptides are widely distributed among the kingdoms of life as they are found in plants [97], jellyfish, sponges, nematodes, crustaceans, arachnids, insects, bivalves, snails, sea urchins, birds [98] and mammals [99], including humans [100]. The two structural classes of defensins are, (i) those exclusive to vertebrate known as α- β- and θ-defensins [101] and (ii), the most extended, possessing a simple structural motif known as the cysteine-stabilized α-helix and β-sheet

(CSαβ) [102] as those depicted in Figure 4A. Defensins have a wide range of biological functions, varying from sweet-tasting proteins to antimicrobial peptides (AMP) [102]. Recruitment of defensins has been reported in scorpion [103,104], snake [105], lizard [106], platypus [107] and spider [108] venom glands. The main function of defensins as animal toxins is to target ion channels [102]. Defensin molecules can also possess multiple biological functions that include ion channel modulation, antimicrobial and antifungal activity, such as crotamine, the β-defensin myotoxin from the rattlesnake *Crotalus durissus terrificus* [109]. In contrast, although isolated from the spider venom of *Ornithoctonus hainana*, the Oh-defensin was shown so far to only possess antimicrobial activity [108].

A single experimentally induced genetic deletion or mutation transforms a non-toxic defensin into a neurotoxin [99], thus, reinforcing the concept that toxic molecules are recruited from ancestral proteins possessing a non-toxic physiological function [2]. This also suggests the evolutionary steps necessary for recruiting defensins in the venom of venomous animals [99]. In agreement with the functional diversity of defensins, it is evident,

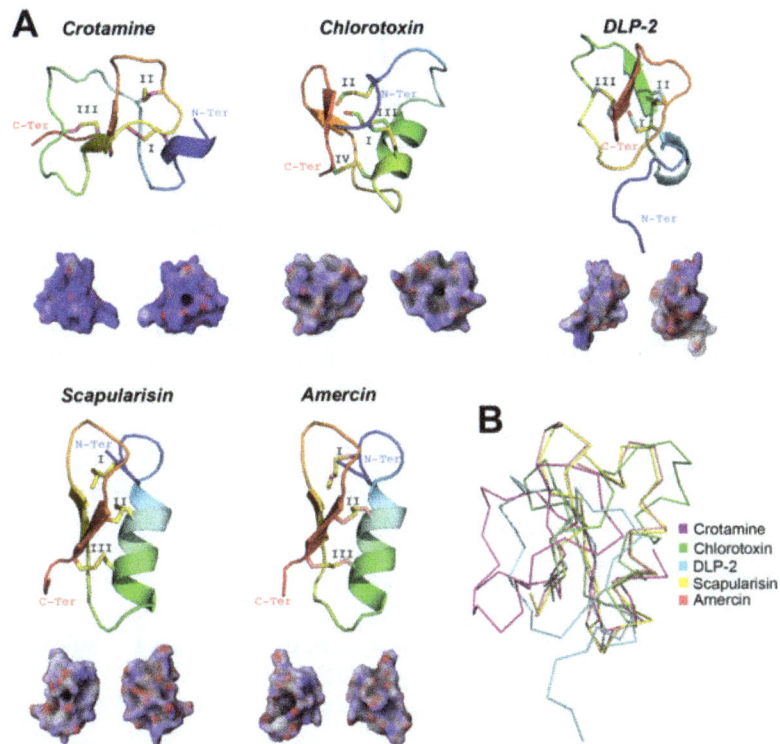

Figure 4 Tertiary structures of tick salivary defensin peptides. Panel **A** displays the crystal structure of toxins from rattlesnake (crotamine; PDB: 1Z99), scorpion (chlorotoxin; PDB: 1CHL) and platypus (DLP-2; PDB: 1D6B), and two predicted tertiary structures from tick salivary glands (GenBank: scapularisin-AAV74387 and amercin-ABI74752). The tertiary structures depict the conserved disulfide bridges (indicated by roman numerals), loops, β-sheets that forms the β-hairpin, and the α-helix. All structures are colored from the N-terminus (blue) to the C-terminus (red). Below each tertiary structure is the respective electrostatic potential in 180° turns (blue = positive; red = negative; white = neutral). A tertiary structural alignment in Panel **B** depicts the Cα protein backbone (color codes for each structure is presented on the right).

from reported crystal structures, that the tertiary structure is highly divergent (Figure 4B; all have ~3.5 Å root mean square deviation compared with crotamine). Defensins in ticks show a diverse expression pattern, thus they have been isolated from tick haemocytes, gut, intestine, ovaries, malpighian tubules and fat body [110,111]. Nevertheless, some of these defensins are exclusively expressed in tick salivary glands [112]. The most widely reported defensin structure in both hard and soft tick species is the CSαβ [113]. The only function assigned to the majority of characterized tick defensins, thus far, is AMP [65,110-112]; however, haemolytic activity was also recently reported for *I. ricinus* and *H. longicornis* defensins [111,114]. This obviously does not exclude the possibility that tick defensins may have other toxic functions in the vertebrate host. Other types of cysteine-rich AMPs from ticks were found to inhibit serine proteases [115], specifically chymotrypsin and elastase [116]. Furthermore, some tick defensins have secondary and tertiary structures similar to membrane potential modulators, such as scorpion neurotoxins, snake safaratoxins and plant γ-thionins [117] suggesting a toxic role for these tick defensins. Another example of

evolutionary convergence between ticks defensins and toxic defensins are the recently discovered multigenic defensin-like peptides, scasin and scapularisin, from the toxicoses-inducing tick *I. scapularis* (scapularisin is shown in Figure 4A) [65]. Of these functionally uncharacterized novel defensin-like peptides, scasins show a strong positive selection acting on the whole molecule [65], an evolutionary pattern observed before in conotoxins from the molluscs of the genera Conus [118] that act as ion channel modulators [119]. Nevertheless, further studies should clarify whether scapularisin and scasins are ion channel effectors, or not. As Figure 4A shows, however, tick defensins have similar electrostatic potentials as those found in snake and scorpion venom.

Lipocalins

Lipocalins are multifunctional proteins with a β-barrel structure that share three conserved domains in their primary structure, namely, motifs 1–3. Lipocalins have been implicated in development, regeneration, and pathological processes, but their specific roles are not known [120]. In reptiles and other venomous taxa, venom systems are enriched through gene duplication

[2,121], thereby increasing its functional divergence to develop a new function or neofunctionalization [47]. Neofunctionalization in tick lipocalins is a good example of functional diversification found in the venom of several venomous taxa [2].

Lipocalin-scaffolds have frequently been recruited as tick salivary components. Examples of toxin recruitment in tick salivary glands are the sand tampan toxins (TSGP) from *O. savignyi*, an abundant protein group that form a phylogenetic cluster with members of the tick lipocalin protein family, suggesting that they originated via gene duplication [21,69-71]. Three TSGPs were isolated from salivary gland extract in the tampan *O. savignyi*: the toxins TSGP2 that produces ventricular tachycardia, TSGP4 that produces Mobitz-type ventricular block, and the non-toxic TSGP3 that inhibits platelet aggregation [21,69,71]. Two other lipocalins closely related to TSGP2/TSGP3 are moubatin (platelet aggregation inhibitor; [122]) and OmCI from the soft tick *O. moubata* (complement inhibitor of C5 activation) [123]. It is worth mentioning that OmCI [124], TSGP2 and moubatin have dual action and triple action was reported for TSGP3 [71]. These tick salivary lipocalins are

depicted in Figure 5. Multifunctionalization is a common trait found in the toxins of *Lonomia obliqua* [125] and, as previously mentioned, for defensin peptides found in some venomous systems.

Due to a recent European upsurge of allergic reactions caused by the pigeon tick *A. reflexus* (e.g. [126]) a major allergen was identified (Arg r1) that is homologous to toxic lipocalins from *O. savignyi* [127,128]. Histamine-binding proteins were also described as lipocalins from the tick saliva of *R. appendiculatus* [46] and recently, a novel group of lipocalins was reported from metastriate tick saliva possessing a modulatory activity on dendritic cells [49]. Another lipocalin was isolated from *A. monolakensis* (AM-33) that binds to cysteinyl leokotrienes with high affinity, avoiding endothelial permeability and formation of edema, thus ensuring the tick to replete an erythrocyte-rich meal [129].

Phospholipase A2

The phospholipase A_2 (PLA_2) superfamily are ubiquitously found throughout the animal kingdom to catalyze the hydrolysis of ester bonds in a variety of different phospholipids producing lysophospholipids and free

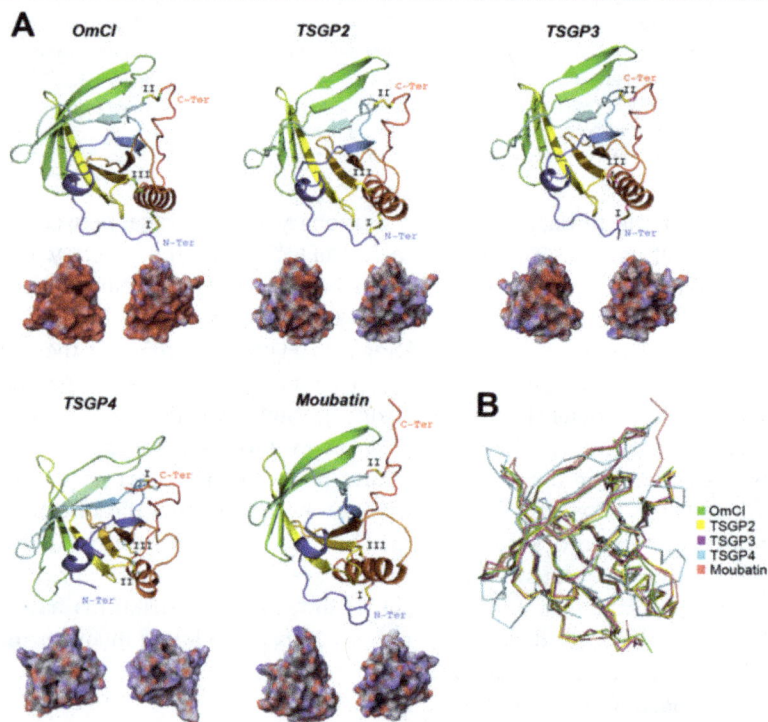

Figure 5 Tertiary structures of tick salivary lipocalins. Panel **A** displays the crystal structure of the tick lipocalin OmCI from *O. moubata* (; PDB: 3ZUO) and four predicted tertiary structures from the tick toxins, namely TSGP2, TSGP3, TSGP4 from *O. savignyi* and moubatin, also from *O. moubata* (respective UniProt: Q8I9U1, Q8I9U0, Q8I9T9 and Q04669). The tertiary structures depict the conserved disulfide bridges (indicated by roman numerals), loops, β-sheets that forms the β-hairpin, and the α-helix. All structures are colored from the N-terminus (blue) to the C-terminus (red). Below each tertiary structure is the respective electrostatic potential in 180° turns (blue = positive; red = negative; white = neutral). A tertiary structural alignment in Panel **B** depicts the Cα protein backbone (color codes for each structure is presented on the right). All structures have <2.2 Å root mean square deviation compared with OmCI.

fatty acids that play important physiological roles [130]. The PLA_2 superfamily includes five distinct enzyme types that are composed of 15 groups with many subgroups depending if they are secreted, cytosolic, calcium-independent or based on their specific target [131]. The PLA_2 superfamily has also been recruited via convergent evolution into the toxic arsenal of cephalopods, cnidarians, insects and arachnids [2]. In the venom of reptiles, PLA_2 appears as an antiplatelet aggregation factor [132], a myotoxin and a neurotoxin [133].

As previously stated, tick toxicoses is related to feeding and feeding cycle. Tick salivary gland PLA_2 activity was found to be higher in engorged *A. americanum* compared with unfed ticks and this increase was correlated with salivary gland secretion [134]. Although the function remains unknown, the PLA_2 activity found in the saliva of *A. americanum* was suggested to play an important role during prolonged tick feeding (10-14 days for *A. americanum*) [73]. The salivary PLA_2 from *A. americanum* is alkaline (pH: 9.5), as previously reported for PLA_2 from bee and snake venom [73], and does not contribute to the anticoagulant activities found in the saliva of *A. americanum* [135], but possess hemolytic activity [73]. The PLA_2 from both tick [136] and rattlesnake [137] possess antibacterial activities, suggesting a functional confluence between these two venomous species. Nevertheless, the PLA_2 from *A. americanum* was

not inhibited by aristolochic acid [73] as previously reported for the PLA_2 from the venomous snake, *Vipera ammodytes meridionalis* [138]. Sousa and colleagues [139] demonstrated the complexity of the toxic effects induced by venomous molecules since the PLA_2-melitin complex in *Apis mellifera* venom acted as a vasoconstrictor on rat aorta; however, no effect was evidenced for the PLA_2 and melitin fractions individually. New methods for testing the molecular functions of tick molecules may contribute to unravel the intricate putative toxic effects of tick salivary PLA_2. For example, a new method for dsRNA delivery was reported using *I. scapularis* eggs and nymphs, incorporating electroporation instead of microinjection. One of the genes that were successfully silenced was a putative PLA_2 from *I. scapularis* [58]. We compared this putative PLA_2 from *I. scapularis* (GenBank: EW812932), and a few other putative PLA_2 from ticks that induce toxicoses (most are depicted in Figure 1) with the crystal structure of the PLA_2 present in *A. mellifera* venom (PDB: 1POC; Figure 6A) and a predicted model from the scorpion toxin, imperatoxin-I that targets the sarcoplasmic reticulum calcium-release channel of skeletal and cardiac muscles [140]. Although these tick PLA_2 have <1.8 Å root mean square deviation compared with *A. mellifera* (Figure 6B), they have lost the disulfide bridge IV that is present in *A. mellifera* PLA_2 (Figure 6A). A recent study reports that, due evolutive

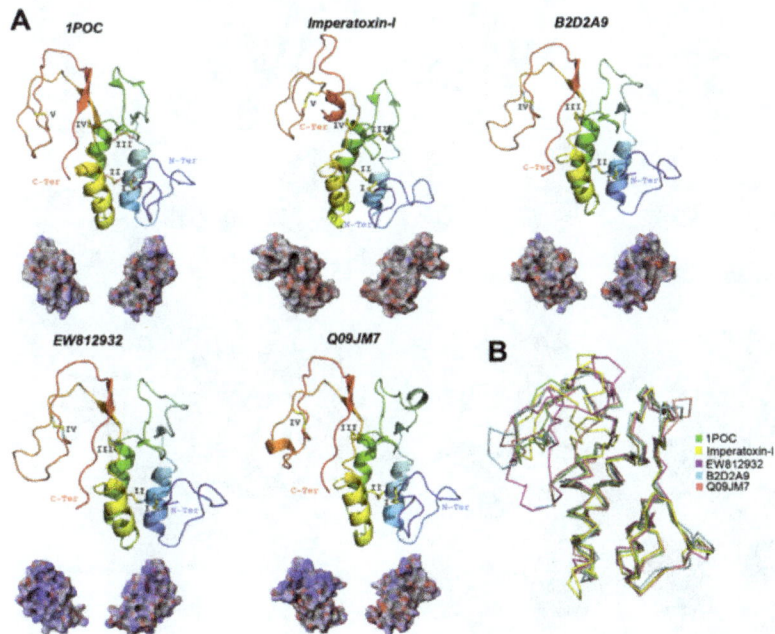

Figure 6 Tertiary structures of tick salivary PLA₂. Panel **A** displays the crystal structure of the PLA_2 from bee venom of *A. mellifera* (PDB: 1POC) and four predicted tertiary structures, three tick PLA_2 (respective UniProt: B2D2A9, Q09JM7 and the translated sequences GenBank: EW812932) and the scorpion toxin, imperatoxin-I (UniProt: P59888). The tertiary structures depict the conserved disulfide bridges (indicated by roman numerals), loops, and the α-helix. All structures are colored from the N-terminus (blue) to the C-terminus (red). Below each tertiary structure is the respective electrostatic potential in 180° turns (blue = positive; red = negative; white = neutral). A tertiary structural alignment in Panel **B** depicts the Cα protein backbone (color codes for each structure is presented on the right).

pressures caused during the arms race with the host(s), ticks express non-canonical variants of highly conserved protein families and these variants possess an altered disulfide bridge pattern that provide functional flexibility – e.g., Kunitz protein family from *I. ricinus* [80].

Lectins

Lectins can be defined as a wide variety of carbohydrate-binding proteins and glycoproteins from viruses, bacteria, fungi, protista, plants, and animals [141]. Lectins found in snake venom mainly affect blood coagulation pathways [142] and can also act as anti-angiogenic compounds [143]. In caterpillars, lectins are known to function as anticoagulants as in the *Lonomia* venom [92], but may also possess a myotoxic effect as in stonefish venom [144]. To date, tick lectin research has mainly focused on its roles in tick innate immunity (for revision see [145,146]). Earlier studies, however, showed that *R. microplus* saliva possesses lectin activity and induces immunosuppression in mice [147]. Additionally, Rego and colleagues [66,148] reported four tick lectins, two from *O. moubata* (Dorin M and OMFREP) and two from *I.*

ricinus (Ixoderin A and Ixoderin B). Based on phylogenetic analysis and expression pattern, a putative role in tick innate immunity was assigned to Ixoderin A and OMFREP. The role of Ixoderin B still remains unknown since it was exclusively expressed in the salivary glands and presented sequence divergence compared with Ixoderin A and OMFREP. Figure 7A depicts these tick lectins compared with ryncolin1 from the venom of the dog-faced water snake, *Cerberus rynchops* (all structures have <1.8 Å root mean square deviation; Figure 7B), which was suggested to induce platelet aggregation and/or initiate complement activation [149]. We note that IxoderinB also (as in Kunitz and PLA$_2$) does not possess the archetypal disulfide bridge pattern (II; Figure 7A). In addition to its sequence divergence to Ixoderin A and OMFREP, the missing disulfide bridge of IxoderinB may cause a more flexible fold thus diversifying its target(s).

A comparison of salivary proteins from humans and ticks
In order to compare the saliva of ticks to the saliva of a non-venomous animal we established a comparison

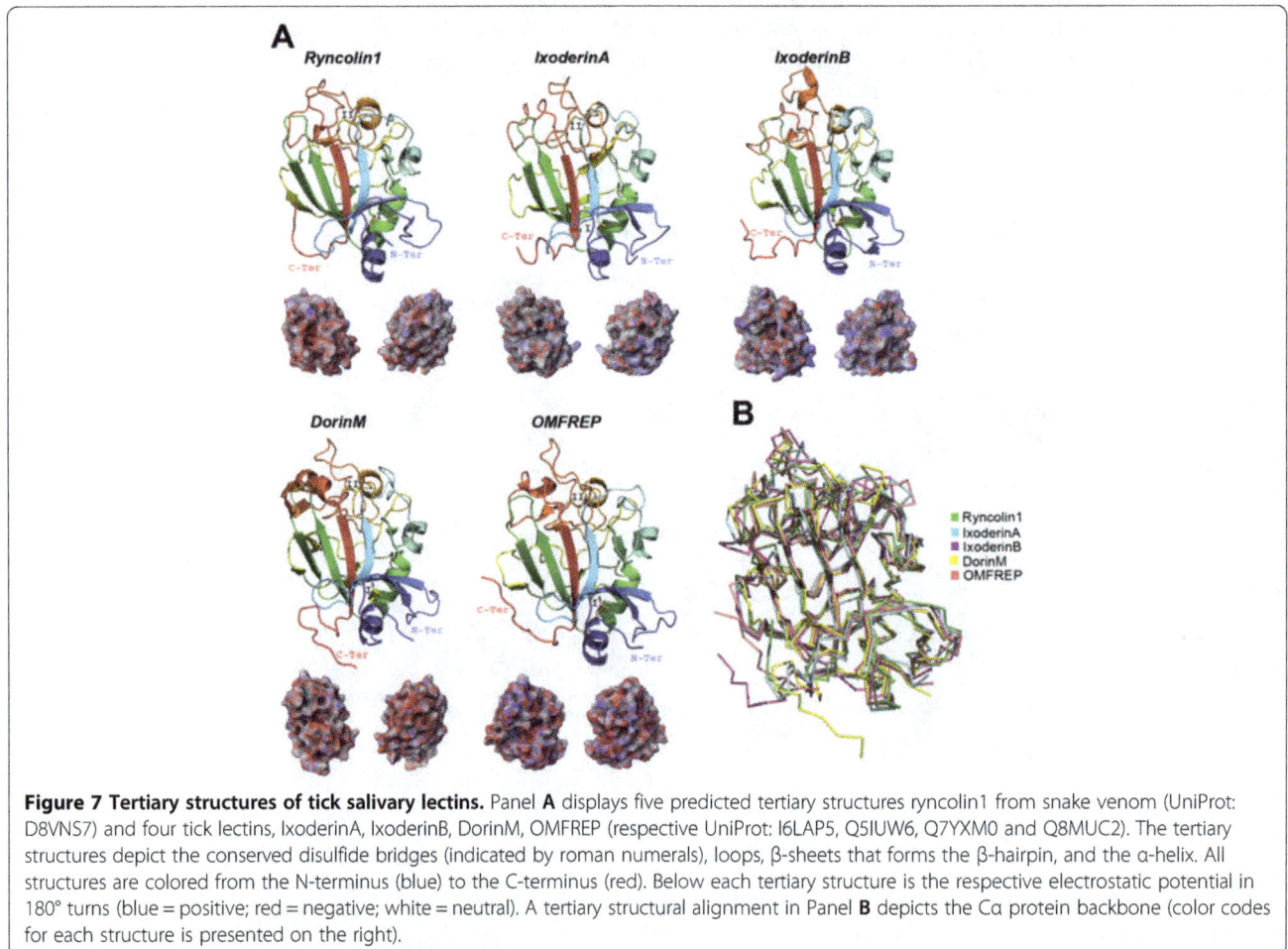

Figure 7 Tertiary structures of tick salivary lectins. Panel **A** displays five predicted tertiary structures ryncolin1 from snake venom (UniProt: D8VNS7) and four tick lectins, IxoderinA, IxoderinB, DorinM, OMFREP (respective UniProt: I6LAP5, Q5IUW6, Q7YXM0 and Q8MUC2). The tertiary structures depict the conserved disulfide bridges (indicated by roman numerals), loops, β-sheets that forms the β-hairpin, and the α-helix. All structures are colored from the N-terminus (blue) to the C-terminus (red). Below each tertiary structure is the respective electrostatic potential in 180° turns (blue = positive; red = negative; white = neutral). A tertiary structural alignment in Panel **B** depicts the Cα protein backbone (color codes for each structure is presented on the right).

between human and tick salivary systems. Recent proteomic studies have identified a total of 2698 proteins in human saliva (revised in [150]). The major protein components of human saliva are amylase, carbonic anhydrase, mucins, cystatins, proline-rich proteins, histatins, statherins and antibodies (revised in [150]), but it also contains defensins, lactoperoxidases, lysozymes and lactoferrins [151]. The complexity of the saliva from non-venomous animals (e.g., humans) is akin to that of ticks, but there are major differences in the molecular functions, the structure and the electrostatic potential of common salivary protein families. The two salivary systems present similar components such as lysozymes, cystatins, lipocalins, defensins and PLA2s. There are also differences, for example, human saliva possesses histatins and statherins, but these proteins have not been found in tick saliva. Despite that human salivary glands and the Kunitz protein family have been scientifically investigated for some time, the authors are unaware of any reports indicating the presence of Kunitz peptides (i.e., the archetypal 60 aa long peptide) in human saliva. The main reports for human Kunitz (also found in saliva) are of domains from larger proteins, e.g., immunoglobulins. Additionally, although the lectin intelectin-1 (UniProt: Q8WWA0) has been found in human saliva, its specific tertiary structure may drastically differ from those currently reported since a PSI-BLAST against the PDB was unable to retrieve a homologous structure and thus we were unable to model this human salivary lectin. In addition, human saliva does not present any allergen-like molecules; as are found in tick saliva. These differences are not surprising if we consider the different alimentary regimes these two species have and the molecular functions these two salivary systems need to perform. The phylogenetic distance between human and ticks maybe the most rational explanation for such differences. The salivary composition of the venomous mammal *D. rotundus*, however, is similar to ticks. *D. rotundus* salivary glands possess both Kunitz proteins and allergen-like molecules, while also possessing humans salivary components, like statherins and lysozymes [44]. As discussed above, this suggests that venomous animals recruit in their salivary glands a special set of proteins with specific venomous functions.

For the protein families we report here, the available tertiary structures for humans (and those we were able to model) are cystatins, defensins, lipocalins and PLA2 (Figure 8). There seems to be a bimodal diversity among these salivary protein families. Figure 8 shows that cystatins and lipocalins are structurally conserved, however, the electrostatic potential differs considerably – as dramatically displayed in OmCI. The opposite is seen for defensins and PLA2s – similar electrostatic potential surface with dissimilar tertiary structures. This bimodal

diversity is also evidenced within these protein families from venomous organisms compared to their human counterparts. The defensin crotamine, for instance (Figure 4), is structurally similar to human β-defensin1 (Figure 8), but their electrostatic potential and landscape differs (crotamine being more basic). It is worth mentioning that the tick salivary proteins depicted in Figure 8 are from some of the toxicoses-inducing ticks represented in Figure 1.

Considerations on the most recent common ancestor of Parasitiformes and Pseudoscorpiones

Given the evidence provided in the previous sections, we find it necessary to reconsider ticks as venomous ectoparasites due its salivary properties and its evolutionary implications. Arthropods represent a major component of the biodiversity of life on Earth and venom systems, as an evolutionary adaptation, appear many times in Arthropoda (e.g., spiders, scorpions, wasp, bees and flies). With a few exceptions in Hexapoda (i.e., bees and some wasps), the main function of venomous molecular systems amongst Arthropods is to assist in predation [47]. Ticks are haematophagous Arachnids and, based on morphological characteristics, it has been argued that Holothyrida is the sister taxa of Ixodida [18]. This evolutionary relationship with Holotryrida, free-living mites that mainly feed on the body fluids of dead arthropods [152], has lead to inferences regarding the feeding habits of the most recent common ancestor (MRCA) of ticks, which has been described as a saprophytic organism [153] or a scavenger in which haematophagy evolved subsequently [81]. Entomophagy was another type of predatory behavior proposed for the MRCA of ticks and the cannibalism observed in some soft tick species was an argument in favor of this hypothesis [81].

Haematophagy has evolved several times from predation in Insecta; for example, triatomine and cimicid bugs evolved from predatory heteropterans and Symphoromyia (and other hemathophagous flies) evolved from predatory ones [154]. The characteristics of the MRCA of ticks are paramount in understanding the evolution of tick salivary constituents. Recently, by using maximum likelihood, Bayesian and parsimony analyses of over 41 kilobases of DNA sequence from 62 single-copy nuclear protein-coding genes, Regier and colleagues [9] presented a strong, revised and actualized phylogeny of Arthropoda. In these phylogenetic analyses, Parasitiformes (Acari) appears as a sister group of Pseudoscorpions (Figure 9A). The phylogenetic position of Pseudoscorpions in Chelicerata has caused large debate in recent years. By using 2907 amino acids from 13 different proteins, another series of phylogenetic analyses by Ovchinnikov and Masta [155] also showed that Pseudoscorpions, although not a sister group, are closely related to Parasitiformes. From both studies [9,155], we could certainly infer that

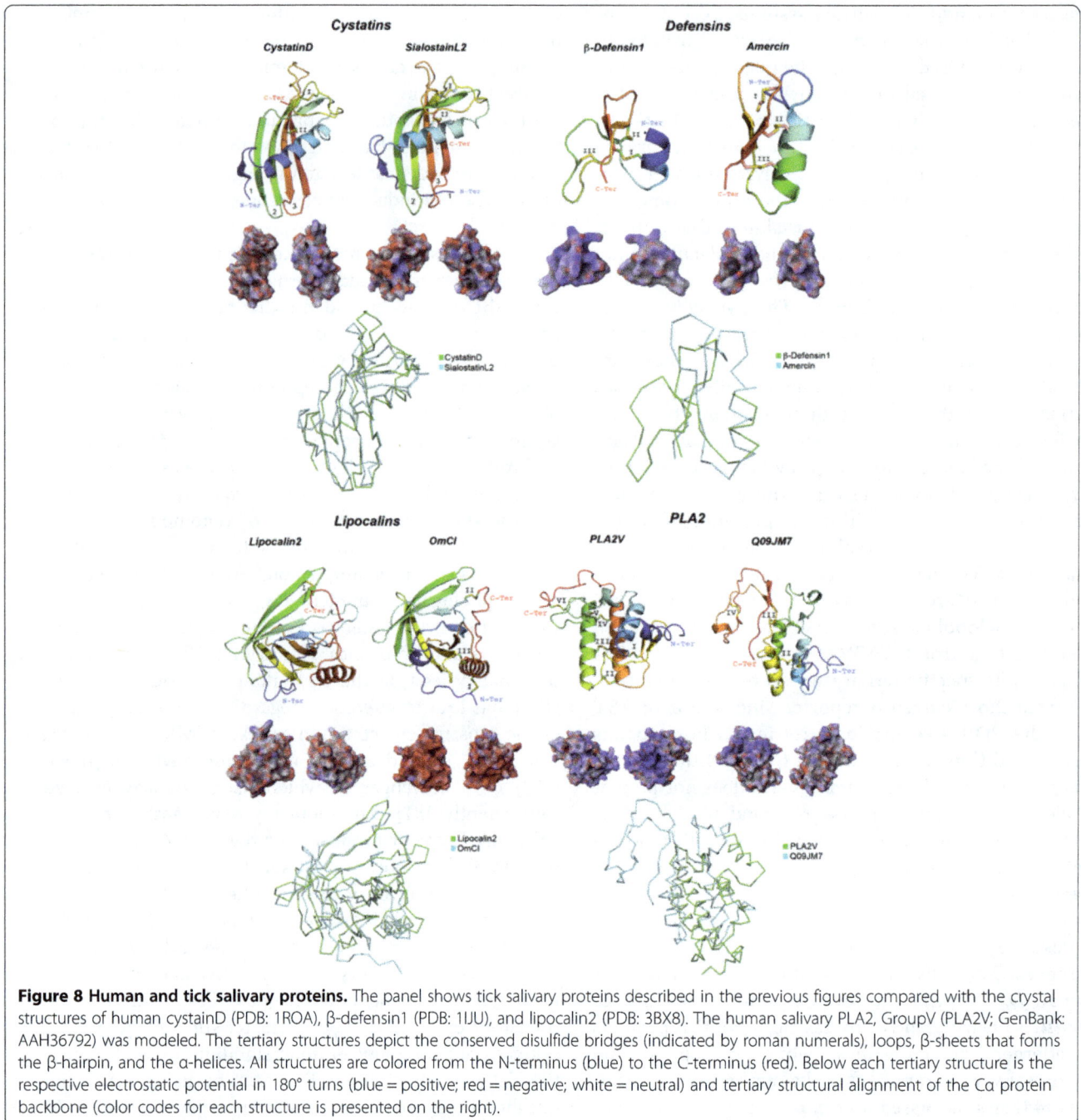

Figure 8 Human and tick salivary proteins. The panel shows tick salivary proteins described in the previous figures compared with the crystal structures of human cystainD (PDB: 1ROA), β-defensin1 (PDB: 1IJU), and lipocalin2 (PDB: 3BX8). The human salivary PLA2, GroupV (PLA2V; GenBank: AAH36792) was modeled. The tertiary structures depict the conserved disulfide bridges (indicated by roman numerals), loops, β-sheets that forms the β-hairpin, and the α-helices. All structures are colored from the N-terminus (blue) to the C-terminus (red). Below each tertiary structure is the respective electrostatic potential in 180° turns (blue = positive; red = negative; white = neutral) and tertiary structural alignment of the Cα protein backbone (color codes for each structure is presented on the right).

Pseudoscorpions and ticks share a recent common ancestor.

Pseudoscorpions are venomous members of Chelicerata [156] that, together with spiders and scorpions [2], constitute well-known examples of venomous animals amongst Chelicerata. Our working hypothesis is that the MRCA of ticks was a venomous predator of smaller preys that later evolved to feed on larger vertebrates. In agreement with this, the feeding plasticity (as a capacity of feeding on several hosts) of the tick ancestor was

recently suggested [157]. The convergence of several protein families between tick saliva and the venom of spiders and scorpions is shown in Figure 9A. This suggests a common origin in the venom systems of these three taxa (pseudoscorpions were not included in the comparison due to a lack of information regarding the venom composition of this species). Additionally, ticks are questing animals that will move actively in order to find their hosts [158-160]. Questing behavior constitute an important trait of tick ecology [158]. The evolution of

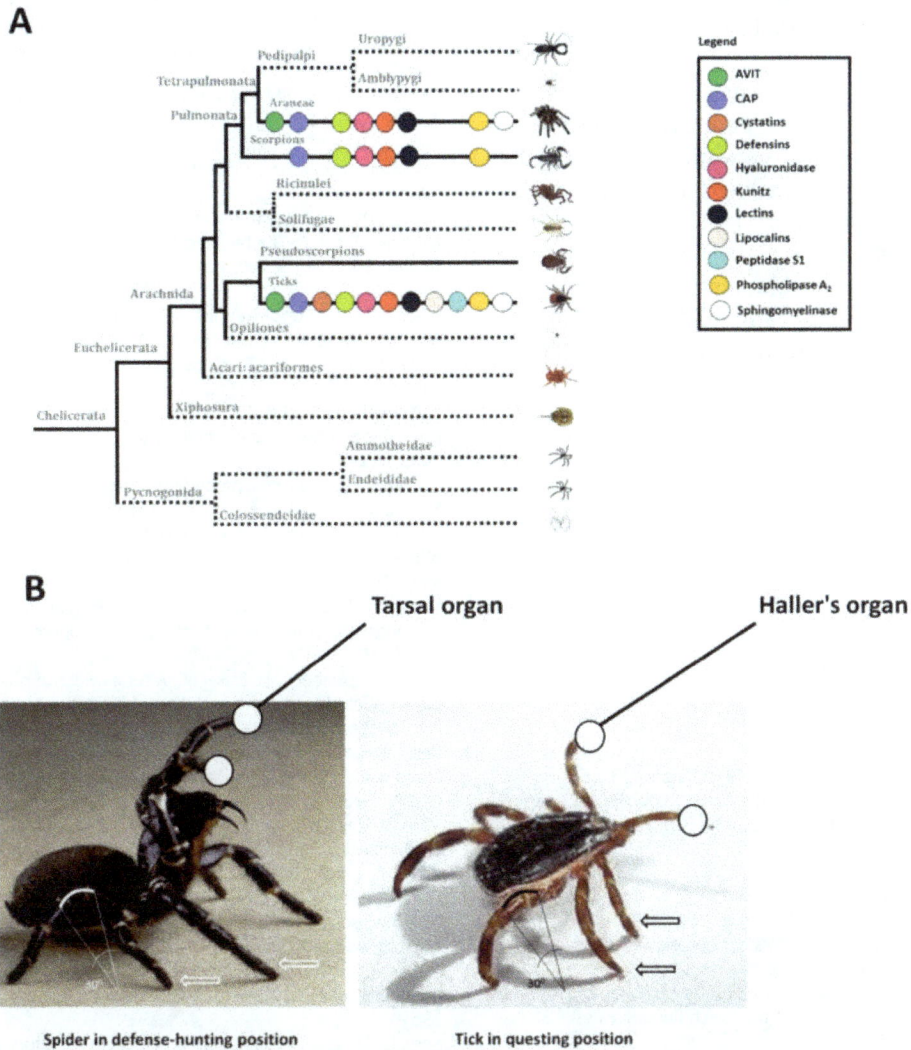

Figure 9 Similarities among venomous animals in chelicerata. Panel **A** shows the phylogenetic tree compiled from published sources [9]. Data regarding protein families in ticks, spiders and scorpions was collected from [2], as well as protein families characterized in recent studies. Panel **B** shows a spider's defense/hunting and a tick's questing positions (photos are not in real dimensional scale). The images depict similarities in attitude, angle of flexion in posterior legs (30°), middle legs (arrows), and first pair of legs exposing the respective sensory organs (white circles).

questing behavior is closely related to the evolution of sensory organs like the Haller's organ [161]. We found striking similarities in the questing behavior and position of sensory organs of ticks compared to the hunting behavior and position of sensory organs [162] of venomous spiders (Figure 9B).

Conclusion

The literature has split the saliva of haematophagous animals into those who consider the saliva as venomous and those that consider it as a special kind of saliva or a complex cocktail of bioactive components. Just last year (2013) the *Journal of Proteomics* published two investigations on the salivary gland components of the haematophagous bat *Desmodus rotundus*, where one refers to it as venomous [44] and the other simply as salivary components [45]. In tick research, the literature predominantly considers tick saliva as a complex cocktail of bioactive components and the toxicoses induced by ticks are mainly discussed in the context of paralysis. These claims narrow the ecological implications of the venomous relationship between ticks and their hosts. To classify ticks as mere ectoparasites is limited and underestimates the venomous structure of multigenic protein families in tick saliva. Therefore, we propose to consider ticks as venomous ectoparasites based on the intrinsic properties of tick saliva.

Methods

Tertiary protein modeling, structural alignment and electrostatic potential

Predicted tertiary models were generated using the Phyre2 server [163]. All predicted models were then refined via minimization and hydrogen-bond network was optimized by means of side chain sampling using the Schrodinger's Maestro Protein Preparation Wizard [164]. The structural deviations (root mean square deviation) were calculated using the protein structural alignment tool, from the Maestro software. Electrostatic potentials were calculated using the Poisson-Boltzmann equation also implemented in the Maestro software.

Abbreviations

AMP: Antimicrobial peptides; CSαβ: Cysteine-stabilized α-helix and β-sheet; Ig: Immunoglobulin; MRCA: Most recent common ancestor; PLA$_2$: Phospholipase A$_2$; PDB: Protein databank; Th2: T helper type 2.

Competing interests

Both authors declare that they have no competing interests.

Authors' contributions

Both authors have contributed equally. Both authors read and approved the final manuscript.

Acknowledgements

JJV was sponsored by project CZ.1.07/2.3.00/30.0032, co-financed by the European Social Fund and the state budget of the Czech Republic. ACC was supported by a grant from the Ministère de l'Education Supérieure et de la Recherche of France.

Author details

[1]Center for Infection and Immunity of Lille (CIIL), INSERM U1019 – CNRS UMR 8204, Université Lille Nord de France, Institut Pasteur de Lille, Lille, France. [2]SaBio. Instituto de Investigación de Recursos Cinegéticos, IREC-CSIC-UCLM-JCCM, Ciudad Real 13005, Spain. [3]Institute of Parasitology, Biology Centre of the Academy of Sciences of the Czech Republic, České Budějovice 37005, Czech Republic.

References

1.   Wikel S: Ticks and tick-borne pathogens at the cutaneous interface: host defenses, tick countermeasures, and a suitable environment for pathogen establishment. *Front Microbiol* 2013, 4:337.
2.   Fry BG, Roelants K, Champagne DE, Scheib H, Tyndall JDA, King GF, Nevalainen TJ, Norman JA, Lewis RJ, Norton RS, Renjifo C, de la Vega RC R: The toxicogenomic multiverse: convergent recruitment of proteins into animal venoms. *Annu Rev Genomics Hum Genet* 2009, 10:483–511.
3.   Mans BJ, Gothe R, Neitz AW: Biochemical perspectives on paralysis and other forms of toxicoses caused by ticks. *Parasitology* 2004, 129:S95–S111.
4.   Estrada-Peña A, Mans BJ: Tick-induced paralysis and toxicoses. In *Biology of ticks*, Volume Volume 2. 2nd edition. Edited by Sonenshine DE, Roe RM. USA: Oxford University Press; 2013:313–332.
5.   Edlow JA, McGillicuddy DC: Tick paralysis. *Infect Dis Clin North Am* 2008, 22:397–413.
6.   Maritz C, Louw AI, Gothe R, Neitz AW: Neuropathogenic properties of *Argas* (*Persicargas*) *walkerae* larval homogenates. *Comp Biochem Physiol A Mol Integr Physiol* 2001, 128:233–239.
7.   Motoyashiki T, Tu AT, Azimov DA, Ibragim K: Isolation of anticoagulant from the venom of tick, *Boophilus calcaratus*, from Uzbekistan. *Thromb Res* 2003, 110:235–241.
8.   Fry BG, Roelants J, Norman JA: Tentacles of venom: toxic protein convergence in the kingdom Animalia. *J Mol Evol* 2009, 68:311–321.
9.   Regier JC, Shultz JW, Zwick A, Hussey A, Ball B, Wetzer R, Martin JW, Cunningham CW: Arthropod relationships revealed by phylogenomic analysis of nuclear protein-coding sequences. *Nature* 2010, 463:1079–1083.
10. Bagnall BG, Doube BM: The Australian paralysis tick *Ixodes holocyclus*. *Aust Vet J* 1975, 51:159–160.
11. Ilkiw JE, Turner DM, Howlett CR: Infestation in the dog by the paralysis tick *Ixodes holocyclus*. I. Clinical and histological findings. *Aust Vet J* 1987, 64:137–139.
12. Pearn J: The clinical features of tick bite. *Med J Aust* 1977, 2:313–318.
13. Barbour AG, Fish D: The biological and social phenomenon of Lyme disease. *Science* 1993, 260:1610–1616.
14. Rydzewski J, Mateus-Pinilla N, Warner RE, Nelson JA, Velat TC: *Ixodes scapularis* (Acari: Ixodidae) distribution surveys in the Chicago metropolitan region. *J Med Entomol* 2012, 49:955–959.
15. Beati L, Nava S, Burkman EJ, Barros-Battesti DM, Labruna MB, Guglielmone AA, Cáceres AG, Guzmán-Cornejo CM, León R, Durden LA, Faccini JL: *Amblyomma cajennense* (Fabricius, 1787) (Acari: Ixodidae), the Cayenne tick: phylogeography and evidence for allopatric speciation. *BMC Evol Biol* 2013, 13:267.
16. Stanek G: Pandora's Box: pathogens in *Ixodes ricinus* ticks in Central Europe. *Wien Klin Wochenschr* 2009, 121:673–683.
17. Gray J, Dantas-Torres F, Estrada-Peña A, Levin M: Systematics and ecology of the brown dog tick, *Rhipicephalus sanguineus*. *Ticks Tick Borne Dis* 2013, 4:171–180.
18. Klompen JSH, Black WC, Keirans JE, Norris DE: Systematic and biogeography of hard ticks, a total evidence approach. *Cladistics* 2000, 16:79–102.
19. Murrel A, Campbell NJH, Barker SC: A total-evidence phylogeny of ticks provides insights into the evolution of life cycles and biogeography. *Mol Phylogenet Evol* 2001, 21:244–258.
20. Neitz AWH, Howell CJ, Potgieter DJJ: Purification of the toxic component in the oral secretion of the sand tampan *Ornithodoros savignyi* Audouin (1827). *J South Afr Chem Ind* 1969, 22:142–149.
21. Mans BJ, Steinmann CM, Venter JD, Louw AI, Neitz AWH: Pathogenic mechanisms of sand tampan toxicoses induced by the tick, *Ornithodoros savignyi*. *Toxicon* 2002, 40:1007–1016.
22. Viljoen GJ, Bezuidenhout JD, Oberem PT, Vermeulen NM, Visser L, Gothe R, Neitz AW: Isolation of a neurotoxin from the salivary glands of female *Rhipicephalus evertsi evertsi*. *J Parasitol* 1986, 72:865–874.
23. Crause JC, van Wyngaardt S, Gothe R, Neitz AW: A shared epitope found in the major paralysis inducing tick species of Africa. *Exp Appl Acarol* 1994, 18:51–59.
24. Negri L, Lattanzi R: Bv8/PK2 and prokineticin receptors: a druggable pronociceptive system. *Curr Opin Pharmacol* 2012, 12:62–66.
25. Hoogstraal H: Argasid and Nuttalliellid ticks as parasites and vectors. *Adv Parasitol* 1985, 24:135–238.
26. Reck J, Marks FS, Guimarães JA, Termignoni C, Martins JR: Epidemiology of *Ornithodoros brasiliensis* (mouro tick) in the southern Brazilian highlands and the description of human and animal retrospective cases of tick parasitism. *Ticks Tick Borne Dis* 2013, 4:101–109.
27. Wikel SK: Tick and mite toxicoses and allergy. In *Handbook of Natural Toxins*, Volume 2. Edited by Tu AT. New York: Marcel Dekker; 1984:371–396.
28. Marufu MC, Chimonyo M, Mans BJ, Dzama K: Cutaneous hypersensitivity responses to *Rhipicephalus* tick larval antigens in pre-sensitized cattle. *Ticks Tick Borne Dis* 2013, 4:311–316.
29. Marufu MC, Dzama K, Chimonyo M: Cellular responses to *Rhipicephalus microplus* infestations in pre-sensitised cattle with differing phenotypes of infestation. *Exp Appl Acarol* 2014, 62:241–522.
30. Mukai LS, Netto AC, Szabo MP, Bechara GH: Hypersensitivity induced in dogs by nymphal extract of *Amblyomma cajennense* ticks (Acari: Ixodidae). *Ann N Y Acad Sci* 2002, 969:184–186.
31. Fourie LJ, Kok DJ: Seasonal dynamics of the Karoo paralysis tick (*Ixodes rubicundus*): a comparative study on Merino and Dorper sheep. *Onderstepoort J Vet Res* 1996, 63:273–276.
32. Patil MM, Walikar BN, Kalyanshettar SS, Patil SV: Tick induced facial palsy. *Indian Pediatr* 2012, 49:57–58.
33. Gürbüz MK, Erdoğan M, Doğan N, Birdane L, Cingi C, Cingi E: Case report: isolated facial paralysis with a tick. *Türkiye Parazitol Derg* 2010, 34:61–64.

34. Venzal JM, Estrada-Peña A, Fernández de Luco D: **Effects produced by the feeding of larvae of *Ornithodoros aff. puertoricensis* (Acari: Argasidae) on laboratory mice.** *Exp Appl Acarol* 2007, **42**:217–223.

35. Berg EA, Platts-Mills TA, Commins SP: **Drug allergens and food-the cetuximab and galactose-α-1,3-galactose story.** *Ann Allergy Asthma Immunol* 2014, **112**:97–101.

36. Hamsten C, Starkhammar M, Tran TA, Johansson M, Bengtsson U, Ahlén G, Sällberg M, Grönlund H, van Hage M: **Identification of galactose-α-1,3-galactose in the gastrointestinal tract of the tick *Ixodes ricinus*; possible relationship with red meat allergy.** *Allergy* 2013, **68**:549–552.

37. Palm N, Rosenstein R, Yu S, Schenten D, Florsheim E, Medzhitov R: **Bee venom phospholipase A2 induces a primary type 2 response that is dependent on the receptor ST2 and confers protective immunity.** *Immunity* 2013, **39**:976–985.

38. Marichal T, Starkl P, Reber LL, Kalesnikoff J, Oettgen HC, Tsai M, Metz M, Galli SJ: **A beneficial role for immunoglobulin E in host defense against honeybee venom.** *Immunity* 2013, **39**:963–975.

39. Heatwole H, Poran NS: **Resistances of sympatric and allopatric eels to sea snake venoms.** *Copeia* 1995, **1**:136–147.

40. Biardi JE, Coss RG: **Rock squirrel (*Spermophilus variegatus*) blood sera affects proteolytic and hemolytic activities of rattlesnake venoms.** *Toxicon* 2011, **57**:323–331.

41. Jansa SA, Voss RS: **Adaptive evolution of the venom targeted vWF protein in opossums that eat pitvipers.** *PLoS One* 2011, **6**:e20997.

42. Qian C, Liu Y, Fang Q, Min-Li Y, Liu SS, Ye GY, Li YM: **Venom of the ectoparasitoid, Nasonia vitripennis, influences gene expression in Musca domestica hemocytes.** *Arch Insect Biochem Physiol* 2013, **83**:211–231.

43. Danneels EL, Rivers DB, de Graaf DC: **Venom proteins of the parasitoid wasp Nasonia vitripennis: recent discovery of an untapped pharmacopee.** *Toxins (Basel)* 2010, **4**:494–516.

44. Low DH, Sunagar K, Undheim EA, Ali SA, Alagon AC, Ruder T, Jackson TN, Pineda Gonzalez S, King GF, Jones A, Antunes A, Fry BG: **Dracula's children: molecular evolution of vampire bat venom.** *J Proteomics* 2013, **89**:95–111.

45. Francischetti IM, Assumpção TC, Ma D, Li Y, Vicente EC, Uieda W, Ribeiro JM: **The "Vampirome": transcriptome and proteome analysis of the principal and accessory submaxillary glands of the vampire bat *Desmodus rotundus*, a vector of human rabies.** *J Proteomics* 2013, **82**:288–319.

46. Paesen GC, Adams PL, Nuttall PA, Stuart DL: **Tick histamine-binding proteins: lipocalins with a second binding cavity.** *Biochim Biophys Acta* 2000, **1482**:92–101.

47. Casewell NR, Wüster W, Vonk FJ, Harrison RA, Fry BG: **Complex cocktails: the evolutionary novelty of venoms.** *Trends Ecol Evol* 2013, **28**:219–229.

48. Fontaine A, Diouf I, Bakkali N, Missé D, Pagès F, Fusai T, Rogier C, Almeras L: **Implication of haematophagous arthropod salivary proteins in host-vector interactions.** *Parasit Vectors* 2011, **4**:187.

49. Preston SG, Majtán J, Kouremenou C, Rysnik O, Burger LF, Cabezas Cruz A, Chiong Guzman M, Nunn MA, Paesen GC, Nuttall PA, Austyn JM: **Novel immunomodulators from hard ticks selectively reprogramme human dendritic cell responses.** *PLoS Pathog* 2013, **9**:e1003450.

50. Palm NW, Rosenstein RK, Medzhitov R: **Allergic host defences.** *Nature* 2012, **484**:465–472.

51. Komegae EN, Grund LZ, Lopes-Ferreira M, Lima C: **The longevity of Th2 humoral response induced by proteases natterins requires the participation of long-lasting innate-like B cells and plasma cells in spleen.** *PLoS One* 2013, **8**:e67135.

52. Kashino SS, Resende J, Sacco AM, Rocha C, Proença L, Carvalho WA, Firmino AA, Queiroz R, Benavides M, Gershwin LJ, De Miranda Santos IK: *Boophilus microplus*: **the pattern of bovine immunoglobulin isotype responses to high and low tick infestations.** *Exp Parasitol* 2005, **110**:12–21.

53. Christe M, Rutti B, Brossard M: **Susceptibility of BALB/c mice to nymphs and larvae of *Ixodes ricinus* after modulation of IgE production with anti-interleukin-4 or anti-interferon-gamma monoclonal antibodies.** *Parasitol Res* 1998, **84**:388–393.

54. Akahoshi M, Song CH, Piliponsky AM, Metz M, Guzzetta A, Abrink M, Schlenner SM, Feyerabend TB, Rodewald HR, Pejler G, Tsai M, Galli SJ: **Mast cell chymase reduces the toxicity of Gila monster venom, scorpion venom, and vasoactive intestinal polypeptide in mice.** *J Clin Invest* 2011, **121**:4180–4191.

55. Wada T, Ishiwata K, Koseki H, Ishikura T, Ugajin T, Ohnuma N, Obata K, Ishikawa R, Yoshikawa S, Mukai K, Kawano Y, Minegishi Y, Yokozeki H, Watanabe N, Karasuyama H: **Selective ablation of basophils in mice reveals their nonredundant role in acquired immunity against ticks.** *J Clin Invest* 2010, **120**:2867–2875.

56. Dai SX, Zhang AD, Huang JF: **Evolution, expansion and expression of the Kunitz/BPTI gene family associated with long-term blood feeding in *Ixodes scapularis*.** *BMC Evol Biol* 2012, **12**:4.

57. Anatriello E, Ribeiro J, de Miranda-Santos I, Brandao L, Anderson J, Valenzuela J, Maruyama S, Silva J, Ferreira B: **An insight into the sialotranscriptome of the brown dog tick, *Rhipicephalus sanguineus*.** *BMC Genomics* 2010, **11**:450.

58. Karim S, Troiano E, Mather TN: **Functional genomics tool: gene silencing in *Ixodes scapularis* eggs and nymphs by electroporated dsRNA.** *BMC Biotechnol* 2010, **10**:1.

59. Ribeiro JM, Anderson JM, Manoukis NC, Meng Z, Francischetti IM: **A further insight into the sialome of the tropical bont tick, *Amblyomma variegatum*.** *BMC Genomics* 2011, **12**:136.

60. Francischetti IM, Anderson JM, Manoukis N, Pham VM, Ribeiro JM: **An insight into the sialotranscriptome and proteome of the coarse bontlegged tick, *Hyalomma marginatum rufipes*.** *J Proteomics* 2011, **74**:2892–2908.

61. Schwarz A, von Reumont BM, Erhart J, Chagas AC, Ribeiro JMC, Kotsyfakis M: **De novo *Ixodes ricinus* salivary gland transcriptome analysis using two next-generation sequencing methodologies.** *FASEB J* 2013, **27**:4745–4756.

62. Mans BJ, Andersen JF, Francischetti IM, Valenzuela JG, Schwan TG, Pham VM, Garfield MK, Hammer CH, Ribeiro JM: **Comparative sialomics between hard and soft ticks: implications for the evolution of blood-feeding behavior.** *Insect Biochem Mol Biol* 2008, **38**:42–58.

63. Francischetti IMB, Mans BJ, Meng Z, Gudderra N, Veenstra TD, Pham VM, Ribeiro JMC: **An insight into the sialome of the soft tick, *Ornithodorus parkeri*.** *Insect Biochem Mol Biol* 2008, **38**:1–21.

64. Francischetti IMB, Meng Z, Mans BJ, Gudderra N, Hall M, Veenstra TD, Pham VM, Kotsyfakis M, Ribeiro JMC: **An insight into the salivary transcriptome and proteome of the soft tick and vector of epizootic bovine abortion, *Ornithodoros coriaceus*.** *J Proteomics* 2008, **71**:493–512.

65. Wang Y, Zhu S: **The defensin gene family expansion in the tick *Ixodes scapularis*.** *Dev Comp Immunol* 2011, **35**:1128–1134.

66. Rego RO, Kovár V, Kopácek P, Weise C, Man P, Sauman I, Grubhoffer L: **The tick plasma lectin, Dorin M, is a fibrinogen-related molecule.** *Insect Biochem Mol Biol* 2006, **36**:291–299.

67. Schwarz A, Valdés JJ, Kotsyfakis M: **The role of cystatins in tick physiology and blood feeding.** *Ticks Tick Borne Dis* 2012, **3**:117–127.

68. Rodriguez-Valle M, Moolhuijzen P, Piper EK, Weiss O, Vance M, Bellgard M, Lew-Tabor A: *Rhipicephalus microplus* **lipocalins (LRMs): genomic identification and analysis of the bovine immune response using in silico predicted B and T cell epitopes.** *Int J Parasitol* 2013, **43**:739–752.

69. Mans BJ, Venter JD, Vrey PJ, Louw AI, Neitz AW: **Identification of putative proteins involved in granule biogenesis of tick salivary glands.** *Electrophoresis* 2001, **22**:1739–1746.

70. Mans BJ, Louw AI, Neitz AW: **The major tick salivary gland proteins and toxins from the soft tick, *Ornithodoros savignyi*, are part of the tick Lipocalin family: implications for the origins of tick toxicoses.** *Mol Biol Evol* 2003, **20**:1158–1167.

71. Mans BJ, Ribeiro JM: **Function, mechanism and evolution of the moubatin-clade of soft tick lipocalins.** *Insect Biochem Mol Biol* 2008, **38**:841–852.

72. Neitz AW, Howell CJ, Potgieter DJ, Bezuidenhout JD: **Proteins and free amino acids in the salivary secretion and haemolymph of the tick *Amblyomma hebraeum*.** *Onderstepoort J Vet Res* 1978, **45**:235–240.

73. Bowman AS, Gengler CL, Surdick MR, Zhu K, Essenberg RC, Sauer JR, Dillwith JW: **A novel phospholipase A2 activity in saliva of the lone star tick, *Amblyomma americanum* (L.).** *Exp Parasitol* 1997, **87**:121–132.

74. Louw E, van der Merwe NA, Neitz AWH, Maritz-Olivier C: **Evolution of the tissue factor pathway inhibitor-like Kunitz domain-containing protein family in *Rhipicephalus microplus*.** *Int J Parasitol* 2013, **43**:81–94.

75. Valdés JJ, Schwarz A, Cabeza de Vaca I, Calvo E, Pedra JHF, Guallar V, Kotsyfakis M: **Tryptogalinin is a tick Kunitz serine protease inhibitor with a unique intrinsic disorder.** *PLoS One* 2013, **8**:e62562.

76. Francischetti IM, Mather TN, Ribeiro JM: **Cloning of a salivary gland metalloprotease and characterization of gelatinase and fibrin(ogen)lytic activities in the saliva of the Lyme disease tick vector *Ixodes scapularis*.** *Biochem Biophys Res Commun* 2003, **305**:869–875.

77. Ribeiro JM, Alarcon-Chaidez F, Francischetti IM, Mans BJ, Mather TN, Valenzuela JG, Wikel SK: An annotated catalog of salivary gland transcripts from *Ixodes scapularis* ticks. *Insect Biochem Mol Biol* 2006, **36**:111–129.

78. Wong ES, Papenfuss AT, Whittington CM, Warren WC, Belov K: A limited role for gene duplications in the evolution of platypus venom. *Mol Biol Evol* 2012, **29**:167–177.

79. Kunitz M, Northrop JH: Isolation from beef pancreas of crystalline trypsinogen, trypsin, a trypsin inhibitor, and an inhibitor-trypsin compound. *J Gen Physiol* 1936, **19**:991–1007.

80. Schwarz A, Cabezas-Cruz A, Kopecký J, Valdés JJ: Understanding the evolutionary structural variability and target specificity of tick salivary Kunitz peptides using next generation transcriptome data. *BMC Evol Biol* 2014, **14**:4.

81. Mans BJ, Neitz AW: Adaptation of ticks to a blood-feeding environment: evolution from a functional perspective. *Insect Biochem Mol Biol* 2004, **34**:1–17.

82. Chen ZY, Hu YT, Yang WS, He YW, Feng J, Wang B, Zhao RM, Ding JP, Cao ZJ, Li WX, Wu YL: Hg1, novel peptide inhibitor specific for Kv1.3 channels from first scorpion Kunitz-type potassium channel toxin family. *J Biol Chem* 2012, **287**:13813–13821.

83. Peng K, Lin Y, Liang SP: Nuclear magnetic resonance studies on huwentoxin-XI from the Chinese bird spider *Ornithoctonus huwena*: 15N labeling and sequence-specific 1H, 15N nuclear magnetic resonance assignments. *Acta Biochim Biophys Sin* 2006, **38**:457–466.

84. Selisko B, Garcia C, Becerril B, Delepierre M, Possani LD: An insect-specific toxin from *Centruroides noxius* Hoffmann. cDNA, primary structure, three-dimensional model and electrostatic surface potentials in comparison with other toxin variants. *Eur J Biochem* 1996, **242**:235–242.

85. Paesen GC, Siebold C, Dallas ML, Peers C, Harlos K, Nuttall PA, Nunn MA, Stuart DI, Esnouf RM: An ion-channel modulator from the saliva of the brown ear tick has a highly modified kunitz/BPTI structure. *J Mol Biol* 2009, **389**:734–747.

86. Jordan SP, Waxman L, Smith DE, Vlasuk GP: Tick anticoagulant peptide: kinetic analysis of the recombinant inhibitor with blood coagulation factor Xa. *Biochemistry* 1990, **29**:11095–11100.

87. van de Locht A, Stubbs MT, Bode W, Friedrich T, Bollschweiler C, Höffken W, Huber R: The ornithodorin-thrombin crystal structure, a key to the TAP enigma? *EMBO J* 1996, **15**:6011–6017.

88. Soares TS, Watanabe RM, Tanaka-Azevedo AM, Torquato RJ, Lu S, Figueiredo AC, Pereira PJ, Tanaka AS: Expression and functional characterization of boophilin, a thrombin inhibitor from *Rhipicephalus* (*Boophilus*) *microplus* midgut. *Vet Parasitol* 2012, **187**:521–528.

89. Paesen GC, Siebold C, Harlos K, Peacey MF, Nuttall PA, Stuart DI: A tick protein with a modified kunitz fold inhibits human tryptase. *J Mol Biol* 2007, **368**:1172–1186.

90. Chen J, Deng M, He Q, Meng E, Jiang L, Liao Z, Rong M, Liang S: Molecular diversity and evolution of cystine knot toxins of the tarantula *Chilobrachys jingzhao*. *Cell Mol Life Sci* 2008, **65**:2431–2444.

91. Richards R, St Pierre L, Trabi M, Johnson LA, de Jersey J, Masci PP, Lavin MF: Cloning and characterisation of novel cystatins from elapid snake venom glands. *Biochimie* 2011, **93**:659–668.

92. Veiga AB, Ribeiro JM, Guimarães JA, Francischetti IM: A catalog for the transcripts from the venomous structures of the caterpillar *Lonomia obliqua*: identification of the proteins potentially involved in the coagulation disorder and hemorrhagic syndrome. *Gene* 2005, **1**:11–27.

93. Evans HJ, Barrett AJ: A cystatin-like cysteine proteinase inhibitor from venom of the African puff adder (*Bitis arietans*). *Biochem J* 1987, **246**:795–797.

94. Imamura S, Konnai S, Yamada S, Parizi LF, Githaka N, Vaz I, Murata S, Ohashi K: Identification and partial characterization of a gut *Rhipicephalus appendiculatus* cystatin. *Ticks Tick Borne Dis* 2013, **4**:138–144.

95. Zhou J, Ueda M, Umemiya R, Battsetseg B, Boldbaatar D, Xuan X, Fujisaki K: A secreted cystatin from the tick *Haemaphysalis longicornis* and its distinct expression patterns in relation to innate immunity. *Insect Biochem Mol Biol* 2006, **36**:527–535.

96. Ibelli AR, Hermance M, Kim T, Gonzalez C, Mulenga A: Bioinformatics and expression analyses of the *Ixodes scapularis* tick cystatin family. *Exp Appl Acarol* 2013, **60**:41–53.

97. Gachomo EW, Jimenez-Lopez JC, Kayodé AP, Baba-Moussa L, Kotchoni SO: Structural characterization of plant defensin protein superfamily. *Mol Biol Rep* 2012, **39**:4461–4469.

98. van Dijk A, Veldhuizen EJ, Haagsman HP: Avian defensins. *Vet Immunol Immunopathol* 2008, **124**:1–18.

99. Zhu S, Peigneur S, Gao B, Umetsu Y, Ohki S, Tytgat J: Experimental conversion of a defensin into a neurotoxin: implications for origin of toxic function. *Mol Biol Evol* 2014, **31**:546–559.

100. Jarczak J, Kościuczuk EM, Lisowski P, Strzałkowska N, Jóźwik A, Horbańczuk J, Krzyżewski J, Zwierzchowski L, Bagnicka E: Defensins: natural component of human innate immunity. *Hum Immunol* 2013, **74**:1069–1079.

101. Menendez A, Finlay BB: Defensins in the immunology of bacterial infections. *Curr Opin Immunol* 2007, **19**:385–391.

102. Zhu S, Gao B, Tytgat J: Phylogenetic distribution, functional epitopes and evolution of the CSalphabeta superfamily. *Cell Mol Life Sci* 2005, **62**:2257–2269.

103. Zhu S, Tytgat J: The scorpine family of defensins: gene structure, alternative polyadenylation and fold recognition. *Cell Mol Life Sci* 2004, **61**:1751–1763.

104. Feng J, Yu C, Wang M, Li Z, Wu Y, Cao Z, Li W, He X, Han S: Expression and characterization of a novel scorpine-like peptide Ev37, from the scorpion *Euscorpiops validus*. *Protein Expr Purif* 2013, **88**:127–133.

105. Oguiura N, Boni-Mitake M, Affonso R, Zhang G: In vitro antibacterial and hemolytic activities of crotamine, a small basic myotoxin from rattlesnake *Crotalus durissus*. *J Antibiot (Tokyo)* 2011, **64**:327–331.

106. Dalla VL, Benato F, Maistro S, Quinzani S, Alibardi L: Bioinformatic and molecular characterization of beta-defensins-like peptides isolated from the green lizard *Anolis carolinensis*. *Dev Comp Immunol* 2012, **36**:222–229.

107. Whittington CM, Papenfuss AT, Bansal P, Torres AM, Wong ES, Deakin JE, Graves T, Alsop A, Schatzkamer K, Kremitzki C, Ponting CP, Temple-Smith P, Warren WC, Kuchel PW, Belov K: Defensins and the convergent evolution of platypus and reptile venom genes. *Genome Res* 2008, **18**:986–994.

108. Zhao H, Kong Y, Wang H, Yan T, Feng F, Bian J, Yang Y, Yu H: A defensin-like antimicrobial peptide from the venoms of spider, *Ornithoctonus hainana*. *J Pept Sci* 2011, **17**:540–544.

109. Yamane ES, Bizerra FC, Oliveira EB, Moreira JT, Rajabi M, Nunes GL, de Souza AO, da Silva ID, Yamane T, Karpel RL, Silva PI Jr, Hayashi MA: Unraveling the antifungal activity of a South American rattlesnake toxin crotamine. *Biochimie* 2013, **95**:231–240.

110. Ceraul SM, Dreher-Lesnick SM, Gillespie JJ, Rahman MS, Azad AF: New tick defensin isoform and antimicrobial gene expression in response to Rickettsia montanensis challenge. *Infect Immun* 2007, **75**:1973–1983.

111. Chrudimská T, Slaninová J, Rudenko N, Růžek D, Grubhoffer L: Functional characterization of two defensin isoforms of the hard tick *Ixodes ricinus*. *Parasit Vectors* 2011, **4**:63.

112. Zhou J, Liao M, Ueda M, Gong H, Xuan X, Fujisaki K: Sequence characterization and expression patterns of two defensin-like antimicrobial peptides from the tick *Haemaphysalis longicornis*. *Peptides* 2007, **28**:1304–1310.

113. Chrudimská T, Chrudimský T, Golovchenko M, Rudenko N, Grubhoffer L: New defensins from hard and soft ticks: similarities, differences, and phylogenetic analyses. *Vet Parasitol* 2010, **167**:298–303.

114. Lu X, Che Q, Lv Y, Wang M, Lu Z, Feng F, Liu J, Yu H: A novel defensin-like peptide from salivary glands of the hard tick, *Haemaphysalis longicornis*. *Protein Sci* 2010, **19**:392–397.

115. Fogaça AC, Almeida IC, Eberlin MN, Tanaka AS, Bulet P, Daffre S: Ixodidin, a novel antimicrobial peptide from the hemocytes of the cattle tick *Boophilus microplus* with inhibitory activity against serine proteinases. *Peptides* 2006, **27**:667–674.

116. Zhang H, Zhang W, Wang X, Zhou Y, Wang N, Zhou J: Identification of a cysteine-rich antimicrobial peptide from salivary glands of the tick *Rhipicephalus haemaphysaloides*. *Peptides* 2011, **32**:441–446.

117. Froy O, Gurevitz M: Membrane potential modulators: a thread of scarlet from plants to humans. *FASEB J* 1998, **12**:1793–1796.

118. Duda TF, Palumbi SR: Molecular genetics of ecological diversification: duplication and rapid evolution of toxin genes of the venomous gastropod Conus. *Proc Natl Acad Sci U S A* 1998, **96**:6820–6823.

119. Terlau H, Olivera BM: Conus venoms: a rich source of novel ion channel-targeted peptides. *Physiol Rev* 2004, **84**:41–68.

120. Carrijo-Carvalho LC, Maria DA, Ventura JS, Morais KL, Melo RL, Rodrigues CJ, Chudzinski-Tavassi AM: A lipocalin-derived Peptide modulating fibroblasts and extracellular matrix proteins. *J Toxicol* 2012, **2012**:325250.

121. Fry BG, Scheib H, de L M Junqueira de Azevedo I, Silva DA, Casewell NR: Novel transcripts in the maxillary venom glands of advanced snakes. *Toxicon* 2012, **59**:696–708.

122. Keller PM, Waxman L, Arnold BA, Schultz LD, Condra C, Connolly TM: Cloning of the cDNA and expression of moubatin, an inhibitor of platelet aggregation. *J Biol Chem* 1993, **268**:5450–5456.

123. Nunn MA, Sharma A, Paesen GC, Adamson S, Lissina O, Willis AC, Nuttall PA: Complement inhibitor of C5 activation from the soft tick *Ornithodoros moubata. J Immunol* 2005, **174**:2084–2091.

124. Roversi P, Ryffel B, Togbe D, Maillet I, Teixeira M, Ahmat N, Paesen GC, Lissina O, Boland W, Ploss K, Caesar JJ, Leonhartsberger S, Lea SM, Nunn MA: Bifunctional lipocalin ameliorates murine immune complex-induced acute lung injury. *J Biol Chem* 2013, **288**:18789–18802.

125. Alvarez FMP, Zannin M, Chudzinski-Tavassi AM: New insight into the mechanism of *Lonomia obliqua* envenoming: toxin involvement and molecular approach. *Pathophysiol Haemost Thromb* 2010, **37**:1–16.

126. Weckesser S, Hilger C, Lentz D, Jakob T: Anaphylactic reactions to bites of the pigeon tick *Argas reflexus. Eur J Dermatol* 2010, **20**:244–245.

127. Hilger C, Bessot JC, Hutt N, Grigioni F, De Blay F, Pauli G, Hentges F: IgE-mediated anaphylaxis caused by bites of the pigeon tick *Argas reflexus*: cloning and expression of the major allergen Arg r 1. *J Allergy Clin Immunol* 2005, **115**:617–622.

128. Mans BJ: Tick histamine-binding proteins and related lipocalins: potential as therapeutic agents. *Curr Opin Investig Drugs* 2005, **6**:1131–1135.

129. Mans BJ, Ribeiro JM: A novel clade of cysteinyl leukotriene scavengers in soft ticks. *Insect Biochem Mol Biol* 2008, **38**:862–870.

130. Burke JE, Dennis EA: Phospholipase A2 biochemistry. *Cardiovasc Drugs Ther* 2009, **23**:49–59.

131. Schaloske RH, Dennis EA: The phospholipase A2 superfamily and its group numbering system. *Biochim Biophys Acta* 2006, **1761**:1246–1259.

132. Chang HC, Tsai TS, Tsai IH: Functional proteomic approach to discover geographic variations of king cobra venoms from Southeast Asia and China. *J Proteomics* 2013, **89**:141–153.

133. Harris JB, Scott-Davey T: Secreted phospholipases A2 of snake venoms: effects on the peripheral neuromuscular system with comments on the role of phospholipases A2 in disorders of the CNS and their uses in industry. *Toxins (Basel)* 2013, **5**:2533–2571.

134. Zhu K, Bowman AS, Dillwith JW, Sauer JR: Phospholipase A$_2$ activity in salivary glands and saliva of the lone star tick (Acari: Ixodidae) during tick feeding. *J Med Entomol* 1998, **35**:500–504.

135. Zhu K, Sauer JR, Bowman AS, Dillwith JW: Identification and characterization of anticoagulant activities in the saliva of the lone star tick, *Amblyomma americanum* (L.). *J Parasitol* 1997, **83**:38–43.

136. Zeidner N, Ullmann A, Sackal C, Dolan M, Dietrich G, Piesman J, Champagne D: A borreliacidal factor in *Amblyomma americanum* saliva is associated with phospholipase A2 activity. *Exp Parasitol* 2009, **121**:370–375.

137. Sampaio SC, Hyslop S, Fontes MR, Prado-Franceschi J, Zambelli VO, Magro AJ, Brigatte P, Gutierrez VP, Cury Y: Crotoxin: novel activities for a classic beta-neurotoxin. *Toxicon* 2010, **55**:1045–1060.

138. Noetzel C, Chandra V, Perbandt M, Rajashankar K, Singh T, Aleksiev B, Kalkura N, Genov N, Betzel C: Enzymatic activity and inhibition of the neurotoxic complex vipoxin from the venom of *Vipera ammodytes meridionalis. Z Naturforsch C* 2002, **57**:1078–1083.

139. Sousa PC, Brito TS, Freire DS, Ximenes RM, Magalhães PJ, Monteiro HS, Alves RS, Martins AM, Toyama DO, Toyama MH: Vasoconstrictor effect of Africanized honeybee (*Apis mellifera* L.) venom on rat aorta. *J Venom Anim Toxins Incl Trop Dis* 2013, **19:**24.

140. Valdivia HH, Kirby MS, Lederer WJ, Coronado R: Scorpion toxins targeted against the sarcoplasmic reticulum Ca(2+)-release channel of skeletal and cardiac muscle. *Proc Natl Acad Sci U S A* 1992, **89**:12185–12189.

141. Vasta GR, Nita-Lazar M, Giomarelli B, Ahmed H, Du S, Cammarata M, Parrinello N, Bianchet MA, Amzel LM: Structural and functional diversity of the lectin repertoire in teleost fish: relevance to innate and adaptive immunity. *Dev Comp Immunol* 2011, **35**:1388–1399.

142. Herrera C, Rucavado A, Warrell DA, Gutiérrez JM: Systemic effects induced by the venom of the snake Bothrops caribbaeus in a murine model. *Toxicon* 2013, **1**:19–31.

143. Momic T, Cohen G, Reich R, Arlinghaus FT, Eble JA, Marcinkiewicz C, Lazarovici P: Vixapatin (VP12), a c-type lectin-protein from *Vipera xantina palestinae* venom: characterization as a novel anti-angiogenic compound. *Toxins (Basel)* 2012, **4**:862–877.

144. Magalhães GS, Junqueira-de-Azevedo IL, Lopes-Ferreira M, Lorenzini DM, Ho PL, Moura-da-Silva AM: Transcriptome analysis of expressed sequence tags from the venom glands of the fish *Thalassophryne nattereri. Biochimie* 2006, **88**:693–699.

145. Kopáček P, Hajdusek O, Buresová V, Daffre S: Tick innate immunity. *Adv Exp Med Biol* 2010, **708**:137–162.

146. Hajdušek O, Síma R, Ayllón N, Jalovecká M, Perner J, de la Fuente J, Kopáček P: Interaction of the tick immune system with transmitted pathogens. *Front Cell Infect Microbiol* 2013, **16:**26.

147. Bautista-Garfias CR, Martínez-Cruz MA, Córdoba-Alva F: Lectin activity from the cattle tick (*Boophilus microplus*) saliva. *Rev Latinoam Microbiol* 1997, **39**:83–89.

148. Rego RO, Hajdusek O, Kovár V, Kopáček P, Grubhoffer L, Hypsa V: Molecular cloning and comparative analysis of fibrinogen-related proteins from the soft tick *Ornithodoros moubata* and the hard tick *Ixodes ricinus. Insect Biochem Mol Biol* 2005, **35**:991–1004.

149. OmPraba G, Chapeaurouge A, Doley R, Devi KR, Padmanaban P, Venkatraman C, Velmurugan D, Lin Q, Kini RM: Identification of a novel family of snake venom proteins Veficolins from *Cerberus rynchops* using a venom gland transcriptomics and proteomics approach. *J Proteome Res* 2010, **9**:1882–1893.

150. Jágr M, Eckhardt A, Pataridis S, Broukal Z, Dušková J, Mikšík I: Proteomics of human teeth and saliva. *Physiol Res* 2014, **63**:S141–S154.

151. Loo JA, Yan W, Ramachandran P, Wong DT: Comparative human salivary and plasma proteomes. *J Dent Res* 2010, **89**:1016–1023.

152. Walter DE, Proctor HC: Feeding behaviour and phylogeny: observations on early derivative Acari. *Exp Appl Acarol* 1998, **22**:39–50.

153. Steen NA, Barker SC, Alewood PF: Proteins in the saliva of the Ixodida (ticks): pharmacological features and biological significance. *Toxicon* 2006, **47**:1–20.

154. Grimaldi D, Engel MS: *Evolution of the Insects.* UK: Cambridge University Press; 2005.

155. Ovchinnikov S, Masta SE: Pseudoscorpion mitochondria show rearranged genes and genome-wide reductions of RNA gene sizes and inferred structures, yet typical nucleotide composition bias. *BMC Evol Biol* 2012, **12**:12–31.

156. Murienne J, Harvey MS, Giribet G: First molecular phylogeny of the major clades of Pseudoscorpiones (Arthropoda: Chelicerata). *Mol Phylogenet Evol* 2008, **49**:170–184.

157. Mans BJ, de Klerk DG, Pienaar R, Latif AA: The host preferences of *Nuttalliella namaqua* (Ixodoidea: Nuttalliellidae): a generalist approach to surviving multiple host-switches. *Exp Appl Acarol* 2014, **62**:233–240.

158. Randolph SE: Tick ecology: processes and patterns behind the epidemiological risk posed by ixodid ticks as vectors. *Parasitol* 2004, **129:**S37–S65.

159. Goddard J: Observations on questing activity of adult *Ixodes brunneus* Koch (Acari: Ixodidae) in Mississippi. *J Parasitol* 2013, **99**:346–349.

160. Bartosik K, Wiśniowski Ł, Buczek A: Questing behavior of *Dermacentor reticulatus* adults (Acari: Amblyommidae) during diurnal activity periods in eastern Poland. *J Med Entomol* 2012, **49**:859–864.

161. Leonovich SA: The main evolutionary trends in sensory organs and questing behavior of parasitiform ticks and mites (Parasitiformes). *Parazitol* 2013, **47**:204–211.

162. Foelix RF, Chu-Wang IW, Beck L: Fine structure of tarsal sensory organs in the whip spider *Admetus pumilio* (Amblypygi, Arachnida). *Tissue Cell* 1975, **7**:331–346.

163. Kelley LA, Sternberg MJE: Protein structure prediction on the Web: a case study using the Phyre server. *Nat Protocols* 2009, **4**:363–371.

164. Li X, Jacobson MP, Zhu K, Zhao S, Friesner RA: Assignment of polar states for protein amino acid residues using an interaction cluster decomposition algorithm and its application to high resolution protein structure modeling. *Proteins Struct Funct Bioinformatics* 2007, **66**:824–837.

5

Proposal for a revised classification of the Demospongiae (Porifera)

Christine Morrow[1] and Paco Cárdenas[2,3*]

Abstract

Background: Demospongiae is the largest sponge class including 81% of all living sponges with nearly 7,000 species worldwide. *Systema Porifera* (2002) was the result of a large international collaboration to update the Demospongiae higher taxa classification, essentially based on morphological data. Since then, an increasing number of molecular phylogenetic studies have considerably shaken this taxonomic framework, with numerous polyphyletic groups revealed or confirmed and new clades discovered. And yet, despite a few taxonomical changes, the overall framework of the *Systema Porifera* classification still stands and is used as it is by the scientific community. This has led to a widening phylogeny/classification gap which creates biases and inconsistencies for the many end-users of this classification and ultimately impedes our understanding of today's marine ecosystems and evolutionary processes. In an attempt to bridge this phylogeny/classification gap, we propose to officially revise the higher taxa Demospongiae classification.

Discussion: We propose a revision of the Demospongiae higher taxa classification, essentially based on molecular data of the last ten years. We recommend the use of three subclasses: Verongimorpha, Keratosa and Heteroscleromorpha. We retain seven (Agelasida, Chondrosiida, Dendroceratida, Dictyoceratida, Haplosclerida, Poecilosclerida, Verongiida) of the 13 orders from *Systema Porifera*. We recommend the abandonment of five order names (Hadromerida, Halichondrida, Halisarcida, lithistids, Verticillitida) and resurrect or upgrade six order names (Axinellida, Merliida, Spongillida, Sphaerocladina, Suberitida, Tetractinellida). Finally, we create seven new orders (Bubarida, Desmacellida, Polymastiida, Scopalinida, Clionaida, Tethyida, Trachycladida). These added to the recently created orders (Biemnida and Chondrillida) make a total of 22 orders in the revised classification. We propose the abandonment of the haplosclerid and poecilosclerid suborders. The family content of each order is also revised.

Summary: The deletion of polyphyletic taxa, the use of resurrected or new names for new clades and the proposal of new family groupings will improve the comparability of studies in a wide range of scientific fields using sponges as their object of study. It is envisaged that this will lead to new and more meaningful evolutionary hypotheses for the end-users of the Demospongiae classification.

Keywords: Taxonomy, Systematics, Sponges, Lithistids, Heteroscleromorpha, Polyphyletic, Monophyletic, Type taxon

* Correspondence: paco.cardenas@fkog.uu.se
[2]Department of Organismal Biology, Division of Systematic Biology, Evolutionary Biology Centre, Uppsala University, Norbyvägen 18D, 752 36 Uppsala, Sweden
[3]Department of Medicinal Chemistry, Division of Pharmacognosy, BioMedical Centre, Husargatan 3, Uppsala University, 751 23 Uppsala, Sweden
Full list of author information is available at the end of the article

Background

The *Systema Porifera* (SP) [1] was the result of a collaboration of 45 researchers from 17 countries led by editors J. Hooper and R. W. M. van Soest. This milestone publication in 2002 provided an updated comprehensive overview of sponge (Porifera) systematics, the largest revision of this group (from genera, subfamilies, families, suborders, orders and class) since the start of spongiology in the mid-19th century. Because before 2002 only a handful of sponge molecular studies were available, the classification of SP is largely based on sponge morphology and re-evaluation of type material, thus providing "a sound platform for the future development of sponge systematics". Since then, an increasing number of molecular phylogenetic studies have considerably shaken the taxonomic framework of SP (for a review, see [2]) especially concerning the Demospongiae. This is the largest class and includes about 81% of all living sponges with nearly 7,000 species and more than 50 new species on average described every year [3,4].

One of main reasons that molecular results contradict traditional taxonomy may be that this classification was essentially based on the morphology and arrangement of spicules, characters which have repeatedly been shown to be highly homoplasic in demosponges (i.e. prone to convergent evolution and secondary loss) [5-8]. 13 years after SP, the various molecular studies have greatly challenged the Demospongiae classification, telling a largely congruent story where numerous polyphyletic groups have been revealed or confirmed and new clades have been identified. And yet, despite a few taxonomic changes (see Cárdenas et al. [3], p. 159 for a review of these changes), the overall framework of SP classification still stands and is mirrored in the World Porifera Database (WPD, http://www.marinespecies.org/porifera). This is the most widely used reference for sponge nomenclature and part of the World Register for Marine Species (WoRMS). This has led to a widening phylogeny/classification gap which creates biases and inconsistencies for the many end-users of this classification (biochemists, microbiologists, ecologists, conservationists, paleontologists, developmental biologists) and ultimately impedes our understanding of today's marine ecosystems and evolutionary processes. In an attempt to bridge this phylogeny/classification gap, the studies of Morrow et al. [5,9] in particular, but also Redmond et al. [10] and the review of Cárdenas et al. [3] started to anticipate and suggest a revised higher taxa classification of Demospongiae. Indeed, we still have too few demosponge taxa sequenced to generate a full revision of Demospongiae classification to genus level, but we do have enough taxon coverage to suggest revisions of the higher taxa. The last sentence of the SP preface [1] states: "The *Systema Porifera* project is not an end – but a sound beginning for this new generation to build on what we propose here". We consider it timely to build on the SP classification and officially propose a revised classification of Demospongiae.

Discussion

Revising the classification

The SP questioned the validity of the subclasses Ceractinomorpha and Tetractinomorpha [11], based on different reproductive strategies. Since then, the polyphyly of the Ceractinomorpha and Tetractinomorpha has been repeatedly confirmed by molecular data. Instead, four well separated Demospongiae clades were identified, often designated under the G1, G2, G3 and G4 clades *sensu* Borchiellini et al. [12]. Since then, these four clades have been considered subclasses and have been named: Keratosa (G1), Myxospongiae (=Verongimorpha) (G2), Haploscleromorpha (G3) and Heteroscleromorpha (G4) [3]. Ceractinomopha and Tetractinomorpha are now officially unaccepted by WPD, but the new four subclasses are not currently implemented in WPD. In this paper, we essentially revisit the current demosponge subclasses, orders and suborders by i) highlighting polyphyletic taxa and the corresponding names that should be abandoned, ii) creating new orders for the newly identified clades and iii) reallocating families to what we believe is their correct order. Doing so, we propose a revised classification of Demospongiae, essentially based on the latest molecular results. To fit the Linnaean rank-based nomenclature, seven new orders (all within the Heteroscleromorpha) have been created to accommodate new groupings of families: Bubarida ord. nov., Desmacellida ord. nov., Polymastiida ord. nov., Scopalinida ord. nov., Clionaida ord. nov., Tethyida ord. nov., and Trachycladida ord. nov.. Other orders not present in SP are upgraded from SP suborders (Spongillida) or resurrected (Axinellida, Merliida, Suberitida, Tetractinellida). Seven SP orders are maintained (Agelasida, Chondrosiida, Dendroceratida, Dictyoceratida, Haplosclerida, Poecilosclerida, Verongiida) and two recently created orders are also included (Chondrillida, Biemnida). Although the naming of orders is not governed by the International Code of Zoological Nomenclature (ICZN), the tradition is to follow a similar rule as for the naming of families (ICZN articles 29.1 and 29.2): adding the suffix –ida to the stem of a genus name (P. Bouchet, pers. comm.). When the genus' stem ends in –ia, this makes an order name ending in –iida which explains why we decided to modify the names of Verongiida and Chondrosiida from their original spelling with one 'i'. We have revised the diagnoses of resurrected orders or orders whose content has changed, by revisiting their morphological and chemical characters. We are well aware that some new definitions might appear too wide, due to the fact that we currently lack morphological synapomorphies for these new clades. For taxa where the morphological characters are ambiguous and molecular data are lacking we have used the qualifier '*incertae sedis*' when allocating them to a particular higher taxa. To avoid the creation of 'orphan' taxa and in order to anticipate the genera re-allocations that will ensue from this proposal, we include a table of Heteroscleromorpha genera (Appendix) with tentative order and family allocations within the framework of our proposal, based on SP and molecular results. We have highlighted where there is supporting molecular data for this allocation and particularly where there is molecular data for the type taxon (Appendix). The allocations of some of the genera are likely to change in the future but we consider this table as a working hypothesis and the necessary first step for the future revision of sponge families and genera. Because of a lack of combined morphological/molecular approaches in Keratosa, Verongimorpha and Haplosclerida *sensu stricto*, the genera content of their families remains to this day unchanged (and is therefore not reviewed in Appendix). Figure 1 represents the *Systema Porifera* Demospongiae classification. Crossed out in red are names that should be abandoned. Figure 2 represents our proposal for a revised Demospongiae classification. Relationships reflect the current knowledge of molecular phylogenetics, resulting from markers *18S, 28S, CO1* (*cytochrome oxidase subunit 1*, usually the Folmer fragment) and almost complete mitochondrial (mt.) genomes. In Figure 2, we have also flagged with an asterisk '*' all the families that are suspected to be non-monophyletic in order to help future systematic studies target problematic groups in need of revision and alert end-users to where contradictory results may arise.

Three versus four subclasses

One of the main discordant points among sponge taxonomists and the higher taxa may be this one: should we create three subclasses (Verongimorpha, Keratosa and Heteroscleromorpha — including the Haplosclerida) or four (Verongimorpha, Keratosa, Haploscleromorpha and Heteroscleromorpha)? The four subclasses classification originates from the first Demospongiae molecular study which named four distinct clades: G1 to G4 [12]. So the issue is whether marine Haplosclerida can be considered part of the Heteroscleromorpha or not and for this, we should first look at molecular studies with the widest taxon sampling which are those issuing from the Porifera Tree of Life (PorToL) project [10,13], and then at those with the highest number of characters; the mt. genome studies of Lavrov et al. [14]. *18S* suggests there are four clades with strong support (>90 bootstrap support (b.s.)), marine Haplosclerida and Heteroscleromorpha group with moderate support (70 b.s.) [10]. *28S* and mt-genome phylogenetic analyses also find the same four strongly supported clades but this time marine Haplosclerida and Heteroscleromorpha group with a stronger support of 90 b.s. [13-15]. So current molecular data support either three or

four subclasses. But skeleton morphology favours three subclasses since Verongimorpha and Keratosa do not have (for the most part) siliceous spicules, and especially do not share the diversity of microscleres present in Heteroscleromorpha and Haplosclerida. By choosing three subclasses we can restrict the order Haplosclerida to the marine Haplosclerida and include it in the Heteroscleromorpha, which becomes by far the largest Demospongiae subclass.

Deleting polyphyletic groups
Abandoning the subclasses Tetractinomorpha and Ceractinomorpha

Tetractinomorpha and Ceractinomorpha are a legacy from the works of Lévi [16,17] that tentatively grouped sponges according to their modes of reproduction (oviparous vs. ovoviviparous). Although early morphological cladistic analysis suggested the polyphyly of these subclasses [18,19], SP followed the classification of Lévi [20] and subdivided the class Demospongiae into three subclasses: Tetractinomorpha, Ceractinomorpha and Homoscleromorpha. The Homoscleromorpha will not be considered here as it was removed from the Demospongiae and is now accepted as a separate sponge class [21]. Shortly after the publication of SP, molecular studies confirmed the polyphyly of Tetractinomorpha and Ceractinomorpha [12,22]. The abandonment of Tetractinomorpha and Ceractinomorpha was officially agreed upon during the 7th International Sponge Symposium (Búzios, Brazil, May 2006), formally published by Boury-Esnault [23] and implemented in WPD.

Abandoning Halisarcida

In SP, Chondrosiida includes four genera: *Chondrilla, Thymosia, Thymosiopsis* and *Chondrosia*. Molecular results have repeatedly suggested the polyphyly of this order with *Chondrilla, Thymosia* and *Thymosiopsis* grouping with *Halisarca* (only genus of Halisarcidae, and of Halisarcida) [10,12,24]. Meanwhile *Chondrosia* was either sister group of Verongiida (very well supported) with ribosomal nuclear markers [10,12] or sister-group of a Verongiida + Chondrillidae clade with *CO1* [25]. It was already suggested that Halisarcida should be abandoned and Halisarcidae reallocated to Chondrosiida [26], a proposal also previously made on morphological grounds [27]. The family Chondrillidae Gray, 1872 (including *Chondrilla, Thymosia* and *Thymosiopsis*) was resurrected to be associated with Halisarcidae in the new order Chondrillida [10]. Despite contradictory results (*18S-28S* vs. *CO1*) with respect to the position of *Chondrosia*, Redmond et al. [10] decided to abandon Chondrosiida and include the resurrected Chondrosiidae Schulze, 1877 in Verongiida. Given the inconsistencies between ribosomal markers and *CO1* with respect to the position of *Chondrosia* we have decided to retain Chondrosiida

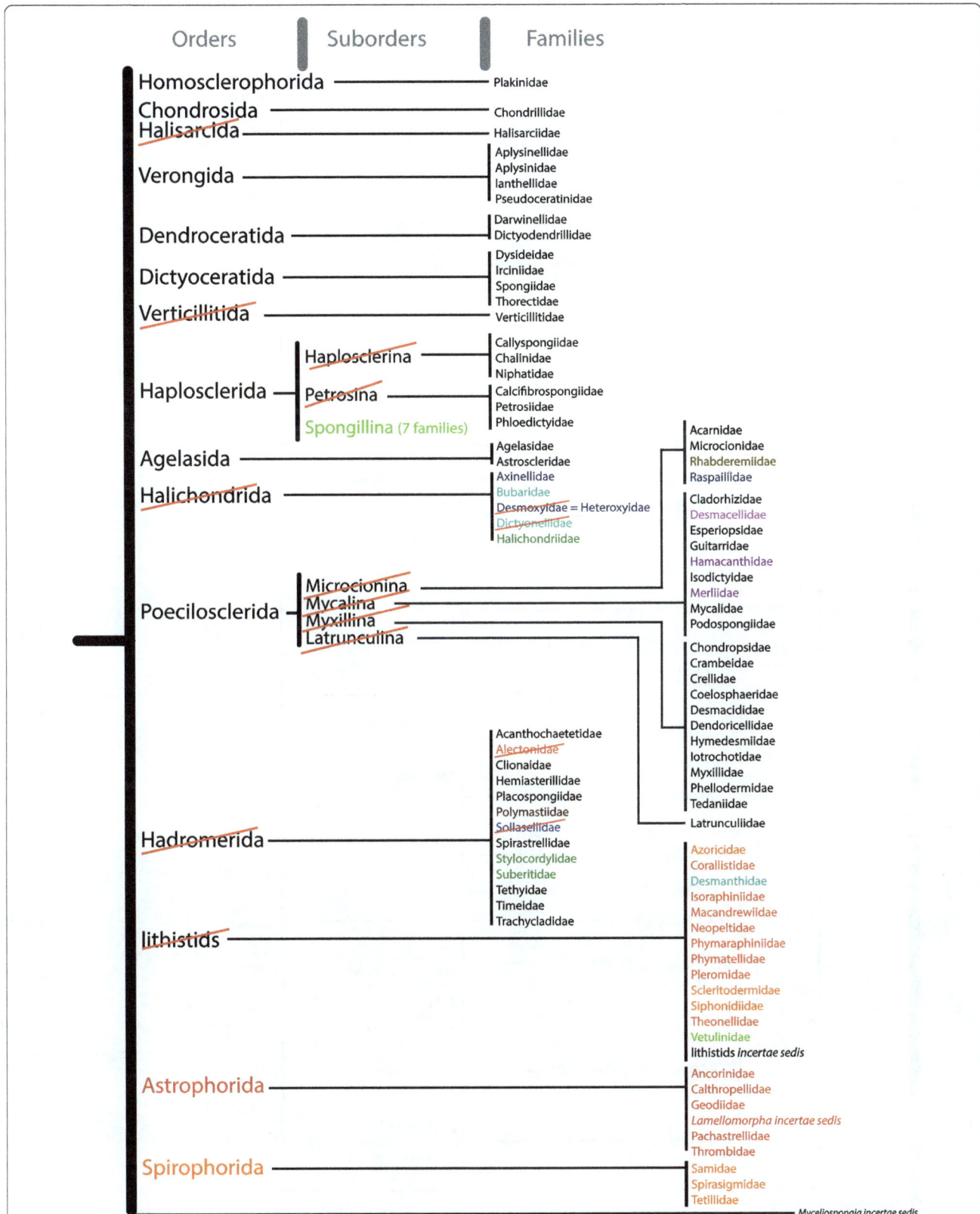

Figure 1 Demospongiae classification from orders to families, as presented in the *Systema Porifera* [1]. Names crossed out in red should be abandoned. Coloured names highlight taxa that should be reallocated; for their new allocation, see Figure 2.

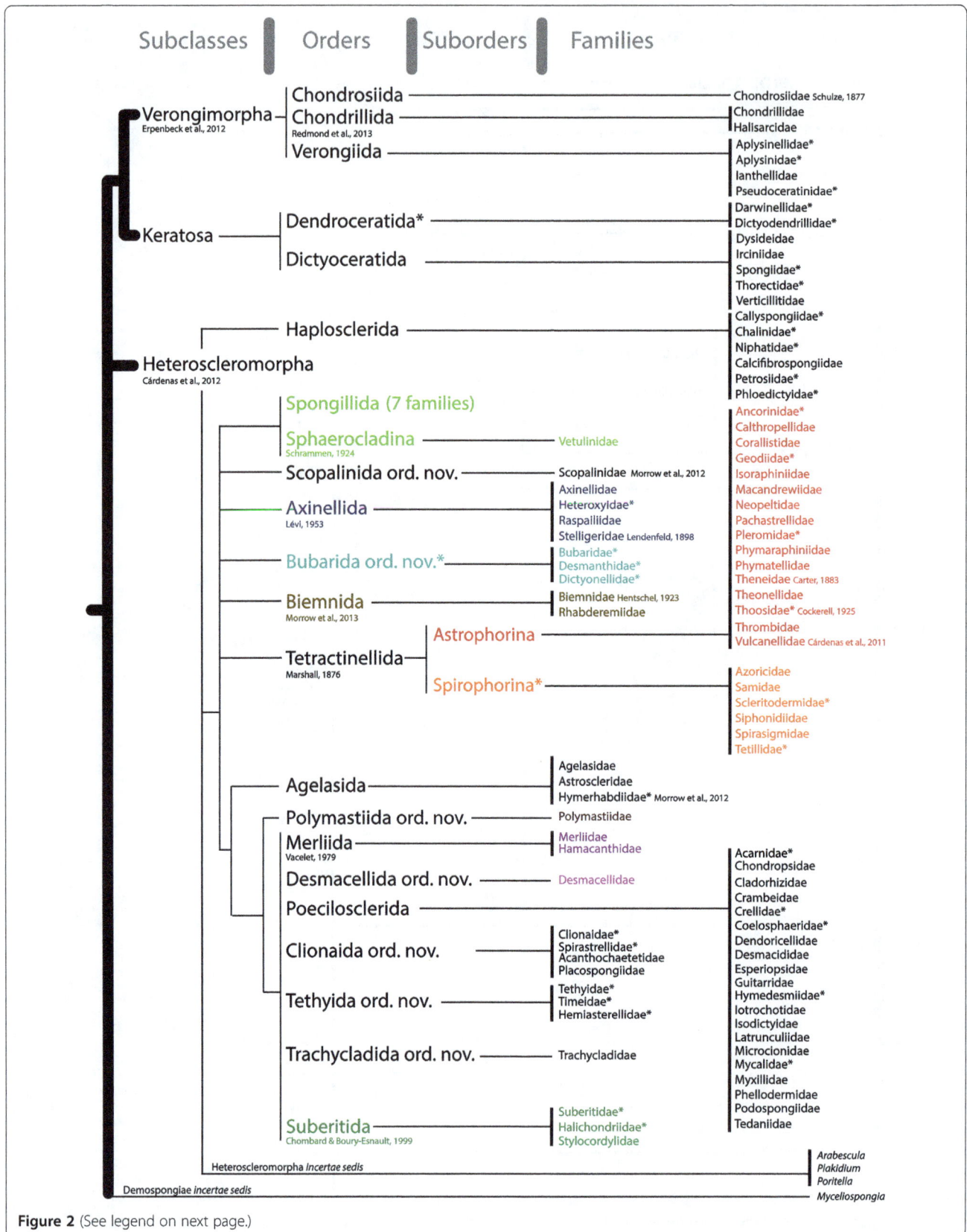

Figure 2 (See legend on next page.)

(See figure on previous page.)
Figure 2 Proposal for a revised classification of the Demospongiae, from subclasses to families. Relationships between the different taxa is deduced from all molecular phylogenetic studies published so far (as of November 2014). Coloured names correspond to the same colour code used in Figure 1. Only the authorships of new taxa or resurrected taxa since the publication of the *Systema Porifera* (2002) are given. An asterisk (*) is placed next to all order and family names suspected to be non-monophyletic, based on molecular phylogenetic results (see text for references).

with its single family Chondrosiidae and its single genus *Chondrosia*.

Abandoning Verticillitida

The calcified sponge order Verticillitida contains a single family Verticillitidae with a single living species: *Vaceletia crypta* (Vacelet, 1977). This species has no obvious morphological affinities with any Demospongiae taxa [28]. However, ribosomal and complete mitochondrial data [29,30] suggested that *V. crypta* belonged to Dictyoceratida, making it the only Keratosa sponge with a mineral skeleton. In the WPD, the order Verticillitida has therefore been synonymized with Dictyoceratida and Verticillitidae has been included in Dictyoceratida. One should note that there are still discussions on whether *V. crypta* is related to the fossil family Verticillitidae [28,31]; it has been proposed to classify *V. crypta* in Vacelitiidae Reitner and Engeser, 1985 instead [31,32].

Abandoning Spongillina, Haplosclerina and Petrosina

In SP, Haplosclerida includes the following three suborders: Haplosclerina, Petrosina (both marine) and Spongillina (freshwater sponges) (Figure 1). The worldwide monophyletic Spongillina has been upgraded to order rank [3], since molecular results (using mt. genomes, *18S* and *28S*) do not support its grouping with marine Haplosclerida (e.g. [10,13,14]). The only two phylogenetic studies disagreeing with these groupings used seven nuclear housekeeping genes (NHKG). In these two studies, the Spongillida group with the marine Haplosclerida, albeit with either no significant or relatively low support (74 b.s.) [24,33]. Furthermore, the taxonomic sampling in these studies was limited: only two species of Spongillida and no species of Vetulinidae — which may be the sister group of Spongillida [10,34] — and therefore a key group that could considerably alter the topology of the NHKG trees. For the rest of Haplosclerida, often referred to as 'marine Haplosclerida', Cárdenas et al. [3] (p. 170) proposed the subclass name Haploscleromorpha. Here we abandon Haploscleromorpha and retain Haplosclerida (with a revised definition) as an order within the subclass Heteroscleromorpha (cf. discussion above on 'Three versus four subclasses').

Although Haplosclerida is a well-supported clade, the suborders Haplosclerina and Petrosina and almost every family therein appear polyphyletic (for a review see [3], p. 192; [10]). A revision of Haplosclerida using new character datasets and implementing a bottom-up approach, studying first the type species of each genus [3] within each of the 5 newly found clades — provisionally called clade A to E [10] — is urgently needed. In the meantime, we propose to abandon these suborders.

Abandoning Lithistida

Lithistida Schmidt, 1870 had been considered an artificial and polyphyletic group long before it was confirmed by molecular results. The only shared character of Lithistida is interlocked spicules called 'desmas' which form a rigid skeleton. Lithistida are easily fossilized and thus have an extremely rich fossil record in comparison with other Demospongiae. Despite their acknowledged polyphyly and after numerous debates during the SP genesis, desma-bearing demosponges were grouped together under the name "lithistid' Demospongiae", mainly for convenience (Figure 1). Even though SP proposed to abandon the order Lithistida [35], this name has remained in the WPD, which can be very misleading for end-users such as biochemists, microbiologists, ecologists or paleontologists. Now that we have molecular support concerning the phylogenetic affinities of most of the desma-bearing families [6,36], we propose to formally reallocate the 13 desma-bearing families to their respective Heteroscleromorpha orders, as was already done by Cárdenas et al. [3,6], and abandon the Lithistida name in WPD. 11 out of the 13 desma-bearing families should be moved to the Tetractinellida: 8 families to the Astrophorina, 3 families to the Spirophorina. The Vetulinidae are now moved to their own order Sphaerocladina (an existing order in the fossil classification) and the Desmanthidae are allocated to Bubarida ord. nov.. 'Lithistids *incertae sedis*' from the SP (*Arabescula*, *Plakidium* and *Poritella*) should now be referred to as 'Heterosclermorpha *incertae sedis*' (see below).

Abandoning Poecilosclerida suborders: Microcionina, Mycalina, Myxillina and Latrunculina

In SP, Poecilosclerida comprised 25 families, distributed in four suborders erected by Hajdu et al. [37]: Microcionina, Mycalina, Myxillina and Latrunculina (Figure 1). These suborders essentially rely on the presence/absence and morphology of chelae microscleres. The SP classification is based on the assumption that chelae can be used to reconstruct phylogeny because of their morphological complexity and presumed selective neutrality but it seems that convergent evolution has brought phylogenetic noise to this hypothesis. Although we are far from understanding the phylogenetic relationships within this large order, molecular studies (using *CO1, 28S* and *18S*) strongly suggest that Microcionina, Mycalina and Myxillina are polyphyletic (Figure 3) [7,10,13,38]. We therefore propose to abandon these suborder names. Latrunculina, which only includes Latrunculiidae, seems to be monophyletic [10,38] but for consistency it is here abandoned along with the other suborders. It is not possible to provide an alternative internal phylogenetic structure for Poecilosclerida since so few taxa have yet been sequenced.

Abandoning Halichondrida

The taxonomic history of this group is long and complex (for a review, see [39]). The SP order Halichondrida contains the following five families: Halichondriidae, Axinellidae, Dictyonellidae, Heteroxyidae and Bubaridae (Figure 1). However, the monophyly of Halichondrida has never been recovered in any morphological, molecular or biochemical cladistics analyses (e.g. [5,39,40]). Halichondrida lack any unambiguous synapomorphic characters and are mainly defined on the basis of shared negative characters. Using the *28S* rDNA marker, Chombard [41]

and Chombard & Boury-Esnault [42] first revealed a close relationship between Halichondriidae and Suberitidae and not with other families assigned to Halichondrida. Chombard & Boury-Esnault [42] proposed the name Suberitina for this new clade. This clade was consistently confirmed in subsequent molecular phylogenetic studies, using more taxa and additional markers (e.g. [5]) and we now consider it should be upgraded to the order rank as Suberitida. At the same time, the other Halichondrida families were distributed amongst other clades: Axinellidae and Heteroxyidae in a well-supported clade for which we use the resurrected order name Axinellida; Dictyonellidae and Bubaridae in another clade here named Bubarida ord. nov. (Figure 2). Finally, *18S* and *CO1* data revealed a new clade (unnamed at this moment) grouping species of *Topsentia, Petromica* and *Axinyssa* [10,36]. Altogether, these well-established results force us to formally propose the abandonment of Halichondrida.

Abandoning Hadromerida

Hadromerida in SP included 13 families (Figure 1), two of which have now been abandoned: Alectonidae (split between Astrophorina and Clionaidae) and the Sollasellidae (now a junior synonym of Raspailiidae). Suberitidae has

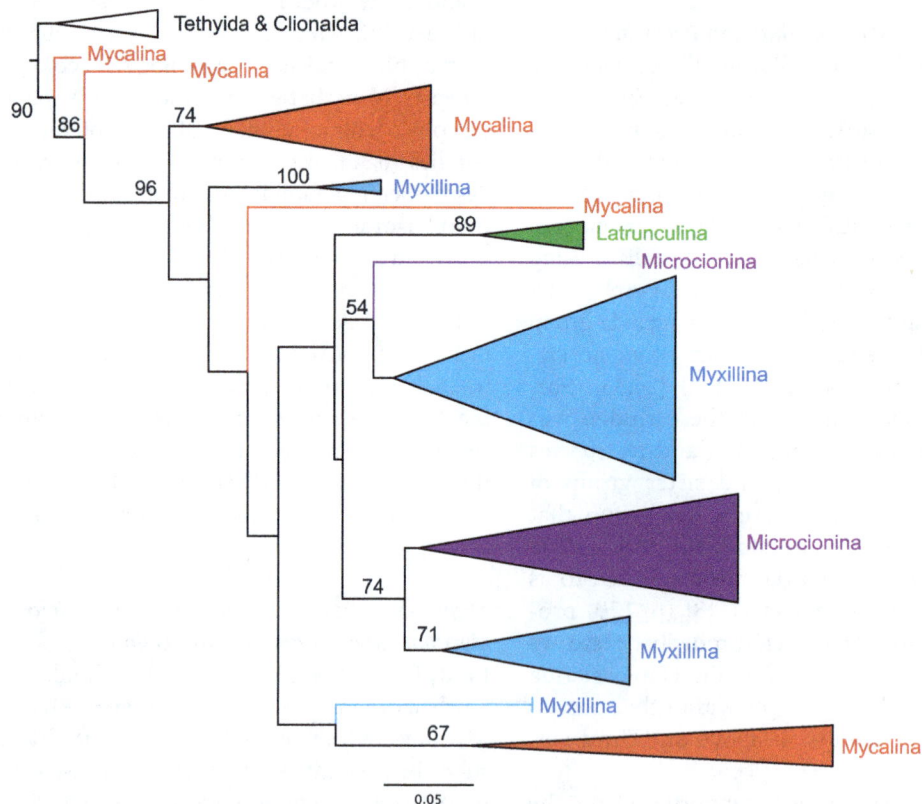

Figure 3 *18S* tree revealing the polyphyly of the Poecilosclerida suborders. PhyML tree with branches collapsed showing the polyphyly of the poecilosclerid suborders (Microcionina, Mycalina, Myxillina), with the exception of Latrunculina; only node bootstrap support > 50 are shown. This is a subset of the data used in Redmond et al. [10].

often been considered as a typical 'hadromerid'. However, since the work of Chombard & Boury-Esnault [42], Suberitidae have been shown to group with Halichondriidae in a well-supported clade. A *CO1* sequence of *Stylocordyla borealis* suggests that Stylocordylidae, which was also considered a 'hadromerid', groups with Suberitidae and Halichondriidae (Morrow and Cárdenas, unpublished results). On the basis of *28S* rDNA data, Chombard [41] had anticipated that the remaining 'hadromerids' grouped in four well-supported clades, later confirmed with larger sampling and additional markers. One contains Spirastrellidae, Clionaidae, Placospongiidae and Acanthochaetetidae; a second Timeidae, Tethyidae and Hemiasterellidae (pars); a third Trachycladidae and a fourth Polymastiidae [5,9,10,13]. Figure 4 is an *18S* ML tree which shows the distribution of former SP 'hadromerid' taxa relative to other Heteroscleromorpha. Lavrov et al. [14] using mitochondrial genomes showed Tethyidae grouping separately to Clionaidae but his analysis did not include Trachycladidae. Some former Hemiasterellidae have also joined some former halichondrids to group in the resurrected Stelligeridae family [5]. Altogether, given the polyphyly of Hadromerida (Figure 4) we propose the abandonment of Hadromerida, the erection of four new

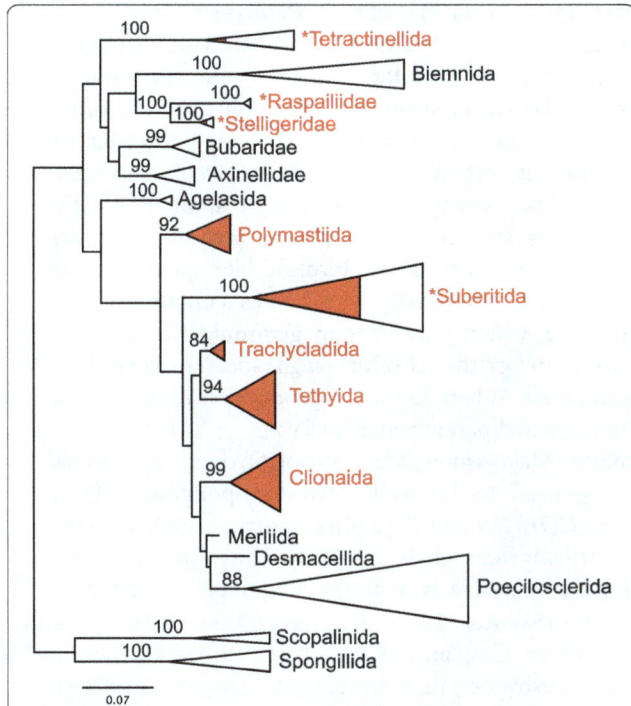

Figure 4 *18S* tree revealing the polyphyly of Hadromerida.
PhyML tree with branches collapsed showing polyphyly of Hadromerida (shown in red) that we propose to abandon; only nodes with > 50 b.s. are shown. This is a subset of the data used in Redmond et al. [10]. *Raspailiidae, *Suberitida, *Stelligeridae and *Tetractinellida include a mix of former 'hadromerid' taxa as well as taxa from other orders.

orders (Clionaida ord. nov., Tethyida ord. nov., Trachycladida ord. nov., Polymastiida ord. nov.) and the elevation of Suberitina to Suberitida Chombard & Boury-Esnault, 1999.

Taxonomy and definitions

For the family composition of each order, see Figure 2. For a tentative generic composition of the orders, see Appendix.

Subclass Verongimorpha Erpenbeck et al., 2012

Definition: Demospongiae without skeleton or with a skeleton made of siliceous asters (*Chondrilla*) or spongin fibres with a laminated bark and a finely fibrillar or granular pith (most of the Verongiida and *Thymosia*). Epithelial cells of the larva have i) a non-perpendicular orientation of the accessory centriole relative to the basal body, ii) a protruding nuclear apex and iii) a Golgi apparatus around the nuclear apex and part of the organelles of the basal apparatus (definition from [3], emended with larva observations from [43]).

Remark: We add here larva cytological characters in the definition but underline that very few Verongimorpha larva have been studied so far [43] so these characters need to be confirmed. Borchiellini et al. [12] resurrected Myxospongiae for their G2 clade (*Halisarca* + *Chondrosia* + Chondrillidae + Verongiida); since then, the name has been used and even erected as a subclass [43]. But Myxospongiae was originally intended for sponges without any skeleton ("*ohne jedes Skelet*" [44]), the so called "slime sponges" by Haeckel [44], which was essentially *Halisarca*. Erpenbeck et al. [25] consider that the G2 clade assemblage is too different from the original content of Myxospongiae and decide to create a replacement name for the Myxospongiae: Verongimorpha. Erpenbeck et al. [25] also consider that most of the sponges in this group are not "slime-sponges" so that the name of the subclass is not descriptive enough of the group. Myxospongiae *sensu* Haeckel, 1866 has been seldom used in the past and so, despite the different assemblage of Myxospongiae (G2), we believe there has not been confusion: the name Myxospongiae has been properly used now for over ten years. However, we agree that the Myxospongiae name poorly reflects the sponges it contains so we recommend using the name Verongimorpha.

Order Chondrillida Redmond et al., [10]

Definition: Verongimorpha in which the skeleton can be absent, but when present is composed of nodular spongin fibers or aster microscleres [10].

Order Chondrosiida Boury-Esnault & Lopès, 1985

Definition: Verongimorpha with a marked ectosome or cortex enriched by a highly organized fibrillar collagen. Collagen is always very abundant. (modified from [3], p. 170).

Remark: This order only includes the genus *Chondrosia*.

Order Verongiida Bergquist, 1978

SP definition emended from Bergquist and Cook [45]: Verongimorpha in which the fibrous skeleton, when present, is either anastomosing or dendritic in construction. Reproduction is always oviparous. They produce complex brominated tyrosine-derived compounds.

Remark: Most relationships are uncertain within this order; Pseudoceratinidae, Aplysinellidae and Aplysinidae are probably not monophyletic [10,13,25,46,47].

Subclass Keratosa Grant, 1861

Definition: Demospongiae with a skeleton made of spongin fibre. Spongin fibres are either homogenous or pithed and strongly laminated with pith grading into bark. One genus has a hypercalcified basal skeleton (*Vaceletia*). ([3], p. 170).

Order Dendroceratida Minchin, 1900

SP definition emended from Bergquist and Cook [48]: Keratosa in which a fibre skeleton is always present but, as compared to Dictyoceratida, is reduced in relation to soft tissue volume. The skeleton arises from a continuous spreading basal plate, and adopts either a dendritic or an anastomosing pattern. In anastomosing forms there is never any clear size distinction between primary and secondary elements. The fibres always contain pith and are strongly laminated, usually quite stout, and in some genera cellular (degenerate spongocytes) elements are incorporated in the bark and to a lesser extent in pith. Free fibrous spicules may supplement the main skeleton. The pith in the fibres is markedly disjunct from the bark, and in structure is close to that of the Verongiida. Biochemically, members of this group are characterized by a moderate sterol content in conjunction with the presence of terpenes, which are always diterpenes.

Remark: This order and both of its families (Darwinellidae and Dictyodendrillidae) may be polyphyletic [10,12,25].

Order Dictyoceratida Minchin, 1900

SP definition emended from Cook and Bergquist [49]: Keratosa in which a spongin fibre skeleton is constructed on an anastomosing plan. The skeleton develops from multiple points of attachment and, except in two genera where primary fibres are absent, is organized as a hierarchy of primary, secondary and sometimes tertiary elements. Fibre construction is homogeneous lacking pith with growth laminae tightly adherent and just detectable, or pithed and strongly laminated with pith grading into bark, consecutive laminae are marked but remain adherent to each other. Pith is structurally and chemically distinct from that seen in fibres of Verongiida and Dendroceratida. One species (*Vaceletia crypta*) has a 'sphinctozoan' grade of organization with a chambered skeleton composed of aragonite with irregular structure, without siliceous spicules.

Remark: The families Spongiidae and Thorectidae are not monophyletic and blend essentially together; some Thorectidae should also be reallocated to the Dysideidae [10,13,25].

Subclass Heteroscleromorpha Cárdenas et al., [3]

Definition: Demospongiae with a skeleton composed of siliceous spicules which can be monaxons and/or tetraxons and when they are present, microscleres are highly diversified (definition of Cárdenas et al. [3] was not modified by the inclusion of Haplosclerida).

Order Haplosclerida Topsent, 1928

Definition: Heteroscleromorpha with an isodictyal anisotropic or isotropic choanosomal skeleton; spicules are diactinal megascleres (oxeas or strongyles), smooth or spined; microscleres, if present, are sigmas and/or toxas, microxeas or microstrongyles (emended definition of Haploscleromorpha Cárdenas et al., [3]).

Remark: Five out of the six families, as well as many genera are not monophyletic [10]. The last family (Calcifibrospongiidae) is monospecific. Currently, five well-supported clades (clades A-E) are recognized [10]. The definition of Cárdenas et al. [3] has been emended here to include *Janulum*, formerly assigned to Raspailiidae, and now included in clade E [10].

Order Spongillida Manconi & Pronzato, [50]

SP definition of Spongillina from Manconi and Pronzato [50]: Exclusively freshwater sponges, with megascleres consisting of oxeas or strongyles, smooth or spined, forming pauci- to multispicular tracts producing irregular to regular meshes, occasionally with large alveolate cavities (a central body cavity in one family); spongin mostly sparse; microscleres present or absent, including smooth or spined oxeas, aster-like or birotule like spicules. Four families with gemmules (resting bodies containing totipotent cells), which may contain gemmuloscleres of diverse morphology that is often diagnostic. Three families lack gemmules. Where known, reproduction is viviparous, with fully ciliated parenchymella larvae.

Remark: Malawispongiidae is polyphyletic and some taxa may need to be reallocated to Spongillidae [51]. Based on *CO1*, *18S* and *ITS2* phylogenetic analyses, Spongillidae are not monophyletic [52,53]. The monotypic family Metschnikowiidae is a doubtful Spongillida and may need to be reallocated to the Haplosclerida. Indeed, it is endemic to the Caspian Sea, thus living in brackish water instead of freshwater; its morphological characters suggest possible affinities with *Janulum* [54,55], recently reallocated to Haplosclerida (clade E) [10].

Order Sphaerocladina Schrammen, 1924

SP definition of Vetulinidae emended from Pisera & Lévi [56]: Heteroscleromorpha with acrepid polyaxial (astro- or sphaeroclone) desmas. No other megascleres or microscleres, only one recent family but several fossil families.

Remark: We decided to use the paleontological order name since this group has a very long and rich fossil record [56]. A sister-group relationship with Spongillida is strongly supported by *18S* [10,34], *CO1* and *28S* [36].

Scopalinida ord. nov.

Definition: Encrusting, massive or erect flabellate growth forms; with smooth or conulose surface supported by prominent spongin fibres cored with styles; megascleres styles, often with telescoped ends; no ectosomal skeleton; tissue contains an unusual cell type filled with refractile granules.

Remark: This order contains only one family, Scopalinidae with genera *Scopalina* and *Svenzea*. *Stylissa*, a former halichondrid, is here tentatively included in this order/family (Appendix) since the type species *Stylissa flabelliformis* grouped within the Scopalinidae [10].

Order Axinellida Lévi, 1953 (resurrected)

Definition: Megascleres are styles or tylostyles and oxea, with acanthostyles in some genera. Surface may be smooth but is usually hispid due to projecting choanosomal styles and these may be surrounded by brushes of fine oxea, anisoxea or styles forming a specialised ectosomal skeleton. Micoscleres when present are asters, acanthoxea or raphides, usually in trichodragmata (emended from [5]). Skeleton in several species comprised of a stiff axial region, usually with abundant spongin and an outer, softer extra-axial region. Colour of living sponge is characteristically orange, yellow or dark brown.

Remark: The order Axinellida Lévi, 1953 was originally erected for the families Axinellidae and Raspailiidae [17]. In subsequent publications Lévi assigned seven other families to this order. Whilst the content of Axinellida has changed from that formally defined by Lévi (1973) [20] and Bergquist (1967; 1970) [57,58], various molecular studies have shown Axinellida to contain three of the nine families that have been assigned to it. Most importantly the studies show that *Axinella polypoides*, the type taxon for Axinellidae clusters here. The principles of the ICZN for family names are based on the name-bearing type genus. Whilst at order level we are not bound by the rules of the ICZN, in keeping with the spirit of the code, we retain the name Axinellida for this order. The Raspailiidae have been moved back from Poecilosclerida to this order based on numerous molecular studies (e.g. [9,10]). The family Stelligeridae was resurrected by Morrow et al. [5] to include former hadromerid (*Stelligera*, *Paratimea*) and halichondrid genera (*Halicnemia*, *Higginsia*). Redmond et al. [10] using *18S* rDNA shows strong support for a Stelligeridae-Raspailiidae clade but no support for Stelligeridae-Raspailiidae + Axinellidae. However in Morrow et al. [9] using *18S + 28S*, Stelligeridae-Raspailiidae + Axinellidae is a strongly supported clade. Unpublished results by

C. Morrow show that some *Heteroxyidae* also cluster with Axinellida.

Order Bubarida ord. nov.

Definition: Heteroscleromorpha built with monactines, diactines of different kinds and different shapes (flexuous, sinuous or vermiculiform; telescoped endings are common). Flexuous or sinuous spicules may be confined to axial skeleton or form a basal layer in encrusting forms. Monocrepidial desmas form a basal skeleton in one genus (*Desmanthus*).

Remark: This order is in need of taxonomic revision: early indications are that there are three clades mixing species of *Acanthella*, *Dictyonella*, *Bubaris*, *Cymbastella*, *Axinyssa*, *Phakellia* and *Phycopsis* [5,10]. In the future it is likely that the family Dictyonellidae van Soest, Diaz & Pomponi, 1990 and Bubaridae Topsent, 1894 will be merged, however Dictyonellidae is retained here as a revision of the order is beyond the scope of this study. It is also possible that older family names such as Desmanthidae Topsent, 1893 may take priority over Bubaridae. The lithistid family Desmanthidae is assigned to this order since the type species *Desmanthus incrustans* joins this clade, based on *18S*, *CO1* and *28S* markers [5,10,59]. However, *CO1* and *18S* studies also suggest that *Petromica* (Desmanthidae), *Topsentia* (Halichondriidae) and some *Axinyssa* (Halichondriidae) group in a very well supported clade (unnamed at the moment) outside of Bubarida [10,36], therefore Desmanthidae and Bubarida are both polyphyletic. For now we retain the family Desmanthidae pending further molecular evidence.

Order Biemnida Morrow et al., 2013

Definition: Megascleres styles, subtylostyles, strongyles, rhabdostyles, or oxeas. Spicules typically enclosed by spongin fibres. Reticulate or plumoreticulate choanosomal skeleton, axially compressed in erect forms. Extra-axial plumose skeleton usually present. Microscleres are microspined sigmas/spirosigmas, toxas, microxeas, raphides, and/or commata. *Biemna* and *Neofibularia* cause a dermatitis-like reaction when in contact with bare skin [9].

Remark: This order includes former Desmacellidae genera (*Biemna*, *Neofibularia* and *Sigmaxinella*), grouped in the resurrected family Biemnidae, as well as the Rhabderemiidae family with a single genus *Rhabderemia*. Grouping of these two families is based on molecular and morphological data, notably the possession of distally microspined "sigmas", which may not be homologous to the sigmas found in poecilosclerid sponges for example [7,9,10,60,61]. In SP, all these genera were in the Poecilosclerida order since they were believed to have lost their chelae.

Order Tetractinellida Marshall, 1876 (resurrected)

Definition: Heteroscleromorpha usually with radial or subradial skeletal arrangement, some genera can be endolithic. Megascleres are monactines and triaenes in various shapes (a synapomorphy of the order, but sometimes secondarily lost). Microscleres include sigmas, asters, sometimes with microrhabds, microxeas and raphides. Desmas are sometimes present.

Remarks: Astrophorida and Spirophorida were included in SP while Tetractinellida was not. However, all molecular studies have shown that Tetractinellida is a strongly supported clade (e.g. [10,62]). We propose to resurrect this order, and recommend the use of the suborders Astrophorina and Spirophorina, as previously suggested [3,5]. Many Astrophorina families and genera seem to be polyphyletic, especially due to the mingling of Ancorinidae and Geodiidae genera [6,63]. Since SP, one family was resurrected (Theneidae) and a new family was created (Vulcanellidae) [6]. Alectonidae, formerly in Hadromerida has been abandoned and the genera *Alectona/Delectona/Thoosa* and *Neamphius* have been artificially kept together in the resurrected Thoosidae, now assigned to Astrophorina, awaiting a clearer appreciation of the phylogenetic relationships of *Neamphius*. The genera *Lamellomorpha* (Astrophorida *incertae sedis* in SP) and *Characella*, are both provisionally assigned to Pachastrellidae, which makes this family polyphyletic [6,10]. Most of the desma-bearing demosponge families have now been assigned to this order: eight and three families are now respectively assigned to Astrophorina and Spirophorina. Studies based on *COI*, *18S* and *28S* suggest that Spirophorina and Tetillidae may not be monophyletic [10,36]. It also seems that Pleromidae and Scleritodermidae may not be monophyletic [36].

Order Agelasida Hartman, 1980

Definition: Megascleres smooth or verticillately spined styles, rhabdostyles or ocassionally oxea, no microscleres. Representatives of all families of the order produce similar pyrrole-2-carboxylic compounds, characteristically with a bromine addition [5].

Remark: The new family Hymerhabdiidae added to this order currently includes the former halichondrid *Hymerhabdia* and the former hadromerid *Prosuberites*, in addition to some former halichondrids belonging to *Stylissa/Axinella/Phycopsis/Cymbastela* [5,10,64]. The former raspailiid genera *Acanthostylotella* and *Amphinomia* are now assigned to Agelasidae [10].

Polymastiida ord. nov.

SP definition by Boury-Esnault [65]: Heteroscleromorpha with a radiating choanoskeleton and a more or less complicated cortex, the outer layer being always a palisade of ectosomal spicules (tylostyles, or oxea and/or exotyles). Megascleres are tylostyles, subtylostyles, strongyloxeas, styles or oxeas; microscleres may include centrotylote microxeas, acanthose microxeas or raphides in trichodragmata.

Remark: This order only includes the family Polymastiidae.

Order Merliida Vacelet, 1979 (resurrected)

Definition: Megascleres are diverse (oxeas, styles, mycalostyles or tylostyles) but associated with unique microscleres (either clavidiscs or diancistra-derivatives: diancistras or cyrtancistras). Raphides, sigmas or small commata-like spicules are also present. One family contains a species with a chaetetid calcareous basal skeleton, and an outer layer of which is filled with sponge tissue and siliceous spicules.

Remarks: Vacelet [66,67] first suggested to isolate *Merlia* in its own order — Merliida *incertae sedis* — but in the SP it was included in the Poecilosclerida, based on microscleres similarities with the Biemnidae and Desmacellidae [68]. The recognition of a separate order was actually confirmed in *18S*, *28S* and *16S* phylogenetic analyses, where it branches separately from the rest of the Poecilosclerida [10,59]. *CO1* phylogenetic analyses further show that *Hamacantha* and *Desmacella* branch before the poecilosclerid clade (Morrow, unpublished results). Topsent [69] was the first to suggest a close relationship between *Hamacantha* and *Merlia* based on the striking similarity between the diancistras (in *Hamacantha*) and the clavidiscs (in *Merlia*). On the basis of this molecular result and also the strong morphological affinities of the microscleres we propose to move Hamacanthidae from Poecilosclerida to Merliida.

Desmacellida ord. nov.

SP definition by Hajdu and van Soest [70]: Heteroscleromorpha with monactinal megascleres arranged in plumose bundles; microscleres sigmas and sometimes raphides.

Remark: Desmacellidae, a family without chelae, seems to diverge before the Poecilosclerida radiation [10] and is here assigned its own order. *Biemna*, *Neofibularia* and *Sigmaxinella* which were allocated to Desmacellidae in SP, were transferred to Biemnida [9]; the remaining genera, *Desmacella*, *Dragmatella* and *Microtylostylifer*, are allocated to Desmacellida. However, it should be noted that there is no molecular data for *Dragmatella* and *Microtylostylifer* species.

Order Poecilosclerida Topsent, 1928

SP definition emended from Hooper & van Soest [71]: both fibre and mineral skeletons always show regional differentiation such that megascleres are often differentiated into distinct ectosomal and choanosomal components; microscleres include chelae (a synapomorphy for the order, but sometimes lost), sigmas and sigmancistra derivatives, and other diverse forms such as toxas, raphides, microxeas, discorhabds or spinorhabds; the order appears to be exclusively viviparous.

Remark: We propose to abandon the suborders (cf. above and Figure 3). The Poecilosclerida included 25 families in the SP, five of which we now propose to reassign to Desmacellida, Merliida, Axinellida and Biemnida. Possible polyphyletic families include Acarnidae, Mycalidae, Coelosphaeridae, Hymedesmiidae and Crellidae [10,13].

Clionaida ord. nov.

Definition: Heteroscleromorpha with tylostyle megascleres; oxeas and styloid spicules are also present in one family. Variety of microscleres including streptasters (spirasters and diplasters), amphiasters, selenasters, microxeas, microrhabds, spiral microstrongyles and derivatives. Microscleres may be lacking altogether. Calcareous basal skeleton present in one family.

Remark: Clionaidae d'Orbigny, 1851 and Spirastrellidae Ridley and Dendy, 1886 seem to mix and may thus not be monophyletic [5,22,72]. Since Clionaidae is an older family name, there is a chance that Spirastrellidae might become invalid in the future so we chose Clionaida over Spirastrellida (which had been suggested by Chombard [41] in her PhD thesis).

Tethyida ord. nov.

Definition: Megascleres may be styles, tylostyles or oxeas arranged in tracts ending as bouquets, at or near the surface. Microscleres are euasters, usually of two sizes.

Remark: Morrow et al. [5] showed that Hemiasterellidae is polyphyletic with some genera grouping closely with some heteroxyid and raspailiid taxa and others with Tethyidae. In the absence of molecular data from the type taxon (*Hemiasterella typus*) we retain Hemiasterellidae in Tethyida. Furthermore, Timeidae and Tethyidae may not be monophyletic, but molecular data on these two families is sparse [5,10,22]. We chose the name Tethyida for this order since we have molecular data for the type taxon of Tethyidae (*Tethya aurantium*) but not for the type taxons of Timeidae or Hemiasterellidae.

Trachycladida ord. nov.

SP definition of Trachycladidae by Hooper & van Soest [73]: Heteroscleromorpha with spined vermiform spinispirae and smooth microrhabds, with a differentiated axial and extra-axial skeleton cored by oxeas, strongyles and/or (tylo-)styles.

Remark: Only one family (Trachycladidae), with 10 species, belongs to this order at the moment.

Order Suberitida Chombard & Boury-Esnault, 1999

Definition: Heteroscleromorpha without an obvious cortex and without microscleres other than microstrongyles/oxeas; megascleres are oxeas, centrotylote oxeas, styles or tylostyles. Choanosomal skeleton usually consisting of a confused arrangement of megascleres, radial arrangement of megascleres in one family. Surface skeleton of paratangential to erect palisade of large or small megascleres. Molecular synapomorphy is a deletion of a small loop of 15 base pairs in the secondary structure of the $28S$ D2 domain with respect to other Heteroscleromorpha (slightly emended from [42] to include Stylocordylidae).

Remark: Suberitidae and Halichondriidae are currently not monophyletic due to, for example, the grouping of *Terpios* with the halichondrids instead of the suberitids, or the early branching of *Homaxinella* with respect to the rest of the Suberitida [5,10,13]. The allocation of *Stylocordyla* to Suberitida is based on previous morphological studies (e.g.

[74,75]) and molecular data (*CO1*) (Morrow and Cárdenas, unpublished results).

Heteroscleromorpha *incertae sedis* may include some of the former 'lithistids' *incertae sedis* listed in SP, which consists of poorly known genera with rhizoclone desmas of uncertain status [76]: *Arabescula*, *Plakidium* and *Poritella*. *Collectella* is clearly a tetractinellid with phyllotriaenes that we tentatively assign to Theonellidae. *Collinella* has been synonymized with *Discodermia* (Theonellidae) in the WPD.

Myceliospongia with its single species *M. araneosa* remains Demospongiae *incertae sedis*, as in the SP [77], awaiting molecular data to assign it to an order.

Concluding remarks

Demospongiae in SP is comprised of 13 orders. In the present proposal, five of these order names are abandoned (Hadromerida, Halichondrida, Halisarcida, lithistids, Verticillitida) and six order names are resurrected or upgraded (Axinellida, Merliida, Spongillida, Sphaerocladina, Suberitida, Tetractinellida) and seven new orders have been erected (Bubarida, Desmacellida, Polymastiida, Scopalinida, Clionaida, Tethyida, Trachycladida). These added to the recently created orders (Biemnida and Chondrillida) make a total of 22 orders in the revised classification. We propose the abandonment of all Haplosclerida and Poecilosclerida suborders and the use of Tetractinellida suborders. Finally, we reassign many families (belonging to Hadromerida, Halichondrida, Halisarcida, Poecilosclerida, Lithistida) to new orders.

When the classification changes, so does the importance of the different groups in terms of species numbers; these numbers are reviewed for the various orders in Figure 5. According to SP, Poecilosclerida was the largest order in term of species: over 2,630 species [4, accessed on the 17th of October 2014]. By reassigning five families Poecilosclerida "loses" about 421 species (essentially Raspailiidae). However the revised Poecilosclerida remains the largest order with 2,209 species. The second largest is the revised Haplosclerida with 1,073 species, and the third largest is Tetractinellida (including 11 former lithistid families) with 1,064 species ([4], accessed on the 17th of October 2014).

The overall aim of this paper is to begin to resolve the growing discrepancy between the classification presented in SP and the body of evidence from molecular phylogenetic studies. Doing so, we hope to convince end-users to 1) abandon the use of artificial groups, and to 2) use the new/resurrected names proposed here when referring to the new Demospongiae clades. This updated classification will undoubtedly facilitate communication between end-users, reduce taxonomically biased results, and ultimately provide a better understanding of Demospongiae evolutionary history. We should however keep in mind that the groupings we propose are new phylogenetic hypotheses that will be challenged by future systematic research.

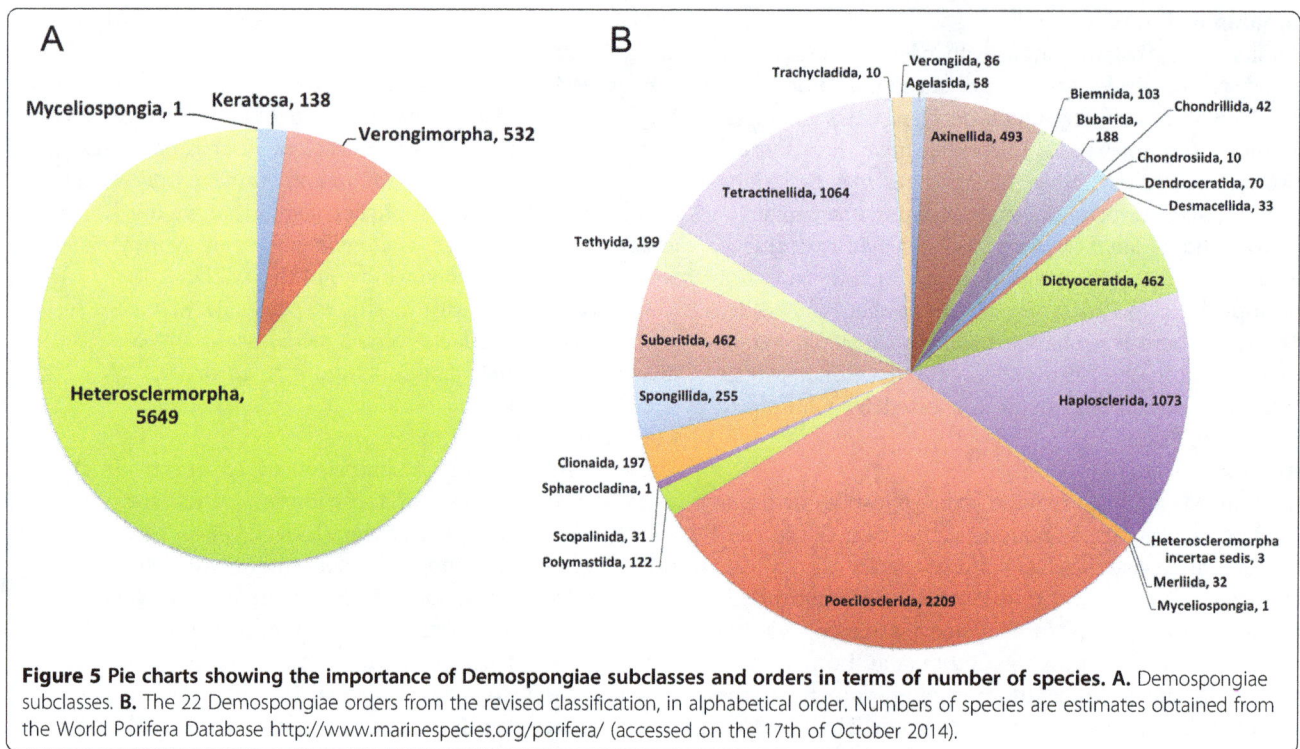

Figure 5 Pie charts showing the importance of Demospongiae subclasses and orders in terms of number of species. A. Demospongiae subclasses. **B.** The 22 Demospongiae orders from the revised classification, in alphabetical order. Numbers of species are estimates obtained from the World Porifera Database http://www.marinespecies.org/porifera/ (accessed on the 17th of October 2014).

Hypotheses regarding sponge phylogenetic relationships will continue to change with the description and sequencing of new species, use of new datasets and improvements in phylogenetic reconstruction methods. New phylogenetic hypotheses will undoubtedly involve further changes to the classification. In other words, absolute nomenclatural stability in a rank-based system is impossible and name changes simply reflect the regular growth of phylogenetic knowledge and understanding [78]. This is certainly frustrating for the many end-users of the sponge classification, but it is a reality that we have to accept and understand if we want our research to rely on updated taxonomic grounds and avoid reaching misleading conclusions. The good news is that end-users and non-sponge specialists can now rely on a large choice of biodiversity web based databases. The WPD is currently the most complete, easily and regularly updated database thanks to a large editorial committee.

Yes, the *Systema Porifera* classification published in 2002 is already partly out of date but we should keep in mind that SP still represents a milestone for sponge researchers, due to its rigorous approach of defining terminal taxa (based on objective evidence from the type species of each genus), and the richness of taxonomic and morphological information it contains. As was certainly the case with SP, we hope the following proposal will stimulate fruitful taxonomic research. In particular, we hope this proposal will help researchers to refocus and revisit clades with a more integrative taxonomic approach [3], combining top-down and bottom-up

phylogenetic strategies [3]. Many of these new groupings require clear morphological diagnoses, we hope this classification will help to reveal hidden and overlooked synapomorphies in the various datasets that sponge biologists now have at their disposal. We invite and welcome comments on our proposal, as well as any suggestions for additional changes.

Appendix

Proposal for Heteroscleromorpha genera assignments (except Haplosclerida) in their respective orders and families of the revised classification. This Table 1 lists the Heteroscleromorpha genera (in alphabetical order) with their current assignment in the World Porifera Database http://www.marinespecies.org/porifera/ (accessed on the 17th of October 2014). It does not include the marine haplosclerids as no changes are proposed for this group. For each of the genera, we give the order and family assignment within the framework of the revised classification. Some of these assignments are tentative and may change. We have included the molecular markers that exist for the type species: *CO1 (cytochrome oxidase subunit 1,* usually the Folmer fragment), *18S (ribosomal small subunit), 28S (ribosomal large subunit), ITS (internal transcribed spacer), AGL11 (aspargine-linked glycosylation 11 protein),* ESTs (Expressed Sequence Tags), mt (mitochondrial genome) and tpm (transcriptome). We have also included absence/presence (yes/no) of molecular markers available for non-type species.

Table 1 Listing of Heteroscleromorpha genera in alphabetical order

Genus	Order	Family	New order (Suborder)	Family	Genes, type species	Genes, non-type
Aaptos	Hadromerida	Suberitidae	Suberitida	Suberitidae	no	yes
Abyssocladia	Poecilosclerida	Cladorhizidae	Poecilosclerida	Cladorhizidae	no	yes
Acalle	Haplosclerida	Metaniidae	Spongillida	Metaniidae	no	no
Acanthancora	Poecilosclerida	Hymedesmiidae	Poecilosclerida	Hymedesmiidae	18S	no
Acanthella	Halichondrida	Dictyonellidae	Bubarida	Dictyonellidae	18S, 28S, COI	yes
Acantheurypon	Poecilosclerida	Raspailiidae	Poecilosclerida	Poecilosclerida *incertae sedis*	18S, 28S, COI	yes
Acanthochaetetes	Hadromerida	Acanthochaetetidae	Clionaida	Acanthochaetetidae	no	yes
Acanthoclada	Halichondrida	Heteroxyidae	Axinellida	Heteroxyidae	no	no
Acanthopolymastia	Hadromerida	Polymastiidae	Polymastiida	Polymastiidae	no	no
Acanthorhabdus	Poecilosclerida	Acarnidae	Poecilosclerida	Acarnidae	no	no
Acanthostylotella	Poecilosclerida	Raspailiidae	Agelasida	Agelasidae	18S, 28S	no
Acanthotetilla	Spirophorida	Tetillidae	Tetractinellida (Spirophorina)	Tetillidae	no	yes
Acanthotriaena	Astrophorida	Pachastrellidae	Tetractinellida (Astrophorina)	Pachastrellidae	no	no
Acanthotylotra	Haplosclerida	Spongillina *incertae sedis*	Spongillida	Spongillida *incertae sedis*	no	no
Acarnus	Poecilosclerida	Acarnidae	Poecilosclerida	AcarnidaeCP	16S, 18S, 28S	no
Acheliderma	Poecilosclerida	Acarnidae	Poecilosclerida	Acarnidae	no	no
Aciculites	'Lithistida'	Scleritodermidae	Tetractinellida (Spirophorina)	Scleritodermidae	no	yes
Adreus	Hadromerida	Hemiasterellidae	Tethyida	Hemiasterellidae	18S, 28S, COI	yes
Agelas	Agelasida	Agelasidae	Agelasida	Agelasidae	28S, COI	yes
Alectona	Astrophorida	Thoosidae	Tetractinellida (Astrophorina)	Thoosidae	COI, 28S	no
Alloscleria	Halichondrida	Heteroxyidae	Axinellida	Heteroxyidae	no	no
Amorphinopsis	Halichondrida	Halichondriidae	Suberitida	Halichondriidae	18S, 28S, COI, EF-1	yes
Amphiastrella	Poecilosclerida	Iotrochotidae	Poecilosclerida	Iotrochotidae	no	no
Amphibleptula	'Lithistida'	Scleritodermidae	Tetractinellida (Spirophorina)	Scleritodermidae	no	no
Amphilectus	Poecilosclerida	Esperiopsidae	Poecilosclerida	Esperiopsidae	no	yes
Amphinomia	Poecilosclerida	Echinodictyinae	Poecilosclerida	Echinodictyinae	18S, 28S	no
Amphitethya	Spirophorida	Tetillidae	Tetractinellida (Spirophorina)	Tetillidae	no	yes
Anaderma	'Lithistida'	Pleromidae	Tetractinellida (Astrophorina)	Pleromidae	no	no
Ancorella	Astrophorida	Pachastrellidae	Tetractinellida (Astrophorina)	Pachastrellidae	no	no
Ancorina	Astrophorida	Ancorinidae	Tetractinellida (Astrophorina)	Ancorinidae	no	yes
Anheteromeyenia	Haplosclerida	Spongillidae	Spongillida	Spongillidae	no	no
Anisocrella	Poecilosclerida	Crellidae	Poecilosclerida	Crellidae	no	no
Annulastrella	Astrophorida	Theneidae	Tetractinellida (Astrophorina)	Theneidae	no	yes
Antho	Poecilosclerida	Ophlitaspongiinae	Poecilosclerida	Microcionidae	18S, 28S	yes
Anthotethya	Hadromerida	Tethyidae	Tethyida	Tethyidae	no	no

Table 1 Listing of Heteroscleromorpha genera in alphabetical order *(Continued)*

Arabescula	'Lithistida'	Lithistida *incertae sedis*	*incertae sedis*		no	no
Artemisina	Poecilosclerida	Ophlitaspongiinae	Poecilosclerida	Microcionidae	no	yes
Asbestopluma	Poecilosclerida	Cladorhizidae	Poecilosclerida	Cladorhizidae	no	yes
Asteropus	Astrophorida	Ancorinidae	Tetractinellida (Astrophorina)	Ancorinidae	no	yes
Astrosclera	Agelasida	Astroscleridae	Agelasida	Astroscleridae	28S,COI,18S, ITS1-5.8S-ITS2-28S	no
Astrotylus	Hadromerida	Polymastiidae	Polymastiida	Polymastiidae	no	no
Atergia	Hadromerida	Polymastiidae	Polymastiida	Polymastiidae	18S	no
Auletta	Halichondrida	Axinellidae	Bubarida	Bubaridae[1]	no	yes
Aulospongus	Poecilosclerida	Raspailiidae	Axinellida	Raspailiidae	no	yes
Awhiowhio	'Lithistida'	Corallistidae	Tetractinellida (Astrophorina)	Corallistidae	no	no
Axechina	Poecilosclerida	Raspailiidae	Axinellida	Raspailiidae	18S, 28S, COI	no
Axinella	Halichondrida	Axinellidae	Axinellida	Axinellidae	18S, 28S	yes
Axinyssa	Halichondrida	Halichondriidae	Bubarida	Dictyonellidae	18S, 28S	yes
Axos	Hadromerida	Hemiasterellidae	Tethyida	Hemiasterellidae	28S, COI	yes
Baikalospongia	Haplosclerida	Lubomirskiidae	Spongillida	Lubomirskiidae	COI, cox2-ATP6, 18S, ITS1-5.8S-ITS2-28S	yes
Balliviaspongia	Haplosclerida	Spongillina *incertae sedis*	Spongillida	Spongillida *incertae sedis*	no	no
Batzella	Poecilosclerida	Chondropsidae	Poecilosclerida	Chondropsidae	no	no
Biemna	Poecilosclerida	Desmacelliae	Biemnida	Biemnidae	18S, 28S, COI	yes
Brachiaster	Astrophorida	Pachastrellidae	Tetractinellida (Astrophorina)	Pachastrellidae	no	no
Bubaris	Halichondrida	Bubaridae	Bubarida	Bubaridae	no	yes
Burtonitethya	Hadromerida	Tethyidae	Tethyida	Tethyidae	no	no
Callipelta	'Lithistida'	Neopeltidae	Tetractinellida (Astrophorina)	Neopeltidae	no	yes
Calthropella	Astrophorida	Calthropellidae	Tetractinellida (Astrophorina)	Calthropellidae	no	yes
Caminella	Astrophorida	Geodiidae	Tetractinellida (Astrophorina)	Geodiidae	no	yes
Caminus	Astrophorida	Geodiidae	Tetractinellida (Astrophorina)	Geodiidae	COI	no
Cantabrina	Poecilosclerida	Echinodictyinae	Poecilosclerida	Echinodictyinae	no	no
Caulospongia	Hadromerida	Suberitidae	Suberitida	Suberitidae	no	yes
Celtodoryx	Poecilosclerida	Coelosphaeridae	Poecilosclerida	Coelosphaeridae	no (contaminant)	no
Ceratoporella	Agelasida	Astroscleridae	Agelasida	Astroscleridae	COI, ITS1-5.8S	no
Ceratopsion	Poecilosclerida	Raspailiidae	Axinellida	Raspailiidae	no	yes
Cerbaris	Halichondrida	Bubaridae	Bubarida	Bubaridae	no	no
Cercicladia	Poecilosclerida	Cladorhizidae	Poecilosclerida	Cladorhizidae	no	no
Cervicornia	Hadromerida	Clionaidae	Clionaida	Clionaidae	18S, 28S	no
Chaetodoryx	Poecilosclerida	Coelosphaeridae	Poecilosclerida	Coelosphaeridae	no	no
Characella	Astrophorida	Pachastrellidae	Tetractinellida (Astrophorina)	(Astrophorina) *incertae sedis*	no	yes
Chelotropella	Astrophorida	Ancorinidae	Tetractinellida (Astrophorina)	Ancorinidae	no	no

Table 1 Listing of Heteroscleromorpha genera in alphabetical order *(Continued)*

Chondrocladia	Poecilosclerida	Cladorhizidae	Poecilosclerida	Cladorhizidae	no	yes
Chondropsis	Poecilosclerida	Chondropsidae	Poecilosclerida	Chondropsidae	no	yes
Cinachyra	Spirophorida	Tetillidae	Tetractinellida (Spirophorina)	Tetillidae	COI, 28S, 18S	yes
Cinachyrella	Spirophorida	Tetillidae	Tetractinellida (Spirophorina)	Tetillidae	no	yes
Ciocalapata	Halichondrida	Halichondriidae	Suberitida	Halichondriidae	no	no
Ciocalypta	Halichondrida	Halichondriidae	Suberitida	Halichondriidae	18S, 28S, COI	yes
Cladorhiza	Poecilosclerida	Cladorhizidae	Poecilosclerida	Cladorhizidae	no	yes
Cladothenea	Astrophorida	Theneidae	Tetractinellida (Astrophorina)	Theneidae	no	no
Clathria	Poecilosclerida	Microcioninae	Poecilosclerida	Microcionidae	no	yes
Cliona	Hadromerida	Clionaidae	Clionaida	Clionaidae	18S, 28S, COI	yes
Clionaopsis	Hadromerida	Clionaidae	Clionaida	Clionaidae	no	no
Cliothosa	Hadromerida	Clionaidae	Clionaida	Clionaidae	no	no
Coelocarteria	Poecilosclerida	Isodictyidae	Poecilosclerida	Isodictyidae	28S	no
Coelodischela	Poecilosclerida	Guitarridae	Poecilosclerida	Guitarridae	no	no
Coelosphaera	Poecilosclerida	Coelosphaeridae	Poecilosclerida	Coelosphaeridae	no	no
Collectella	'Lithistida'	Lithistida incertae sedis	Tetractinellida (Astrophorina)	Theonellidae	no	no
Columnitis	Hadromerida	Tethyidae	Tethyida	Tethyidae	no	no
Corallistes	'Lithistida'	Corallistidae	Tetractinellida (Astrophorina)	Corallistidae	no	yes
Cornulella	Poecilosclerida	Acarnidae	Poecilosclerida	Acarnidae	no	yes
Cornulum	Poecilosclerida	Acarnidae	Poecilosclerida	Acarnidae	no	no
Cortispongilla	Haplosclerida	Malawispongiidae	Spongillida	Malawispongiidae	COI, 18S, 5.8S-ITS2-28S	no
Corvoheteromeyenia	Haplosclerida	Spongillidae	Spongillida	Spongillidae	no	no
Corvomeyenia	Haplosclerida	Metaniidae	Spongillida	Metaniidae	no	yes
Corvospongilla	Haplosclerida	Spongillidae	Spongillida	Spongillidae	no	no
Costifer	'Lithistida'	Isoraphiniidae	Tetractinellida (Astrophorina)	Isoraphiniidae	no	no
Crambe	Poecilosclerida	Crambeidae	Poecilosclerida	Crambeidae	18S, 28S	yes
Craniella	Spirophorida	Tetillidae	Tetractinellida (Spirophorina)	Tetillidae	no	yes
Crella	Poecilosclerida	Crellidae	Poecilosclerida	Crellidae	18S, 28S, tpm	yes
Crellastrina	Poecilosclerida	Crellidae	Poecilosclerida	Crellidae	no	no
Crellomima	Poecilosclerida	Crellidae	Poecilosclerida	Crellidae	no	no
Cryptax	Halichondrida	Halichondriidae	Suberitida	Halichondriidae	no	no
Cryptosyringa	Astrophorida	Ancorinidae	Tetractinellida (Astrophorina)	Ancorinidae	no	no
Cyamon	Poecilosclerida	Raspailiidae	Axinellida	Raspailiidae	no	no
Cyclacanthia	Poecilosclerida	Latrunculiidae	Poecilosclerida	Latrunculiidae	no	no
Cymbastela	Halichondrida	Axinellidae	Bubarida	Dictyonellidae	28S	yes
Daedalopelta	'Lithistida'	Neopeltidae	Tetractinellida (Astrophorina)	Neopeltidae	no	no
Damiria	Poecilosclerida	Acarnidae	Poecilosclerida	Acarnidae	no	no
Damiriopsis	Poecilosclerida	Myxillidae	Poecilosclerida	Myxillidae	no	no

Table 1 Listing of Heteroscleromorpha genera in alphabetical order *(Continued)*

Delectona	Astrophorida	Thoosidae	Tetractinellida (Astrophorina)	Thoosidae	no	no
Dendoricella	Poecilosclerida	Dendoricellidae	Poecilosclerida	Dendoricellidae	no	no
Dercitus	Astrophorida	Ancorinidae	Tetractinellida (Astrophorina)	Ancorinidae	COI, 18S	yes
Desmacella	Poecilosclerida	Desmacellidae	Desmacellida	Desmacellidae	18S, 28S	yes
Desmacidon	Poecilosclerida	Desmacididae	Poecilosclerida	Desmacididae	no	no
Desmanthus	'Lithistida'	Desmanthidae	Bubarida	Bubaridae	18S, 28S, COI	no
Desmapsamma	Poecilosclerida	Desmacididae	Poecilosclerida	Desmacididae	16S, 18S, 28S, COI	no
Desmascula	'Lithistida'	Azoricidae	Tetractinellida (Spirophorina)	Azoricidae	no	no
Desmoxya	Halichondrida	Heteroxyidae	Axinellida	Heteroxyidae	no	no
Diacarnus	Poecilosclerida	Podospongiidae	Poecilosclerida	Podospongiidae[3]	18S, 28S, COI	no
Dictyonella	Halichondrida	Dictyonellidae	Bubarida	Dictyonellidae	18S, 28S	yes
Didiscus	Halichondrida	Heteroxyidae	Axinellida	Raspailiidae	no	yes
Diplastrella	Hadromerida	Spirastrellidae	Clionaida	Spirastrellidae	no	yes
Diplopodospongia	Poecilosclerida	Podospongiidae	Poecilosclerida	Podospongiidae	no	no
Discodermia	'Lithistida'	Theonellidae	Tetractinellida (Astrophorina)	Theonellidae	no	yes
Discorhabdella	Poecilosclerida	Crambeidae	Poecilosclerida	Crambeidae	no	no
Disyringa	Astrophorida	Ancorinidae	Tetractinellida (Astrophorina)	Ancorinidae	18S, 28S	no
Dolichacantha	Poecilosclerida	Acarnidae	Poecilosclerida	Acarnidae	no	no
Dosilia	Haplosclerida	Spongillidae	Spongillida	Spongillidae	no	no
Dotona	Hadromerida	Clionaidae	Clionaida	Clionaidae	no	no
Dragmacidon	Halichondrida	Axinellidae	Axinellida	Axinellidae	no	yes
Dragmatella	Poecilosclerida	Desmacellidae	Desmacellida	Desmacellidae	no	no
Dragmaxia	Halichondrida	Axinellidae	Axinellida	Axinellidae	no	yes
Drulia	Haplosclerida	Metaniidae	Spongillida	Metaniidae	5.8S-ITS2-28S	no
Duosclera	Haplosclerida	Spongillidae	Spongillida	Spongillidae	no	no
Dyscliona	Hadromerida	Clionaidae	Clionaida	Clionaidae	no	no
Echinochalina	Poecilosclerida	Microcioninae	Poecilosclerida	Microcionidae	no	yes
Echinoclathria	Poecilosclerida	Ophlitaspongiinae	Poecilosclerida	Microcionidae	no	yes
Echinodictyum	Poecilosclerida	Echinodictyinae	Poecilosclerida	Echinodictyinae	no	yes
Echinospongilla	Haplosclerida	Potamolepidae	Spongillida	Potamolepidae	COI, 18S, ITS1-5.8S-ITS2-28S	no
Echinostylinos	Poecilosclerida	Phellodermidae	Poecilosclerida	Phellodermidae	no	no
Ecionemia	Astrophorida	Ancorinidae	Tetractinellida (Astrophorina)	Ancorinidae	18S	yes
Ectyonopsis	Poecilosclerida	Myxillidae	Poecilosclerida	Myxillidae	no	yes
Ectyoplasia	Poecilosclerida	Raspailiidae	Axinellida	Raspailiidae	18S, 28S	yes
Endectyon	Poecilosclerida	Raspailiidae	Axinellida	Raspailiidae	no	yes
Eospongilla	Haplosclerida	Palaeospongillidae	Spongillida	Palaeospongillidae	no	no
Ephydatia	Haplosclerida	Spongillidae	Spongillida	Spongillidae	COI, 18S, 5.8S-ITS2-28S, mt, ESTs	yes
Epipolasis	Halichondrida	Axinellidae	Axinellida	Axinellidae	no	yes
Erylus	Astrophorida	Geodiidae	Tetractinellida (Astrophorina)	Geodiidae	COI, 28S	yes

Table 1 Listing of Heteroscleromorpha genera in alphabetical order *(Continued)*

Esperiopsis	Poecilosclerida	Esperiopsidae	Poecilosclerida	Esperiopsidae	no	no
Euchelipluma	Poecilosclerida	Guitarridae	Poecilosclerida	Guitarridae	no	no
Eunapius	Haplosclerida	Spongillidae	Spongillida	Spongillidae	COI, ITS2, 18S	yes
Eurypon	Poecilosclerida	Raspailiidae	Axinellida	Raspailiidae	no	yes
Exsuperantia	'Lithistida'	Phymaraphiniidae	Tetractinellida (Astrophorina)	Phymaraphiniidae	no	yes
Fangophilina	Spirophorida	Tetillidae	Tetractinellida (Spirophorina)	Tetillidae	no	yes
Fibulia	Poecilosclerida	Dendoricellidae	Poecilosclerida	Dendoricellidae	no	no
Forcepia	Poecilosclerida	Coelosphaeridae	Poecilosclerida	Coelosphaeridae	no	yes
Gastrophanella	'Lithistida'	Siphonidiidae	Tetractinellida (Spirophorina)	Siphonidiidae	no	no
Geodia	Astrophorida	Geodiidae	Tetractinellida (Astrophorina)	Geodiidae	no	yes
Goreauiella	Agelasida	Astroscleridae	Agelasida	Astroscleridae	no	no
Guitarra	Poecilosclerida	Guitarridae	Poecilosclerida	Guitarridae	28S	no
Halichondria	Halichondrida	Halichondriidae	Suberitida	Halichondriidae	18S, 28S, 5.8S, COI	yes
Halicnemia	Halichondrida	Heteroxyidae	Axinellida	Stelligeridae	no	yes
Halicometes	Hadromerida	Tethyidae	Tethyida	Tethyidae	no	no
Hamacantha	Poecilosclerida	Hamacanthidae	Merliida	Hamacanthidae	no	no
Hamigera	Poecilosclerida	Hymedesmiidae	Poecilosclerida	Hymedesmiidae	18S, 28S	yes
Hemiasterella	Hadromerida	Hemiasterellidae	Tethyida	Hemiasterellidae	no	yes
Hemimycale	Poecilosclerida	Hymedesmiidae	Poecilosclerida	Hymedesmiidae	18S, 28S	no
Hemitedania	Poecilosclerida	Tedaniidae	Poecilosclerida	Tedaniidae	no	no
Herengeria	'Lithistida'	Corallistidae	Tetractinellida (Astrophorina)	Corallistidae	no	no
Heteromeyenia	Haplosclerida	Spongillidae	Spongillida	Spongillidae	no	no
Heterorotula	Haplosclerida	Spongillidae	Spongillida	Spongillidae	no	yes
Heteroxya	Halichondrida	Heteroxyidae	Axinellida	Heteroxyidae	no	no
Higginsia	Halichondrida	Heteroxyidae	Axinellida	Stelligeridae	no	yes
Hispidopetra	Agelasida	Astroscleridae	Agelasida	Astroscleridae	no	no
Histodermella	Poecilosclerida	Coelosphaeridae	Poecilosclerida	Coelosphaeridae	no	no
Holopsamma	Poecilosclerida	Microcioninae	Poecilosclerida	Microcionidae	no	yes
Holoxea	Astrophorida	Ancorinidae	Tetractinellida (Astrophorina)	Ancorinidae	no	yes
Homaxinella	Hadromerida	Suberitidae	Suberitida	Suberitida *incertae sedis*	no	yes
Homophymia	'Lithistida'	Neopeltidae	Tetractinellida (Astrophorina)	Neopeltidae	no	yes
Houssayella	Haplosclerida	Metaniidae	Spongillida	Metaniidae	no	no
Hymedesmia	Poecilosclerida	Hymedesmiidae	Poecilosclerida	Hymedesmiidae	no	yes
Hymenancora	Poecilosclerida	Myxillidae	Poecilosclerida	Myxillidae	no	no
Hymeniacidon	Halichondrida	Halichondriidae	Suberitida	Halichondriidae	18S, 28S, COI	yes
Hymeraphia	Poecilosclerida	Raspailiidae	Axinellida	Raspailiidae	18S, 28S	yes
Hymerhabdia	Agelasida	Hymerhabdiidae	Agelasida	Hymerhabdiidae	18S, 28S, COI	no
Hymetrochota	Poecilosclerida	Iotrochotidae	Poecilosclerida	Iotrochotidae	no	no
Inflatella	Poecilosclerida	Coelosphaeridae	Poecilosclerida	Coelosphaeridae	18S	no
Iophon	Poecilosclerida	Acarnidae	Poecilosclerida	Acarnidae	no	yes

Table 1 Listing of Heteroscleromorpha genera in alphabetical order *(Continued)*

Iotroata	Poecilosclerida	Iotrochotidae	Poecilosclerida	Iotrochotidae	no	no
Iotrochopsamma	Poecilosclerida	Iotrochotidae	Poecilosclerida	Iotrochotidae	no	no
Iotrochota	Poecilosclerida	Iotrochotidae	Poecilosclerida	Iotrochotidae	18S, 28S, mt	no
Isabella	'Lithistida'	Corallistidae	Tetractinellida (Astrophorina)	Corallistidae	no	no
Isodictya	Poecilosclerida	Isodictyidae	Poecilosclerida	Isodictyidae	no	yes
Janulum	Poecilosclerida	Raspailiidae	Haplosclerida	Phloeodictyidae	18S	no
Jaspis	Astrophorida	Ancorinidae	Tetractinellida (Astrophorina)	Ancorinidae	no	yes
Jereicopsis	'Lithistida'	Azoricidae	Tetractinellida (Spirophorina)	Azoricidae	no	no
Johannesia	Halichondrida	Halichondriidae	Suberitida	Halichondriidae	no	no
Julavis	Halichondrida	Heteroxyidae	Axinellida	Heteroxyidae	no	no
Kaliapsis	'Lithistida'	Phymaraphiniidae	Tetractinellida (Astrophorina)	Phymaraphiniidae	no	no
Kirkpatrickia	Poecilosclerida	Hymedesmiidae	Poecilosclerida	Hymedesmiidae	no	no
Lamellomorpha	Astrophorida	Pachastrellidae	Tetractinellida (Astrophorina)	(Astrophorina) *incertae sedis*	18S	no
Laminospongia	Halichondrida	Halichondriidae	Suberitida	Halichondriidae	no	no
Latrunculia	Poecilosclerida	Latrunculiidae	Poecilosclerida	Latrunculiidae	no	yes
Laxotethya	Hadromerida	Tethyidae	Tethyida	Tethyidae	28S	no
Leiodermatium	'Lithistida'	Azoricidae	Tetractinellida (Spirophorina)	Azoricidae	no	no
Lepidosphaera	Poecilosclerida	Coelosphaeridae	Poecilosclerida	Coelosphaeridae	no	no
Lepidothenea	'Lithistida'	Phymaraphiniidae	Tetractinellida (Astrophorina)	Phymaraphiniidae	no	no
Leptosastra	Hadromerida	Hemiasterellidae	Tethyida	Hemiasterellidae	no	no
Liosina	Halichondrida	Dictyonellidae	Tethyida	Hemiasterellidae	28S, COI	no
Lipastrotethya	Halichondrida	Dictyonellidae	Bubarida	Dictyonellidae	no	no
Lissodendoryx	Poecilosclerida	Coelosphaeridae	Poecilosclerida	Coelosphaeridae	no	yes
Lithobactrum	'Lithistida'	Siphonidiidae	Tetractinellida (Spirophorina)	Siphonidiidae	no	no
Lithochela	Poecilosclerida	Crambeidae	Poecilosclerida	Crambeidae	no	no
Lithoplocamia	Poecilosclerida	Raspailiidae	Poecilosclerida	Raspailiidae	no	no
Lollipocladia	Poecilosclerida	Cladorhizidae	Poecilosclerida	Cladorhizidae	no	no
Lubomirskia	Haplosclerida	Lubomirskiidae	Spongillida	Lubomirskiidae	65 sequences, mt, ESTs	yes
Lutetiospongilla	Haplosclerida	Palaeospongillidae	Spongillida	Palaeospongillidae	no	no
Macandrewia	'Lithistida'	Macandrewiidae	Tetractinellida (Astrophorina)	Macandrewiidae	no	no
Makedia	Haplosclerida	Spongillina *incertae sedis*	Spongillida	Spongillida *incertae sedis*	no	no
Malawispongia	Haplosclerida	Malawispongiidae	Spongillida	Malawispongiidae	no	no
Manihinea	'Lithistida'	Theonellidae	Tetractinellida (Astrophorina)	Theonellidae	18S	no
Megaciella	Poecilosclerida	Acarnidae	Poecilosclerida	Acarnidae	no	no
Melonanchora	Poecilosclerida	Myxillidae	Poecilosclerida	Myxillidae	no	no
Melophlus	Astrophorida	Geodiidae	Tetractinellida (Astrophorina)	Geodiidae	no	yes
Merlia	Poecilosclerida	Merliidae	Merliida	Merliidae	16S, 18S, 28S	no

Table 1 Listing of Heteroscleromorpha genera in alphabetical order *(Continued)*

Metania	Haplosclerida	Metaniidae	Spongillida	Metaniidae	no	no
Metschnikowia	Haplosclerida	Metschnikowiidae	Spongillida	Metschnikowiidae	no	no
Microscleroderma	'Lithistida'	Scleritodermidae	Tetractinellida (Spirophorina)	Scleritodermidae	no	yes
Microtylostylifer	Poecilosclerida	Desmacellidae	Desmacellida	Desmacellidae	no	no
Microxistyla	Halichondrida	Heteroxyidae	Axinellida	Heteroxyidae	no	no
Monanchora	Poecilosclerida	Crambeidae	Poecilosclerida	Crambeidae	28S, COI	yes
Monocrepidium	Halichondrida	Bubaridae	Bubarida	Bubaridae	no	no
Mycale	Poecilosclerida	Mycalidae	Poecilosclerida	Mycalidae	no	yes
Myrmekioderma	Halichondrida	Heteroxyidae	Axinellida	Heteroxyidae	18S, 28S, COI	yes
Myxilla	Poecilosclerida	Myxillidae	Poecilosclerida	Myxillidae	28S	yes
Myxodoryx	Poecilosclerida	Hymedesmiidae	Poecilosclerida	Hymedesmiidae	no	no
Neamphius	Astrophorida	Thoosidae	Tetractinellida (Astrophorina)	(Astrophorina) *incertae sedis*	COI, 28S	yes
Negombata	Poecilosclerida	Podospongiidae	Poecilosclerida	Podospongiidae	18S, COI	yes
Negombo	Halichondrida	Heteroxyidae	Axinellida	Heteroxyidae	no	no
Neoaulaxinia	'Lithistida'	Phymatellidae	Tetractinellida (Astrophorina)	Phymatellidae	no	no
Neocladia	Poecilosclerida	Cladorhizidae	Poecilosclerida	Cladorhizidae	no	yes
Neofibularia	Poecilosclerida	Desmacellidae	Biemnida	Biemnidae	18S, 28S, COI	yes
Neopelta	'Lithistida'	Neopeltidae	Tetractinellida (Astrophorina)	Neopeltidae	no	no
Neophrissospongia	'Lithistida'	Corallistidae	Tetractinellida (Astrophorina)	Corallistidae	no	yes
Neopodospongia	Poecilosclerida	Podospongiidae	Poecilosclerida	Podospongiidae	no	yes
Neoschrammeniella	'Lithistida'	Corallistidae	Tetractinellida (Astrophorina)	Corallistidae	no	no
Neosiphonia	'Lithistida'	Phymatellidae	Tetractinellida (Astrophorina)	Phymatellidae	no	no
Nethea	Astrophorida	Pachastrellidae	Tetractinellida (Astrophorina)	Pachastrellidae	no	yes
Nucleotethya	Hadromerida	Tethyidae	Tethyida	Tethyidae	no	no
Nudospongilla	Haplosclerida	Spongillidae	Spongillida	Spongillidae	no	yes
Ochridaspongia	Haplosclerida	Malawispongiidae	Spongillida	Malawispongiidae	no	no
Ohridospongilla	Haplosclerida	Spongillina *incertae sedis*	Spongillida	Spongillida *incertae sedis*	no	yes
Oncosclera	Haplosclerida	Potamolepidae	Spongillida	Potamolepidae	no	yes
Onotoa	Hadromerida	Placospongiidae	Clionaida	Placospongiidae	no	no
Ophiraphidites	Halichondrida	Axinellidae	Bubarida	Bubaridae	no	no
Ophlitaspongia	Poecilosclerida	Ophlitaspongiinae	Poecilosclerida	Microcionidae	16S, 18S	no
Oxytethya	Hadromerida	Tethyidae	Tethyida	Tethyidae	no	no
Pachastrella	Astrophorida	Pachastrellidae	Tetractinellida (Astrophorina)	Pachastrellidae	no	yes
Pachydictyum	Haplosclerida	Malawispongiidae	Spongillida	Malawispongiidae	COI, 18S, ESTs	yes
Pachymatisma	Astrophorida	Geodiidae	Tetractinellida (Astrophorina)	Geodiidae	COI, 28S, 18S, ITS1-5.8S-ITS2	yes
Pachyrotula	Haplosclerida	Spongillidae	Spongillida	Spongillidae	no	no
Palaeospongilla	Haplosclerida	Palaeospongillidae	Spongillida	Palaeospongillidae	no	no
Pandaros	Poecilosclerida	Raspailiidae	Axinellida	Raspailiidae	COI	no

Table 1 Listing of Heteroscleromorpha genera in alphabetical order *(Continued)*

Paracornulum	Poecilosclerida	Acarnidae	Poecilosclerida	Acarnidae	COI	no
Paradesmanthus	'Lithistida'	Desmanthidae	Bubarida	Bubaridae	no	no
Parahigginsia	Halichondrida	Heteroxyidae	Axinellida	Heteroxyidae	no	no
Pararhaphoxya	Halichondrida	Axinellidae	Bubarida	Bubaridae	no	yes
Paratetilla	Spirophorida	Tetillidae	Tetractinellida (Spirophorina)	Tetillidae	no	yes
Paratimea	Hadromerida	Stelligeridae	Axinellida	Stelligeridae	no	yes
Pectispongilla	Haplosclerida	Spongillidae	Spongillida	Spongillidae	no	no
Penares	Astrophorida	Geodiidae	Tetractinellida (Astrophorina)	Geodiidae	5.8S-ITS2-28S	yes
Penicillus	Hadromerida	Polymastiidae	Polymastiida	Polymastiidae	no	no
Petromica	'Lithistida'	Desmanthidae	*incertae sedis*		no	yes
Phakellia	Halichondrida	Axinellidae	Bubarida	Bubaridae	18S, 28S, COI	yes
Phakettia	Halichondrida	Dictyonellidae	Bubarida	Dictyonellidae	no	yes
Phelloderma	Poecilosclerida	Phellodermidae	Poecilosclerida	Phellodermidae	no	yes
Phlyctaenopora	Poecilosclerida	Mycalidae	Poecilosclerida	Mycalidae	no	no
Phorbas	Poecilosclerida	Hymedesmiidae	Poecilosclerida	Hymedesmiidae	18S	yes
Phoriospongia	Poecilosclerida	Chondropsidae	Poecilosclerida	Chondropsidae	no	no
Phycopsis	Halichondrida	Axinellidae	Bubarida	Bubaridae	no	yes
Pione	Hadromerida	Clionaidae	Clionaida	Clionaidae	no	yes
Pipestela	Halichondrida	Axinellidae	Bubarida	Bubaridae	no	no
Placospherastra	Hadromerida	Placospongiidae	Clionaida	Placospongiidae	no	no
Placospongia	Hadromerida	Placospongiidae	Clionaida	Placospongiidae	18S, COI	yes
Plakidium	'Lithistida'	Lithistida *incertae sedis*	*incertae sedis*		no	no
Pleroma	'Lithistida'	Pleromidae	Tetractinellida (Astrophorina)	Pleromidae	no	no
Plicatellopsis	Hadromerida	Suberitidae	Suberitida	Suberitidae	no	no
Plocamiancora	Poecilosclerida	Myxillidae	Poecilosclerida	Myxillidae	no	yes
Plocamione	Poecilosclerida	Raspailiidae	Poecilosclerida	Raspailiidae	no	no
Plocamionida	Poecilosclerida	Hymedesmiidae	Poecilosclerida	Hymedesmiidae	18S, 28S, COI	yes
Podospongia	Poecilosclerida	Podospongiidae	Poecilosclerida	Podospongiidae	no	no
Poecillastra	Astrophorida	Vulcanellidae	Tetractinellida (Astrophorina)	Vulcanellidae	COI, 28S	yes
Polymastia	Hadromerida	Polymastiidae	Polymastiida	Polymastiidae	no	yes
Pomelia	'Lithistida'	Scleritodermidae	Tetractinellida (Spirophorina)	Scleritodermidae	no	no
Poritella	'Lithistida'	Lithistida *incertae sedis*	*incertae sedis*		no	no
Potamolepis	Haplosclerida	Potamolepidae	Spongillida	Potamolepidae	no	yes
Potamophloios	Haplosclerida	Potamolepidae	Spongillida	Potamolepidae	no	no
Pottsiela	Haplosclerida	Spongillidae	Spongillida	Spongillidae	no	no
Pozziella	Poecilosclerida	Hamacanthidae	Merliida	Hamacanthidae	no	no
Prosuberites	Hadromerida	Suberitidae	Agelasida	Hymerhabdiidae	28S	yes
Proteleia	Hadromerida	Polymastiidae	Polymastiida	Polymastiidae	18S	no
Protosuberites	Hadromerida	Suberitidae	Suberitida	Suberitidae	no	yes
Psammastra	Astrophorida	Ancorinidae	Tetractinellida (Astrophorina)	Ancorinidae	no	no

Table 1 Listing of Heteroscleromorpha genera in alphabetical order *(Continued)*

Psammochela	Poecilosclerida	Myxillidae	Poecilosclerida	Myxillidae	no	no
Psammoclema	Poecilosclerida	Chondropsidae	Poecilosclerida	Chondropsidae	no	yes
Pseudohalichondria	Poecilosclerida	Hymedesmiidae	Poecilosclerida	Hymedesmiidae	no	no
Pseudospongosorites	Hadromerida	Suberitidae	Suberitida	Suberitidae	tpm	no
Pseudosuberites	Hadromerida	Suberitidae	Suberitida	Suberitidae	no	yes
Pseudotrachya	Hadromerida	Polymastiidae	Polymastiida	Polymastiidae	no	no
Ptilocaulis	Halichondrida	Axinellidae	Axinellida	Raspailiidae	18S, 28S, mt	yes
Pyloderma	Poecilosclerida	Dendoricellidae	Poecilosclerida	Dendoricellidae	no	no
Quasillina	Hadromerida	Polymastiidae	Polymastiida	Polymastiidae	18S	no
Racekiela	Haplosclerida	Spongillidae	Spongillida	Spongillidae	no	no
Racodiscula	'Lithistida'	Theonellidae	Tetractinellida (Astrophorina)	Theonellidae	no	no
Radiella	Hadromerida	Polymastiidae	Polymastiida	Polymastiidae	no	no
Radiospongilla	Haplosclerida	Spongillidae	Spongillida	Spongillidae	no	yes
Raspaciona	Poecilosclerida	Raspailiidae	Axinellida	Raspailiidae	18S, 28S, COI	no
Raspailia	Poecilosclerida	Raspailiidae	Axinellida	Raspailiidae	no	yes
Reidispongia	'Lithistida'	Phymatellidae	Tetractinellida (Astrophorina)	Phymatellidae	no	no
Reniochalina	Halichondrida	Axinellidae	Axinellida	Raspailiidae	18S, 28S	yes
Rezinkovia	Haplosclerida	Lubomirskiidae	Spongillida	Lubomirskiidae	18S-5.8S-ITS2-28S, mt	yes
Rhabdastrella	Astrophorida	Ancorinidae	Tetractinellida (Astrophorina)	Geodiidae	no	yes
Rhabderemia	Poecilosclerida	Rhabderemiidae	Biemnida	Rhabderemiidae	no	yes
Rhabdeurypon	Poecilosclerida	Raspailiidae	Axinellida	Raspailiidae	no	no
Rhabdobaris	Halichondrida	Bubaridae	Bubarida	Bubaridae	no	no
Rhaphidhistia	Hadromerida	Trachycladidae	Trachycladida	Trachycladidae	no	no
Rhaphoxya	Halichondrida	Dictyonellidae	Bubarida	Dictyonellidae	no	no
Rhizaxinella	Hadromerida	Suberitidae	Suberitida	Suberitidae	no	yes
Ridleia	Hadromerida	Polymastiidae	Polymastiida	Polymastiidae	no	no
Rotuloplocamia	Poecilosclerida	Iotrochotidae	Poecilosclerida	Iotrochotidae	no	no
Samus	Spirophorida	Samidae	Tetractinellida (Spirophorina)	Samidae	no	no
Sanidastra	Haplosclerida	Spongillidae	Spongillida	Spongillidae	no	no
Sarcomella	Halichondrida	Halichondriidae	Suberitida	Halichondriidae	no	no
Saturnospongilla	Haplosclerida	Spongillidae	Spongillida	Spongillidae	no	no
Sceptrella	Poecilosclerida	Latrunculiidae	Poecilosclerida	Latrunculiidae	no	yes
Sceptrintus	Poecilosclerida	Podospongiidae	Poecilosclerida	Podospongiidae	no	no
Scleritoderma	'Lithistida'	Scleritodermidae	Tetractinellida (Spirophorina)	Scleritodermidae	no	no
Scolopes	Hadromerida	Clionaidae	Clionaida	Clionaidae	no	no
Scopalina	Halichondrida	Scopalinidae	Scopalinida	Scopalinidae	18S, 28S, COI	yes
Semisuberites	Poecilosclerida	Esperiopsidae	Poecilosclerida	Esperiopsidae	no	no
Setidium	'Lithistida'	Scleritodermidae	Tetractinellida (Spirophorina)	Scleritodermidae	no	no
Sigmaxinella	Poecilosclerida	Desmacellidae	Biemnida	Biemnidae	no	yes
Sigmeurypon	Poecilosclerida	Ophlitaspongiinae	Poecilosclerida	Microcionidae	no	no
Sigmosceptrella	Poecilosclerida	Podospongiidae	Poecilosclerida	Podospongiidae	no	no

Table 1 Listing of Heteroscleromorpha genera in alphabetical order *(Continued)*

Siliquariaspongia	'Lithistida'	Theonellidae	Tetractinellida (Astrophorina)	Theonellidae	no	no
Siphonidium	'Lithistida'	Siphonidiidae	Tetractinellida (Spirophorina)	Siphonidiidae	no	no
Sollasella	Poecilosclerida	Raspailiidae	Axinellida	Raspailiidae	no	yes
Spanioplon	Poecilosclerida	Hymedesmiidae	Poecilosclerida	Hymedesmiidae	18S, 28S	no
Sphaerotylus	Hadromerida	Polymastiidae	Polymastiida	Polymastiidae	no	yes
Spheciospongia	Hadromerida	Clionaidae	Clionaida	Clionaidae	18S	yes
Spinospongilla	Haplosclerida	Malawispongiidae	Spongillida	Malawispongiidae	no	no
Spinularia	Hadromerida	Polymastiidae	Polymastiida	Polymastiidae	no	no
Spirasigma	Spirophorida	Spirasigmidae	Tetractinellida (Spirophorina)	Spirasigmidae	no	no
Spirastrella	Hadromerida	Spirastrellidae	Clionaida	Spirastrellidae	no	yes
Spirorhabdia	Poecilosclerida	Crellidae	Poecilosclerida	Crellidae	no	no
Spiroxya	Hadromerida	Clionaidae	Clionaida	Clionaidae	no	no
Spongilla	Haplosclerida	Spongillidae	Spongillida	Spongillidae	COI,18S, 28S, ITS1-5.8S-ITS2-28S, tpm	yes
Spongosorites	Halichondrida	Halichondriidae	Suberitida	Halichondriidae	no	yes
Stelletta	Astrophorida	Ancorinidae	Tetractinellida (Astrophorina)	Ancorinidae	COI, 28S, 18S	yes
Stellettinopsis	Astrophorida	Ancorinidae	Tetractinellida (Astrophorina)	Ancorinidae	no	yes
Stelligera	Hadromerida	Stelligeridae	Axinellida	Stelligeridae	18S, 28S, COI	yes
Stellitethya	Hadromerida	Tethyidae	Tethyida	Tethyidae	no	yes
Stelodoryx	Poecilosclerida	Myxillidae	Poecilosclerida	Myxillidae	no	no
Sterrastrolepis	Haplosclerida	Potamolepidae	Spongillida	Potamolepidae	no	no
Stratospongilla	Haplosclerida	Spongillidae	Spongillida	Spongillidae	no	no
Stromatospongia	Agelasida	Astroscleridae	Agelasida	Astroscleridae	no	no
Strongylacidon	Poecilosclerida	Chondropsidae	Poecilosclerida	Chondropsidae	no	yes
Strongylamma	Poecilosclerida	Tedaniidae	Poecilosclerida	Tedaniidae	no	no
Strongylodesma	Poecilosclerida	Latrunculiidae	Poecilosclerida	Latrunculiidae	no	no
Stryphnus	Astrophorida	Ancorinidae	Tetractinellida (Astrophorina)	Ancorinidae	no	yes
Stylissa	Halichondrida	Scopalinidae	Scopalinida	Scopalinidae	18S, 28S, COI	yes
Stylocordyla	Hadromerida	Stylocordylidae	Suberitida	Stylocordylidae	COI (submitted)	no
Suberites	Hadromerida	Suberitidae	Suberitida	Suberitidae	18S, 28S, mt, ESTs	yes
Sulcastrella	'Lithistida'	Desmanthidae	*incertae sedis*		no	no
Svenzea	Halichondrida	Scopalinidae	Scopalinida	Scopalinidae	18S, 28S, COI	yes
Swartschewskia	Haplosclerida	Lubomirskiidae	Spongillida	Lubomirskiidae	COI, 18S, 5.8S-ITS2-28S, mt	no
Tectitethya	Hadromerida	Tethyidae	Tethyida	Tethyidae	no	yes
Tedania	Poecilosclerida	Tedaniidae	Poecilosclerida	Tedaniidae	no	yes
Tedaniphorbas	Poecilosclerida	Acarnidae	Poecilosclerida	Acarnidae	no	no
Tentorina	Spirophorida	Spirasigmidae	Tetractinellida (Spirophorina)	Spirasigmidae	no	no
Tentorium	Hadromerida	Polymastiidae	Polymastiida	Polymastiidae	18S	no
Terpios	Hadromerida	Suberitidae	Suberitida	Suberitida *incertae sedis*	no	yes

Table 1 Listing of Heteroscleromorpha genera in alphabetical order *(Continued)*

Tethya	Hadromerida	Tethyidae	Tethyida	Tethyidae	18S, 28S, COI	yes
Tethyastra	Hadromerida	Tethyidae	Tethyida	Tethyidae	no	no
Tethycometes	Hadromerida	Tethyidae	Tethyida	Tethyidae	no	no
Tethyopsis	Astrophorida	Ancorinidae	Tetractinellida (Astrophorina)	Ancorinidae	no	yes
Tethyspira	Halichondrida	Dictyonellidae	Axinellida	Raspailiidae	18S, 28S, COI	no
Tethytimea	Hadromerida	Tethyidae	Tethyida	Tethyidae	no	yes
Tetilla	Spirophorida	Tetillidae	Tetractinellida (Spirophorina)	Tetillidae	no	yes
Tetrapocillon	Poecilosclerida	Guitarridae	Poecilosclerida	Guitarridae	no	yes
Thenea	Astrophorida	Theneidae	Tetractinellida (Astrophorina)	Theneidae	COI, 28S	yes
Theonella	'Lithistida'	Theonellidae	Tetractinellida (Astrophorina)	Theonellidae	COI, 28S, 18S	yes
Thoosa	Astrophorida	Thoosidae	Tetractinellida (Astrophorina)	Thoosidae	no	no
Thrinacophora	Poecilosclerida	Raspailiidae	Axinellida	Raspailiidae	no	yes
Thrombus	Astrophorida	Thrombidae	Tetractinellida (Astrophorina)	Thrombidae	28S	no
Timea	Hadromerida	Timeidae	Tethyida	Tethyidae	no	yes
Topsentia	Halichondrida	Halichondriidae	Suberitida	Suberitida *incertae sedis*	no	yes
Trachostylea	Poecilosclerida	Raspailiidae	Axinellida	Raspailiidae	no	no
Trachycladus	Hadromerida	Trachycladidae	Trachycladida	Trachycladidae	18S, 28S	yes
Trachyteleia	Hadromerida	Polymastiidae	Polymastiida	Polymastiidae	no	no
Tribrachium	Astrophorida	Ancorinidae	Tetractinellida (Astrophorina)	Ancorinidae	no	no
Trikentrion	Poecilosclerida	Raspailiidae	Axinellida	Raspailiidae	no	yes
Triptolemma	Astrophorida	Pachastrellidae	Tetractinellida (Astrophorina)	Pachastrellidae	no	yes
Trochospongilla	Haplosclerida	Spongillidae	Spongillida	Spongillidae	no	yes
Tsitsikamma	Poecilosclerida	Latrunculiidae	Poecilosclerida	Latrunculiidae	18S, COI	no
Tylexocladus	Hadromerida	Polymastiidae	Polymastiida	Polymastiidae	no	no
Ulosa	Poecilosclerida	Esperiopsidae	Poecilosclerida[2]	Esperiopsidae[2]	no	yes
Umborotula	Haplosclerida	Spongillidae	Spongillida	Spongillidae	no	no
Uritaia	Halichondrida	Halichondriidae	Suberitida	Halichondriidae	no	no
Uruguaya	Haplosclerida	Potamolepidae	Spongillida	Potamolepidae	no	no
Uruguayella	Haplosclerida	Spongillidae	Spongillida	Spongillidae	no	no
Vetulina	'Lithistida'	Vetulinidae	Sphaerocladina	Vetulinidae	18S	yes
Volzia	Hadromerida	Clionaidae	Clionaida	Clionaidae	no	no
Vosmaeria	Halichondrida	Halichondriidae	Suberitida	Halichondriidae	28S	no
Vulcanella	Astrophorida	Vulcanellidae	Tetractinellida (Astrophorina)	Vulcanellidae	no	yes
Waltherarndtia	Poecilosclerida	Raspailiidae	Axinellida	Raspailiidae	no	no
Weberella	Hadromerida	Polymastiidae	Polymastiida	Polymastiidae	no	no
Wigginsia	Poecilosclerida	Acarnidae	Poecilosclerida	Acarnidae	no	no
Willardia	Hadromerida	Acanthochaetetidae	Clionaida	Acanthochaetetidae	no	no
Xenospongia	Hadromerida	Tethyidae	Tethyida	Tethyidae	18S, 28S	no

Table 1 Listing of Heteroscleromorpha genera in alphabetical order (Continued)

Yucatania	Astrophorida	Thrombidae	Tetractinellida (Astrophorina)	Thrombidae	no	no
Zyzzya	Poecilosclerida	Acarnidae	Poecilosclerida	Acarnidae	no	yes

[1] One sequence of *Auletta* in Erpenbeck et al. [39] groups with *Acanthella* and *Phakellia ventilabrum*.
[2] *28S* sequences of *Ulosa digitata* suggest that this genus could belong to the Suberitida, as it groups with Halichondriidae [5].
[3] Redmond et al. [10] show *D. spinipoculum* (type species) clustering with Raspailiidae whilst *D. bismarckensis* groups with Podospongiidae. It is possible that the *18S* sequence of *D. spinipoculum* in Redmond et al. [10] is a result of contamination and therefore we retain *Diacarnus* in Podospongiidae.

Competing interests
The authors declare that they have no competing interests.

Authors' contributions
CM and PC equally contributed in conceiving and writing the paper. Both authors read and approved the final manuscript.

Authors' information
CM and PC are both sponge taxonomists and phylogeneticists publishing species descriptions and group revisions combining molecular and morphological data.

Acknowledgements
We warmly thank the organizers of the 9th International Sponge Conference (Fremantle, Western Australia, 4-8 Nov. 2013) for facilitating an informal and yet fruitful discussion on the need for a revised classification of Demospongiae. We also thank the organizers (Hans Tore Rapp and Joana Xavier (University of Bergen)) and participants of the 2nd International Workshop on Taxonomy of Atlanto-Mediterranean Deep-Sea Sponges (Bergen, Norway, 2-6 June 2014) for their comments on a first draft of this proposal. Andrzej Pisera is thanked for pointing out at this workshop the availability of the order name Sphaerocladina to welcome the Vetulinidae. Christine Schönberg is thanked for suggesting the name Clionaida. We thank Rob van Soest, Jean Vacelet, Nicole Boury-Esnault, John Hooper and one anonymous reviewer for their comments, which improved this manuscript. PC received support from the 'Swedish Taxonomy Initiative' project (dha 154/2011) and from the EU project 'BlueGenics' KBBE-2012-6. CM received funding from the Beaufort Marine Biodiscovery Research Award under the Sea Change Strategy for Science Technology and Innovation (2006-2013).

Author details
[1]Queen's University Belfast, Marine Laboratory, Portaferry BT22 1PF, Northern Ireland, UK. [2]Department of Organismal Biology, Division of Systematic Biology, Evolutionary Biology Centre, Uppsala University, Norbyvägen 18D, 752 36 Uppsala, Sweden. [3]Department of Medicinal Chemistry, Division of Pharmacognosy, BioMedical Centre, Husargatan 3, Uppsala University, 751 23 Uppsala, Sweden.

References
1. Hooper JNA, van Soest RWM. Systema Porifera: A guide to the classification of sponges. New York: Kluwer Academic / Plenum Publishers; 2002.
2. Wörheide G, Dohrmann M, Erpenbeck D, Larroux C, Maldonado M, Voigt O, et al. Deep Phylogeny and Evolution of Sponges (Phylum Porifera). Adv Mar Biol. 2012;61:1–78.
3. Cárdenas P, Pérez T, Boury-Esnault N. Sponge Systematics facing new challenges. Adv Mar Biol. 2012;61:79–209.
4. World Porifera database, http://www.marinespecies.org/porifera [http://www.marinespecies.org/porifera]
5. Morrow CC, Picton BE, Erpenbeck D, Boury-Esnault N, Maggs CA, Allcock AL. Congruence between nuclear and mitochondrial genes in Demospongiae: A new hypothesis for relationships within the G4 clade (Porifera: Demospongiae). Mol Phylogen Evol. 2012;62:174–90.
6. Cárdenas P, Xavier JR, Reveillaud J, Schander C, Rapp HT. Molecular phylogeny of the Astrophorida (Porifera, *Demospongiae*) reveals an unexpected high level of spicule homoplasy. PLoS One. 2011;6:e18318.

7. Vargas S, Erpenbeck D, Göcke C, Hall KA, Hooper JNA, Janussen D, et al. Molecular phylogeny of *Abyssocladia* (Cladorhizidae: Poecilosclerida) and *Phelloderma* (Phellodermidae: Poecilosclerida) suggests a diversification of chelae microscleres in cladorhizid sponges. Zool Scr. 2013;42:106–16.
8. Erpenbeck D, Breeuwer JAJ, Parra-Velandia FJ, van Soest RWM. Speculation with spiculation? — Three independent gene fragments and biochemical characters versus morphology in demosponge higher classification. Mol Phylogen Evol. 2006;38:293–305.
9. Morrow CC, Redmond NE, Picton BE, Thacker RW, Collins AG, Maggs CA, et al. Molecular Phylogenies Support Homoplasy of Multiple Morphological Characters Used in the Taxonomy of Heteroscleromorpha (Porifera: Demospongiae). Integr Comp Biol. 2013;53:428–46.
10. Redmond NE, Morrow CC, Thacker RW, Diaz MC, Boury-Esnault N, Cárdenas P, et al. Phylogeny and Systematics of Demospongiae in Light of New Small Subunit Ribosomal DNA (18S) Sequences. Integr Comp Biol. 2013;53:388–415.
11. Hooper JNA, van Soest RWM. Class Demospongiae Sollas, 1885. In: Hooper JNA, van Soest RWM, editors. Systema Porifera: A guide to the classification of sponges. Volume 1. New York: Kluwer Academic/Plenum Publishers; 2002. p. 15–8.
12. Borchiellini C, Chombard C, Manuel M, Alivon E, Vacelet J, Boury-Esnault N. Molecular phylogeny of Demospongiae: implications for classification and scenarios of character evolution. Mol Phylogen Evol. 2004;32:823–37.
13. Thacker RW, Hill AL, Hill MS, Redmond NE, Collins AG, Morrow CC, et al. Nearly Complete 28S rRNA Gene Sequences Confirm New Hypotheses of Sponge Evolution. Integr Comp Biol. 2013;53:373–87.
14. Lavrov DV, Wang X, Kelly M. Reconstructing ordinal relationships in the Demospongiae using mitochondrial genomic data. Mol Phylogen Evol. 2008;49:111–24.
15. Kayal E: The evolution of the mitochondrial genomes of calcareous sponges and cnidarians. PhD thesis. Iowa State University, 2012.
16. Lévi C. Ontogeny and Systematics in Sponges. Syst Zool. 1957;6:174–83.
17. Lévi C. Sur une nouvelle classification des Démosponges. Compte rendu hebdomadaire des séances de l'Académie des sciences. 1953;236:853–5.
18. van Soest RWM. Toward a phylogenetic classification of sponges. In: Rützler K, editor. New Perspectives in Sponge Biology. Washington: Smithsonian Institution; 1990. p. 344–8.
19. van Soest RWM. Demosponge higher taxa classification re-examined. In: Reitner J, Keupp H, editors. Fossil and recent sponges. Berlin Heidelberg: Springer Verlag; 1991. p. 54–71.
20. Lévi C. Systématique de la classe des Demospongiaria (Démosponges). In: Grassé PP, editor. Traité de Zoologie Spongiaires, vol. 3. Paris: Masson & Co; 1973. p. 577–632.
21. Gazave E, Lapébie P, Ereskovsky A, Vacelet J, Renard E, Cárdenas P, et al. No longer Demospongiae: Homoscleromorpha formal nomination as a fourth class of Porifera. Hydrobiologia. 2012;687:3–10.
22. Nichols SA. An evaluation of support for order-level monophyly and interrelationships within the class Demospongiae using partial data from the large subunit rDNA and cytochrome oxidase subunit I. Mol Phylogen Evol. 2005;34:81–96.
23. Boury-Esnault N. Systematics and evolution of Demospongiae. Can J Zool. 2006;84:205–24.
24. Hill MS, Hill AL, Lopez J, Peterson KJ, Pomponi S, Diaz MC, et al. Reconstruction of Family-Level Phylogenetic Relationships within Demospongiae (Porifera) Using Nuclear Encoded Housekeeping Genes. PLoS One. 2013;8:e50437.
25. Erpenbeck D, Sutcliffe P, Cook SC, Dietzel A, Maldonado M, van Soest RWM, et al. Horny sponges and their affairs: On the phylogenetic relationships of keratose sponges. Mol Phylogen Evol. 2012;63:809–16.

26. Ereskovsky AV, Lavrov DV, Boury-Esnault N, Vacelet J. Molecular and morphological description of a new species of *Halisarca* (Demospongiae: Halisarcida) from Mediterranean Sea and a redescription of the type species *Halisarca dujardini*. Zootaxa. 2011;2768:5–31.

27. Vacelet J, Donadey C. A new species of Halisarca (Porifera, Demospongiae) from the Caribbean, with remarks on the cytology and affinities of the genus. In: Jones WC, editor. European contributions to the taxonomy of sponges. Midleton: Litho Press Co; 1987. p. 5–12.

28. Vacelet J, Recent 'Sphinctozoa'. Family Verticillitidae Steinmann, 1882. In: Hooper JNA, van Soest RWM, editors. Systema Porifera: A Guide to the Classification of Sponges, vol. 1. New York: Kluwer Academic/Plenum Publishers; 2002. p. 1109–10.

29. Wörheide G. A hypercalcified sponge with soft relatives: *Vaceletia* is a keratose demosponge. Mol Phylogen Evol. 2008;47:433–8.

30. Wang X, Lavrov DV. Seventeen New Complete mtDNA Sequences Reveal Extensive Mitochondrial Genome Evolution within the Demospongiae. PLoS One. 2008;3:e2723.

31. West RR, Vacelet J, Wood RA, Willenz P, Hartman WD. Hypercalcified extant and fossil chaetetid-type and Post- Devonian stromatoporoid-type Demospongiae: Systematic descriptions. In: Treatise on Invertebrate Paleontology On Line, number 58 Part E, Revised. Lawrence, Kansas, USA: The University of Kansas, Paleontological Institute; 2013. p. 1–95.

32. Reitner J, Engeser TS. Revision der Demospongier mit einem thalamiden aragonitischen basaskelett und trabekulärer internstruktur ('Sphinctozoa' pars). Geowissenschaftliche Abhandlungen Reihe A (Geologie und Paläontologie). 1985;60:151–93.

33. Sperling EA, Peterson KJ, Pisani D. Phylogenetic-Signal Dissection of Nuclear Housekeeping Genes Supports the Paraphyly of Sponges and the Monophyly of Eumetazoa. Mol Biol Evol. 2009;26:2261–74.

34. Addis JS, Peterson KJ. Phylogenetic relationships of freshwater sponges (Porifera, Spongillina) inferred from analyses of 18S rDNA, COI mtDNA, and ITS2 rDNA sequences. Zool Scr. 2005;34:549–57.

35. Pisera A, Lévi C. 'Lithistid' Demospongiae. In: Hooper JNA, van Soest RWM, editors. Systema Porifera: A Guide to the classification of Sponges. New York: Kluwer Academic / Plenum Publishers; 2002. p. 299–301.

36. Schuster A, Erpenbeck D, Pisera A, Hooper JNA, Bryce M, Fromont J, et al. Deceptive desmas: molecular phylogenetics suggests a new classification and uncovers convergent evolution of lithistid demosponges. PLoS One. 2015;10(1):e116038.

37. Hajdu E, van Soest RWM, Hooper JNA. Proposal for a phylogenetic subordinal classification of poecilosclerid sponges. In: van Soest RWM, van Kempen TMG, Braekman J-C, editors. Sponges in Time and Space. Rotterdam: Balkema; 1994. p. 123–39.

38. Hajdu E, de Paula TS, Redmond NE, Cosme B, Collins AG, Lôbo-Hajdu G. Mycalina: Another Crack in the Poecilosclerida Framework. Integr Comp Biol. 2013;53:462–72.

39. Erpenbeck D, Hall K, Alvarez B, Büttner G, Sacher K, Schätzle S, et al. The phylogeny of halichondrid demosponges: past and present re-visited with DNA-barcoding data. Org Divers Evol. 2012;12:57–70.

40. Erpenbeck D, van Soest RWM. A survey for biochemical synapomorphies to reveal phylogenetic relationships of halichondrid demosponges (Metazoa: Porifera). Biochem Syst Ecol. 2005;33:585–616.

41. Chombard C: Les Demospongiae à asters: phylogénie moléculaire et homologie morphologique. Ph.D thesis. Muséum National d'Histoire Naturelle, 1998.

42. Chombard C, Boury-Esnault N. Good congruence between morphology and molecular phylogeny of Hadromerida, or how to bother sponge taxonomists [abstract]. Mem Queensl Mus. 1999;44:100.

43. Maldonado M. Embryonic development of verongid demosponges supports the independent acquisition of spongin skeletons as an alternative to the siliceous skeleton of sponges. Biol J Linn Soc. 2009;97:427–47.

44. Haeckel E. Generelle Morphologie der Organismen. Allgemeine grundzüge der organischen formen-wissenschaft, mechanisch begründet durch die von Charles Darwin reformirte descendenztheorie. Band 2. Berlin: Geork Reimer; 1866.

45. Bergquist PR, Cook SDC. Order Verongida Bergquist, 1978. In: Hooper JNA, van Soest RWM, editors. Systema Porifera: A Guide to the Classification of Sponges, vol. 1. New York: Kluwer Academic/Plenum Publishers; 2002. p. 1081.

46. Erwin PM, Thacker RW. Phylogenetic analyses of marine sponges within the order Verongida: a comparison of morphological and molecular data. Invertebr Biol. 2007;126:220–34.

47. Diaz MC, Thacker RW, Redmond NE, Matterson KO, Collins AG. Phylogenetic Novelties and Geographic Anomalies among Tropical Verongida. Integr Comp Biol. 2013;53:482–94.

48. Bergquist PR, Cook SDC. Order Dendroceratida Minchin, 1900. In: Hooper JNA, van Soest RWM, editors. Systema Porifera: A Guide to the Classification of Sponges, vol. 1. New York: Kluwer Academic/Plenum Publishers; 2002. p. 1067.

49. Cook SDC, Bergquist PR. Order Dictyoceratida Minchin, 1900. In: Hooper JNA, van Soest RWM, editors. Systema Porifera: A Guide to the Classification of Sponges, vol. 1. New York: Kluwer Academic/Plenum Publishers; 2002. p. 1021.

50. Manconi R, Pronzato R. Suborder Spongillina subord. nov.: Freshwater Sponges. In: Hooper JNA, van Soest RWM, editors. Systema Porifera: A Guide to the classification of Sponges, vol. 1. New York: Kluwer Academic/Plenum Publishers; 2002. p. 921–1020.

51. Itskovich V, Kaluzhnaya O, Ostrovsky I, McCormack G. The number of endemic species of freshwater sponges (Malawispongiidae; Spongillina; Porifera) from Lake Kinneret is overestimated. J Zool Syst Evol Res. 2013;51:252–7.

52. Harcet M, Bilandžija H, Bruvo-Mađarić B, Ćetković H. Taxonomic position of *Eunapius subterraneus* (Porifera, Spongillidae) inferred from molecular data – A revised classification needed? Mol Phylogen Evol. 2010;54:1021–7.

53. Erpenbeck D, Weier T, de Voogd NJ, Wörheide G, Sutcliffe P, Todd JA, et al. Insights into the evolution of freshwater sponges (Porifera: Demospongiae: Spongillina): Barcoding and phylogenetic data from Lake Tanganyika endemics indicate multiple invasions and unsettle existing taxonomy. Mol Phylogen Evol. 2011;61:231–6.

54. Lundbeck W. Porifera. (Part I.) Homorrhaphidae and Heterorrhaphidae. In: The Danish Ingolf-Expedition 6(1). Copenhagen: Bianco Luno; 1902.

55. Vacelet J. Éponges de la Roche du Large et de l'étage bathyal de Méditerranée (Récoltes de la soucoupe plongeante Cousteau et dragages). Mémoires du Muséum national d'Histoire naturelle (A, Zoologie). 1969;59:145–219. pls I–IV.

56. Pisera A, Lévi C. Family Vetulinidae Lendenfeld, 1903. In: Hooper JNA, van Soest RWM, editors. Systema Porifera: A Guide to the classification of Sponges, vol. 1. New York: Kluwer Academic/Plenum Publishers; 2002. p. 363–5.

57. Bergquist PR. Additions to the Sponge Fauna of the Hawaiian Islands. Micronesica. 1967;3:159–74.

58. Bergquist PR. The Marine Fauna of New Zealand: Porifera, Demospongiae, Part 2 (Axinellida and Halichondrida). New Zealand Department of Scientific and Industrial Research Bulletin [New Zealand Oceanographic Institute Memoir 51]. 1970;197:1–85.

59. Hall KA, Ekins MG, Hooper JNA. Two new desma-less species of *Theonella* Gray, 1868 (Demospongiae: Astrophorida: Theonellidae), from the Great Barrier Reef, Australia, and a re-evaluation of one species assigned previously to *Dercitus* Gray, 1867. Zootaxa. 2013;3814:451–77.

60. Cedro VR, Hajdu E, Correia MD. Three new intertidal sponges (Porifera: Demospongiae) from Brazil's fringing urban reefs (Maceió, Alagoas, Brazil), and support for *Rhabderemia*'s exclusion from Poecilosclerida. J Nat Hist. 2013;47:2151–74.

61. van Soest RWM, Hooper JNA. Taxonomy, phylogeny and biogeography of the marine sponge genus Rhabderemia Topsent, 1890 (Demospongiae, Poecilosclerida). Scientia Marina. 1993;57:319–51.

62. Chombard C, Boury-Esnault N, Tillier S. Reassessment of homology of morphological characters in Tetractinellid sponges based on molecular data. Syst Biol. 1998;47:351–66.

63. Cárdenas P, Rapp HT, Schander C, Tendal OS. Molecular taxonomy and phylogeny of the Geodiidae (Porifera, *Demospongiae*, Astrophorida) — combining phylogenetic and Linnaean classification. Zool Scr. 2010;39:89–106.

64. Gazave E, Carteron S, Chenuil A, Richelle-Maurer E, Boury-Esnault N, Borchiellini C. Polyphyly of the genus *Axinella* and of the family Axinellidae (Porifera: Demospongiaep). Mol Phylogen Evol. 2010;57:35–47.

65. Boury-Esnault N. Family Polymastiidae Gray, 1867. In: Hooper JNA, van Soest RWM, editors. Systema Porifera: A Guide to the classification of Sponges, vol. 1. New York: Kluwer Academic/Plenum Publishers; 2002. p. 201–19.

66. Vacelet J. Description et affinités d'une éponge sphinctozoaire actuelle. In: Lévi C, Boury-Esnault N, editors. Biologie des Spongiaires - Sponge Biology Colloques Internationaux du Centre National de la Recherche Scientifique, 291. Paris: Editions du CNRS; 1979. p. 483–93.

67. Vacelet J. Eponges hypercalcifiées ("Pharétronides", "Sclérosponges") des cavités des récifs coralliens de Nouvelle-Calédonie. Bulletin du Muséum National d'Histoire Naturelle. 1981;3:313–51.

68. Hajdu E, van Soest RWM. Family Merliidae Kirkpatrick, 1908. In: Hooper JNA, van Soest RWM, editors. Systema Porifera: A Guide to the classification of Sponges, vol. 1. New York: Kluwer Academic / Plenum Publishers; 2002. p. 691–3.

69. Topsent E. Spongiaires de l'Atlantique et de la Méditerranée provenant des croisières du Prince Albert Ier de Monaco. Résultats des campagnes scientifiques accomplies par le Prince Albert I Monaco. 1928;74:1–376. pls I-XI.

70. Hajdu E, van Soest RWM. Family Desmacellidae Ridley & Dendy, 1886. In: Hooper JNA, van Soest RWM, editors. Systema Porifera: A Guide to the classification of Sponges, vol. 1. New York: Kluwer Academic/Plenum Publishers; 2002. p. 642–50.

71. Hooper JNA, van Soest RWM. Order Poecilosclerida Topsent, 1928. In: Hooper JNA, van Soest RWM, editors. Systema Porifera: A Guide to the classification of Sponges, vol. 1. New York: Kluwer Academic/Plenum Publishers; 2002. p. 403–8.

72. Barucca M, Azzini F, Bavestrello G, Biscotti M, Calcinai B, Canapa A, et al. The systematic position of some boring sponges (Demospongiae, Hadromerida) studied by molecular analysis. Mar Biol. 2007;151:529–35.

73. Hooper JNA, van Soest RWM. Family Trachycladidae Hallmann, 1917. In: Hooper JNA, van Soest RWM, editors. Systema Porifera: A Guide to the classification of Sponges, vol. 1. New York: Kluwer Academic/Plenum Publishers; 2002. p. 268–74.

74. Ridley SO, Dendy A, Report on the Monaxonida collected by H.M.S. 'Challenger' during the years 1873-1876. Report on the Scientific Results of the Voyage of HMS 'Challenger', 1873-1876 Zoology. 1887;20(59):i-lxviii–1-275, pls I-LI.

75. Vosmaer GCJ: Klassen und Ordnungen der Spongien (Porifera). In Die Klassen und Ordnungen des Thierreichs 2. Edited by Bronn HG: Leipzig & Heidelberg; 1887: i-xii, 1-496, pls I-XXXIV

76. Pisera A, Lévi C. "Lithistids" Incertae Sedis. In: Hooper JNA, van Soest RWM, editors. Systema Porifera: A Guide to the classification of Sponges. New York: Kluwer Academic / Plenum Publishers; 2002. p. 384–7.

77. Vacelet J, Perez T, Hooper JNA. Demospongiae incertae sedis: Myceliospongia Vacelet and Perez, 1998. In: Hooper JNA, van Soest RWM, editors. Systema Porifera: A Guide to the classification of Sponges, vol. 1. New York: Kluwer Academic / Plenum Publishers; 2002. p. 1099–101.

78. Dominguez E, Wheeler QD. Taxonomic Stability is Ignorance. Cladistics. 1997;13:367–72.

6

The nervous system of *Paludicella articulata* - first evidence of a neuroepithelium in a ctenostome ectoproct

Anna V Weber, Andreas Wanninger and Thomas F Schwaha[*]

Abstract

Introduction: Comparatively few data are available concerning the structure of the adult nervous system in the Ectoprocta or Bryozoa. In contrast to all other ectoprocts, the cerebral ganglion of phylactolaemates contains a central fluid-filled lumen surrounded by a neuroepithelium. Preliminary observations have shown a small lumen within the cerebral ganglion of the ctenostome *Paludicella articulata*. Ctenostome-grade ectoprocts are of phylogenetic relevance since they are considered to have retained ancestral ectoproct features. Therefore, the ctenostome *Paludicella articulata* was analyzed in order to contribute to the basal neural bauplan of ctenostomes and the Ectoprocta in general.

Results: The presence of a lumen and a neuroepithelial organization of the nerve cells within the cerebral ganglion are confirmed. Four tentacle nerves project from the cerebral ganglion into each tentacle. Three of the tentacle nerves (one abfrontal and two latero-frontal nerves) have an intertentacular origin, whereas the medio-frontal nerve arises from the cerebral ganglion. Six to eight visceral nerves and four tentacle sheath nerves are found to emanate from the cerebral ganglion and innervate the digestive tract and the tentacle sheath, respectively.

Conclusions: The situation in *P. articulata* corresponds to the situation found in other ctenostomes and supports the notion that four tentacle nerves are the ancestral configuration in Ectoprocta and not six as proposed earlier. The presence of a lumen in the ganglion represents the ancestral state in Ectoprocta which disappears during ontogeny in all except in adult Phylactolaemata and *P. articulata*. It appears likely that it has been overlooked in earlier studies owing to its small size.

Introduction

Bryozoa or Ectoprocta are widespread colonial suspension-feeders and predominantly marine animals mainly attached to hard substrates. They represent a large lophotrochozoan phylum with over 6.000 recent and about 15.000 extinct species. In various phylogenetic analyses ectoprocts were placed at different positions within the Bilateria [1-3]. Owing to several morphological similarities of the adults, in particular the lophophore, ectoprocts were traditionally united with phoronids and brachiopods in the clade Lophophorata [4], which also receives support in some recent molecular analyses [5,6]. Monophyly of Brachiopoda and Phoronida is often found in molecular phylogenies, but to the exclusion of Ectoprocta [1,2,7-9].

Within Ectoprocta, three major subtaxa are commonly recognized: Phylactolaemata, Stenolaemata and Gymnolaemata, the latter comprising the Ctenostomata and Cheilostomata. An ectoproct colony (zooarium) is composed of single individuals called zooids. Each zooid has a tough, often calcified reinforced body wall, the cystid, in which the soft part, named polypide, may retract into. The polypide mainly consists of the lophophore and digestive tract and is used for food uptake [4,10].

Phylactolaemata is often regarded as the sister-taxon to all remaining Ectoprocta. However, the relationships within the ectoproct subtaxa remain controversial. The "Ctenostomata" are currently regarded as paraphyletic [11,12]. They are little investigated and can be divided

* Correspondence: thomas.schwaha@univie.ac.at
Department of Integrative Zoology, University of Vienna, Althanstraße 14, 1090 Vienna, Austria

into two subclades, Carnosa and Stolonifera [4]. Ctenostomes lack a mineralized cystid and, thus, complex calcified structures. Concerning bryozoan soft body morphology only little data and mostly old monographs are available [13-16] with only a small amount of recent studies focusing on specific organ systems [17-20]. As a consequence, there is limited data available concerning the adult ectoproct nervous system. Some of the earliest notes on the nervous system are available in old monographs (e.g. [14]). Only a few studies specifically addressing the nervous system were conducted in the early 20th century (e.g. [21,22]). The most profound knowledge on the nervous system was gained by a series of studies of Lutaud (e.g. [23-26]), also summarized in Mukai et al. [4]. In the last decades only a few notable studies had a focus on adult nervous systems (e.g. [20,27]).

In adult Phylactolaemata the serotonergic nervous system is concentrated in the cerebral ganglion, from which a serotonergic neurite extends to each tentacle base [28]. Investigations showed that the cerebral ganglion of adult Phylactolaemata bears a small fluid-filled lumen [14]. An organization of the nervous cells as a neuroepithelium that bears interconnections of neurons via adherens junctions was described [27]. The ontogenetic origin of the cerebral ganglion has been described as an invagination of the inner layer of the bilayered bud, i.e. ultimately derived from the epidermis of the mother zooid, in all bryozoan taxa investigated so far. Consequently, in the ganglion of early developmental stages, there is a lumen which is described to disappear during development in all clades except in the Phylactolaemata [20,29,30].

Taken together, the data currently available show that the adult ectoproct nervous system is rather simple and mainly consists of a cerebral ganglion at the base of the lophophore, a circum-oral/circum-pharyngeal nerve ring and nerves emerging from the cerebral ganglion - which constitute the tentacular and tentacle sheath innervation - as well as some nerve fibers that project to the gut. Simultaneous response to environmental cues of several zooids (or heterozooids) raise questions concerning neural communication between individual zooids within a colony. In the cheilostome Electra pilosa [25] as well as in other malacostegine cheilstome ectoprocts [31] a chain of perikarya at the base of the longitudinal and transverse parts of the cystid was described. These perikarya are associated with parietal filaments and are linked along the tentacle sheath with the cerebral ganglion, and the filaments of neighboring zooids are periodically connected through pore plates [24]. In the cheilostome Electra pilosa and the ctenostome Hislopia malayensis 4 basiepidermal tentacle nerves were described [23,32]. In the cheilostome Cryptosula pallasiana two additional basiperitoneal tentacle nerves are present [18]. In the Phylactolaemata all tentacle nerves have an intertentacular origin and form intertentacular forks [21]. Conversely, in Gymnolaemata tentacle nerves branch off directly from the cerebral ganglion/nerve ring. In the ctenostome H. malayensis the medio-frontal nerve emerges directly from the cerebral ganglion, whereas the abfrontal and the latero-frontal nerves emanate from intertentacular forks [32]. In the cheilostomate E. pilosa the medio-frontal as well as the abfrontal nerve branches off directly from the cerebral ganglion [23]. This trend reflects the current phylogenetic view that the Phylactolaemata represent a basal ectoproct offshoot and that the Ctenostomata are paraphyletic [11,12,32,33].

In the current study special focus was on the innervation of the tentacles in order to assess whether the tentacle nerves show an intertentacular origin or whether they emanate directly from the cerebral ganglion. Preliminary data also showed a lumen within the cerebral ganglion of Paludicella articulata (Ctenostomata) (T. Schwaha, pers. observation). This has never been described in the Ctenostomata before. In order to test for a neuroepithelial organization of the nervous cells within the cerebral ganglion (see [27]), to analyze the hollow structure of the cerebral ganglion in more detail, and to gain more insight into the neural anatomy of P. articulata, the zooids were studied at histological and ultrastructural level. In addition, confocal microscopy analyses of specimens stained for the pan-neural marker α-tubulin were performed.

Materials and methods
Fixation, staining and image acquisition
Adult colonies of Paludicella articulata were sampled at the Laxenburg pond, Lower Austria, at a depth of 0.1-0.5 m and were transferred to the laboratory. Some of the zooids were anesthetized by adding few drops of chloral hydrate and subsequently fixed either for immunocytochemical staining, transmission electron microscopy or light microscopy. For comparison, some specimens were fixed with retracted polypides (without chloral hydrate for relaxation). Fixation for immunocytochemistry was done in 4% paraformaldehyde (PFA) in 0.1 M phosphate buffer (PB) with 0.01% NaN_3 for 1 hour at room temperature. Subsequently, samples were rinsed three times for 20 minutes and stored in PB with 0.01% NaN_3 added. To permeabilize the tissues, colony branches were dissected and washed three times for 10 minutes in PB to remove the NaN_3. Next, the samples were blocked overnight at room temperature in 6% normal goat serum in PB with 4% Triton X-100 (Sigma Aldrich, St. Louis, MO, USA). A monoclonal mouse anti-acetylated α tubulin primary antibody (ImmunoStar, Hudson, USA) was applied in a concentration of 1:500 blocking solution in PB. In the following, the specimens were rinsed in the blocking solution three times for two hours at room temperature. Then, the secondary goat anti-mouse

antibody conjugated to the fluorescent dye Alexa Fluor 488 (Invitrogen, Carlsbad, CA, USA) was added in a 1:500 dilution to the blocking solution. Nuclear counterstaining was performed with 4´6-diamidino-2-phenylindole (DAPI, Sigma, Buchs SG, Switzerland) in a concentration of 1:300. The specimens were incubated overnight and finally washed three times for a total of 1.5 hours in PB without NaN$_3$. All samples were embedded either in Fluoromount G (Southern Biotech, Birmingham, AL, USA) or Vectashield (Vector Laboratories, Burlingame, CA, USA) on glass slides.

Stained samples were scanned with a Leica TCS SP5 II (Leica Microsystems, Wetzlar, Germany) confocal laser scanning microscope (CLSM) with a step size of 0.5 μm-1 μm along the Z-axis. Image stacks were merged into maximum projections or rendered three-dimensionally with the aid of the reconstruction software Amira 5.4 software (FEI Visualization Science Group, Hillsborow, OR, USA).

Transmission electron microscopy, light microscopy and 3D reconstruction

Zooids or colony parts were prefixed according to Gruhl and Bartolomaeus [27] in a 2.5% glutaraldehyde solution in 0.01 M PB (pH 7.4) at room temperature for one hour. Specimens were then rinsed three times for 20 minutes in PB. For postfixation, the samples were treated with 1% osmium tetroxide in PB for 1 hour at room temperature. Dehydration was done via acidified dimethoxypropane and afterwards embedded in Agar low-viscosity resin (LVR, Agar Scientific, Stanstead, Essex, UK) using acetone as intermediate. After overnight polymerization at 60°C, semi-thin (1 μm thickness) and ultrathin (60 nm thickness) serial sections were generated with Diatome diamond knifes (Diatome, Biel, Switzerland) on a Leica UC6 ultramicrotome (Leica Microsystems, Wetzlar, Germany). Ultra-thin sections were collected on mesh copper grids or on formvar-coated single-slot copper grids. Staining was carried out with uranylacetate for 35–40 minutes and lead citrate for 5 minutes. Ultrathin sections were examined on a Zeiss Libra 120 and a Zeiss EM 902 transmission electron microscope (Carl Zeiss AG, Oberkochen, Germany). Images were adjusted for brightness and contrast using Adobe Photoshop CS5 (Adobe, San Jose, CA, USA).

Semi-thin sections were mounted in LVR after staining with toluidine blue for 5–10 seconds at 60°C. Photographs were taken with a Nikon Eclipse E800 light microscope (Nikon, Chiyoda, Tokio, Japan) equipped with a Nikon Fi2-U3 microscope camera. Images were transformed into grayscale format with Adobe Photoshop CS5 and then imported into the reconstruction and visualization software Amira 5.4. Alignment of the image stack was conducted with the Align Slices tool. Structures of interest (tentacle nerves, cerebral ganglion, peritoneum, epidermis) were manually segmented. Based on the segmentation data, surfaces were generated and were optimized via successive alternating steps of polygon-reduction and additional surface smoothing. Snapshots of the 3D models were taken with the Amira software (see also [34]).

Results

Gross morphology of *Paludicella articulata*

Paludicella articulata forms very simple thread-like or runner-like colonies on various substrates in quiet freshwater streams. Within the colony the individual zooids are arranged one after another and form branches [35]. Colonies of *P. articulata* consist of sometimes creeping but more often elongated, mostly erect, slender zooids. There are always 3 adjacent zooids: one distally and two lateral ones. The cuticle (ectocyst) is uncalcified and chitinous. The cystid essentially is the protective body-wall into which the entire polypide may retract if disturbed. The polypide mainly consists of the digestive tract and the lophophore which carries the tentacles. The digestive tract can be divided into the pharynx, esophagus, cardia, cecum, and intestine (Figure 1; Figure 2a). In the current investigation intertentacular pits were found between adjacent tentacles of the lophophore in all semi- and ultrathin sections as a proximal indentation of the tentacle epidermis. The average depth of the intertentacular pits is about 20 μm and the diameter is 7 μm (Figure 3a). Between the epidermal layer of the tentacles and the peritoneum a prominent extracellular matrix is situated (Figure 4b, Figure 5b-d). The tentacle coeloms extend from the lophophoral ring coelom at the lophophoral base into each tentacle.

Gross morphology of the cerebral ganglion

The most complex part of the polypide is the lophophoral base, where also the main components of the nervous system are situated. The highest concentration of nervous cells is found in the cerebral ganglion from where the circum-oral nerve ring and a circum-pharyngeal nerve plexus emerge. The circum-oral nerve ring surrounds the mouth opening and thus the most distal part of the pharynx. This nerve ring is closed on the side opposite the ganglion by only a few nerve strands. As a result there is a considerable increase in the concentration of perikarya within the circum-oral nerve ring towards the ganglion. The circum-pharyngeal nerve plexus surrounds the pharynx proximal to the circum-oral nerve ring. It mainly consists of 6 nerve strands that project orally. Three nerves emerge on each side of the cerebral ganglion. In contrast to the circum-oral nerve ring the circum-pharyngeal plexus is open at the oral side (Figure 2; Figure 3).

As mentioned above the majority of nerve cells can be found in the cerebral ganglion, which is located on the anal side between the lophophoral base and the first

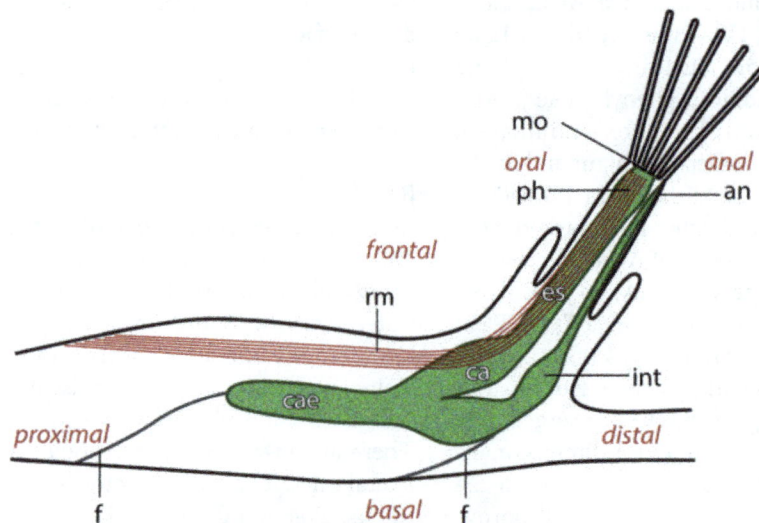

Figure 1 Schematic representation of a lateral view through a zooid of *Paludicella articulata* after Allman [36]. Abbreviations: an, anus; ca, cardia; cae, cecum; es, esophagus; f, funiculus; int, intestine; mo, mouth opening; rm, retractor muscle; ph, pharynx.

esophageal subdivision of the digestive tract at the level of the pharynx (Figure 6a, b). It is completely enveloped by an extracellular matrix which clearly separates it from the pharyngeal epithelium. Analyses of semi- and ultra-thin sections indicate that the cerebral ganglion contains a central fluid-filled lumen (Figure 6b, d-f).

Fine structure of the cerebral ganglion

The cerebral ganglion contains neurites, with characteristic neurotubules, and numerous perikarya. The latter are centrally arranged, surrounding and facing the lumen. All cells within the cerebral ganglion show a similar subcellular constitution, i.e. no difference between the ganglionic cells concerning the distribution and abundance of their cell organelles was recognizable. All nerve cells were of the same size and no difference in staining (light microscope) and contrast (TEM) appeared (Figure 6c). TEM investigations showed interconnections of ganglionic cells by adherens junctions in the area surrounding the lumen within the cerebral ganglion (Figure 6e, f). This finding is typical for an organization of the nervous tissue as a neuroepithelium.

Nerves projecting from the cerebral ganglion
Tentacle innervation

The lophophore is innervated by perioral nerve tracts. The gut and the tentacle sheath are innervated by further nerves branching off the cerebral ganglion. Each tentacle is innervated by four basiepidermal nerve fibers, which were identified in all microscopic analyses (CLSM, TEM, light microscopy). Each tentacle possesses one abfrontal nerve, one medio-frontal nerve and one pair of latero-frontal nerves. Three tentacle nerves emerge from

an intertentacular fork, located between two adjacent tentacles. The abfrontal nerve arises from the fusion of the two abfrontal nerve roots (Figure 2a-d; Figure 3; Figure 4, Figure 5a-d). Only the medio-frontal nerve branches off directly from the cerebral ganglion/circum-oral nerve ring. Instead of separate nerve strands described in other species (see above), two basiperitoneal cells are present in *P. articulata*. Within these basiperitoneal cells individual microtubules were observed in some ultrathin sections, but these never appeared as distinct as the neurotubules in the four tentacle nerves (Figure 5e, f). No basiperitoneal nerves could be detected in the α-tubulin staining. Thus, it appears likely that the basiperitoneal cells do not constitute nerves.

Visceral innervation

In *Paludicella articulata* six to eight visceral nerves branch off the cerebral ganglion (Figure 2d-f). There is one conspicuous median visceral nerve, which follows the esophagus in proximal direction. This nerve strand is the most prominent and always larger than the remaining visceral nerves. The medio-lateral visceral nerves are not bilaterally symmetrical and the number of nerves can differ between zooids. They appear as three to five nerve strands flanking the medio-frontal visceral nerve. One to two medio-lateral visceral nerves lie adjacent to the medio-frontal nerve, which adjoins one to three medio-lateral visceral nerves. The two latero-visceral nerves, however, are bilaterally symmetric and can be found in every each individual (Figure 2d-f).

Tentacle sheath nerves

Each zooid possesses four main tentacle sheath nerves that arise from the cerebral ganglion. The number of

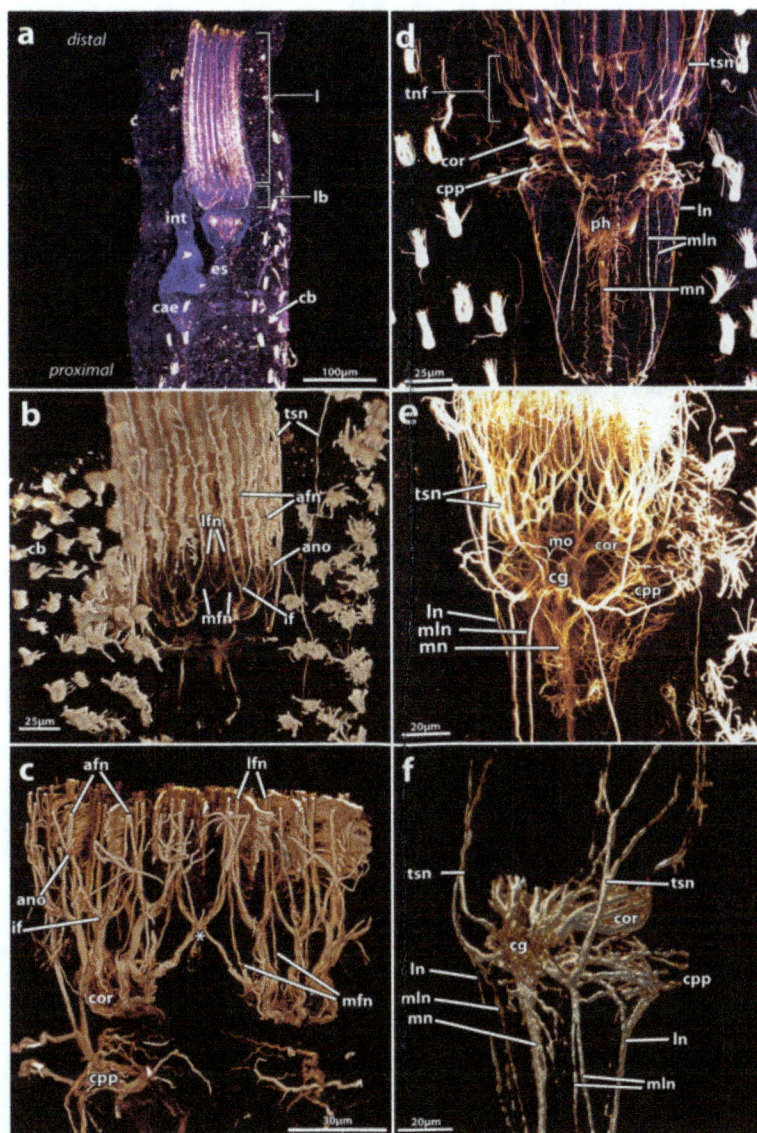

Figure 2 Adult *Paludicella articulata*. Maximum projections of confocal image stacks. Fluorescence staining with antibodies against acetylated α-tubulin (red) and in part staining of nucleic acids by DAPI (blue). **a)** overlay of acetylated α-tubulin and DAPI signal for an overview of the nervous system. **b)** overview of the central nervous system with focus on the tentacle innervation; volume rendering based on confocal image stack. **c)** Detail of the tentacle innervation by 3 different nerve types; medio-frontal tentacle nerves arise directly from the cerebral ganglion; the abfrontal and the paired latero-frontal tentacle nerves emerge from a intertentacular origin (intertentacular fork and abfrontal nerve origin); volume rendering based on confocal image stack. **d)** Overview of the acetylated α-tubulin-immunoreactive elements of the central nervous system and from the nerve strands that emerge from the cerebral ganglion. **e)** Detailed view of all nerves that arise from the cerebral ganglion: visceral nerves, tentacle sheath nerves, and tentacle innervation. **f)** Oblique view of the visceral nerves and tentacle sheath nerves; volume rendering based on confocal image stack. Abbreviations: ano, abfrontal nerve origin; afn, abfrontal nerve; c, cystid; ca, cardia; cae, cecum; cb, ciliary bundle; cg, cerebral ganglion; cor, circum-oral nerve ring; cpp, circum-pharyngeal plexus; es, esophagus; if, intertentacular forks; int, intestine; l, lophophore; lb, lophophore base; ln, latero visceral nerve; lfn, latero-frontal nerve; mfn, medio-frontal nerve; mln, medio-lateral visceral nerve; mn, medio visceral nerve; mo, mouth opening; ph, pharynx; tnf, area of tentacle nerve forks; ts, tentacle sheath; tsn, tentacle sheath nerve; asterisks, abanal closing of the circum-oral nerve ring via fine neurites.

nerves (or tentacle sheath nerve roots), however, differ between zooids, and the nerves are usually asymmetrical. In most cases two to three nerves fuse, after the emergence from the ganglion, into a single neurite bundle. In particular, at the proximal side, close to the ganglion, some interconnections between the nerves are apparent. However, no specific pattern was found in the number of these interconnections and other ramifications (Figure 2d, e). The four tentacle nerves project through the tentacle sheath towards the aperture of the zooid from where they continue into the body wall/cystid. Within the cystid hardly any nerves are stained (Figure 2d-f).

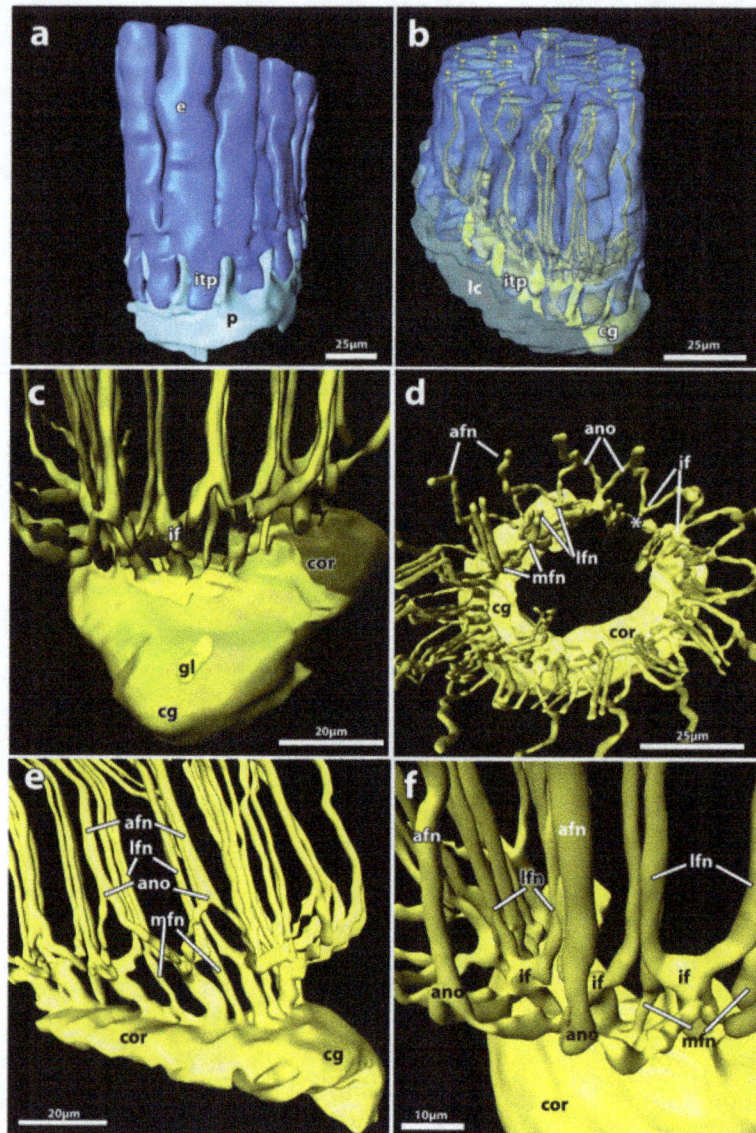

Figure 3 **3D-reconstruction based on semi-thin cross-section series of an adult zooid of *Paludicella articulata*. a)** Lateral view of the base of the lophophore, showing the epidermis and the peritoneum. **b)** Lateral view, epidermis and peritoneum displayed transparently, tentacle innervation visible below the epidermis. **c)** Lateral view of the cerebral ganglion, parts of the circum-oral nerve ring and the tentacle nerves from the oral side, nervous structures displayed transparently to allow for visibility of the lumen within the cerebral ganglion. **d)** Distal view of the central nervous system, showing the circum-oral nerve ring, the tentacle nerves and the cerebral ganglion. **e)** Slightly oblique lateral view of the circum-oral nerve ring, the tentacle innervation and the cerebral ganglion on the anal side to demonstrate the increase of nervous mass towards the anal side. **f)** Oblique view of the tentacle nerves, showing the direct origin of the medio-frontal tentacle nerve and the intertentacular origin of the other tentacle nerves as intertentacular forks and abfrontal nerve origins. Abbreviations: ano, abfrontal nerve origin; afn, abfrontal nerve; cg, cerebral ganglion; cor, circum-oral nerve ring; e, epidermis; gl, ganglion lumen; if, intertentacular forks; itp, intertentacular pit; lc, lophophore coelom; lfn, latero-frontal nerve; mfn, medio-frontal nerve; p, peritoneum; asterisks, abanal closing of the circum-oral nerve ring via fine neuritis Colors: dark blue, epidermal layer of the lophophoral base; light blue, peritoneum; yellow, nervous system.

Discussion
Tentacle innervation
With the exception of the Phylactolaemata [4], most adult ectoprocts have four to six tentacle nerves. In ultrastructural investigations of the tentacles four basiepidermal tentacle nerves per tentacle were found in the cheilostome *Electra pilosa* [23] and the ctenostome

Flustrellidra hispida [37]. In the cheilostome *Cryptosula pallasiana* two additional basiperitoneal tentacle nerves were detected [18]. Less information about the tentacle innervation is available for the Stenolaemata. In the cyclostome *Crisia eburnea* (Stenolaemata) also four tentacle nerves were reported [38]. Accordingly, mainly four tentacle nerves per tentacle appear to be present in

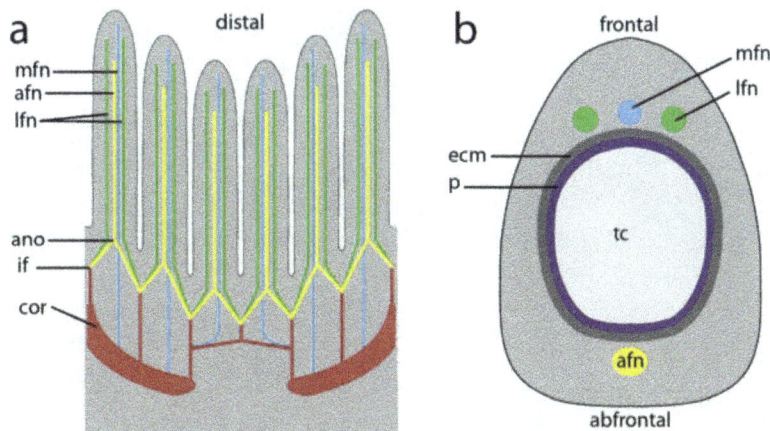

Figure 4 Schematic representations of the tentacle innervation and the central nervous system of *Paludicella articulata*. a) Sagittal view of the lophophore and the lophophoral base. **b)** Cross-section through one tentacle of the lophophore, demonstrating the tentacle innervations by 4 tentacle nerves. Abbreviations: ano, abfrontal nerve origin; afn, abfrontal tentacle nerve; cor, circum-oral nerve ring; ecm, extracellular matrix; if, intertentacular fork; lfn, latero-frontal tentacle nerve; mfn, medio-frontal tentacle nerve; p, peritoneum; tc, tentacle coelom.

ectoprocts. These usually comprise one abfrontal nerve, one medio-frontal nerve and paired latero-frontal nerves. The position of the four tentacle nerves is usually described as basiepidermal. Only *Flustrellidra hispida* (Ctenostomata), *Membranipora membranacea* (Cheilostomata) [22], *Farella repens* (Ctenostomata) and *Alcyonidium* sp. (Ctenostomata) [31] and *Cryptosula pallasiana* (Cheilostomata) [18] seem to possess two basiperitoneal (motoric) and two basiepidermal (sensory) tentacle nerves. As already mentioned, *E. pilosa* (Cheilostomata), which is closely related to *M. membranacea* (both Malacostega), also possesses four basiepidermal tentacle nerves per tentacle. In *E. pilosa* the latero-frontal nerves were found to have an intertentacular origin. They branch off from the circum-oral nerve ring and bifurcate via an intertentacular fork from where one nerve projects into a tentacle. The abfrontal and medio-frontal nerves branch off directly from the circum-oral nerve ring [23]. In adult *Hislopia malayensis* (Ctenostomata) four basiepidermal tentacle nerves were found. The abfrontal and latero-frontal tentacle nerves have intertentacular origins, whereas the medio-frontal tentacle nerves branch off directly from the circum-oral nerve ring in *H. malayensis* [32]. The situation in *Paludicella articulata* (present study), which also possesses four basiepidermal tentacle nerves, mostly resembles the situation found in *H. malayensis*. In both species three tentacle nerves have intertentacular origins, with the medio-frontal tentacle nerves directly emanating from the circum-oral nerve ring (i.e., the cerebral ganglion). The abfrontal nerve origin in *P. articulata* represents also a point of amalgamation with nerves arising from the intertentacular fork similar to the conspicuous abfrontal nerve bodies in *H. malayensis* from where also a single abfrontal nerve extends [32].

Until now *Cryptosula pallasiana* (cheilostome) remains the only species for which six tentacle nerves have been described [18]. The presence of the additional two basiperitoneal tentacle nerves does not apply to *P. articulata* in the current study. Instead of two basiperitoneal tentacle nerve strands, basiperitoneal cells were found. These cells include numerous microtubules, but no bundles of neurotubules or nerve fibres were found. Therefore, and because they never appear as distinct fibers or strands, it appears likely that they constitute basiperitoneal cells instead of nerves. These cells have been described in virtually all ultrastructural investigations: in the cheilostome *E. pilosa* [23] and the phylactolaemates *Fredericella sultana*, *Plumatella emarginata*, *Lophopus crystallinus* [17] and *Asajirella gelatinosa* [4].

The structure of the cerebral ganglion

The nervous system of *Paludicella articulata* essentially consists of a cerebral ganglion, a circumoral nerve ring, tentacle innervation, tentacle sheath nerves and visceral innervation. The main part of the nervous system is restricted to the basis of the lophophore. This is in line with all other ectoprocts investigated, which also possess a circum-oral or circum-pharyngeal nerve ring. In all ectoproct species investigated so far the central nervous system originates from an invagination of the prospective anal side of the pharyngeal epithelium during the budding process. Therefore, a lumen within the cerebral ganglion in the early developmental stages appears [29,30,39]. This has also been described for *P. articulata* [35]. Until now, the prevailing opinion was that the only ectoproct group which as adults still bears a lumen within their cerebral ganglion is the Phylactolaemata [4]. Several studies have described a vesicle-like cerebral ganglion in

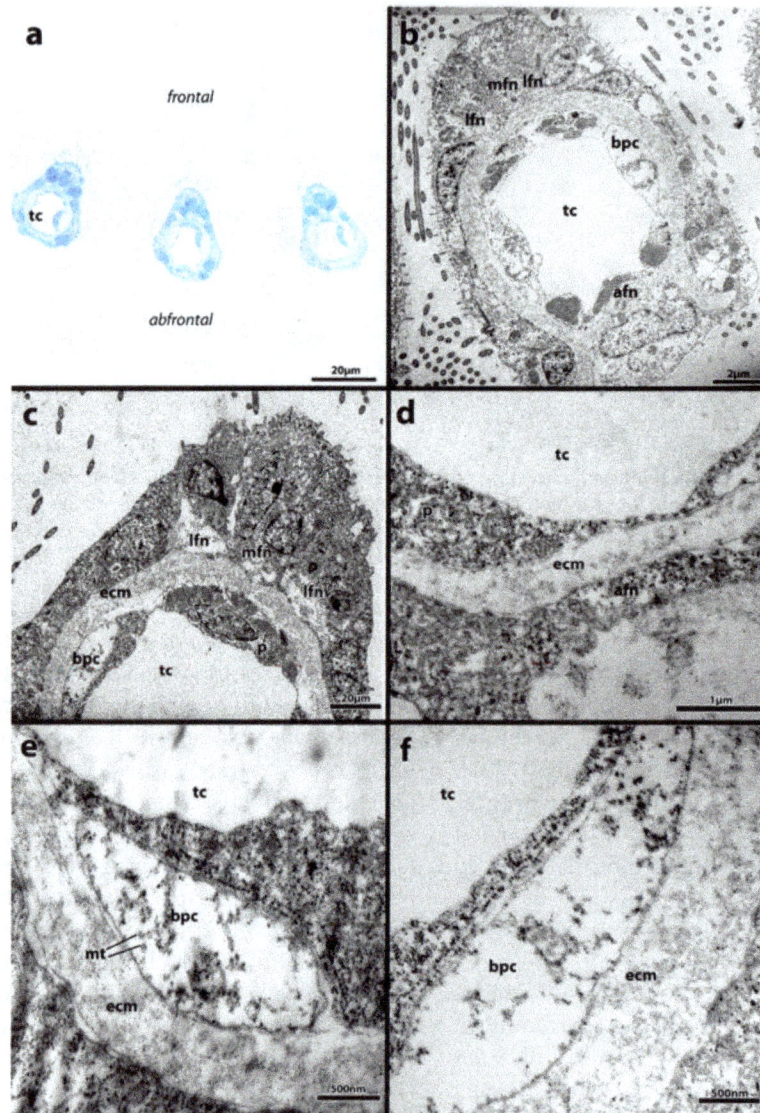

Figure 5 Cross-sections of tentacles of *Paludicella articulata*. a) Semi-thin section through 3 tentacles. **b)** TEM-Overview of a tentacle cross-section, with the 4 tentacle nerves, the 2 basiperitoneal cells and the centrally situated tentacle coelom visible. **c)** Detail of the 3 frontal tentacle nerves. **d)** Detail of the ultrastructure of an abfrontal tentacle nerve. **e)** Detailed view of a basiperitoneal cells with inherent sections of microtubules. **f)** TEM-Image of a basiperitoneal cell, also with sections through inherent microtubules. Abbreviations: afn, abfrontal tentacle nerve; bpc, basiperitoneal cell; ecm, extracellular matrix; lfn, latero-frontal tentacle nerve; mfn, medio-frontal tentacle nerve; mt – microtubules, p, peritoneum; spc – subperitoneal cell; tc, tentacle coelom.

adult Phylactolaemata [14,20,21,30,39]. The first ctenostome species where a lumen was found is *P. articulata* (current study). The phylactolaemates *Fredericella sultana* and *Plumatella emarginata* also bear a fluid-filled cavity within their cerebral ganglion. Furthermore, a neuroepithelial organization of the cells lining the lumen with apical adherens junctions is present in these two species [27]. In the ctenostome *P. articulata* (present study) a neuroepithelial organization of the ganglion cells was also found. These interconnections of the nerve cells are also predominately located around the lumen. However, in *H. malayensis*, another ctenostome representative, the

ganglion never contains a lumen but is compact from late budding stages onwards [32].

In the phylactolaemates *Lophopus crystallinus* [22], *F. sultana* and *P. emarginata* [27], as well as in the ctenostome *P. articulata* (current study), the perikarya of the cerebral ganglion face the lumen. In the cerebral ganglion of *H. malayensis* one conspicuous prominent perikaryon was described based on histological sections. It is situated centrally within the ganglion and is larger than the other nerve cells [32]. In the cerebral ganglion of *P. articulata* all cells appear morphologically similar. No difference concerning the abundance and the distribution of cell

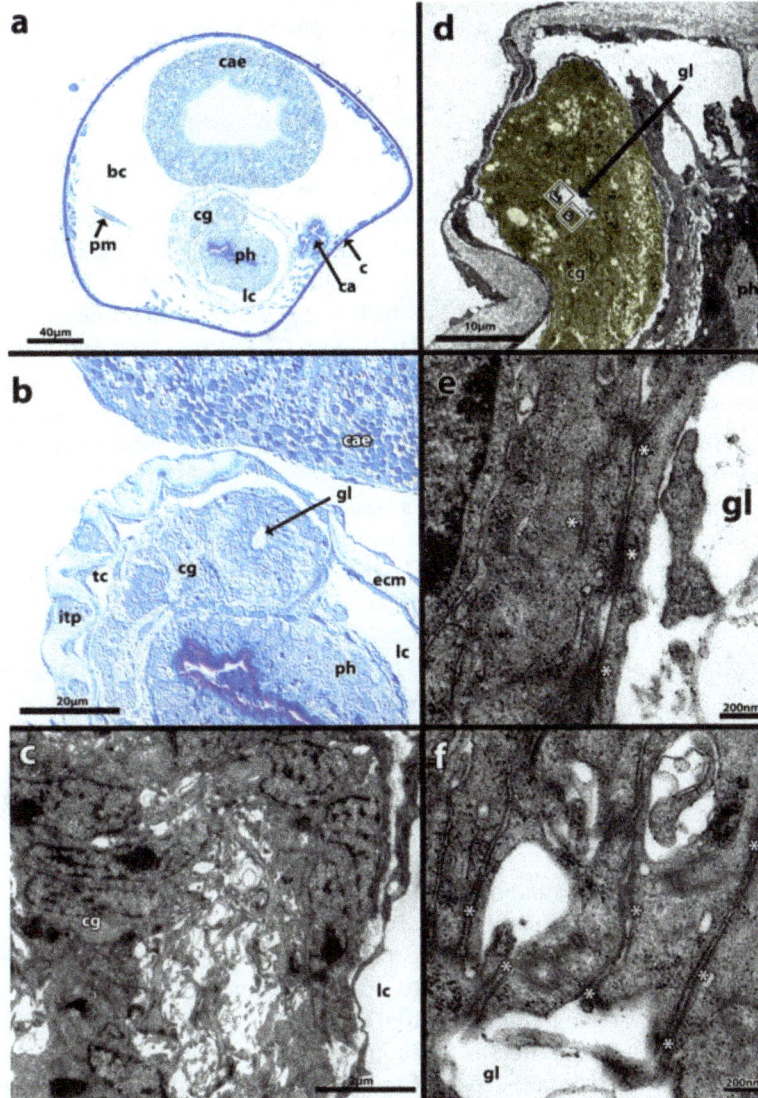

Figure 6 Semi-thin (a, b) and ultra-thin sections (c-f) of the cerebral ganglion of *Paludicella articulata*. a) Semi-thin cross section through the entire zooid, demonstrating the situation of the cerebral ganglion within the zooid **b)** Detailed view of the ganglion lumen in a retracted situation. **c)** TEM micrographs of the ganglionic tissue. **d)** Overview of the cerebral ganglion in a retracted polypide with its small lumen. **e)** Cell connections via adherens junctions in the area of the lumen within the cerebral ganglion. **f)** Adherens junctions of a nerve cell adjoining the fluid-filled lumen. Abbreviations: bc, body coelom; c, cystid; ca, cardia; cae, cecum; cg, cerebral ganglion; ecm, extracellular matrix; gl, ganglion lumen; itp, intertentacular pit; lc, lophophore coelom; ph, pharynx; pm, parietal muscle; tc, tentacle coelom; asterisks, cell-cell connections via adherens junctions.

organelles within these ganglionic cells were found. This is in contrast to the ctenostome *H. malayensis* [32]. In the phylactolametes *F. sultana* and *P. emarginata* additional neurosecretory cells are present which appear to be missing in the Ctenostomata [27]. In the ctenostome *Flustrellidra hispida* two different types of ganglion cells were described histologically, which differ in size [22]. In the cheilostome *E. pilosa* three different categories of neurons were found by ultrastructural studies. These are neurons with characteristic neurotubules, secretory and glial cells [24]. In ultrastructural investigations of

the phylactolaemates *F. sultana* and *P. emarginata* [27] no difference between the ganglionic cells were found, which most likely resembles the situation found in *P. articulata* (current study).

Visceral innervation

The visceral innervation has been analyzed in several ectoproct species [21,22,24]. In malacostegine cheilostomes ectoprocts two pairs of visceral nerves were described. A thick median visceral nerve is reported to project to the esophagus and lateral visceral nerves emerge from a separate visceral ganglion and project from the apex of the esophagus to the anus. These visceral nerves are connected via annular ramifications between the subdivisions of the gut [24]. *Electra pilosa*, *E. verticilliata* and *Membranipora membranacea* (Cheilostomata) also possess a prominent medio-visceral nerve, two medio-lateral visceral nerves and two lateral visceral nerves [26]. Only one major nerve is described to descend at the frontal side of the pharynx in *Cryptosula pallasiana* [18]. In *Zoobotryon verticillatum* (Ctenostomata) a frontal and lateral esophageal nerve were detected [21]. In *Flustrellidra hispida* the situation is similar, with one median visceral nerve and two fine medio-lateral nerves which are interconnected via a commissure and an asymmetrical tripolar nerve cell [22]. Most visceral nerves arise either from the cerebral ganglion or from a visceral ganglion, a small accessory ganglion located proximally of the cerebral ganglion. In *P. articulata* all nerves emerge from the cerebral ganglion and a comparable, prominent medio-visceral nerve was also detected. The three to five medio-lateral visceral nerves were found to be asymmetrically arranged. However, the two lateral visceral nerves appear bilaterally symmetrical. Taken together, the visceral innervation of *P. articulata* most closely resembles that of the cheilostomes with its three types of visceral nerves [26], all emerging from the cerebral ganglion.

Tentacle sheath nerves

In the phylactolaemate *Cristatella mucedo*, two pairs of prominent tentacle sheath nerves – frontal and basal - that arise from the cerebral ganglion were described [21]. In *Flustrellidra hispida* (Ctenostomata) two tentacle sheath nerves arise from the cerebral ganglion and bifurcate to form four distinct tentacle sheath nerves. The distal pair of the tentacle sheath nerves projects towards the aperture and finally ends as distal plexus. The other two nerves surround the attachment of the tentacle sheath [22]. In the current investigation, the tentacle sheath nerves were detectable among the main nerves that branch off the cerebral ganglion. The roots of the tentacle sheath nerves always fuse into four distinct nerves. However, the tentacle nerve roots that arise from the cerebral ganglion can differ in number and appear asymmetrically positioned. In *E. pilosa* (Cheilostomata) also two pairs of main tentacle sheath nerves emerge from the cerebral ganglion. The more laterally situated and three-branched nerves, called "trifid nerves", are thin bundles that merge at the aperture [31]. These trifid nerves bend towards the esophagus before they again bend in distal direction and project towards the tentacle sheath. The inner median "direct nerves" are thick strands without branches until they merge with the axial branches of the trifid nerves. The axial branches of the trifid nerve extend distally, almost parallel to the "direct nerves". The merged neurite

bundle finally form the compound tentacle sheath nerves [23]. A paired parietal plexus or "cystidial nerves", (cf. [25]) is situated parallel to the compound tentacle sheath nerves in the cystid. The parietal plexus is also connected to the cerebral ganglion via the tentacle sheath [24,40]. The tentacle sheath innervation of *P. articulata* most closely resembles that of cheilostomes [23]. One pair of "direct nerves" with their characteristic bending in distal and afterwards again in apical direction appears also in the ctenostome *P. articulata*. The so called "trifid nerves", which merge with the direct nerves, are present in a similar pattern. The triple-branching of these nerves is not clearly recognizable. The merged "direct" and comparable "trifid" nerve appears to be similar to the great tentacle sheath nerve [4]. As mentioned before, the number of all tentacle sheath nerves arising from the cerebral ganglion is odd and the nerves are arranged asymmetrically. An asymmetrical arrangement of nerves has so far not been described for any other ectoproct species.

Conclusions

The anatomy of the adult ectoproct nervous system, which is concentrated at the lophophoral base, has been subject of several investigations in the Gymnolaemata and Phylactolaemata. However, the nervous system of the Stenolaemata remains almost unstudied. Based on the available data, a trend is recognizable concerning the tentacle innervation from intertentacular origins in the basal Phylactolaemata towards direct innervation from the cerebral ganglion in Gymnolaemata (cf. [31]). The state in the paraphyletic Ctenostomata shows three out of four nerves to originate from the ganglion [32], whereas the Cheilostomata have only two of the four [23]. The current study supports the situation of the tentacle innervation in the Ctenostomata as observed in the *Hislopia malayensis* [32]. In order to verify this trend, additional cheilostome and cyclostome species need to be studied.

The lumen within the cerebral ganglion in shape of a small vesicle is found in all ectoprocts during ontogeny. The differentiation of the adult ganglion seems to lead to the obliteration of the ganglion lumen in most ectoprocts (cf. [32]), whereas its persistence into the adult is so far only documented for the Phylactolaemata [20,27]. The present study is the first to assess a lumen in the adult of *P. articulata*, a non-phylactolaemate ectoproct. Owing to its small size, it is quite possible that this lumen is more widespread and persistent than previously thought. It seems that *P. articulata* has retained the lumen like the basal Phylactolaemata, or, instead, has independently evolved in the adult condition. For a proper evaluation, however, additional species need to be investigated.

Competing interests
The authors declare that they have no competing interests.

Authors' contributions
AVW performed most of the practical work, data analysis and drafted the manuscript, AW supervised the study and contributed significantly to the writing of the manuscript, TS designed the study, contributed to data interpretation and writing of the manuscript. All authors read and approved the final manuscript.

Acknowledgements
We are indebted to the Cell Imaging & Ultrastructure Core Facility of the Faculty of Life Sciences, University of Vienna, for the use of the electron microscopes and support during TEM-analyses. Special thanks to two anonymous reviewers for improving the manuscript.

References

1. Dunn CW, Hejnol A, Matus DQ, Pang K, Browne WE, Smith SA, Seaver E, Rouse GW, Obst M, Edgecombe GD, Sorensen MV, Haddock SHD, Schmidt-Rhaesa A, Okusu A, Kristensen RM, Wheeler WC, Martindale MQ, Giribet G: Broad phylogenomic sampling improves resolution of the animal tree of life. *Nature* 2008, 452:745–U745.
2. Hejnol A, Obst M, Stamatakis A, Ott M, Rouse GW, Edgecombe GD, Martinez P, Baguna J, Bailly X, Jondelius U, Wiens M, Muller WEG, Seaver E, Wheeler WC, Martindale MQ, Giribet G, Dunn CW: Assessing the root of bilaterian animals with scalable phylogenomic methods. *Proc R Soc B* 2009, 276:4261–4270.
3. Nesnidal MP, Helmkampf M, Bruchhaus I, Hausdorf B: Compositional heterogeneity and phylogenomic inference of metazoan relationships. *Mol Biol Evol* 2010, 27:2095–2104.
4. Mukai H, Terakado K, Reed CG: Bryozoa. In *Microscopic anatomy of invertebrates. Volume* 13. Edited by Harrison FW, Woollacott RM. New York, Chichester: Wiley-Liss; 1997: 45–206.
5. Nesnidal M, Helmkampf M, Meyer A, Witek A, Bruchhaus I, Ebersberger I, Hankeln T, Lieb B, Struck T, Hausdorf B: New phylogenomic data support the monophyly of Lophophorata and an Ectoproct-Phoronid clade and indicate that Polyzoa and Kryptrochozoa are caused by systematic bias. *BMC Evol Biol* 2013, 13:253.
6. Nesnidal MP, Helmkampf M, Bruchhaus I, Ebersberger I, Hausdorf B: Lophophorata monophyletic - after all. In *Deep Metazoan Phylogeny: The Backbone of the Tree of Life: New Insights from Analyses of Molecules, Morphology, and Theory of Data Analysis.* Edited by Wagele JW, Bartolomaeus T. 2014:127–1426.
7. Halanych KM, Bacheller JD, Aguinaldo AMA, Liva SM, Hillis DM, Lake JA: Evidence from 18S ribosomal DNA that lophophorates are protostome animals. *Science* 1995, 267:1641–1643.
8. Giribet G, Distel DL, Polz M, Sterrer W, Wheeler WC: Triploblastic relationships with emphasis on the acoelomates and the position of Gnathostomulida, Cycliophora, Plathelminthes, and Chaetognatha: a combined approach of 18S rDNA sequences and morphology. *Syst Biol* 2000, 49:539–562.
9. Helmkampf M, Bruchhaus I, Hausdorf B: Phylogenomic analyses of lophophorates (brachiopods, phoronids and bryozoans) confirm the Lophotrochozoa concept. *Proc Roy Soc B* 2008, 275:1927–1933.
10. Ryland JS: *Bryozoans.* London: Hutchinson University Library; 1970.
11. Todd JA: The central role of ctenostomes in bryozoan phylogeny. In *Proceedings of the 11th International Bryozoology Association Conference.* Edited by Herrera Cubilla A, Jackson JBC. Balboa: Smithsonian Tropical Research Institute 2000: 104–135.
12. Waeschenbach A, Taylor PD, Littlewood DTJ: A molecular phylogeny of bryozoans. *Mol Phyl Evol* 2012, 62:718–735.
13. Kraepelin K: Die deutschen Süßwasser-bryozoen. 1. Anatomisch-systematischer Teil. *Abh Geb Naturw Ver Hamburg* 1887, 10:168p, 167 pl.
14. Braem F: Untersuchungen über die Bryozoen des süßen Wassers. *Zoologica* 1890, 6:1–134.
15. Borg F: Studies on recent cyclostomatous Bryozoa. *Zool Bidr Uppsala* 1926, 10:181–507.
16. Silen L: Zur Kenntnis des Polymorphismus der Bryozoen. Die Avicularien der Cheilostomata Anasca. *Zool Bidr Uppsala* 1938, 17:149–366.
17. Gruhl A, Wegener I, Bartolomaeus T: Ultrastructure of the body cavities in Phylactolaemata (Bryozoa). *J Morphol* 2009, 270:306–318.
18. Gordon DP: Microarchitecture and function of the lophophore in the bryozoan *Cryptosula pallasiana. Mar Biol* 1974, 27:147–163.
19. Ostrovsky AN, Gordon DP, Lidgard S: Independent evolution of matrotrophy in the major classes of Bryozoa: transitions among reproductive patterns and their ecological background. *Mar Ecol Prog Ser* 2009, 378:113–124.
20. Schwaha T, Handschuh S, Redl E, Walzl M: Organogenesis in the budding process of the freshwater bryozoan *Cristatella mucedo* Cuvier 1789 (Bryozoa, Phylactolaemata). *J Morphol* 2011, 272:320–341.
21. Gerwerzhagen A: Beiträge zur Kenntnis der Bryozoen, I. Das Nervensystem von *Cristatella mucedo. Z wiss Zool* 1913, 107:309–345.
22. Graupner H: Zur Kenntnis der feineren Anatomie der Bryozoen. *Z wiss Zool* 1930, 136:38–77.
23. Lutaud G: L'innervation du lophophore chez le Bryozaire Chilostome *Electra pilosa* (L.). *Z Zellforsch Mikroskop Anatom* 1973, 140:217–234.
24. Lutaud G: The bryozoan nervous system. In *Biology of bryozoans.* Edited by Woollacott RM, Zimmer RL. New York: Academic press; 1977: 377–410
25. Lutaud G: Étude ultrastructurale du "plexus colonial" et recherche de connexions nerveuses interzoidales chez le Bryozaire Chilostome *Electra pilosa* (Linné). *Cah Biol Mar* 1979, 20:315–324.
26. Lutaud G: L'innvervation sensorielle du lophophore et de la région orale chez les Bryozaires Cheilostomes. *Ann Sci Nat Zool Ser 13* 1993, 14:137–146.
27. Gruhl A, Bartolomaeus T: Ganglion ultrastructure in phylactolaemate Bryozoa: Evidence for a neuroepithelium. *J Morphol* 2008, 269:594–603.
28. Schwaha T, Wanninger A: Myoanatomy and serotonergic nervous system of plumatellid and fredericellid Phylactolaemata (Lophotrochozoa, Ectoprocta). *J Morphol* 2012, 273:57–67.
29. Nitsche H: Beiträge zur Anatomie und Entwicklungsgeschichte der Phylactolaemen Süßwasserbryozoen insbesondere von *Alyconella fungosa* PALL. In *Inaug.-Dissert. Philos. Fakult. Berlin.* 1868.
30. Nielsen C: Entoproct life-cycles and the Entoproct/Ectoproct relationship. *Ophelia* 1971, 9:209–341.
31. Bronstein G: Étude du système nerveux de quelques Bryozaires Gymnolémides. *Trav Stat Biol Roscoff* 1937, 15:155–174.
32. Schwaha T, Wood TS: Organogenesis during budding and lophophoral morphology of *Hislopia malayensis* Annandale, 1916 (Bryozoa, Ctenostomata). *BMC Dev Biol* 2011, 11:23.
33. Banta WC: Origin and early evolution of cheilostome Bryozoa. In *Bryozoa 1974.* Edited by Pouyet S. Lyon: Université Claude Bernard; 1975:565–582.
34. Ruthensteiner B: Soft Part 3D visualization by serial sectioning and computer reconstruction. *Zoosymposia* 2008, 1:63–100.
35. Davenport CB: Observations on budding in *Paludicella* and some other Bryozoa. *Bull Mus Comp Zool* 1891, 22:1–114, 112 plates.
36. Allman GJ: A monograph of the Fresh-water Polyzoa. *Ray Soc Lond* 1856, 28:1–119.
37. Smith LW: Ultrastructure of the tentacles of *Flustrellidra hispida* (Fabricius). In *Living and Fossil Bryozoa.* Edited by Larwood GP. London: Academic Press; 1973: 335–342.
38. Nielsen C, Riisgard HU: Tentacle structure and filter-feeding in *Crisia eburnea* and other cyclostomatous bryozoans, with a review of upstream-collecting mechanisms. *Mar Ecol Prog Ser* 1998, 168:163–186.
39. Davenport CB: *Cristatella*: The Origin and Development of the Individual in the Colony. *Bull Mus Comp Zool* 1890, 20:101–151.
40. Hiller S: The so-called "colonial nervous system" in Bryozoa. *Nature* 1939, 143:1069–1070.

7

Sensing deep extreme environments: the receptor cell types, brain centers, and multi-layer neural packaging of hydrothermal vent endemic worms

Shuichi Shigeno[1*], Atsushi Ogura[2], Tsukasa Mori[3], Haruhiko Toyohara[4], Takao Yoshida[1], Shinji Tsuchida[1] and Katsunori Fujikura[1]

Abstract

Introduction: Deep-sea alvinellid worm species endemic to hydrothermal vents, such as *Alvinella* and *Paralvinella*, are considered to be among the most thermotolerant animals known with their adaptability to toxic heavy metals, and tolerance of highly reductive and oxidative stressful environments. Despite the number of recent studies focused on their overall transcriptomic, proteomic, and metabolic stabilities, little is known regarding their sensory receptor cells and electrically active neuro-processing centers, and how these can tolerate and function in such harsh conditions.

Results: We examined the extra- and intracellular organizations of the epidermal ciliated sensory cells and their higher centers in the central nervous system through immunocytochemical, ultrastructural, and neurotracing analyses. We observed that these cells were rich in mitochondria and possessed many electron-dense granules, and identified specialized glial cells and serial myelin-like repeats in the head sensory systems of *Paralvinella hessleri*. Additionally, we identified the major epidermal sensory pathways, in which a pair of distinct mushroom bodies-like or small interneuron clusters was observed. These sensory learning and memory systems are commonly found in insects and annelids, but the alvinellid inputs are unlikely derived from the sensory ciliary cells of the dorsal head regions.

Conclusions: Our evidence provides insight into the cellular and system-wide adaptive structure used to sense, process, and combat the deep-sea hydrothermal vent environment. The alvinellid sensory cells exhibit characteristics of annelid ciliary types, and among the most unique features were the head sensory inputs and structure of the neural cell bodies of the brain, which were surrounded by multiple membranes. We speculated that such enhanced protection is required for the production of normal electrical signals, and to avoid the breakdown of the membrane surrounding metabolically fragile neurons from oxidative stress. Such pivotal acquisition is not broadly found in the all body parts, suggesting the head sensory inputs are specific, and these heterogenetic protection mechanisms may be present in alvinellid worms.

Keywords: Deep-sea, Sensory cells, Brain, Nervous system, Glia, Annelids, Evolution

* Correspondence: sushigeno@jamstec.go.jp
[1]Department for Marine Biodiversity Research, Japan Agency for Marine-Earth Science and Technology, 2-15 Natsushima-cho, Yokosuka 237-0061, Kanagawa, Japan
Full list of author information is available at the end of the article

Introduction

The alvinellid worms are annelids that are generally found on microbial mats closely inhabiting the smokers extruding from the active chimneys of deep-sea hydrothermal vents [1,2]. The fauna inhabiting these hot spring fields are exposed to highly fluctuating physico-chemical conditions, high levels of heavy metals, sulfide, and carbon dioxide, and harmful compounds such as hydrogen peroxide and hydroxyl radicals [3,4]. The emblematic characteristic of these alvinellids is thus their exceptional tolerance to high temperatures and the toxicity of acidic and reducing fluids [5-8]. Indeed, the alvinellid thermostabilization, detoxification, and anti-oxidative stress capacities have been attributed to a number of biochemical, physiological, and structural properties [4,9-13], supported by deep sequencing analysis of the transcriptomic and proteomic level stability [14-16].

Despite these extensive studies, little attention has been given to the sensory and nervous systems, particularly their behavioral ecology. Studies of sensory systems of animals endemic to hydrothermal vents are important for two main reasons. First, the sensory ecology of vent-endemic species is largely unknown, with the exception of some classic pioneering work on alvinellid larval settling [17], and crustacean vision and olfaction (e.g., [18-21]). Second, the neural cells are expected to be highly sensitive to toxic and redox fluids, since the neuronal cells primarily carry out electronic and active chemical signal transduction via synapses (e.g., [22]); therefore, the sensory receptors and neural tissues may possess specific tolerance mechanisms.

Classic anatomical studies have revealed that the overall body, tissue, and cellular organization of *Alvinella pompejana* exhibits a typical polychaete ground plan, without visual and gravity sense organs [23]. The sensory receptor cells identified on the branchial crown and feeding appendages are ciliate cells, mitochondria-rich, and of the bipolar type with single long axon, often associated with supportive cells [24,25]. Thus far the nuchal organs, usually a paired epidermal ciliary structure in most polychaetes [26], have not been found in the prostomium of alvinellids. The structure of the central nervous system has been extensively studied in polychaetes and other annelid taxa [27-31], including a few chemosynthetic siboglinid species [32-34]; however, the ultrastructural organization of the brain and nerves, as well as regional specialization, including inter- and intra-lobe connective patterns, is largely unknown in the deepsea annelids, including terebelliform alvinellids.

Since the first discovery of alvinellids on the East Pacific Rise [23], 11 species have been described in the order Terebellida and family Alvinellidae, which comprises two genera, *Alvinella* and *Paralvinella* [35]. Among the alvinellid species, we examined *Paralvinella hessleri*, which is abundant in the hydrothermal communities of active chimneys. These chimneys can be easily accessed with remotely operated vehicles in the fields of the Izu-Ogasawara Arc or the Okinawa Trough of Japan. Compared to the well-studied *A. pompejana*, *P. hessleri* is smaller in size (total 10 mm or less), thus more individuals can be maintained in restricted laboratory space, and examined with whole-mount, three-dimensional (3D) analysis without dissection. In this study, the detailed physiology of *P. hessleri*, including mechanisms of thermotolerance, were not examined, but our preliminary experiments showed that the thermotolerance of these species is similar to those of *A. pompejana* and *Paralvinella sulfincola* from the North Pacific; *P. hessleri* prefers temperatures between 40-50°C, and endure temperatures as high as 55°C ([15,36]; see also [16]; Shigeno et al., unpublished). Using this species, we sought to provide the first comprehensive maps of the distribution of ciliated sensory cells, neural projections in the higher brain centers, and newly identified cellular components, to explain mechanisms for tolerating the hydrothermal vent environment.

Results

Alvinellid body plans

The external cylindrical body of *Paralvinella hessleri* is divided into three parts, as defined in most species also known as polychaetes: a head or prostomium, a segmented trunk with many appendages, and a terminal pygidium (Figures 1 and 2; [2,37]). The head part includes a protrusible head-end with a divided shield, and numerous buccal tentacles extending from the rostro-ventral mouth opening. The body wall consists of longitudinal, circular, and oblique body wall musculature used in locomotion. The segmented trunk is the widest part of the body. A few polychaete genera have a large branchial crown on the upper side of the head part that is used in feeding and probably also as "feelers" [38,39]. In the Terebellidae, as in many sessile polychaetes, no traces of sensory palps are found [30]. The epidermal sensory cell types and their ciliary organizations identified in this study are summarized in Figure 2B and Table 1. Some neritic burrowing and tube-dwelling polychaetes have various small eyespots and statocysts (tilt and balance sensors), but these were not found in *P. hessleri*. The central nervous system is composed of the supraesophageal mass or brain, subesophageal mass, and the segmented ventral nerve cords (Figure 2C). The head contains the brain. A pair of nuchal grooves is situated in the dorsal head part, as has been observed in other polychaetes. The details of sensory innervation, including the basic organization of the central nervous system, are schematized in Figure 2C.

The sensory receptor cell types

We first focused on the main alvinellid epidermal ciliary sensory cells, which are used for chemical and mechanical sensing, as proposed by neritic polychaete studies [38]. The six cell types were identified and categorized, and some subtypes were defined by subtle structural differences of

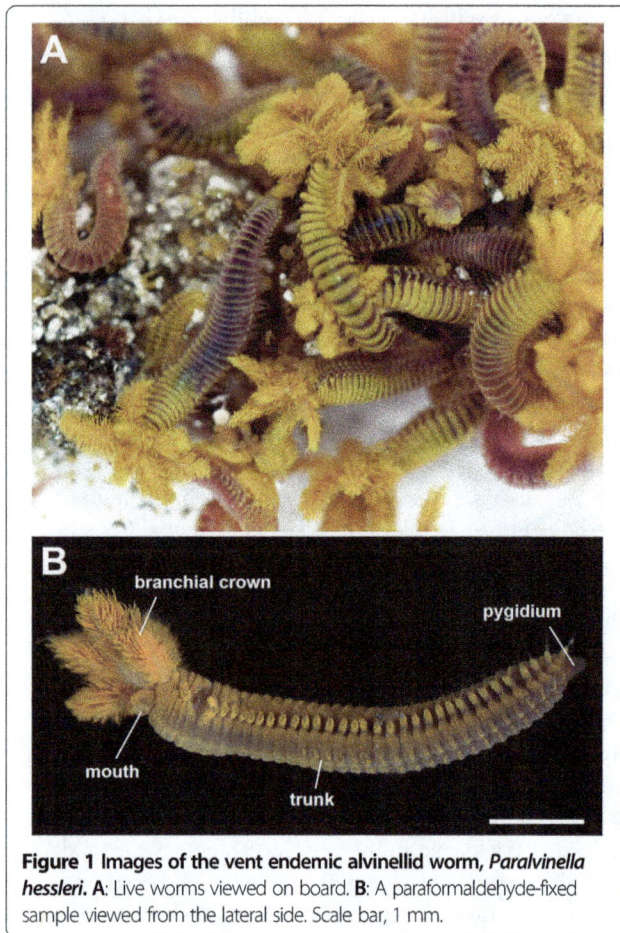

Figure 1 Images of the vent endemic alvinellid worm, *Paralvinella hessleri*. A: Live worms viewed on board. **B**: A paraformaldehyde-fixed sample viewed from the lateral side. Scale bar, 1 mm.

analysis identified subtle repeated gaps among these cells. The cilia are generally long and similar in length (10–12 μm).

3) **The tuft type (Figure 3K):** The cells are multiciliate penetrative type, namely the cilia penetrate the cuticle. The ciliary shape is round and tuft-like. One major subtype is distributed only in the dorsal, lateral, but not in the ventral side of the head. The cilia are short and densely packed compared with those of the patch ciliary type. In contrast to the branch surfaces, many pores and granules are found in the epidermis of the head part (Figure 3F).

4) **The bud type (Figure 3L-N):** These penetrative multiciliate cells are situated in their characteristic positions on the head and trunk. As in polychaetes, the bud may correspond to the papilla [30,38]. The cilia are sharp and roughly arranged in a short line.

5) **The patch type (Figure 3O):** This multiciliate cell type may have motor functions for producing directed water flow or removing waste matter from the body surface, but chemical or mechanical sensation also may be possible. The axonal projection is lacking and cells are tightly connected to the epidermal cells by the gap junction (not shown). These cells are most dense on the lateral sides of each branch, and are broadly distributed over the entire body. The cilia are not sharp as in the bud type and usually exhibiting loop-like form as a possible artifact (Figure 3O).

6) **The ventral patch type at the trunk (Figure 2B):** The ciliary structure is almost identical to the patch ciliary type, with ciliary density and length as stated above. Yet the distribution of cells displays a stereotypic linear pattern, and repeated positions are only observed in the ventral and ventro-lateral sides of the trunk (see the trunk part).

The sensory cells were detected using at least three independent methods: scanning electron microscopy (SEM), transmission electron microscopy (TEM), and light microscopy using labeling for acetylated alpha-tubulin, a widely used cytoskeleton marker for cilia and neural axons (shown later). Phalloidin fluorescent staining was also helpful for detecting F-type actin in the ciliary cells and neurons in differential contrast to the tubulin labeling (not shown).

As revealed with TEM, the typical sensory ciliary cells include multiciliate bipolar types, which are situated in the single epidermal cell layer under the thick cuticle (see Figure 4A for a representative cell). As described previously, the mitochondria structure, including the outer and inner membrane with the cristae and matrix, are typical (Figure 4B; [24]). The size range is from 0.5 to 1.0 μm in diameter. The number of mitochondria per cell varies by tissue type; many epithelial

cilia and microvilli (Table 1). The distribution patterns of ciliary cells and pores for excretion are different in each body part; namely, the branchial crown, the head, and the trunk with appendages (Figure 3A-F).

1) **The sensilla type (Figure 3G-I):** These cells are ciliate sensory cells with a short distal process, also known as typical bipolar receptor cells [38]. Three to five cilia penetrate the cuticle, and the sensory cells send long axons directly to the central nervous system. This cell type is broadly distributed on the head, branches, buccal tentacles, and the trunk. The three (the short, intermediate, and long) ciliary subtypes were detected. The short and intermediate types seem to be developing cell, but here we distinguished as a sensilla type due to their shapes, ciliary numbers, and thickness of cilia.

2) **The line type (Figure 3J):** These are penetrative multiciliate cells, and the cilia and cell bodies exhibit bilateral longitudinal lines running inside of the dorsal grooves extending from the head end. The ciliary cells are superficially lined, but more detailed

A. GENERAL BODY PLAN

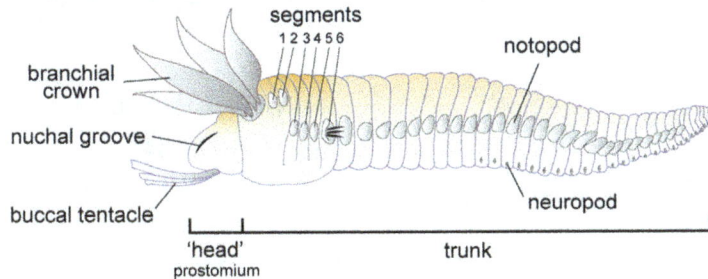

segments
1 2 3 4 5 6

branchial crown

notopod

nuchal groove

buccal tentacle

neuropod

'head'
prostomium

trunk

B. SENSORY SYSTEM

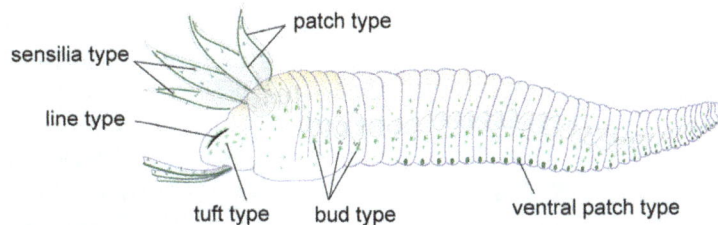

patch type

sensilia type

line type

tuft type   bud type   ventral patch type

C. NERVOUS SYSTEM

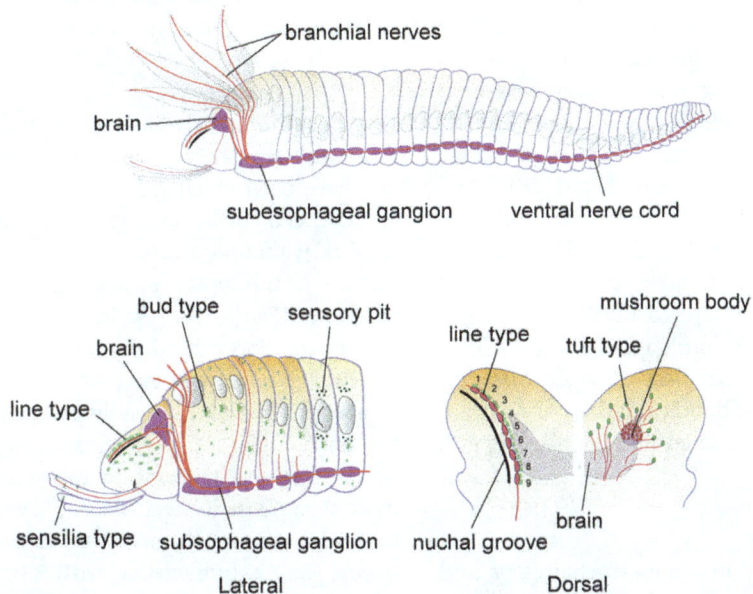

branchial nerves

brain

subesophageal gangion   ventral nerve cord

bud type   sensory pit   mushroom body

brain   line type   tuft type

line type

sensilia type   subesophageal ganglion   nuchal groove   brain

Lateral   Dorsal

Figure 2 Schemes of body plans of *Paralvinella hessleri*: the sensory and neural structure. **A**: Lateral view of the general body plan. Most of the parapodia setae are omitted. **B**: The epidermal sensory systems (green), major epidermal cilia. The patch type cilia on the branchial crown are represented as lines, and most of the cilia on trunk are omitted for clarity. **C**: The central nervous system and some peripheral nerves. The cell bodies (purple), major neural axons (red lines), and sensory organs (green patches). Divisions between the subesophageal ganglia and ventral nerve cords are not clear. Below left: The enlargement of rostral parts of the body to show the details of neural projections and epidermal ciliary sensory cells, side view. Below right: Dorsal view of the bi-lobed prostomium and the brain. On the left-hand side, the repeated features of glial and ciliary sensory cells (green) are located inside of the nuchal groove. The position of small globuli cell clusters or the mushroom body and distributed sensory cells are shown with axonal projection to the brain.

cells, including primary sensory cells, typically have several thousand mitochondria. The electron-dense granules are often observed in the intracellular spaces.

The myelin-like glial repeats in the head sensory neurons
One of the most striking features in the sensory system of *P. hessleri* is the placement of a multiply repeated glial

Table 1 Classification of the sensory and motor ciliated cell types of _Paralvinella hessleri_

Cell types	Body localities	Projections in CNS	Specific characters _ciliary length/width_	Possible homologies
Sensilla type	branchial crown	anterior VNC	chemical sensory cells	penetrative sensory cells
	whole bodies	VNC, each segment	chemical sensory cells _1-12/0.3µm_	
Line type	dorsal prostomium	anterior VNC	forming lines on the dorsal head _4-13/0.2µm_	neck ciliary bands?
Tuft type	whole prostomium	brain, MB	round and short cilia _1-3/0.2µm_	penetrative sensory cells
Bud type	ventral area of notopods	VNC	roughly lined with pods _1-21/0.1-0.2µm(sharp)_	olfactory organ-like
Patch type	lateral branch and whole bodies	-	typical motor cilia _1-12/0.2µm_	motor ciliary cells
Ventral patch type	ventral abdomen	-	ventrally localized _1-11/0.2µm_	sensory receptor cells?
				lateral organs

Possible homologies are based on Purschke [38]. _MB_, Mushroom body; _VNC_, Ventral nerve cord.

structure for the long axons extending from the longitudinal head ciliary lines at the lateral sides of the granule cell-rich nuchal grooves of the head part. Based on the topographical position of the prostomium or head (Figure 3D shows the surface, and Figure 5A, B show granule cells), the longitudinal ciliary lines can be compared to those of nuchal sensory organs of neritic polychaetes (Figure 5C; [38]), or modifications. In the whole-mount specimens, the lipophilic dye tracer experiments revealed the structure of primary sensory cells spreading over the head surface (Figure 5D, E; Additional file 1: Movie S1), and the axonal fibers derived from the sensory cells of the longitudinal ciliary bands extending from the rostral head ends (Figure 5F-I). At least 10 distinct repeated swellings were morphologically identified using concanavalin A membrane staining (Figure 5F). Their afferents do not directly enter into the brain neuropils, but run along the lateral sides of the brain (Figure 5F). The ultrastructural analyses of the axonal bundles show the position of the glial cell body and multi-layered membranes (about 70–100 nm thickness with a 20 nm gap, Figure 6A-C). In these bundles, we identified differentially arranged types, including loosely organized thick bundles (type A: 1–2 µm), and tightly organized and fasciculated fine bundles (type B: 0.2–0.6 µm) (Figure 6D, E).

Arealization

The combined whole-mount immunocytochemistry and the serial optically dissected sections delineated gross features of the brain of _P. hessleri_ (Figure 7A, B). The cord-like brain is situated dorsally in the head, and is connected to the subesophageal cords by thick connectives (summarized in Figure 2C). The acetylated alpha-tubulin positive axons and lipophilic neurotracing studies confirmed that most identified sensory inputs from the surface of the head flow into the overall brain neuropils (Figure 7C). In the brain and associated neuropils, three centers are distinct in the dorsal to ventral regions: 1) a tiny pair of mushroom bodies or globuli cell clusters, characterized by small interneuron pools, receive the sensory inputs from the head tuft type cells, and form a small neuropil with

numerous fine and less fasciculated axons (Figure 7B-G); 2) the antero-dorsally located entry area of axons of head sensory cells (Figure 7H); and 3) the commissural or central region, situated at the basal region (Figure 7I; Additional file 2: Movie S2). Those regions may be comparable with the brain of some mobile polychaetes such as _Nereis diversicolor_; however, the brain or the brain of the latter contains subdivisions such as the optic neuropil, central neuropil, and more small neuropils [40,41], but such obvious subdivisions are lacking in _P. hessleri_ (summarized in Figure 8A).

The mushroom body-like higher brain centers

The mushroom bodies traditionally called corpora pedunculata have been extensively studied as higher chemosensory learning centers in arthropods and annelids [42,43]. In _P. hessleri_, we found a candidate pair of mushroom body cell clusters in the typical position in the dorso-lateral sides of the brain (Figure 8A), which receive inputs from sensory cells in the dorsal head regions (see Figure 7C). The ultrastructural studies revealed that the small interneurons of the mushroom bodies are histologically distinct compared with the surrounding intermediate-size interneurons (Figure 8B). In the small interneurons, there are a few mitochondria and more than 40 to 80 electronically dense particles (ca. smaller than 100 nm in diameter) in a single ultrathin section (Figure 8C). More characteristically, we found a unique support system for the neural cell bodies covered with glial cells (Figure 8D), and granular-rich axonal bundles in the mushroom body region (Figure 8E), although such glial membranes were not widespread in the neuropils (Figure 8F, G). The spatial distribution of the glial cell bodies could not be precisely determined due to their complexity, but most of neuronal cell bodies are tightly covered with thick membrane comprised of two to three layered glial membranes (Figure 8H). Characteristic fine dense granules are also distributed in some axons and cell bodies (Figure 8E), and may be neurosecretory cells (as described in the brain of other polychaetes [29,44]), but larger granules of a different type are also seen (Figure 8I).

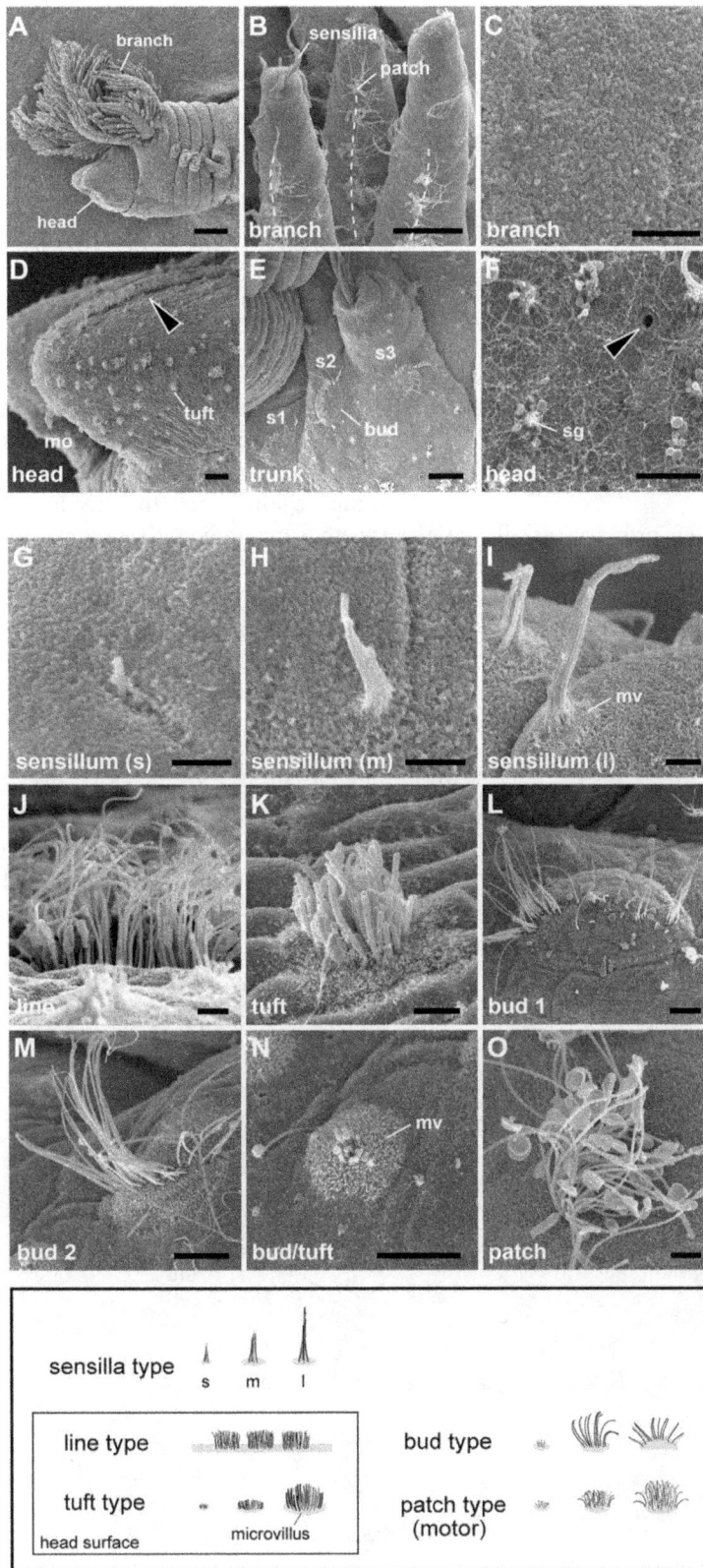

Figure 3 (See legend on next page.)

(See figure on previous page.)
Figure 3 Scanning electron microscopy images of epidermal ciliary sensory cells. The cell types are summarized below for each body part.
A: An overall side view of the rostral part. **B**: Branchial crown and the dotted lines show lines of patch type cilia along the lateral sides. **C**: Enlargement
of a branch to show the no-pore condition of the epidermis. **D**: The head part and nuchal groove (arrowhead). **E**: The setigerous segments to show
the distribution of the bud type cilia. **F**: The pores (arrowhead) and granules on the epidermis of the head. **G - O**: The epidermal cilia. mo, mouth
opening; mv, microvilli; s1-3, setigerous segment 1–3; sg, secretory granules. Scale bar in **A**: 100 μm; **B, D, E**: 20 μm; **M-O**, 5 μm; **C, F, G-L**: 2 μm.

The sensory systems of the head branchial crown and buccal tentacles

We next focused on the branchial crown (Figure 9A, B) and buccal tentacles (Figure 9C) in the anterior body part. These are important sensory organs for accessing outer fluids and food sources [25], and a previous study showed that the branches exhibit higher anaerobic enzyme activity than abdominal tissues [45].

The branchial crown

Most of the sensory cells are the primary sensilla type (summarized in Table 1). In each branchial leaflet, there are numerous patch type cells along the lateral sides, and broadly distributed sensilla types (see Figure 3A). Whole-mount immunocytochemistry identifies the ciliary position as well as the axonal origins in the branches (Figure 9D, E; Additional file 3: Movie S3). Notably, we failed to detect the multilayer glial membranes that were found in the head axons in any branchial sensory cells (data not shown).

The buccal tentacles

The tubulin immunochemistry distinguished the ciliary sensory cell types, and showed that most of the sensory cells are sensilla and patch type cells (Figure 9F, G). As in the case of the branchial crown, the multilayer glial membranes surrounding the axons are not distinguished, as described in previous studies [24].

The trunk nervous system

Although we could not detect specific supportive glial systems in the branchial crown and buccal tentacles, such systems might have widely established components in the trunk region, since the myelin—a lipid-rich multilayer sheath surrounding the axons—is found in the ventral nerve cords of some annelid worms [46]. We explored this hypothesis by examining the peripheral nervous systems of the trunk region of alvinellids. Each segment includes a pair of paddle-like parapodia, which consist of setigerous notopods used for locomotory behavior and possibly as sense organs. Many ciliary penetrative sensory cells, including the sensilla, patch, bud, and ventral patch type cilia were found (Figure 10A, B; summarized in Table 1; Additional file 4: Movie S4, Additional file 5). As in most polychaetes, their sensory cells are connected to the ganglionated ventral nerve cords. The giant fibers exist in the typical anterior position of the ventral nerve cords (not shown), but myelinated glial repeats (as observed in

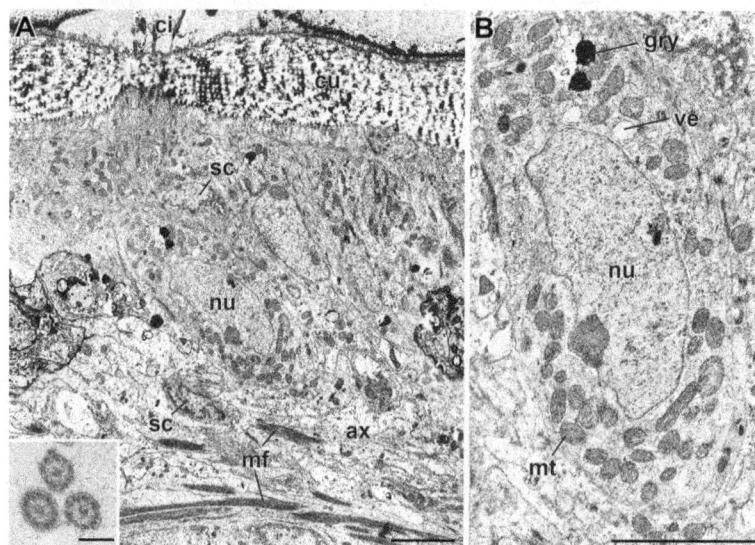

Figure 4 A primary sensilla cell type and the intracellular organella. Transmission electron microscopy (TEM) images. **A**: Cross section of the single layer epidermis of the branchial crown. **B**: Enlarged view of the same cell to show the organella. Inset: a cross-section showing the microtubules of cilia outside of the cuticle. ax, axon; cu, cuticle; gry; yellow-color granule; mf, muscle fiber; mt, mitochondrion; nu, nucleus; sc, supporting cells; ve, vesicle. Scale bar in **A, B**: 5 μm; Inset, 200 nm.

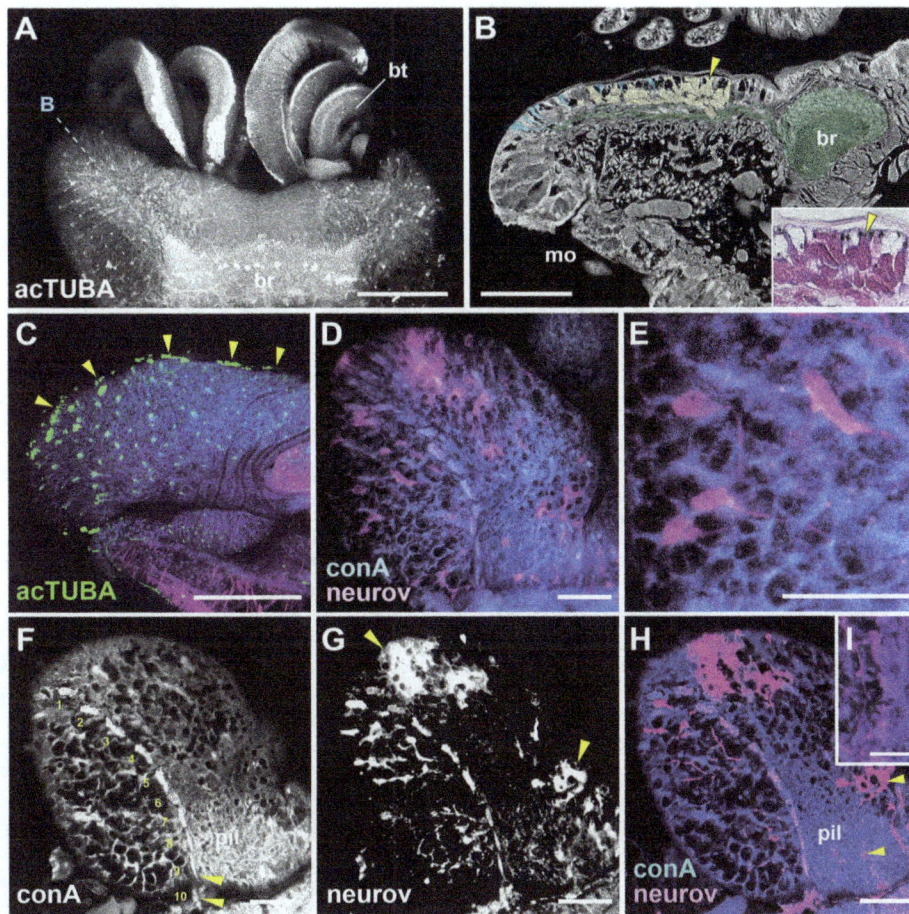

Figure 5 The brain, ciliary sensory cells, and glial repeats of the head or prostomium: whole-mount laser confocal microscopy images.
A: Dorsal view of the prostomium with buccal tentacles. **B**: A semi-sagittal histological section cut as shown in A, showing a negative image of the hematoxylin and eosin stained section. Pseudocolors were used to enhance the position of line type cilia (light blue), axonal bundles and the brain (green), and secretory cells in the nuchal groove (yellow). Inset: granules stained with eosin. The arrowheads indicate the same position.
C: A side view of the prostomium showing the distribution of acetylated alpha-tubulin (acTUBA) positive ciliary cells. The arrowheads indicate the line type cilia. The concanavalin A (conA, blue) and phalloidin (actin, red) cell membrane or muscle markers used for counterstaining. **D, E**: The sensory cell bodies and their axons coupled with dye (neurov: NeuroVue®Red) and repeated myelin-like glia stained with conA membrane marker (blue): a dorsal view of whole-mount prostomium and the enlargement. **F-G**: The sensory cells and the axonal projection patterns in the image series of conA, neurov, and merge. The numbers indicate the glial repeats and the arrowheads are positions filled with dye, and some labeled axons in the brain. **I**: Enlarged view showing the axons in the glial repeats. br, brain; bt, buccal tentacle; mo, mouth opening; pil, neuropil. Scale bar in **A-C**: 100 μm; **D-H**: 40 μm; **I**: 20 μm.

the head parts) were not identified in the neuropils by our 3D confocal and ultrastructural analyses (Figure 10C-E). A pair of ventral nerve cords is medially centralized; additionally, most of the neural cells are monopolar, and their axons usually form the neuropils, as revealed by acetylated alpha-tubulin and serotonin-like immunoreactivity (Figure 10D). The neuropils have no obvious compartments by glia (Figure 10E). Some large fibers, possibly longitudinally projected motor axons running along the antero-posterior axis, were detected by the ultrastructural analysis (Figure 10E). The most distinguishing feature is the multi-layered glial membranes in the cell body layer, which are distinct and better developed than in the brain (Figure 10F, G). Nearly all cell bodies examined are covered with glial membranes.

Discussion
Unique biological systems are often expected to be discovered in animals living in unusual environmental conditions. In order to contribute to the discussion of the sensory and neural systems animals from hydrothermal vent environments, we studied the cellular, intracellular, and histological structure in *Paralvinella hessleri*, a member of the vent-endemic alvinellids, using electron microscopy, immunohistochemistry, and neurotracing in combination with laser confocal microscopy.

The sensory receptor cell types
We identified a total of four sensory cell types with possible homologies to other annelids based on the structure

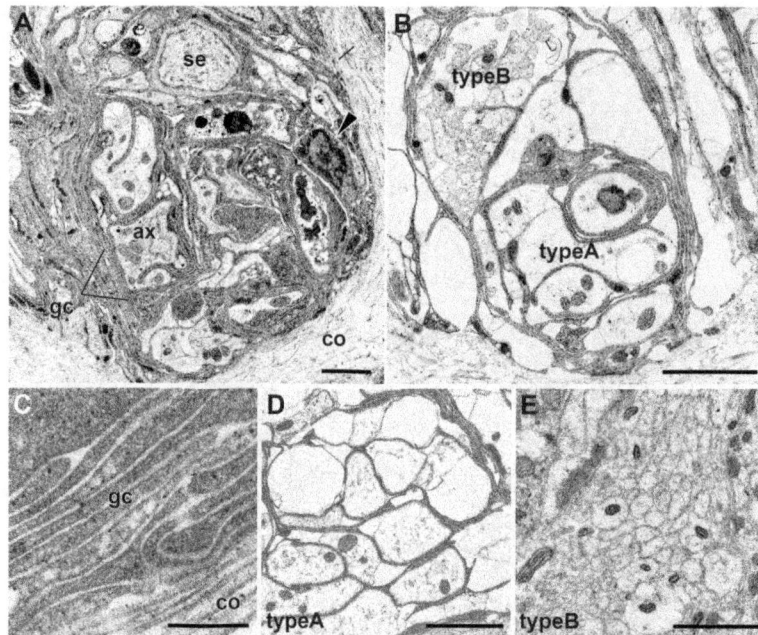

Figure 6 The myelin-like glial wrappers. The transverse sections, TEM images. **A**: The axonal bundles cut with sensory cell level, and **B**. the glial inter-repeat region with two different axonal types. The arrowhead indicates possible glial cell body. **C**: The detailed structure of glial membrane. **D**: Two axonal bundles. The type A and B run along the ventral and dorsal sides, respectively. **E**: Type B axonal bundles showing details of the membrane and mitochondria. ax, axonal fibers; co, collagen fibers; gc, glial cell wrappers; se, line type sensory cell. Scale bar in **A**, **B**, **D**: 2 μm; **C**, 300 nm; **E**: 1 μm.

of the cilia, patterning of the cell clusters, and position on the body (Table 1). The forms of most alvinellid cells are rather simple, and do not display extreme modification such as the chambered or balloon-like multi-ciliary dendrites of the nuchal organs of lugworms [26]. The sensory receptor cells of alvinellid worms are generally found within the epithelium, and connect to the adjoining cells by the anchoring membrane junction. The alvinellid sensory cells are bipolar, with long axonal fibers, and penetrative cilia exhibiting 9 + 2 patterns of axonemal microtubules with additional microvilli, similar to those found in most polychaetes [26,28]. However, the most striking morphological features of the alvinellid sensory cells are as seen in Figure 4. 1) The thick outer cuticle, often without penetrative pores, in the branchial crown; 2) many yellow-colored and electron-dense granules distributed in the cells; 3) mitochondria-rich sensory cells (e.g., more than 50 non-branched mitochondria (n >6 cells) were observed in the single ultrathin section of the sensory cell, compared with less than five to ten mitochondria in the lugworm *Arenicola marina* [25]). Furthermore, there are few Golgi apparatus and endoplasmic reticulum compared with the ciliary sensory cells of lugworms [26]; and 4) the axons from the sensory cells are tightly covered with multi-layered myelinated glial membranes, although this varies among the different body parts and functional subsystems for chemical sensing (Figure 11).

The glial membrane supporters

The types of glial systems identified in *P. hessleri* have not been described in any animals, except for a scale-worm also endemic to hydrothermal vents (Shigeno et al., submitted). The neural cells of many animals are more or less protected by glia, and the specialized myelinating glia for the long axons are found in many vertebrates, crustaceans, and annelids [46]. In the case of well-studied earthworms, the median and lateral giant axons of the ventral nerve cords are encompassed by concentrically wrapped lamellar sheets of insulating plasma membrane [47]. In this study, the glial system found in the alvinellid worms is considerably different. First, the myelin structure is present in the head sensory input systems and not found in the ventral nerve cords. Second, in addition to the axonal bundles, the cell bodies of the brain themselves are covered with multilamellar sheets.

The glial cells in vertebrates and some invertebrates serve diverse functions, including as a nutrition source, and for immunity, peptide signaling, neural development, as a resistor in fine electronic signal modifiers. They also provide protection by filtering or blocking toxic substances, including oxidative stressors and heavy metals, through maintenance of the chemical environment used to conduct electrical impulses [48-50]. If the cells are myelinating, the conduction speed of electrical impulses increases, thus providing an adaptive advantage for rapid behavioral responses

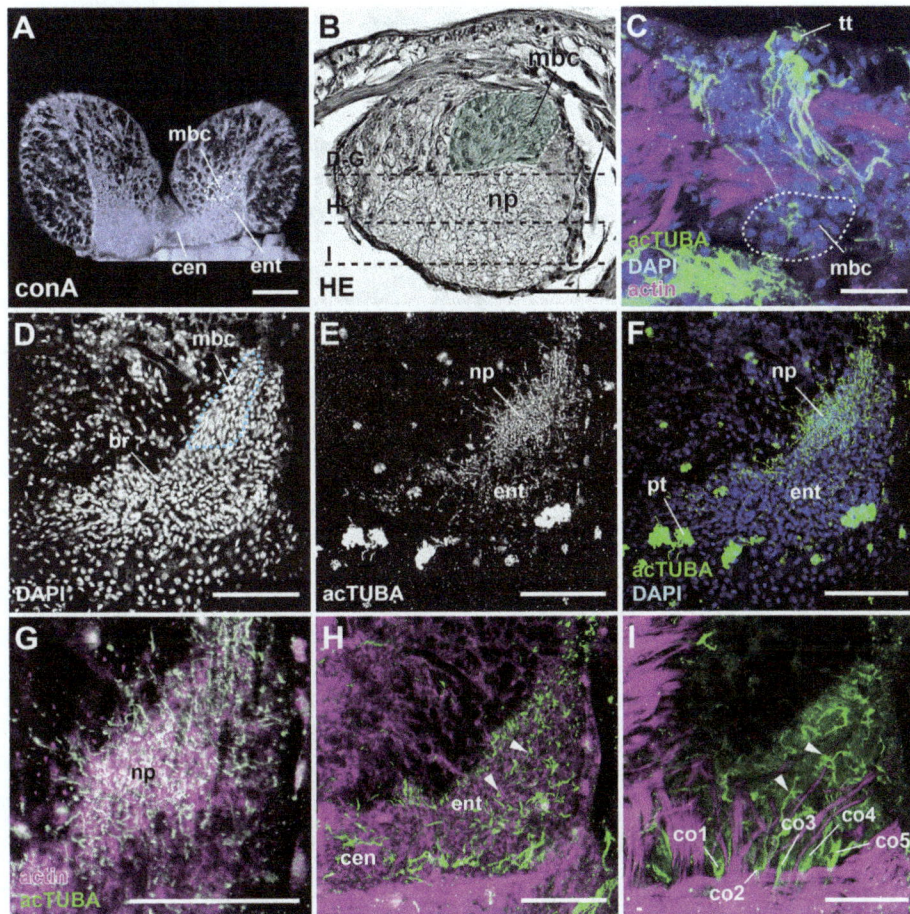

Figure 7 The brains and higher sensory centers. A: A horizontal confocal optical section of the whole head region. **B**: A sagittal histological section of the head to show the brain subdivisions. The white dotted line covered with pseudo-green color indicates the position of the mushroom body-like structure in a section stained with hematoxylin-eosin. The dotted lines indicate optical cutting sites for confocal microscopy. **C**: The direct sensory pathways from tuft type cilia to the mushroom body, a sagittal section of the head. **D-F**: The nuclei distribution and neuropil structure. Horizontal optical sections viewed from the dorsal side at the level of the neuropil of the mushroom body (encircled with dotted red line). **G-I**: The highly mingled axons from small mushroom body neurons (**G**), the brain neuropil (**H**), and more ventral region (**I**). Nuclei stained with DAPI and tubulin positive fibers to show the distinct neuropils and axonal bundle patterns. The horizontal optical section series of confocal microscopy viewed from the dorsal (globuli cluster level) to ventral side. Arrowheads indicate the axonal fiber patterns running from dorsal to ventral, and to connectives. The phalloidin (actin) was used for counterstaining of neural fibers and muscles. br, brain; cen, central neuropil; co 1–5, connective to ventral nerve cords; ent, sensory axon entry region of the neuropil; pt, patch type cilia; mbc, mushroom body cells; np, neuropil; tt, tuft type cilia. Scale bar in **A**, **D-I**: 100 μm; **B**: 50 μm; **C**: 20 μm.

to stimuli from alarm cues [51]. In the environment of hydrothermal vent fields, it is likely that specific glial protectors are required, since the environment is rich in heavy metals, hydrogen sulfide, and other toxic chemicals and metabolites produced by high temperature and pressure.

In addition, moderately hypoxic conditions are ubiquitous close to the vent field; therefore, some mechanism of protection is required to prevent the breakdown of glial membranes. This breakdown leads to irreversible cell death due to the production of reactive oxygen substances via hypoxia, which is well-studied in mammalian brains affected by ischemic stroke and integrity loss of the blood–brain barrier [52-54]. The actual functions of glial cells in alvinellids remain largely speculative, and it is not known how many glial cell types are present in these worms; however, our findings provide the first evidence for the role of such specialized glial systems in the hydrothermal endemic animals. Our findings also showed that not all of the sensory and neural cells are covered with multi-glial membranes, perhaps due to the presence of an alternative protector, such as the thick, non-pored collagenous walls in the epithelium of branches (summarized in Figure 11). This discovery suggests that unexpected biochemical heterogeneity and protection mechanisms may be present in the internal body spaces of these worms.

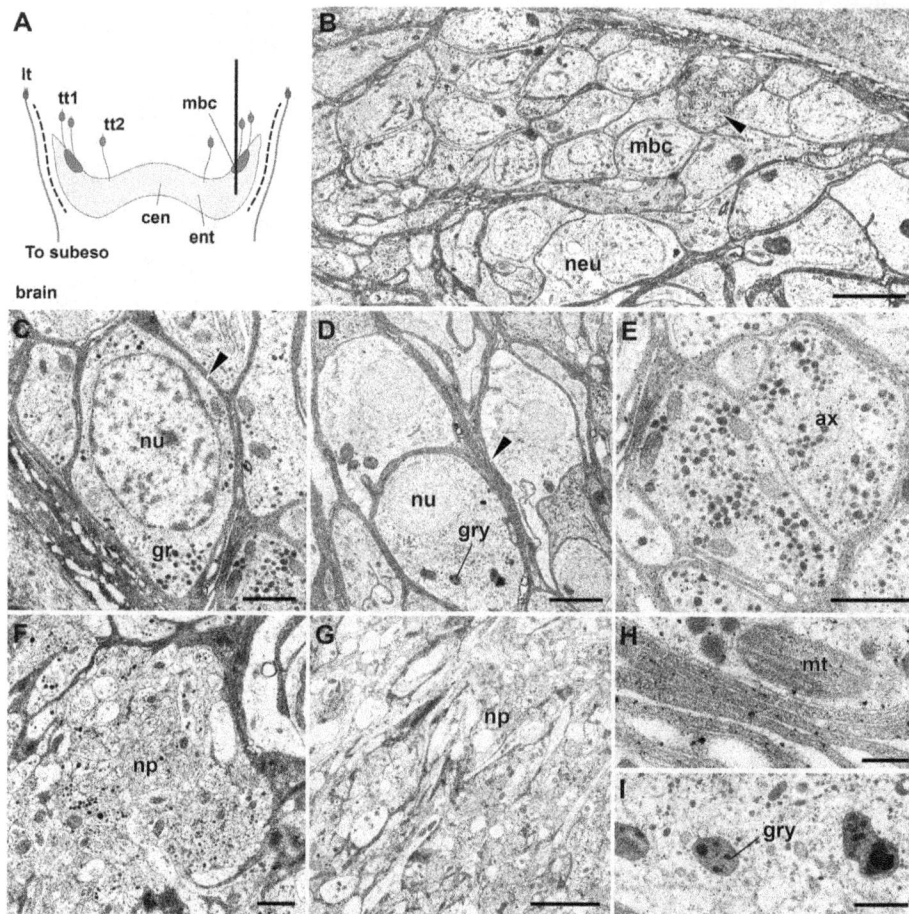

Figure 8 The mushroom body-like structure and the cell body wrappers. A schematic figure and TEM images. **A**: Schematic figure showing the brain organization and associated nerves, dorsal view. The thick longitudinal line indicates the cutting site for Figure 8B. **B**: TEM images of small interneurons and a large neighbor neuron. The arrowhead indicates granule-rich cell. **C, D**: The small or large interneurons covered with glia (an arrowhead). **E**: The granule-rich axons from the small cells. **F**: The neuropil of mushroom body. **G**: Details of the mushroom body neuropil. **H**: The enlarged view of glial membranes of small cells. **I**: Granules in the neurons. ax, axons; ent, sensory axon entry region of the neuropil; gr, granules; gry, yellow color granules; lt, line type sensory neurons; mt, mitochondrion; neu, a large neighbor neuron; np, neuropil; nu, nucleus; tt1 an tt2, tuft type cell 1 or 2; mbc, mushroom body cells. Scale bar in **B, D, G**: 4 μm; **C, E, F**: 1 μm, **H, I**: 0.1 μm.

Specialization of the central nervous system

Based on a comparative analysis of polychaete brains, the overall organization of the alvinellid brain can be compared to that of *Serpula vermicularis* and *Pista cristata*, both of which are burrowing species and use the numerous elongate branches or buccal tentacles for deposit feeding [30,55]. The brains of the alvinellids and these species are externally flat in shape, and are attached to the subesophageal mass and the ventral nerve cords with connective bundles. Many axonal tracts are broadly distributed without fasciculation in the neuropils, and in contrast with nereids and scale-worms, their compartments are not specialized [28,56]. Despite such similar basic morphology, the tract pathways related to the functional organization are dissimilar (Figure 12). First, the nerves from the branchial crown usually construct the major anterior components of the brain, as in the calcareous tubeworm *Serpula*;

however, in alvinellids they run directly to the subesophageal mass or ventral nerve cords. Second, the nuchal organs are more centrally located and extend into the anterior region of the brain in *Serpula* and *Pista*, but in alvinellids the main pathways for the head line type cells are positioned similarly as in the nuchal organs, and run directly to the subesophageal mass or ventral nerve cords.

The mushroom body-like structure

One of the most distinguished areas in the polychaete brains is the corpora pedunculata, also known as mushroom bodies, a pair of neuropils with associated small somata called globuli cells. These may function in the chemosensory learning and memory centers found in many polychaetes [28,42], but are reduced [31] or not developed in sessile species, including the calcareous tubeworm *Serpula* [30]. However, this does not mean that all

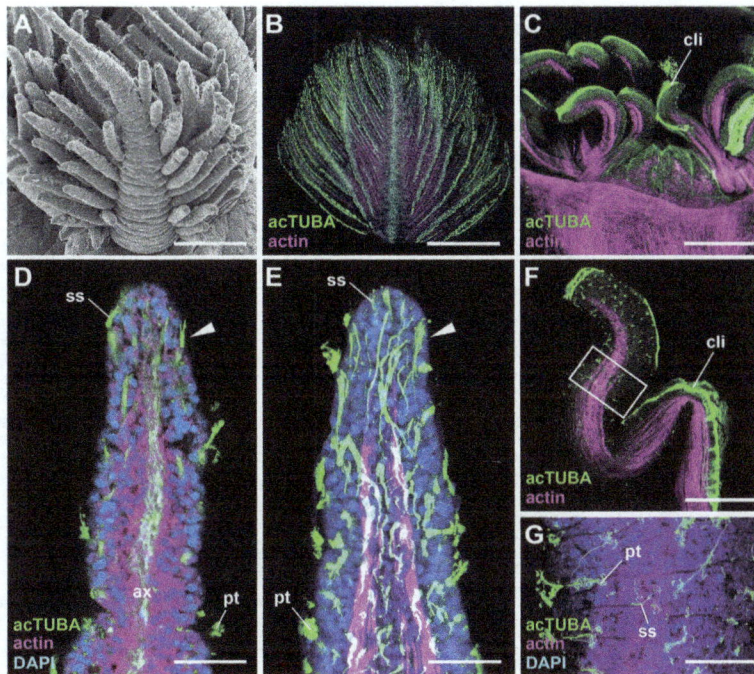

Figure 9 The ciliary epidermal cells of the branchial crown and buccal tentacles. A: Scanning electron microcopy image of a branch. **B-G**: Laser confocal microscopy images. **B**: The whole view of the single branch with many patch type cilia. **C**: The buccal tentacles with dense patch type cilia arranged along each ventral side. **D**: An optical section of the leaflet showing the sensillia type cells (arrowhead) and axonal bundles. **E**: The 3D reconstruction images of optical sections showing the pathways from the sensilla type cells. **F, G**: A buccal tentacle and enlarged view to show the distribution of cilia. ax, axon; cli, ciliary line; pt, patch type cilia; ss, sensilla type cell. Scale bar in **A-C**, **F**: 100 μm; **D**, **E**: 10 μm; **G**: 25 μm.

sessile species lack mushroom bodies and globuli cells. In this study, we observed a pair of interneuron clusters in the anterolateral sides of the brain, indicating that these clusters might be positionally and functionally compared to the annelid mushroom bodies as homologous structures for the higher-order processing centers [31]. Alternatively, higher-order neurons are often developed as distinct interneurons for a type of chemosensory system such as in the brains of aplacophoran molluscs [57,58]. This indicates that the interneuron pools of alvinellids may have developed independently to act as specialized sensory receptors unique to hydrothermal vents.

One could assume that the highly developed branchial crown functions as a sensory organ due to the direct exposure of the bacteria-covered tubes to hot vent fluids. In this study, most sensory cells in the branches were of the ciliate type, and long axons extend into the subesophageal mass or the ventral nerve cords where there were no clearly identifiable lobes or neuropil compartments. Additionally, the buccal tentacles may possibly be used for the bacterial feeding (Shigeno, unpublished data), and have two sensory ciliated cell types and specific neuropil compartments that were not observed in the brain. If specialization of the neuropil compartments of chemosensory or gustatory centers are related to their chemical receptor diversity, as in the case of the glomeruli

centers in annelid brains [31], we may expect that the sensing capacity of the branchial crown and buccal tentacles of the alvinellid worms is less specialized, and chemicals detected are not processed as environmental information. Whether or not the chemical receptors of alvinellids are specialized remains to be determined, and continued molecular studies of ionotropic, gustatory, and olfactory receptors are needed to better understand the chemosensory systems of animals endemic to hydrothermal vents.

Comparative evolutionary frameworks

As in free-living marine ragworms and sessile calcareous tubeworms, the sensory information collected by alvinellid sensory cells is processed by the mushroom bodies or comparable higher-order sensory centers, or the brain centers or the ventral nerve cords, as illustrated in Figure 11. This scheme for the sensory input and output signaling emphasizes the specific characteristics of alvinellid sensory processing systems.

First, the axonal projections of the line type sensory cells or the nuchal organ of *P. hessleri* extend directly into the subesophageal ganglia. This situation is dissimilar to that of the calcareous tubeworm *Serpula*, where the nuchal pathways extend into the brain (Figure 11; [59]). Additionally, the nuchal organ related centers of alvinellids are less specialized than those of the bloodworm *Glycera rouxii,*

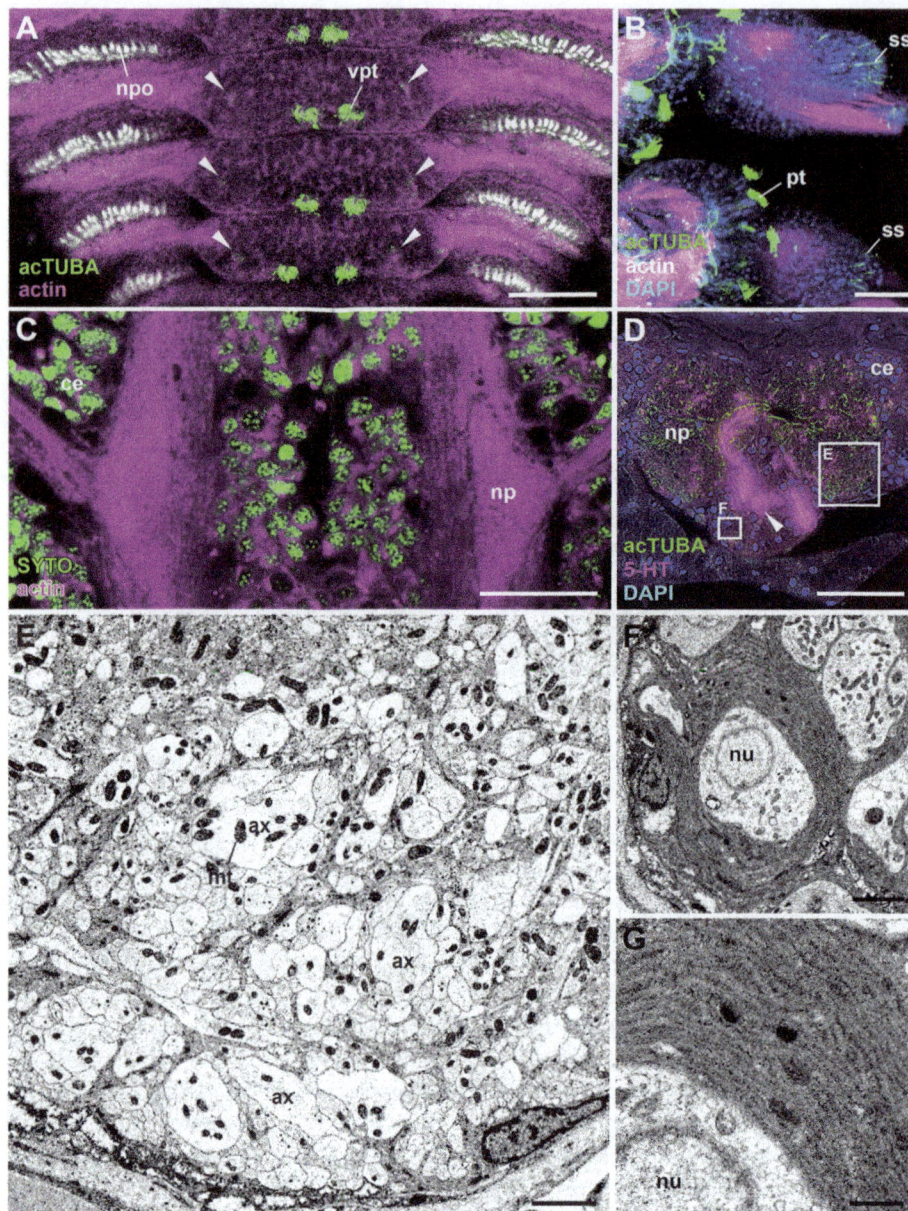

Figure 10 The trunk nervous system with the sensory cilia and ventral nerve cord. A-D: The laser scanning microscopy views. **A**: Distribution of ventral patch type cilia with smaller cilia allayed along the ventro-lateral sides (arrowheads) with autofluorescence of the neuropodia. **B**: Notopodia and sensilla distributed at the tips. The axonal projections are seen from sensilla of the tips (ss), but such projections are not seen from patch type cilia (pt). **C**: Cellular distribution and neuropils of the ventral nerve cord. **D**: A cross-section of the ventral nerve cord to show the position of tubulinergic and serotonergic cells (arrowheads) within the cell body layers and axons in the neuropils. The two rectangles indicate the position of analysis for Figures E and F. The TEM images. **E**: Neuropil without obvious wrapping structure. The large axons are possibly motor neurons. **F**: The wrapped cell body of the ventral nerve cords. **G**: An enlarged view of glial membranes. ax, a relatively giant axonal bundle; ce, cell body layer; ss, sensilla type cell; pt, patch type cell; mt, mitochondrion; np, neuropil; nu, nucleus; npo, neruopodium. Scale bar in **A-D**: 50 μm; **E**, **F**: 2 μm; **G**: 0.1μm.

which has distinct centers known as annexed ganglia and associated giant cells [30], suggesting that the alvinellid cilliary cells might have a simple receptor capacity. Second, the signals from the branchial crown of *P. hessleri* are processed only in the subesophageal ganglia, whereas the sensory cells of the branchial crown in calcareous tubeworms project into both the brain and

subesophageal ganglia [59]. Third, the mushroom bodies are composed of small interneurons and as in the rag-worms, are located at the antero-dorsal sides of the brain [31]; however, the alvinellid mushroom bodies receive inputs from the broadly distributed single sensory cell types of the head (see Figure 11). In addition, the alvinellid sensory inputs to the brain through the mushroom bodies

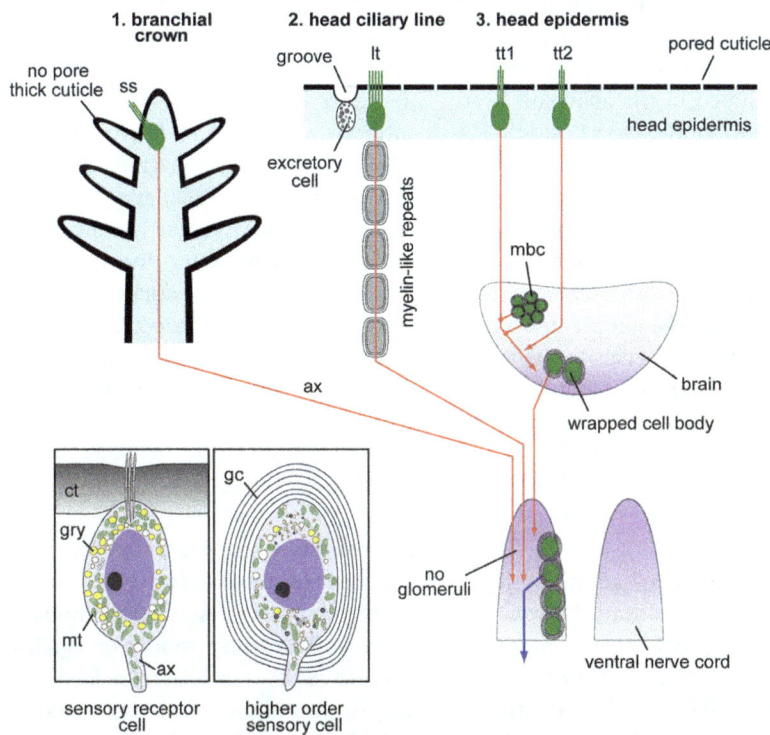

Figure 11 Schemes of the sensory pathways and the differentially specialized protection systems. In contrast to the primary sensory cells situated on the epidermis, the sensory neurons are multiply covered by glial wrappers. The myelin-like repeats are only formed in the head ciliary lines and not in the ventral nerve cords. Insets: the schemes show two distinct structural types of primary sensory receptor cells of epidermis and neuronal cells of the brain and ventral nerve cords. ax, axonal fiber; ct, cuticle; gc, glial cell membranes; gry, yellow color granule; lt, line type cell; mt, mitochondrion; ss, sensilla type cell; tt1 an tt2, tuft type cell 1 or 2.

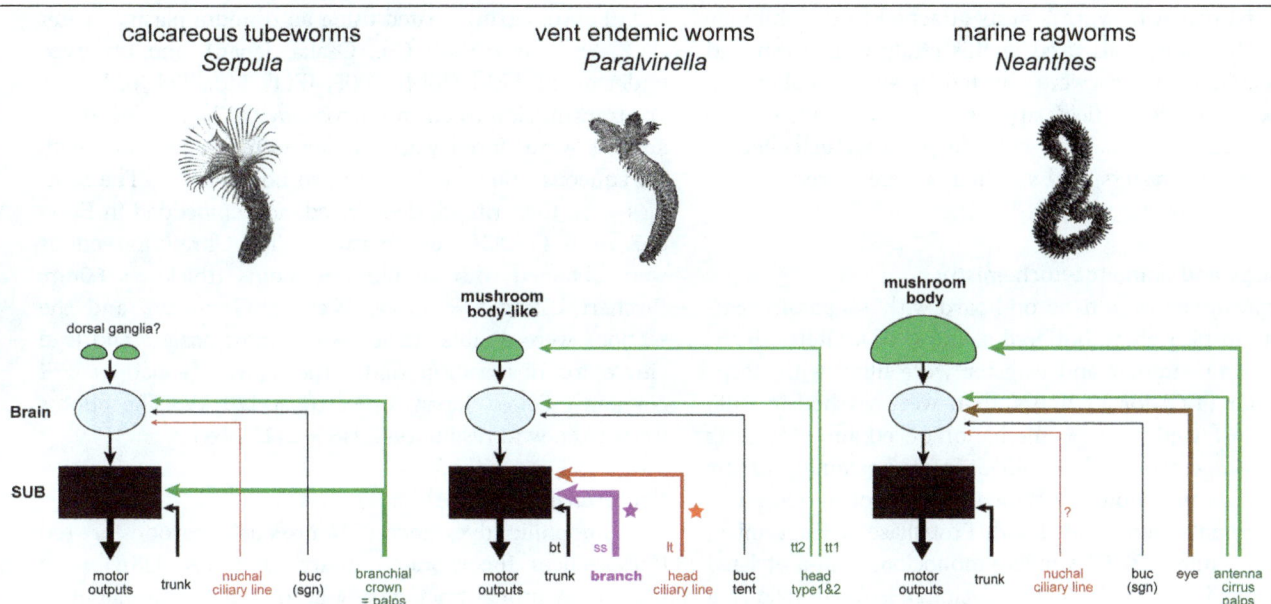

Figure 12 Comparative schemes of the sensory inputs and motor outputs to emphasize the functional subdivisions. The tuft type sensory cells and their inputs are characterized for alvinellid worms. The two distinct tuft type cells on the head or prostomium are identified. Note that the branchial crown of Terebellida (*Paralvinella*) is not homologous to those of Serpulidae and palps of Nereididae. Some neural pathways are not precisely identified. Data are simplified from Orrhage and Müller [30]. bt, bud type cell; lt, line type; ss, sensilla type; tt, tuft type; VNC, subesophageal ganglia with ventral nerve cord/ganglia.

are similar to those of free-living ragworms in the genus *Neanthes* [30]. We therefore propose that the input systems from the eyes, antenna, cirrus, and pulps of the ragworms, which are not developed in the alvinellid head parts, could be compared to those of the type 1 and type 2 sensory cells. We further suggest that alvinellids utilize two distinct sub-systems for sensory signaling: (1) the brain, which serves as a primary sensory processing system, and (2) several "short-cut" pathways from the head and branchial sensory cells to the subesophageal and ventral nerve cords, presumably for the simple and rapid transduction of environmental sensory signals to the trunk motor control networks or any adaptive organs, which regulate homeostasis through the endocrine and circulatory systems in the trunk region.

Materials and methods
Animal samples
More than three hundred individuals of adults or juveniles of alvinellids, *Paralvinella hessleri* [60], Polychaeta, Sedentaria, and Alvinellidae, were collected aboard the research vessel "Natsushima" during research cruise NT12-10 (31°53.049′ N, 139°58.104′ E, 907 m depth, off Myojin-sho submarine caldera, onboard ID 1374–9; or 32°06.214′ N, 139°52.05′ E, 1294 m depth, off Myojin Knoll, onboard ID 1377–4), in the Izu-Ogasawara Arc, Japan. Samples were collected on the 25th or 29th of April 2012 with Hyper-Dolphin 3000 remotely operated vehicles (HPD#1374 or #1377). Worms were collected using a suction sampler from the surface of the white microbial mat with worm's nests attached to the chimney walls. The individuals used in this study were fresh and active animals, which were selected by eye for collection. Following the ROV field surveys, the live animals were maintained in non-aerated cold deep-sea water collected in the same canisters, and specimens were stored at 4–8° C for a few days.

Histology and immunocytochemistry
The specimens were fixed on board with 4% paraformaldehyde in phosphate buffered saline (PBS); alternatively, the suction sampler and canister were filled with deep-sea water (4°C) for 12 hours, then were washed in PBS, and transferred to 80% methanol or ethanol for long term storage at −30°C or −80°C. Immunostaining of the whole-mounts followed a standard protocol (n >30). The whole-mounts were treated with Proteinase K (5 µg/ml in PBS) for 5 min at 37°C. A mouse monoclonal anti-acetylated alpha-tubulin antibody (6-11B-1 clone) isolated following immunization with sea urchin flagella proteins (acTUBA, Sigma Chemical, T6793, 1:3000) in PBST (PBS with 1% Tween 20 and 1% BSA), or rabbit polyclonal anti-serotonin (5-hydroxytryptamine, 5-HT, 1:500) antibody (Sigma, S5545, 1:500) was used to detect 5-HT positive selected neurons.

These primary antibodies have been widely employed in invertebrate neuroanatomy (e.g., [41]). A CF™ 488A goat anti-mouse IgG or CF™ 594 goat anti-rabbit IgG antibody (Biotium Incorporated, 1:400) was used as a secondary antibody. DAPI (4′-6-Diamidino-2-phenylindole; Sigma, 0.1µg/ml), SYTO®13, a green-fluorescent nucleic acid stain dye (Invitrogen, 1:5000), and rhodamine-conjugated Concanavalin A (ConA, Vector Laboratories, 1:1000) and CF™ 594 phalloidin (Cytoskeleton Incorporated, 1:600) were used for counterstaining of the nuclei, the cell membrane, or F-actin rich muscle fibers, respectively. Samples were examined using confocal laser scanning microscopy as described below. Some paraformaldehyde fixed samples were embedded in paraffin and cut with a rotatory microtome to a 5–10 µm thickness. These sections were stained in Mayer's hematoxylin and eosin solution.

Electron microscopy ultrastructural analysis
The electron microscopy analyses were conducted according to previously described methods [61], with some modifications. The whole-mount or dissected specimens (n >10) were fixed in a solution of 2.5% glutaraldehyde in cold deep-sea water for at least one week. After extensive washes with 0.22 mm-filtered seawater, samples were postfixed with 2% osmium tetroxide in filtered seawater for two hours at 4°C. For field-emission scanning microscopy (FE-SEM), samples were stained with 1% aqueous tannic acid (pH 6.8) for one hour and treated with 1% aqueous osmium tetroxide for one hour at 4°C. Samples were dehydrated in a graded ethanol series, critical point-dried (JCPD-5; JEOL Ltd., Tokyo, Japan), coated using an osmium plasma coater (POC-3; Meiwa Shoji Co., Osaka, Japan), and observed under an FE-SEM (JSM-6700F; JEOL Ltd., Tokyo, Japan). For transmission electron microscopy (TEM), the postfixed animals were rinsed with distilled water, and stained with 1% aqueous uranyl acetate for two hours at 4°C. The samples were then rinsed, dehydrated, and embedded in Epon 812 resin (TAAB, Aldermaston, UK). Ultrathin sections were obtained with an ultramicrotome (thickness 60nm; Reichert Ultracut S; Leica, Wetzlar, Germany), and the sections were double stained with uranyl acetate and lead citrate for observation under the TEM (Tecnai 20; FEI company, Tokyo, Japan) operated at 120 kV. The photos were taken with resolution 2048 × 2115 pixel.

Neurotracing analysis
The lipophilic dye tracers NeuroVue® maroon or red (Polysciences Incorporated, Warrington, PA, USA) were used for neuronal tract tracing according to the manufacturer's protocol, with some modifications. The dyes transfer into plasma membranes in formaldehyde fixed tissues and diffuses laterally within the membrane, finally labeling the cell bodies, axons, and allowing visualization of neuronal processes. More than twenty worms were fixed in 4%

paraformaldehyde in cold deep-sea water overnight, and washed in PBS without any polysorbate detergents and alcohol. The fine pieces of coated dye filters were directly applied to the head part with the fine tips of capillary tubes, and they were stored in 0.1% paraformaldehyde in PBS, and at 37°C for one to three weeks. The dye tracers transferred into the epidermal tissues and diffused in the neuronal membranes, allowing visualization of the selected or accidentally labeled cell bodies and long axonal arborization. ConA and Alexa Fluor® 350 or 594 were used as a versatile lectin probe for the membrane counterstaining to illuminate dye-traced cells. The whole-mount samples were stained with other dyes and viewed using confocal microscopy as described below.

Image processing

Samples were examined as whole-mounts or sections using the confocal laser scanning microscope Fluoview, FV500 ver4.3c (Olympus, Lake Success, NY, USA) with the fluorescent microscope IX-71 automated inverted microscope platform (Olympus). Laser power was employed with UV, Argon, and HeNe, and the appropriate filter set was selected according to the fluorescent markers. The pseudo-colors were used to enhance the structural images according to the manufacturer's protocol. Additional processing of images for contrast, brightness, and color balance was performed as needed with Adobe Photoshop CS5 (Adobe Systems Incorporated, San Jose, CA, USA). Schematic diagrams were created with Adobe Illustrator CS5 (Adobe Systems Incorporated).

Terminology

The terms used for the alvinellid sensory cells and nervous systems are based on the definitions given by previous studies [29,37,38], unless otherwise specified. Richter et al. [62] was also used as a nomenclatural reference for invertebrate neuroanatomical terms. Desbruyeres et al. [2] was used for alvinellid anatomical terms.

Additional files

Additional file 1: Video S1. A movie of whole-mount confocal Z-projection image stacks of a left head part of *Paralvinella hessleri*. See Figure 5D-H for details. Format: MP4 Size: 921 KB. The video can be viewed with Quick Time Player, Apple Inc. Mac OSX, ver.10.3.

Additional file 2: Video S2. A movie of whole-mount confocal Z-projection image stacks of a right side of the brain of *Paralvinella hessleri*. See Figure 7E-I for details. Format: MP4 Size: 1.3 MB.

Additional file 3: Video S3. A movie of whole-mount confocal Z-projection image stacks of a part of branchial crown of *Paralvinella hessleri*. See Figure 9D and E for details. Format: MP4 Size: 954 KB.

Additional file 4: Video S4. A movie of whole-mount confocal Z-projection image stacks of a part of the notopodia and sensilla of *Paralvinella hessleri*. See text and Figure 10B for details. Format: MP4 Size: 1.4 MB.

Additional file 5: Figure S1. Details of the trunk and ciliary types. A. Side view of the trunk showing the distribution of bud type (arrowheads) and possible developing or different type of tuft type cilia (arrows), a SEM image. B. Enlarged view of tuft type cilia (arrows) with the ventral patch type. C. A confocal optical section of the trunk, ventral view. The acetylated alpha-tubulin (acTUBA) positive cilia are seen. The DAPI (nuclei, blue) and phalloidin (actin, red) cell membrane markers used for counterstaining. D and E. Enlarged views of a part of the C showing the ventral patch type cilia and possible tuft types (arrows). vpt, ventral patch type cilia. Scale bar in A, C: 50 μm; B, D, E: 10 μm.

Competing interests
The authors declare that they have no competing interests.

Authors' contributions
SS designed the study, conducted the experimental studies, and wrote the manuscript draft. TY, SM, HT, and ST contributed substantially to the collection of deep-sea animal samples on board, and helped write the manuscript. TM, AO, and KF contributed to the data interpretation and to writing the manuscript. All authors read and approved the final manuscript.

Acknowledgments
The authors would like to thank the captain and crew of R/V "Natsushima" and the operation teams of the ROV "Hyper Dolphin 3000". We are grateful to research scientists on board during cruise NT12-10 for their kind assistance in obtaining the study materials. We particularly thank Koji Inoue, Masatomi Hosoi, Retori Hiraoka, and the Chief Scientist of this cruise, Motohiro Shimanaga. For the electron microscopy studies, we thank Katsuyuki Uematsu and Akihiro Tame. We also thank an anonymous reviewer for reading the manuscript and providing critical comments. Grant sponsor: Japanese Society for the Promotion of Science (No. 90360560) and Japan Agency for Marine-Earth Science and Technology Internal Research & Development Award to SS.

Author details
[1]Department for Marine Biodiversity Research, Japan Agency for Marine-Earth Science and Technology, 2-15 Natsushima-cho, Yokosuka 237-0061, Kanagawa, Japan. [2]Nagahama Institute of Bio-Science and Technology, Institute of Bio-Science and Technology, 1266 Tamura-Cho, Nagahama 526-0829, Shiga, Japan. [3]Nihon University, 1866 Kameino, Fujisawa 252-0880, Kanagawa, Japan. [4]Division of Applied Biosciences, Kyoto University, Graduate School of Agriculture, Laboratory of Marine Biological Function, Kitashirakawa Oiwake-cho, Sakyo-ku, Kyoto 606-8602, Japan.

References
1. Desbruyeres D, Laubier L: **Systematics, phylogeny, ecology and distribution of the Alvinellidae (Polychaeta) from deep-sea hydrothermal vents.** *Ophelia* 1991, 5:31–45.
2. Desbruyeres D, Chevaldonne P, Alayse AM, Jollivet D, Lallier FH, Jouin-Toulmond C, Zal F, Sarradin PM, Cosson R, Caprais JC, Arndt C, O'Brien J, Guezennec J, Hourdez S, Riso R, Gaill F, Laubier L, Toulmond A: **Biology and ecology of the "Pompeii worm"** (*Alvinella pompejana* Desbruyeres and Laubier), **a normal dweller of an extreme deep-sea environment: a synthesis of current knowledge and recent developments.** *Deep-Sea Res Part I* 1998, 45:383–422.
3. Luther GW, Rozan TF, Taillefert M, Nuzzio DB, Di Meo C, Shank TM, Lutz RA, Cary SC: **Chemical speciation drives hydrothermal vent ecology.** *Nature* 2001, 410:813–816.
4. Mary J, Rogniaux H, Rees JF, Zal F: **Response of** *Alvinella pompejana* **to variable oxygen stress: a proteomic approach.** *Proteomics* 2010, 10:2250–2258.
5. Chevaldonne P, Fisher CR, Childress JJ, Desbruyeres D, Jollivet D, Zal F, Toulmond A: **Thermotolerance and the "pompeii worms".** *Mar Ecol Prog Ser* 2000, 208:293–295.
6. Le Bris N, Gaill F: **How does the annelid** *Alvinella pompejana* **deal with an extreme hydrothermal environment?** *Rev Environ Sci Biotechnol* 2007, 6:197–221.
7. Ravaux J, Hamel G, Zbinden M, Tasiemski AA, Boutet I, Leger N, Tanguy A, Jollivet D, Shillito B: **Thermal limit for metazoan life in question: in vivo heat tolerance of the Pompeii worm.** *PLoS One* 2013, 8:64074.

8. Genard B, Marie B, Loumaye E, Knoops B, Legendre P, Zal F, Rees JF: **Living in a hot redox soup: antioxidant defences of the hydrothermal worm** *Alvinella pompejana*. *Aqua Biol* 2013, **18:**217–228.

9. Gaill F, Bouligand Y: **Supercoil of collagen fibrils in the integument of** *Alvinella*, **an abyssal annelid**. *Tissue Cell* 1987, **19:**625–642.

10. Toulmond A, Slitine FE, Defrescheville J, Jouin C: **Extracellular hemoglobins of hydrothermal vent annelids: structural and functional characteristics in three alvinellid species**. *Biol Bull* 1990, **179:**366–373.

11. Sicot FX, Mesnage M, Masselot M, Exposito JY, Garrone R, Deutsch J, Gaill F: **Molecular adaptation to an extreme environment: origin of the thermal stability of the pompeii worm collagen**. *J Mol Biol* 2000, **302:**811–820.

12. Marie B, Genard B, Rees JF, Zal F: **Effect of ambient oxygen concentration on activities of enzymatic antioxidant defences and aerobic metabolism in the hydrothermal vent worm,** *Paralvinella grasslei*. *Mar Biol* 2006, **150:**273–284.

13. Shin DS, Didonato M, Barondeau DP, Hura GL, Hitomi C, Berglund JA, Getzoff ED, Cary SC, Tainer JA: **Superoxide dismutase from the eukaryotic thermophile** *Alvinella pompejana*: **structures, stability, mechanism, and insights into amyotrophic lateral sclerosis**. *J Mol Biol* 2009, **385:**1534–1555.

14. Gagniere N, Jollivet D, Boutet I, Brelivet Y, Busso D, Da Silva C, Gaill F, Higuet D, Hourdez S, Knoops B, Lallier F, Leize-Wagner E, Mary J, Moras D, Perrodou E, Rees JF, Segurens B, Shillito B, Tanguy A, Thierry JC, Weissenbach J, Wincker P, Zal F, Poch O, Lecompte O: **Insights into metazoan evolution from** *Alvinella pompejana* **cDNAs**. *BMC Genomics* 2010, **11:**634.

15. Dilly GF, Young CR, Lane WS, Pangilinan J, Girguis PR: **Exploring the limit of metazoan thermal tolerance via comparative proteomics: thermally induced changes in protein abundance by two hydrothermal vent polychaetes**. *Proc R Soc B* 2012, **279:**3347–3356.

16. Holder T, Basquin C, Ebert J, Randel N, Jollivet D, Conti E, Jékely G, Bono F: **Deep transcriptome-sequencing and proteome analysis of the hydrothermal vent annelid** *Alvinella pompejana* **identifies the CvP-bias as a robust measure of eukaryotic thermostability**. *Biol Direct* 2013, **8:**2.

17. Rittschof D, Forward RB, Cannon G, Welch JM, McClary M, Holm ER, Clare AS, Conova S, McKelvey LM, Bryan P, Van Dover CL: **Cues and context: larval responses to physical and chemical cues**. *Biofouling* 1998, **12:**31–44.

18. Van Dover CL, Szuts EZ, Chamberlain SC, Cann JR: **A novel eye in 'eyeless' shrimp from hydrothermal vents of the Mid-Atlantic Ridge**. *Nature* 1989, **337:**458–460.

19. Nuckley DJ, Jinks RN, Battelle BA, Herzog ED, Kass L, Renninger GH, Chamberlain SC: **Retinal anatomy of a new species of bresiliid shrimp from a hydrothermal vent field on the Mid-Atlantic Ridge**. *Biol Bull* 1996, **190:**98–110.

20. Renninger GH, Kass L, Gleeson RA, Van Dover CL, Battelle BA, Jinks RN, Herzog ED, Chamberlain SC: **Sulfide as a chemical stimulus for deep-sea hydrothermal vent shrimp**. *Biol Bull* 1995, **189:**69–76.

21. Charmantier-Daures M, Segonzac M: **Organ of Bellonci and sinus gland in three decapods from Atlantic hydrothermal vents:** *Rimicaris exoculata*, *Chorocaris chacei*, **and** *Segonzacia mesatlantica*. *J Crust Biol* 1998, **18:**213–223.

22. Richter-Landsberg C, Goldbaum O: **Stress proteins in neural cells: functional roles in health and disease**. *Cell Mol Life Sci* 2003, **60:**337–349.

23. Desbruyeres D, Laubier L: *Alvinella pompejana* **gen. sp. nov., aberrant Ampharetidae from East Pacific Rise hydrothermal vents**. *Oceanol Acta* 1980, **3:**267–274.

24. Storch V, Gaill F: **Ultrastructural observations on feeding appendages and gills of** *Alvinella pompejana* **(Annelida, Polychaeta)**. *Helgoländer Meeresuntersuch* 1986, **40:**309–319.

25. Jouin C, Gaill F: **Gills of hydrothermal vent annelids: structure, ultrastructure and functional implications in two alvinellid species**. *Prog Oceanog* 1990, **24:**59–69.

26. Verger-Bocquet M: **Polychaeta: sensory structures**. In *Microscopic Anatomy of Invertebrates. Vol. 7 Annelida*. Edited by Harrison FW, Gardiner SL. New York: Wiley-Liss; 1992:181–196.

27. Hanström B: *Vergleichende Anatomie des Nervensystems der Wirbellosen Tiere unter Berücksichtigung seiner Funktion*. Berlin, Heidelberg, New York: Julius Springer; 1928.

28. Bullock TH, Horridge GA: *Structure and Function in the Nervous System of Invertebrates. Vol. II*. London: Freeman and Company; 1965.

29. Golding DW: **Polychaeta: nervous system**. In *Microscopic Anatomy of Invertebrates*. Edited by Harrison FW, Gardiner SL. New York: Wiley-Liss; 1992:153–179.

30. Orrhage L, Müller MCM: **Morphology of the nervous system of Polychaeta (Annelida)**. *Hydrobiol* 2005, **535/536:**79–111.

31. Heuer CM, Muller CHG, Todt C, Loesel R: **Comparative neuroanatomy suggests repeated reduction of neuroarchitectural complexity in Annelida**. *Front Zool* 2010, **7:**13.

32. Jones ML, Gardiner SL: **On the early development of the vestimentiferan tube worm** *Ridgeia* sp. **and observations on the nervous system and trophosome of** *Ridgeia* sp. **and** *Riftia pachyptila*. *Biol Bull* 1989, **177:**254–276.

33. Miyamoto N, Shinozaki A, Fujiwara Y: **Neuroanatomy of the vestimentiferan tubeworm** *Lamellibrachia satsuma* **provides insights into the evolution of the polychaete nervous system**. *PLoS One* 2013, **8:**55151.

34. Eichinger I, Hourdez S, Bright M: **Morphology, microanatomy and sequence data of** *Sclerolinum contortum* **(Siboglindae, Annelida) of the Gulf of Mexico**. *Org Div Evol* 2013, **13:**311–329.

35. Read G, Fauchald K: **Polychaeta. Annelida. World Polychaeta database**. [http://www.marinespecies.org/polychaeta/aphia.php?p=taxdetails&id=882], accessed 3/28/2014.

36. Girguis PR, Lee RW: **Thermal preference and tolerance of alvinellids**. *Science* 2006, **312:**231–231.

37. Gardiner SL: **Polychaeta: external anatomy**. In *Microscopic Anatomy of Invertebrates Vol. 7. Annelida*. Edited by Harrison FW, Gardiner SL. New York: Wiley-Liss; 1992:11–17.

38. Purschke G: **Sense organs in polychaetes (Annelida)**. In *Morphology, Molecules, Evolution and Phylogeny in Polychaeta and Related Taxa*, Hydrobiol. 535/536th edition. Edited by Bartolomaeus T, Purschke G. ; 2005:53–78.

39. Grelon D, Morineaux M, Desrosiers G, Juniper SK: **Feeding and territorial behavior of** *Paralvinella sulfincola*, **a polychaete worm at deep-sea hydrothermal vents of the Northeast Pacific Ocean**. *J Exp Mar Biol Eco* 2006, **329:**174–186.

40. Heuer CM, Loesel R: **Immunofluorescence analysis of the internal brain anatomy of** *Nereis diversicolor* **(Polychaeta, Annelida)**. *Cell Tissue Res* 2008, **331:**713–724.

41. Engelhardt RP, Dhainautcourtois N, Tramu G: **Immunohistochemical demonstration of a cck-like peptide in the nervous-system of a marine annelid worm,** *Nereis diversicolor* **Muller, CF**. *Cell Tissue Res* 1982, **227:**401–411.

42. Loesel R, Heuer CM: **The mushroom bodies–prominent brain centres of arthropods and annelids with enigmatic evolutionary origin**. *Acta Zool* 2010, **91:**29–34.

43. Strausfeld NJ: *Arthropod Brains: Evolution, Functional Elegance, and Historical Significance*. Harvard: Belknap Press; 2012.

44. Conzelmanna M, Williams EA, Tunaru S, Randel N, Shahidi R, Asadulina A, Berger J, Offermanns S, Jékely G: **Conserved MIP receptor–ligand pair regulates** *Platynereis* **larval settlement**. *Proc Natl Acad Sci U S A* 2013, **110:**8224–8229.

45. Rinke C, Lee RW: **Pathways, activities and thermal stability of anaerobic and aerobic enzymes in thermophilic vent paralvinellid worms**. *Mar Ecol Prog Ser* 2009, **382:**99–112.

46. Hartline DK, Colman DR: **Rapid conduction and the evolution of giant axons and myelinated fiber**. *Curr Biol* 2007, **17:**29–35.

47. Günther J: **Impulse conduction in the myelinated giant fibers of the earthworm. structure and function of the dorsal nodes in the median giant fiber**. *J Comp Neurol* 1976, **168:**505–532.

48. Schweigreiter R, Roots BI, Bandtlow CE, Gould RM: **Understanding myelination through studying its evolution**. *Int Rev Neurobiol* 2006, **73:**219–273.

49. Roots BI, Cardone B, Pereyra P: **Isolation and characterization of the myelin-like membranes ensheathing giant axons in the earthworm nerve cord**. *Ann NY Acad Sci* 1991, **633:**559–561.

50. Allen NJ, Barres BA: **Neuroscience: glia—more than just brain glue**. *Nature* 2009, **457:**675–677.

51. Zalc B, Goujet D, Colman D: **The origin of the myelination program in vertebrates**. *Curr Biol* 2008, **18:**R511–R512.

52. Witt KA, Mark KS, Hom S, Davis TP: **Effects of hypoxia-reoxygenation on rat blood–brain barrier permeability and tight junctional protein expression**. *Am J Physiol Heart Circ Physiol* 2003, **285:**2820–2831.

53. Del Zoppo GJ: **The neurovascular unit in the setting of stroke**. *J Intern Med* 2010, **267:**156–171.

54. Ronaldson PT, Davis TP: **Blood–brain barrier integrity and glial support: mechanisms that can be targeted for novel therapeutic approaches in stroke**. *Curr Pharm Des* 2012, **18:**3624–3644.

55. Orrhage L: **On the anatomy of the central nervous system and the morphological value of the anterior end appendages of Ampharetidae, Pectinariidae and Terebellidae (Polychaeta)**. *Acta Zool* 2001, **82:**57–71.

56. Korn H: Vergleichend-embryologische Untersuchungen an *Harmothoe* Kinberg (Polychaeta, Annelida). *Z Wiss Zool* 1958, **161**:346–443.

57. Shigeno S, Sasaki T, Haszprunar G: **Central nervous system of *Chaetoderma japonicum* (Caudofoveata, Aplacophora): implications for diversified ganglionic plans in early molluscan evolution.** *Biol Bull* 2007, **213**:122–134.

58. Faller S, Rothe BH, Todt C, Schmidt-Rhaesa A, Loesel R: **Comparative neuroanatomy of Caudofoveata, Solenogastres, Polyplacophora, and Scaphopoda (Mollusca) and its phylogenetic implications.** *Zoomorph* 2012, **131**:149–170.

59. Orrhage L: **On the structure and homologues of the anterior end of the polychaete families Sabellidae and Serpulidae.** *Zoomorph* 1980, **96**:113–168.

60. Desbruyeres D, Laubier L: *Paralvinella hessleri*, **new species of Alvinellidae (Polychaeta) from the Mariana Back-Arc Basin Hydrothermal Vents.** *Proc Biol Soc Wash* 1989, **102**:761–767.

61. Tsuchida S, Suzuki Y, Fujiwara Y, Kawato M, Uematsu K, Yamanaka T, Mizota C, Yamamoto H: **Epibiotic association between filamentous bacteria and the vent-associated galatheid crab, *Shinkaia crosnieri* (Decapoda: Anomura).** *J Mar Biol Ass UK* 2011, **91**:23–32.

62. Richter S, Loesel R, Purschke G, Schmidt-Rhaesa A, Scholtz A, Stach T, Vogt L, Wanninger A, Brenneis G, Doring C, Faller S, Fritsch M, Grobe P, Heuer CM, Kaul S, Moller OS, Muller CHG, Rieger V, Rothe BH, Stegner MEJ, Harzsch S: **Invertebrate neurophylogeny: suggested terms and definitions for a neuroanatomical glossary.** *Front Zool* 2010, **7**:29.

8

Overcoming the fragility – X-ray computed micro-tomography elucidates brachiopod endoskeletons

Ronald Seidel[1,2]* and Carsten Lüter[2]

Abstract

Introduction: The calcareous shells of brachiopods offer a wealth of informative characters for taxonomic and phylogenetic investigations. In particular scanning electron microscopy (SEM) has been used for decades to visualise internal structures of the shell. However, to produce informative SEM data, brachiopod shells need to be opened after chemical removal of the soft tissue. This preparation occasionally damages the shell. Additionally, skeletal elements of taxonomic/systematic interest such as calcareous spicules which are loosely embedded in the lophophore and mantle connective tissue become disintegrated during the preparation process.

Results: Using a nondestructive micro-computed tomography (μCT) approach, the entire fragile endoskeleton of brachiopods is documented for the first time. New insights on the structure and position of tissue-bound skeletal elements (spicules) are given as add ons to existing descriptions of brachiopod shell anatomy, thereby enhancing the quality and quantity of informative characters needed for both taxonomic and phylogenetic studies. Here, we present five modern, articulated brachiopods (*Rectocalathis schemmgregoryi* n. gen., n. sp., *Eucalathis* sp., *Gryphus vitreus, Liothyrella neozelanica* and *Terebratulina retusa*) that were X-rayed using a Phoenix Nanotom XS 180 NF. We provide links to download 3D models of these species, and additional five species with spicules can be accessed in the Supplemental Material. In total, 17 brachiopod genera covering all modern articulated subgroups and 2 inarticulated genera were X-rayed for morphological analysis. *Rectocalathis schemmgregoryi* n. gen., n. sp. is fully described.

Conclusion: Micro-CT is an excellent non-destructive tool for investigating calcified structures in the exo- and endoskeletons of brachiopods. With high quality images and interactive 3D models, this study provides a comprehensive description of the profound differences in shell anatomy, facilitates the detection of new delicate morphological characters of the endoskeleton and gives new insights into the body plan of modern brachiopods.

Keywords: Brachiopoda, Endoskeleton, Spicules, X-ray, Micro-computed tomography (μCT), Interactive 3D model, *Rectocalathis schemmgregoryi* n. gen., n. sp.

Introduction

The morphology of brachiopod shells has been of interest to scientists for more than two centuries. In articulated brachiopods such as Rhynchonellida, Thecideida and Terebratulida the shell develops through calcite biomineralisation processes. Both dorsal and ventral valves usually consist of two (sometimes three) calcite layers comprising an outer, hard and protective nanocrystalline layer and a much thicker, inner secondary layer built from a hybrid organic/inorganic fibre composite material [1]. As outgrowth from this secondary shell layer of the dorsal valve [2], the brachidium forms a lophophore support [3] consisting of the crura and the loop-forming brachidia (Figure 1A). It is ensheathed in outer epithelium which controls growth by simultaneous secretion and resorption [4]. However, there are also calcareous structures within the soft tissue in many exclusively articulated brachiopods which are never associated with the two calcareous valves. Those structures are referred to

---

* Correspondence: ronald.seidel@mpikg.mpg.de
[1]Current address: Max Planck Institute of Colloids and Interfaces, Potsdam-Golm Science Park, Am Mühlenberg 1 OT Golm, 14476 Potsdam, Germany
[2]Museum für Naturkunde, Leibniz Institute for Evolution and Biodiversity Science, Invalidenstrasse 43, 10115 Berlin, Germany

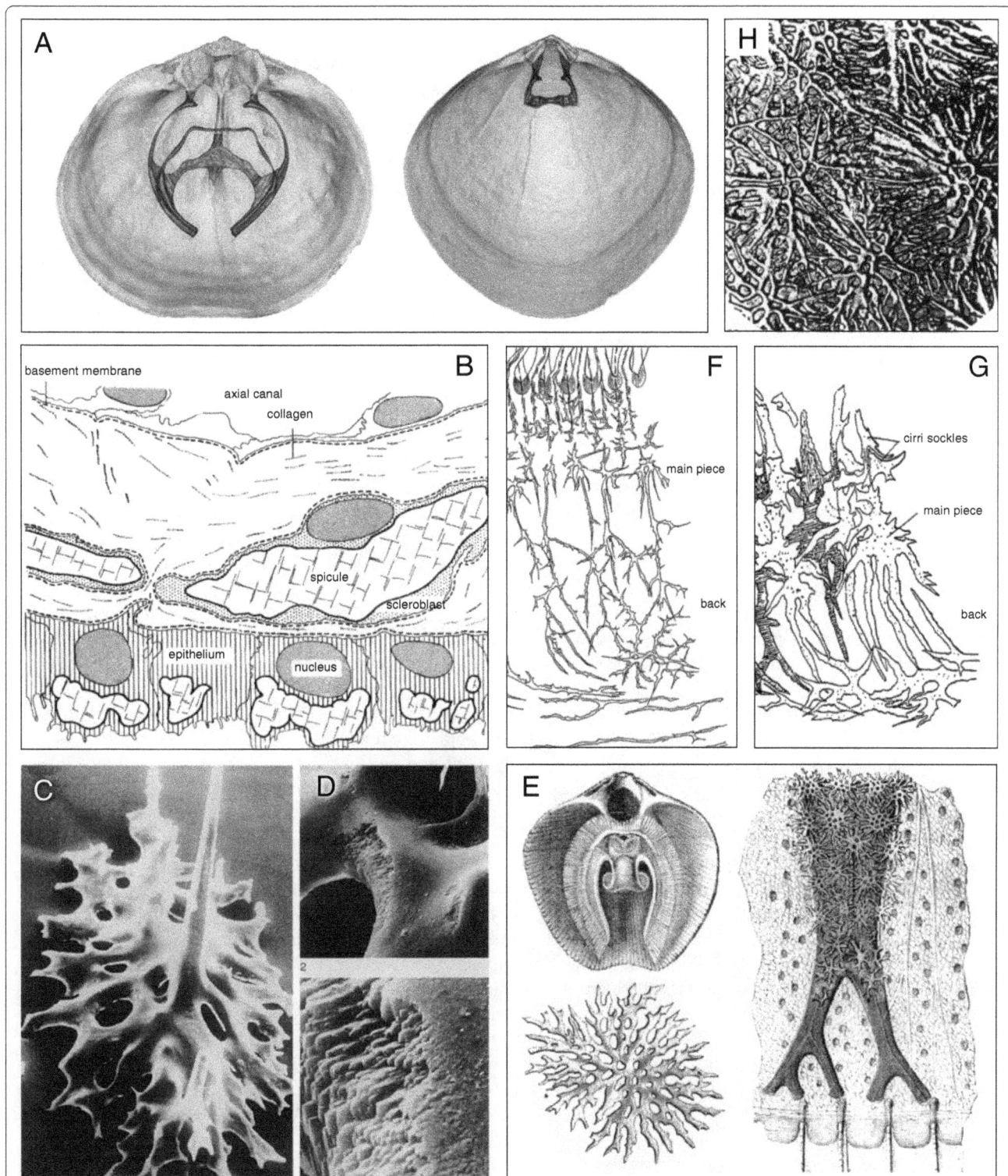

Figure 1 Mineralized endo- and exoskeleton of brachiopods. A) Interior of dorsal valves with "long loop" of *Calloria inconspicua* (Sowerby, 1846) (left) and "short loop" of *Liothyrella neozelanica* Thomson, 1918 (right) (R. Seidel). **B)** Stylised section showing the distribution of spicules in the epithelium and connective tissue of the lophophore of *Terebratulina retusa* (Linnaeus, 1758) according to Williams [4]. **C/D)** SEM of a spicule from the mantle of *T. retusa* with **(D)** the inner layer structure exposed in an area of recent biomineralisation [3]. **E)** The lophophoral endoskeleton and spicules in the mantle covering the pallial sinus of *T. retusa* by Eudes-Deslongchamps [7]. **F/G)** Spicules in the lophophore **F)** without and **G)** with cirri sockles at the bases of the tentacles [8]. **H)** Spicules of the dorsal body wall of *L. neozelanica* [12].

as spicules and are defined as small irregular bodies of calcite secreted by scleroblasts within the connective tissue of the mantle and lophophore [4] (Figure 1B). The spicules are products of endodermal-mesenchymal cells [5] and are therefore a true endoskeleton. Inside a scleroblast, a spicule is separated from all other cell contents, presumably by a protein layer. The surface of a spicule is a rather smooth and thin calcite layer, but the actual body consists of highly declined laminae [4] (Figure 1C, D) which in turn are composed of calcitic spherules about 200 nm in diameter [6].

A true spicular endoskeleton is found only in articulated brachiopods where it is strongly developed in all species of the Terebratulidae, Platidiidae and Kraussinidae. In the families Megathyrididae, Dallinidae, Laqueidae and Thecideidae it is only developed in some species, and it does not occur at all in the Terebratellidae [4]. Spicules were first described in detail by Hancock [9] in *Terebratulina retusa* (Linnaeus, 1758), by Lacaze-Duthiers [10] in *Lacazella mediterranea* (Risso 1826) and extensively by Eudes-Deslongchamps [8] in *T. retusa* (Figure 1E), *Kraussina rubra* (Pallas, 1766), *Gryphus vitreus* (Born, 1778), *Megerlia truncata* (Linnaeus, 1767) and *Platidia anomioides* (Scacchi & Philippi in Philippi, 1844). Based on comprehensive investigations of the material obtained from the Valdivia- (1898-1899) and Gauß-Expeditions (1901-1903), F. Blochmann [8,11] tested the usefulness of the calcareous bodies for species diagnostics when other shell characters fail (Figure 1F, G). Despite existing illustrations of minute single spicules [12] (Figure 1H), the fragility of the endoskeleton limits the ability to visualise and therefore understand the anatomy of the endoskeleton as a whole. However, understanding the anatomy of the endoskeleton is essential for taxonomic and systematic analyses, as well as for any hypotheses on the functional properties of the endoskeleton of brachiopods.

Scanning electron microscopy (SEM) has been used for decades to visualise biological structures of interest in brachiopod shells, such as muscle scars, punctae, teeth, crura and the brachidium [4,13] and references therein. However, this method is limited by the fact that it is only surface-sensitive. Studying hard tissue of modern brachiopods with SEM requires treatment with chemical dissolvers to remove the soft tissue, which can lead to the damage of the shell. Additionally, endoskeletal elements such as calcareous spicules which are loosely embedded within the lophophore and mantle connective tissue disintegrate during the preparation process. Despite exhaustive research on hard tissue of brachiopods using high resolution SEM-imaging, a pictured overview of the occurrence, abundance and exact spatial distribution of endoskeletal spicules within the soft body has been lacking – until now.

In order to understand the inner shell morphology of either fossil and modern brachiopod shells X-ray radiographs of animals have been used since the 1970's [14-16]. Hagadorn [17] was the first to use micro-computed tomography (µCT) and reconstructed 3D models to visualise the coiled lophophore support structures inside pyritized *steinkerns* of the Silica Shale brachiopod *Paraspirifer bownockeri* (Stewart, 1927). With increasing resolution and practicability of X-ray devices within the last decade computed tomography has become a valuable adjunct technique. A few studies have been carried out since on brachiopod shell morphology using micro-computed tomography ([18-22] as well as synchrotron radiation X – ray tomographic microscopy (SRXTM), which provides an even higher image fidelity [23,24].

The majority of analyses of fossil brachiopods using a microtomograph were carried out by Pakhnevich [18-21]. In 2010 he included 6 modern species in his investigations showing developmental series of the brachidium of *Macandrevia cranium* (Müller, 1776), the spiculation of the mantle connective tissue of the ventral valve of *T. retusa* and scans of *Eucalathis murrayi* (Davidson, 1878). Unfortunately, details of the scanning parameters and the scanning set up are missing (e.g. scans performed wet or dry, in ddH$_2$O or EtOH). The presented images do not include all the information µCT scans actually are able to provide. For instance, data on spiculation within the lophophore and mantle *of T. retusa* or *E. murrayi* is limited, although micro computed tomography appears highly applicable for such tasks.

By using the non-invasive µCT approach, we were able to document the internal shell anatomy and to visualise the shells of modern brachiopods. We added structural and positional data on tissue-bound skeletal elements (e.g. spicules) to existing descriptions of brachiopod shell anatomy. We scanned long-term preserved specimens from the historical wet collection of the Museum für Naturkunde Berlin, Germany as well as recently collected samples which represent 19 brachiopod genera covering all articulated subgroups and two genera of (inarticulated) craniids. Here, we illustrate whole mount scans of five representative species to underline the capability of the non-invasive µCT approach for visualising hard tissue of brachiopods and present new characters of the endoskeleton with a SEM-comparable resolution of approximately 10 µm. Comprehensive descriptions of the spicule arrangements within the brachiopod soft tissue are given in the results chapter (for results on the remaining 5 species with spicules see Supplemental Material). One of the five species illustrated in detail was found during the deep-sea expedition SO 205 "Mangan" on manganese nodules in the Central East Pacific and is described as a new species *Rectocalathis schemmgregoryi* n. gen., n. sp.

Results

Modern brachiopods exhibit manifold spiculation. The spicules vary in shape and spatial distribution within the soft tissue, covering almost the entire visceral cavity or only parts of it. The spicules are readily visible within reconstructed 2D image stacks and in rendered three-dimensional visualizations. Scan parameters have to be set specifically for each specimen due to shell differences in thickness and density. The spiculation varies within the tissue of single specimens (e.g. in *P. anomioides* and *Pajaudina atlantica* Logan, 1988). With respect to enhanced character visualisation, the valves and internal characters of all scanned species are coloured. Please note that the unified colouration of brachidia, crura and other lophophoral supportive structures such as median lamellae or peribrachial ridge refers to the functional properties of lophophoral support, not to homology. The terminology used in this study is in accordance with the terminology of the Treatise of Invertebrate Paleontology, Part H, Brachiopoda Revised, Vol. 1. Please use the Hand Tool in Adobe Acrobat to activate the 3D models and to use all intended functions (see tool bar > Enable virtual sectioning). For specimens used in this study see Table 1 and Additional file 1: Table S1. Please see Table 2 for volume analyses and morphometric measurements of the species shown in this paper and Additional file 2: Table S2 for volume analyses and morphometric measurements of all species of the initial study.

Terebratulina retusa (Cancellothyrididae) – ZMB Bra 2253

Spiculation - Spicules develop within the lophophore, and both the dorsal and ventral mantle (Figure 2A, Additional file 3). The spiculation throughout the soft tissue is strong. Within the plectolophous lophophore both lateral and spiral brachia are strongly spiculated at the bases of the tentacles, the brachial lip and the proximal part of the tentacles (Figure 2A, C, D). Between the posterior part of the spiral brachia the primarily isolated long and slender spicules are bigger and multibranched, building a network, which is most dense just in front of the mouth (Figure 2F). The brachial canals are almost completely surrounded by spicules (Figure 2E). The spiculation within the dorsal and ventral mantle is equally strong (Figure 2B, C), and consists of star-shaped spicules (Figure 2G, I, K). Towards the margin the size of the spicules decreases noticeably but the gaps between single spicules increase slightly, thus the entire mantle cavity is sheltered in an endoskeleton compartment. The rear of the spiculated body wall is triple keeled or ridged (Figure 2H, J). Along with the ridges, long and branched spicules reach into the visceral cavity (Figure 2C arrows).

Rectocalathis schemmgregoryi n. gen., n. sp. (Chlidonophoridae) – ZMB Bra 2254

Systematic affiliation:

Phylum BRACHIOPODA Duméril, 1806
Subphylum RHYNCHONELLIFORMEA Williams & others, 1996
Class RHYNCHONELLATA Williams & others, 1996
Order TEREBRATULIDA Waagen, 1883
Suborder TEREBRATULIDINA Waagen, 1883
Superfamily CANCELLOTHYRIDOIDEA Thomson, 1926
Family CHLIDONOPHORIDAE Muir-Wood, 1959
Subfamily EUCALATHINAE Muir-Wood, 1965
Genus *Rectocalathis* n. gen.

Type species - *Rectocalathis schemmgregoryi* n. sp.

Etymology - Genus named after its anatomically correct orientation (from Latin: *rectus* = upright, straight) when attached to its natural substrate due to the unusual position of the foramen.

Diagnosis - Small eucalathine brachiopod with a smooth to faintly costate shell and an extreme form of epithyrid (subumbonal) pedicle opening towards the ventral side of the animal.

Species *Rectocalathis schemmgregoryi* n. sp.

Type material - Holotype: ZMB Bra 2254, Paratypes: ZMB Bra 2350, 2352 (total: 3 specimens).

Type locality - NE Pacific, Clipperton Region, 11°47′56″N 116°49′85″W, Cruise SO 205 MANGAN, Station EBS 40, depth: 3954 m, May 5th 2010.

Etymology - Species named after the late brachiopod researcher Mena-Daniela Schemm-Gregory, who unexpectedly died in July 2013, much too early in her promising career.

Distribution - So far only known from the type locality.

Diagnosis - Punctate and small, subcircular to pentagonal shell with complete oval foramen perforating the ventral valve. Dorsal valve with long, slightly medially incurved, thick crura with blunt crural processes, no loop. Heavy spiculation within the lophophore and tentacles, and fragile median spiculation in the dorsal mantle.

Description - From ventral, shell subcircular, well rounded anterior margin, lateral margin posteriorly straight and curved towards the front. Anterior commissure rectimarginate or slightly uniplicate, lateral commissure straight (Figure 3A, B, Additional file 4). Dorsal umbo projects beyond the hinge line (Figure 3A), beak long and erect (Figure 3B). From posterior, ventral valve with wide, well differentiated symphytium, triangular, with growth lines (Figure 3E right). Dorsal valve convex, ventral valve wedge-shaped (Figure 3E middle). Both valves posteroaxially keeled, but dorsal valve less sharp than ventral valve (Figure 3E left). Surface with faint costae (n = 14-16), fine growth lines, punctae (anterior margin ~134/mm², in

Table 1 Species presented within this article

Species	Location	Date	GPS	Depth (m)	Collector	ZMB
T. retusa	Norway	-	-	-	Sars	Bra 2253
R. schemmgregoryi n. gen., n. sp.	SO205/St. 40, NE Pacific	05.05.2010	011°47′56″N 116°49′85″W	3954	P. Martinez Arbizu	Bra 2254
Eucalathis sp.	RV MSM19/1076-1, S-Atlantic, SW of Africa	04.12.2011	040°22.46′S 014°53.77′E	2018	N. Furchheim	Bra 2351
G. vitreus	Naples, Italy, Mediterranean Sea	-	-	-	F. Blochmann	Bra 2261
L. neozelanica	Doubtful Sound, Bauza Island, NZ	12.06.1996	045°22′03″S 167°00′70″E	-	P. Meredith	Bra 2255

For the remaining species of the initial study (see Additional file 1: Table S1).

front of protegulum ~69/mm^2). Shell size in mm L × W × H (dorsal/ventral valve): 4/4.5 × 4/4 × 0.8/1.

Dorsal valve interior (Figure 3A, C). – Hinge line almost straight and wide (0.57 × valve width), sockets large and both ridges almost parallel to hinge line. Outer socket ridges long, narrow, submarginal, inner socket ridges minute, fused with large cardinal flange. Dorsal umbo projects beyond the hinge line, cardinal process large, anteroaxially labiate, laterally fused with cardinal flange. Crura bases broad, crura thick, straight or slightly convergent to the front, from ventral view axes of crural processes slightly rotated from the axis of the crura and descending braches. Surface with punctae, no tubercles, faint median septum in the posterior half, and faint depression posterior of the centre of the valve where the dorsal mantle spiculation is located.

Inside of ventral valve (Figure 3B). – Deltidial plates conjoint, no conjunction line, symphytium abruptly curved posteriorly. Hinge teeth with elongate base, almost parallel to the hinge line, with outer, lateral base submarginal, inner base elongated towards the foramen, distal ends pointed. Surface with punctae, no tubercles. Symmetrical muscle scars within depressed area of the keel between foramen and a thick median groove (mg) within the middle of the posterior half.

Spiculation - Spicules develop within the lophophore, the tentacles and the dorsal mantle (Figure 3C). The spiculation within the schizolophous lophophore is strong, and takes up almost the entire volume of the mantle cavity (Figure 3F, G). It consists of a smaller,

centred, radiating network of thick and multibranched spicules and two outer "rings" which support the lophophore with its two arms bearing the strongly spiculated tentacles (Figure 3D). The radiating structure is positioned in between the lophophoral arms as a symmetrical web of spicules supporting all parts of the arms of the lophophore. On the anterior side, the arms of the lophophore are curled ventrally and separated. Posteriorly, they are fused leaving a large opening for the mouth. The spicules of the tentacles are half-cylinder shaped. The lophophoral endoskeleton is positioned on the anterior third of the ventral side of the crura. Posteriorly, two excrescences of the endoskeleton extend next to the broad crural processes to support the endoskeleton's orientation (Figure 3D). The dorsal mantle exhibits only a faint but distinct spiculation, which is positioned medially just below the tips of the crura, and slopes anteriorly (Figure 3G). The spicules are slender and delicate, and too fragile to support the loop holding the weight of the lophophoral endoskeleton. In contrast to the spiculation in *Eucalathis* sp., these spicules show a distinct morphology composed of two parts (Figure 3G1 and G2). The anterior part is in one plane with the anterior body wall and the posterior part is oriented dorsoventrally, reaching axially into the visceral cavity. While the spicules of the anterior part are short and form a dense network, the spicules of the posterior part are rather long and less branched. The mantle spiculation is not connected to the floor of the valve or to the lophophoral endoskeleton.

Table 2 Volume analyses and morphometric measurements using VG Studio Max 2.1

Species	Vol. Total	Vol. DV	Vol. VV	Vol. Spic Loph.	Vol. Spic. DV	Vol. Spic VV	DV Length	VV Lenght	DV Width	VV Width	DV Heigh	VV Heigh
T. retusa	194,2	88,4	104,5	1,8	1,1	1,23	16,4	18,2	14,7	14,7	3,6	5
R. schemmgregoryi n. gen., n. sp.	3,8	1,5	2	0,2	0,01	×	4	4,5	4,4	4,4	0,8	1
Eucalathis sp.	na	na	na	na	na	na	3,8	4,3	4	4	1,1	1,5
G. vitreus	682,2	223,3	441,2	5	7,1	1,7	29,3	32,6	28,5	28,5	7,6	10,8
L. neozelanica	1399	551	772,7	7,6	34	26	38,7	42,3	40	40	7,6	13

Volumes are in mm^3 and length in mm. Abbr.: DV = dorsal valve, *na* = not applicable, VV = ventral valve, Spic Loph = spiculation within lophophore, *Spic* DV/VV spiculation within mantle of dorsal/ventral valve, Vol = Volume, Vol. total = Vol. VV + DV + Loph + Spic, × = no spiculation. For volume analyses and morphometric measurements of all species of the initial study see Additional file 2: Table S2.

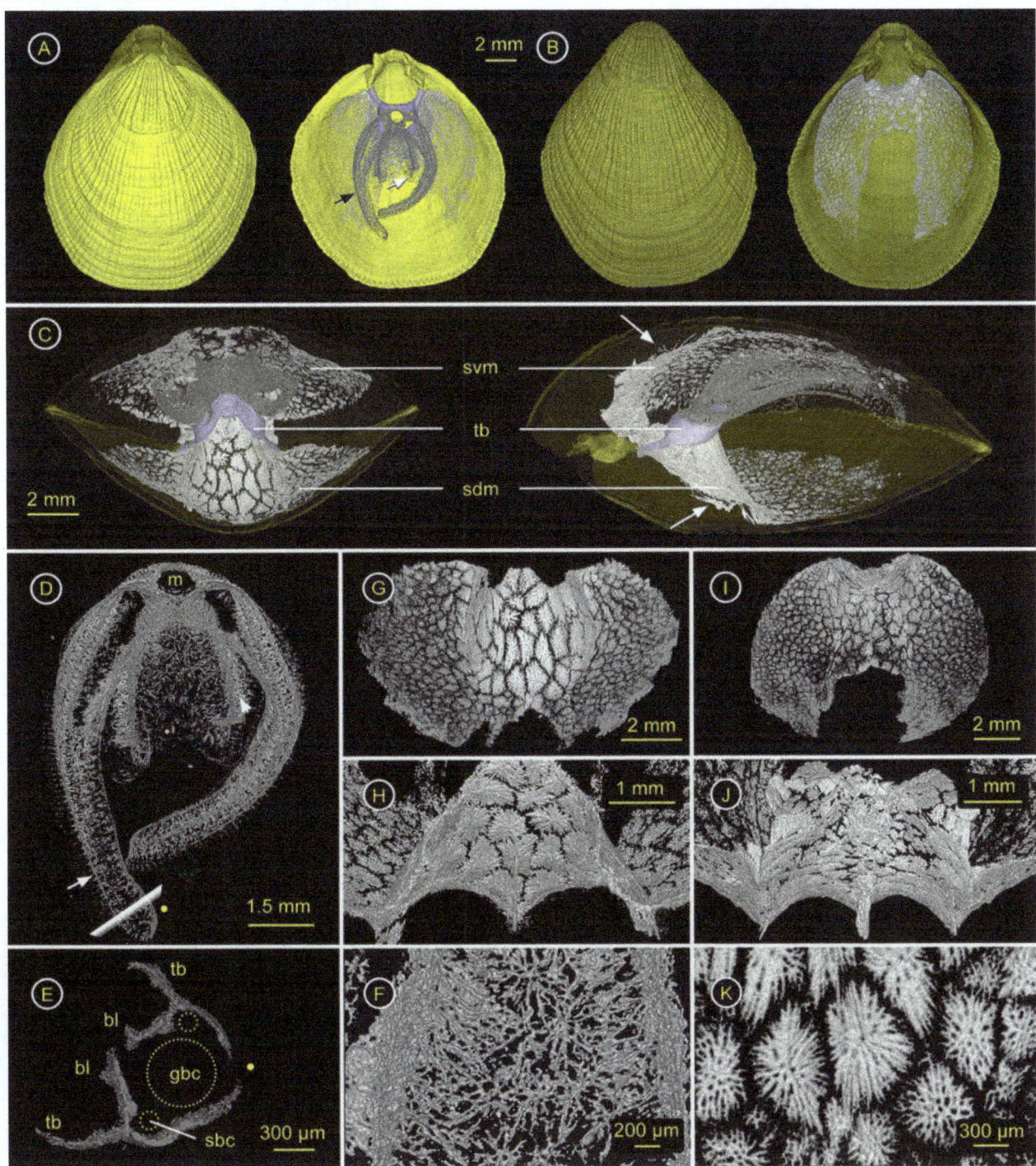

Figure 2 (See legend on next page.)

(See figure on previous page.)
Figure 2 *Terebratulina retusa* (**Cancellothyrididae**) – **ZMB Bra 2253.** Ventral valve = top & dorsal valve = bottom. Click **here** to download a 3D model of *T. retusa* (interactive PDF). **A)** left: Dorsal valve with ventral valve in the background; right: Interior of the dorsal valve with brachidium (purple), endoskeleton of the plectolophous lophophore (grey, black arrow = left lateral brachium, white arrow = right spiral brachium) and spiculation of the dorsal mantle. **B)** left: Ventral valve; right: Inside of ventral valve showing the incomplete, mesothyrid foramen and mantle spiculation. **C)** Anterior view through transparent shell with ventral valve on top, transverse band (tb) of the brachidium (purple) supports heavy spiculated lophophore (dark gray), ventral (svm) and dorsal mantle spiculation (sdm); right: Lateral view through transparent shell showing the short brachidium (purple), the endoskeleton of the ventrally ascending, plectolophous lophophore (dark gray), and the heavy mantle spiculation with excrescences on the posterior side into the visceral cavity (arrows). **D)** Ventral view of plectolophous lophophoral endoskeleton, left lateral brachium (arrow), right spiral brachium (arrow head), mouth (m). **E)** Cross section through lateral brachium (see **D)** for orientation), tentacle bases (tb), brachial lip (bl), and location of great (gbc) and small brachial channel (sbc). **F)** Spicules between the spiral brachia in front of the mouth. **G)** Anterior view of spicules within the dorsal body wall, smaller spicules (~0.6 mm in diameter) towards the shell margin and thick and large (~1.2 mm) star-shaped spicules in the center forming a posteriorly triple keeled skeleton, with **(H)** excrescences of single, elongated and branched spicules along the ridges, extending into the visceral cavity. **(I)** Anterior and posteroventrally **(J)** view of spiculation in the ventral mantle. **(K)** Magnification of star-shaped spicules from the median part of the dorsal mantle, max. size ~1.2 mm in diameter and max. 150 μm in thickness.

Remarks - The spiculation pattern of *Rectocalathis schemmgregoryi* n. gen, n. sp. easily identifies it as a member of Eucalathinae. Especially the spicules of the dorsal mantle epithelium, which are not connected to any other spicules of the internal skeleton, are a characteristic of this group (see also *Eucalathis* sp.). Within Eucalathinae, four genera *Eucalathis* Fischer & Oehlert, 1890, *Bathynanus* Foster, 1974, *Notozyga* Cooper, 1977, and *Nanacalathis* Zezina, 1981, are described [25]. All species within these genera have disjunct deltidial plates resulting in a hypothyrid foramen, except *Nanacalathis* which has conjunct deltidial plates forming a mesothyrid foramen. *Rectocalathis schemmgregoryi* n. gen, n. sp. also has conjunct deltidial plates, but an extreme form of an epithyrid foramen with the pedicle opening lying in a subumbonal position on the ventral side of the animal. Comparable to *Nanacalathis* spp., *Rectocalathis schemmgregoryi* n. gen., n. sp. lacks a brachidial loop, but it clearly differs from *Nanacalathis* through its nearly smooth shell. So far, *Rectocalathis schemmgregoryi* n. gen., n. sp. has only been found in the deep sea at a single station during the MANGAN-Expedition of the German research vessel SONNE (SO 205). Same as *Bathynanus* and *Nanacalathis*, it is expected that *Rectocalathis* n. gen. is also restricted to deep water.

Eucalathis sp. (Chlidonophoridae) – ZMB Bra 2351

Spiculation - Spicules develop within the lophophore, the tentacles (Figure 4A, Additional file 5) and within the dorsal mantle (Figure 4C, D). No spicules were found in the ventral valve (Figure 4B). The lophophoral endoskeleton is massive, dense and large, occupying a considerable volume within the mantle cavity (Figure 4C). It consists of a smaller, inner supportive structure made of large, thick and long, branched spicules and two outer rings which support the lophophore with its two arms bearing the tentacles (Figure 4D). The inner structure with its semicircle-shape fits with a minimal distance on the short looped brachidium and is connected posteriorly to

the outer rings of the two arms. Because it fits on the inner side of the brachidium, the heavy lophophoral endoskeleton maintains its position but is also capable of movement due to the gap to the loop. Additionally, posterior outgrowths extend below the broad crural processes to support the endoskeleton's orientation (Figure 4A). The lophophore bears an alternating row of tentacles. Whereas, the outer tentacles show a rich spiculation throughout (Figure 4C middle), the inner tentacles lack spicules. The conjoint spicules of the outer tentacles are approximately ~30 μm in width and half-cylinder shaped (Figure 4E). The dorsal mantle exhibits only a faint but distinct spiculation, which is positioned medially just below the brachidium at the apical loop, sloping anteriorly (Figure 4C, D). The spicules are slender and delicate, and too fragile to support the loop, which holds the weight of the lophophoral endoskeleton. The mantle spiculation is not connected to the floor of the valve or to the lophophoral endoskeleton.

Gryphus vitreus (Terebratulidae) – ZMB Bra 2261

Spiculation - Spicules develop within the lophophore, and the dorsal and ventral mantle (Figure 5A, B, Additional file 6). Spicules at the tentacle bases and within the tentacles are absent in this specimen. The spiculation within the lophophore is generally faint (Figure 5D), and restricted more or less to the area below the bases of the tentacles, with elongated spicules called basal crystals (Figure 5G). In front of the crural processes the lophophoral spiculation exhibits a compact, plate-shaped structure with a transverse projection (Figure 5E). The spiculation within the ventral valve is faint too, and only few axial spicules were found (Figure 5C, F). In contrast, the spiculation within the dorsal mantle is strong below the short loop of the brachidium (Figure 5C), with plate-shaped, star-shaped and perforated spicules (Figure 5D). No spiculation within the dorsal mantle covering the periphery of the visceral cavity and the pallial vessels towards the margin of the valve could be found.

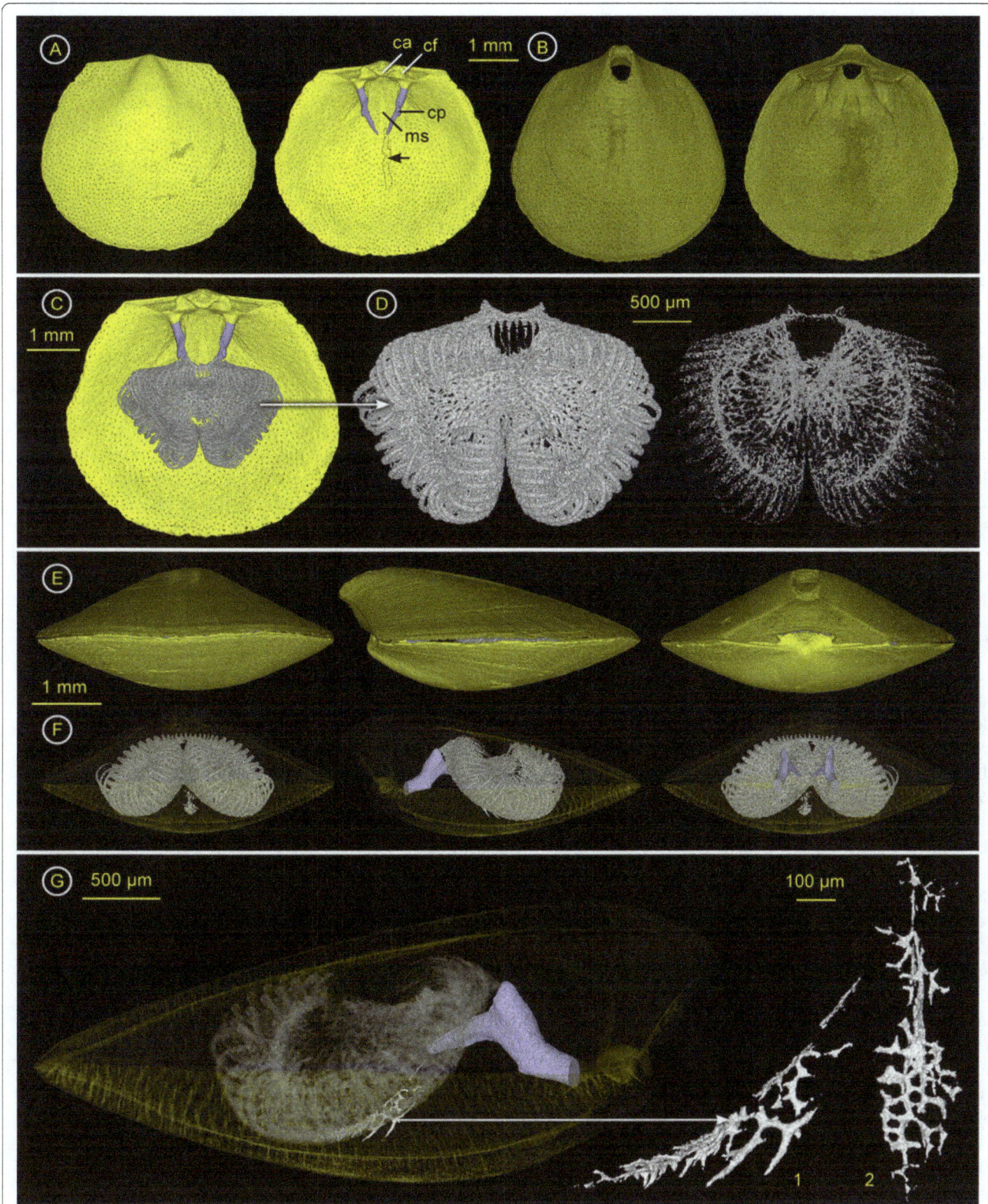

Figure 3 (See legend on next page.)

(See figure on previous page.)
Figure 3 *Rectocalathis schemmgregoryi* **n. gen., n. sp. (Chlidonophoridae) – ZMB Bra 2254.** Ventral valve = top & dorsal valve = bottom. Click **here** to download a 3D model of *R. schemmgregoryi* n. gen., n. sp. (interactive PDF). **A) Left**: Outside of the dorsal valve. **Right**: Inside of dorsal valve without the lophophoral endoskeleton. Shell with faint depression posteriorly of the center where the dorsal mantle spiculation is located. Crura purple. **B) Left**: Outside of the ventral valve. **Right**: Inside of ventral valve. **C)** Inside of the dorsal valve with lophophoral endoskeleton and spicules of the tentacles. **D) Left**: Ventral view of lophophoral endoskeleton, showing spiculated tentacles and heavy spiculation inside the loops. Posteriorly of the mouth, two symmetrical extensions project towards the crural processes. **Right**: Ventral view of "strongly purged" lophophoral endoskeleton to visualize the geometrical nature. Strongest speculation in the centre of the lophophore, spreading out from anterior of the mouth, to support the lophophoral arms, their strongest spiculation at the base of the tentacles. **E) Left**: Anterior view of whole specimen. **Middle**: Lateral view of whole specimen. **Right**: Posterior view of whole specimen. **F) Left**: Anterior view through transparent shell. **Middle**: Lateral view through transparent shell, **Right**: Posterior view through transparent shell. **G) Left**: Lateral view through transparent shell. Spicules of the dorsal body wall (white) are visible due to transparency of the lophophoral endoskeleton (arrow). **Right**: Magnification of median spicules of the dorsal body wall. 1) lateral view, 2) anterior view.

Liothyrella neozelanica (Terebratulidae) – ZMB Bra 2255

Spiculation - *L. neozelanica* develops spicules within the lophophore, ventral and dorsal body wall (Figure 6A, B, Additional file 7). The mineralisation within the plectolophous lophophore is minimal (Figure 6D, E). Only in the posteroaxial part of the lophophoral endoskeleton more spicules are developed which cluster. Here, as in *G. vitreus*, a compact, plate-shaped structure with a transverse projection can be found (Figure 6E). The shape of single spicules within the arms of the lophophore could not be identified unambiguously. The spiculation within the dorsal and ventral valve is unequal (Figure 6C). While the ventral valve exhibits a very faint and loose network of more slender, branched, and irregularly shaped spicules, the body wall of the dorsal valve shows a rich spiculation (Figure 6D). Large and thick, star-shaped and closely spaced spicules cover the visceral cavity below the brachidium (Figure 6F). Especially laterally, close to the brachidium, the spiculation is significantly stronger than anywhere else within the mantle. Spiculation within the dorsal mantle covering the periphery of the visceral cavity and the pallial vessels towards the margin of the valve, could not be found.

Full data sets and additional material

Figures and interactive 3D models of additional 5 species with spicules (*Laqueus rubellus* (Sowerby, 1846) – Additional files 8 and 9, *M. truncata* – Additional files 10 and 11, *P. anomioides* – Additional files 12 and 13, *Pumilus antiquatus* Atkins, 1958 – Additional files 14 and 15, *P. atlantica* – Additional files 16 and 17) can be accessed as Online Supplemental Material, as well as a phylogenetic tree and tables concerning the scan parameters and morphometric analyses of all species of the initial study.

Discussion

Applicability of micro-computed tomography for the study of modern brachiopods

Scanning modern brachiopods with an X-ray microcomputed tomograph (μCT) creates naturally aligned virtual sections (2D) and interactive 3D models. μCT is a unique tool for obtaining and visualising structural

and positional data of the spiculation and the inner shell morphology of modern brachiopods. The present μCT data visualises the endoskeleton and the shell with high accuracy and natural orientation. In a short period of time, the endoskeletal spiculation in 10 out of 19 investigated genera was detected by analyzing stacks of 2 dimensional images. The scanning and reconstruction is a clean and quick procedure, whereas the subsequent virtual sectioning and colouration necessary to create 3D models is much more time consuming, but is particularly important for understanding the three dimensional architecture of structures, linking colours to regions of interest and it results in an unlimited set of views. Hard parts of the animals can be easily defined and marked, whereas soft tissue visualization needs staining.

With the achieved resolution of 10 microns the anatomy of single spicules and differences in the surface texture (e.g. muscle scars) can be visualised. However, the quality of images depends significantly on the rendering algorithm that is applied, which is in turn linked to the computing power and total operation time. In this study the volume renderer PHONG and an isosurface renderer were used. The latter does not affect the CPU as much as the volume render and yields good results for simulating transparency. The chosen scan parameters represent appropriate balances between so called "fast scans" (~15 minutes) and scans of longer duration (>50 minutes) which resulted in data of more than 20 GB / file). The presence of minute structures such as spicules and punctae canals could easily be detected with the scan parameters that were applied, in whole mount scans of any specimen size.

μCT (Phoenix Nanotom XS 180 NF) produces high resolution images of brachiopods visualising their mineralized exo- and endoskeleton and allows rotational analysis of the specimen. Depending on the condition of the animal and the scan duration (scan parameters, such as number of images, timing, "average and skip") μCT images are comparable to SEM images by showing the full set of informative characters of the shell, e.g. muscle scars, punctae, growth lines, costae, tubercles, septae,

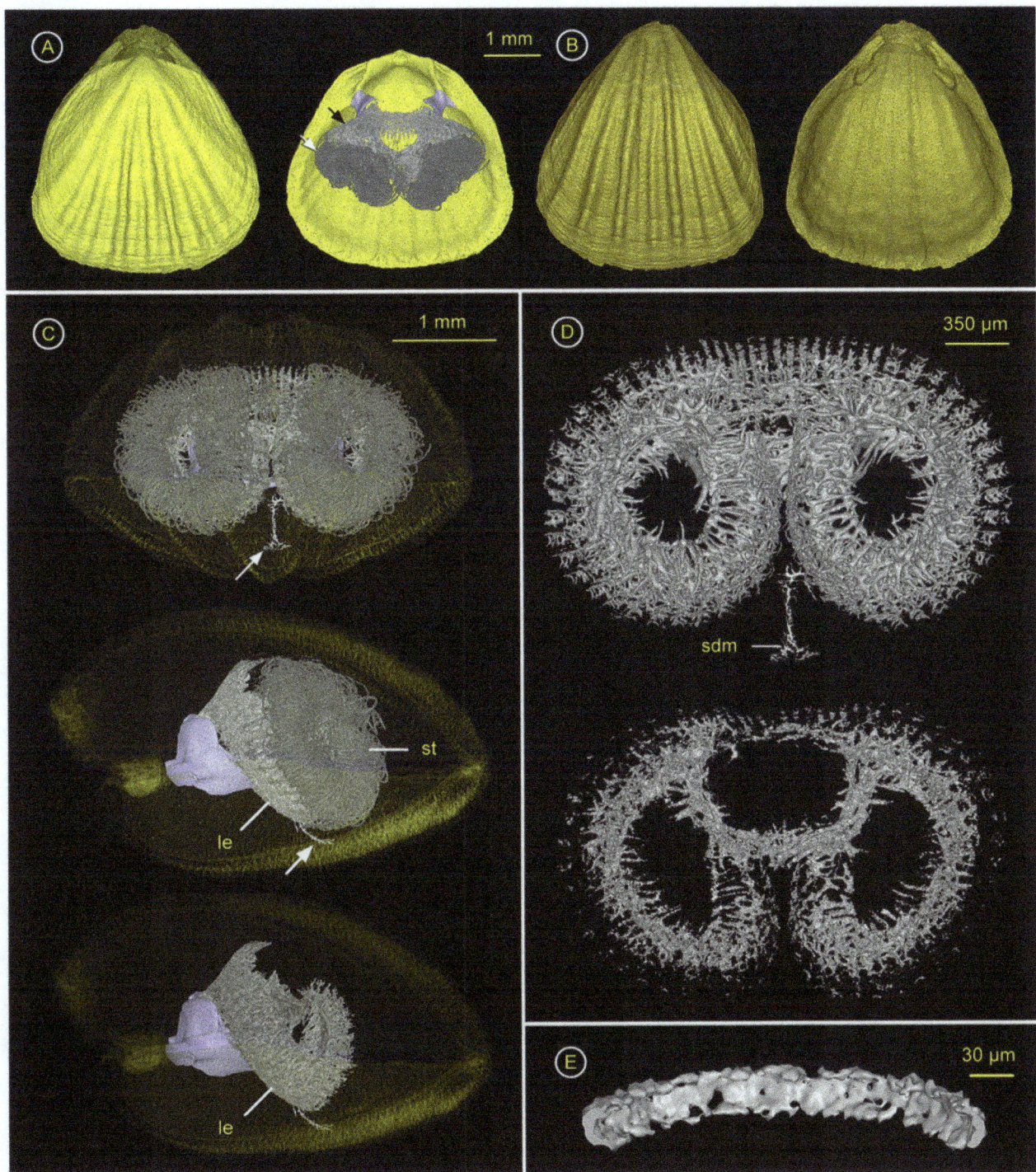

Figure 4 *Eucalathis* **sp. (Chlidonophoridae) – ZMB Bra 2351.** Ventral valve = top & dorsal valve = bottom. Click **here** to download a 3D model of *Eucalathis* sp. (interactive PDF). **A)** left: Dorsal valve with ventral valve in the background; right: Interior of the dorsal valve, with the short looped brachidium (purple), the endoskeleton of the lophophore (black arrow) with spicules of the tentacles (white arrow). **B)** left: Ventral valve; right: Interior of ventral valve showing the incomplete foramen. **C)** Anterior view (top) and lateral views (middle and bottom) through transparent shell with ventral valve on top, showing the dense lophophoral endoskeleton of the schizolophous lophophore (le), spicules within the tentacles (st) and a few median, slender and rather loose spicules with dorsoventral orientation in the dorsal mantle (arrow). **D)** Anteroventral view of lophophoral endoskeleton with faint spiculation of the dorsal mantle (sdm), bottom: "virtually purged" lophophoral endoskeleton to visualise the geometrical nature, the smaller, inner ring rests on the short looped brachidium. **E)** Cross section through spicules of a tentacle.

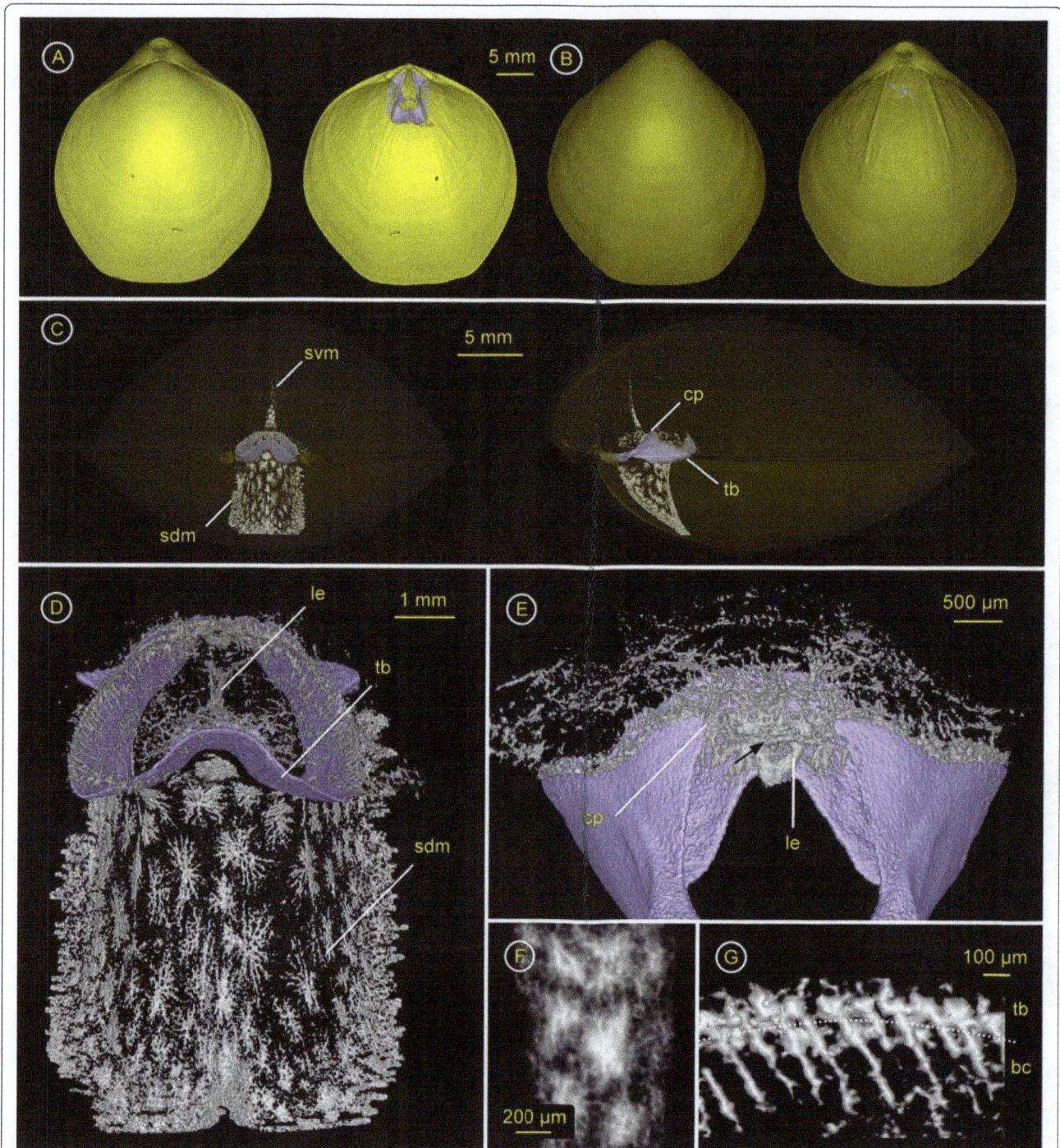

Figure 5 *Gryphus vitreus* (**Terebratulidae**) – **ZMB Bra 2261.** Ventral valve = top & dorsal valve = bottom. Click **here** to download a 3D model of *G. vitreus* (interactive PDF). **A)** left: Dorsal valve with ventral valve in the background; right: Interior of the dorsal valve, with the short looped brachidium (purple), the endoskeleton of the lophophore and spiculation in the dorsal mantle. **B)** left: Ventral valve; right: Inside of ventral valve with faint spiculation in the ventral mantle. **C)** Anterior view (left) and lateral view (right) through transparent shell with ventral valve on top, showing the short looped brachidium (purple) with crural processess (cp) and transverse band (tb), dorsal (sdm) and ventral mantle spiculation (svm). **D)** Anteroventral view of lophophoral endoskeleton (le) supported by the brachidium with a ventrally arched transverse band (tb), and the axial posteriorly keeled dorsal mantle spiculation (sdm) with star-shaped spicules (max. ~1.3 mm in diameter). **E)** Posterior view of the dorsal loop (purple) with triangular, and anteriorly converging crural processes (cp) and the lophophoral endoskeleton (le) with a plate-shaped structure and transverse projection (arrow). **F)** Magnification of faint spiculation within the ventral mantle, showing few axial spicules in a loose network. **G)** Magnification of spiculation of the lateral lophophoral endoskeleton with elongated spicules, basal crystals (bc) and tentacle bases (tb).

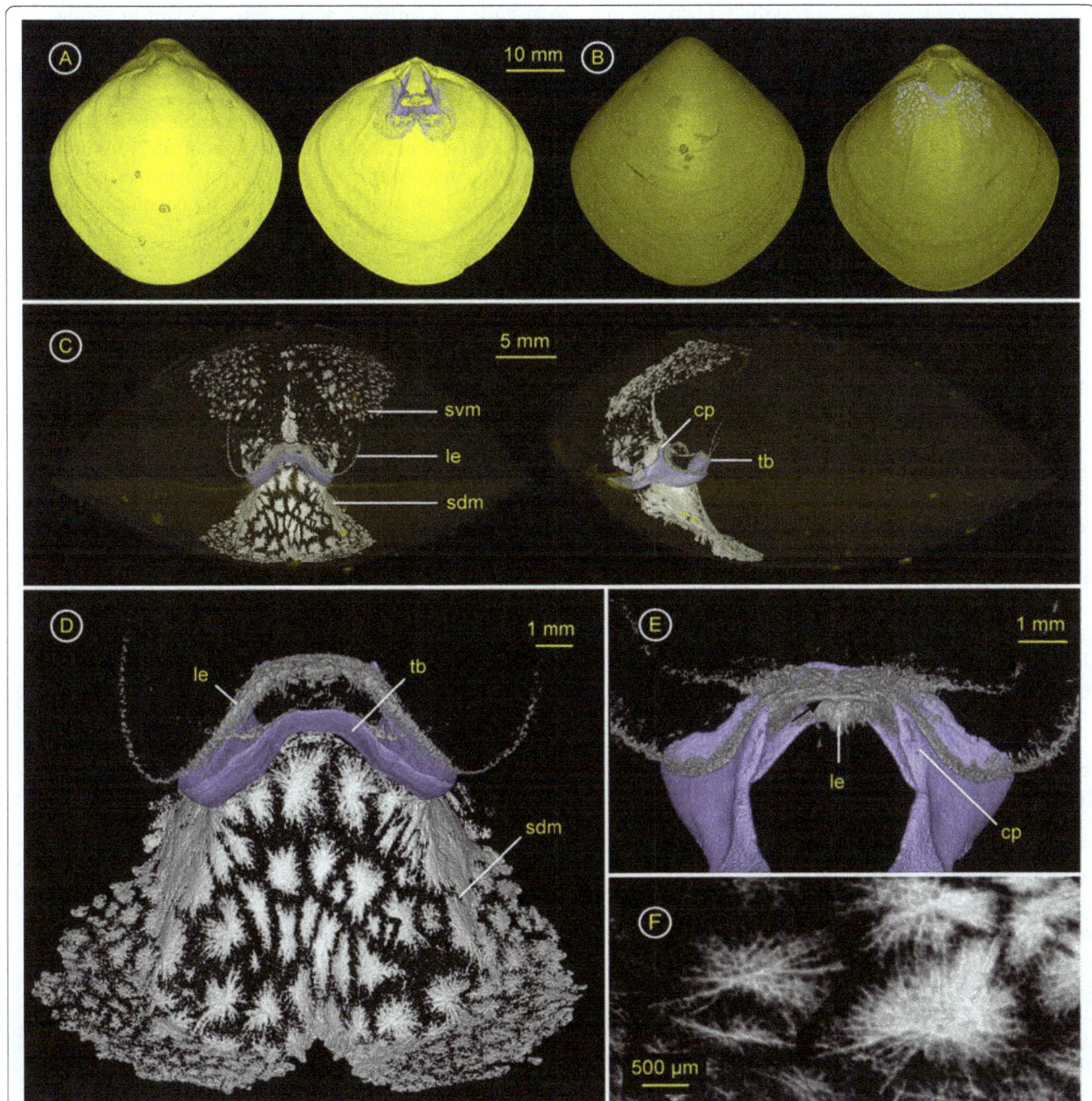

Figure 6 *Liothyrella neozelanica* **(Terebratulidae) – ZMB Bra 2255.** Ventral valve = top & dorsal valve = bottom. Click **here** to download a 3D model of *L. neozelanica* (interactive PDF). **A)** left: Dorsal valve with Ventral valve in the background; right: Interior of the dorsal valve, with the short looped brachidium (purple), the endoskeleton of the lophophore and spiculation in the dorsal mantle. **B)** left: Ventral valve; right: Inside of ventral valve with spiculation in the ventral mantle. **C)** Anterior view (left) and lateral view (right) through transparent shell with ventral valve on top, showing the short looped brachidium (purple) with crural processess (cp) and transverse band (tb), dorsal (sdm) and ventral mantle spiculation (svm). **D)** Anteroventral view of lophophoral endoskeleton (le) supported by the brachidium with a ventrally arched transverse band (tb), and the axial posteriorly keeled dorsal mantle spiculation (sdm) with star-shaped spicules (max. ~1.3 mm in diameter). **E)** Posterior view on the brachial loop (purple) with triangular, and anteriorly converging crural processes (cp) and the lophophoral endoskeleton (le) with a plate-shaped structure and transverse projection (arrow). **F)** Magnification of spicules from the anterior part of the dorsal body wall, showing the starlike nature, sizes range from ~0.5 mm to ~2.5 mm in diameter and max. ~150 µm in thickness.

spicules, teeth and sockets, crura and brachidium. However, in regard to the resolution it is less effective than scanning electron microscopy (SEM), hence SEM would be the first method of choice for surface-sensitive investigations, especially when resolutions of only a few microns or less are needed. None of the specimens investigated in this study were physically damaged through µCT scanning. As a result, after being scanned the specimen of *Eucalathis* sp. was successfully included in a DNA sequencing analysis which underlines the compatibility of µCT and molecular analyses. However, there is no certainty about a maximum radiation dose for successful post-scan DNA sequencing and therefore, rather short scans and immediate DNA amplification are recommended.

What is the function of a brachiopod's endoskeleton?

Speculation about the function of spiculation has been fuelled by a general debate concerning the vulnerability of brachiopods [26]. Their generally poor soft tissue yield results in little predation pressure, especially when the tissue contains spicules or toxins [27,28]. However, it is known that modern echinoderms, gastropods and crabs certainly penetrate the shell and do feed on brachiopods [29,30 and references therein]. The strength of the brachiopod shell is more affected by the thickness and ribbing, rather than by the biconvexity or presence of punctae [31]. However, organic matter is inaccessible to predators when placed within the punctae and similarly the internal tissue is not palatable when containing high amounts of inorganic spicules. Analyses of tissue of modern brachiopods and their value to predators revealed a higher inorganic content in the tissue of punctate than the impunctate brachiopods [32], which is in accordance with the observations in this study. Therefore, the co-occurence of spiculation and punctation may be referred to as non-chemical responses to predatory pressure.

Another criterion is the mechanical support for the soft tissue, especially of the lophophore. Mechanical tests with *Terebratulina unguicula* (Carpenter, 1864) revealed that a mean elastic modulus for spiculated tissues is more than three times greater than for artificially decalcified tissues [33]. In this study, all of the "short looped" specimens with little exoskeletal support of the lophophore (brachidium) exhibit spicules within the lophophore. Certain load-bearing features of the endoskeleton can be identified. The design of these structures suggests a perfect cooperation with the brachidium to maintain function and orientation of the lophophore with its protruding arms, as observed in the posterior excrescences of the endoskeleton in *Eucalathis* sp. and

Rectocalathis schemmgregoryi n. gen., n. sp. or in the distinct thick and plate-shaped structures for load-bearing in *G. vitreus* and *L. neozelanica*. The conspicuous load-bearing structure of *G. vitreus* and *L. neozelanica* can be addressed as an additional synapomorphic character of the family Terebratulidae (see Additional file 18: Figure S6. Phylogeny), identified due to this nondestructive µCT approach. A massive, area-wide spiculation within soft parts offers either mechanical support or fulfils a potent anti-predator function, or both. It is questionable whether the weak spiculation in the dorsal body wall of *Eucalathis* sp. and *Rectocalathis schemmgregoryi* n. gen., n. sp. serves as mechanical support. The anatomy suggests a regulation of the coelomic cavity pressure in interaction with muscles. Nonetheless, the weak spiculation in the dorsal body wall as well as in the specialized lophophoral endoskeleton indicates the close phylogenetic relationship of both species (see Additional file 16: Figure S6. Phylogeny)

Conclusion

The aim of this µCT study was to depict the profound differences of the hard parts and shell anatomy of modern brachiopods and to establish µCT scan and subsequent data processing protocols suitable for specimens preserved in alcohol. Micro-CT is an excellent non-destructive tool for visualizing the calcified exo- and endoskeleton of brachiopods. Acquired data on spiculation delivers a substantial contribution to formerly available documentations of brachiopod hard parts (endoskeleton) due to its visualization in its entirety. The phylogenetic relevance of spiculation has been indicated, with promising new informative characters. However extensive sampling is necessary to verify these findings and to discover intra-specific variability of spiculation.

Materials and methods

Representatives of 17 brachiopod genera covering all modern articulated subgroups and two representatives of inarticulated craniids were scanned for morphological analysis. The results from specimens with spicules are shown in this paper or can be accessed in the Supplemental Material. All specimens were selected from either the historical wet collection of the Museum für Naturkunde

Table 3 Scan parameters used in this study

Species	Images	Voltage (kV)	Current (µA)	Timing value (ms)	Averaging & Skip	Magnification	Scan Time ~ (min)
T. retusa	900	90	200	750	2 – 1	5,7	45
R. schemmgregoryi n. gen., n. sp.	900	60	100	750	2 – 1	17,6	45
Eucalathis sp.	900	60	130	750	2 – 1	13,2	45
G. vitreus	900	80	280	750	2 – 1	3,1	45
L. neozelanica	900	80	280	750	2 – 1	2,4	45

For scan parameters of the remaining species of the initial study (see Additional file 19: Table S3).

Berlin, Germany or from recently collected samples. For specimens used in this study see Table 1 and Additional file 1: Table S1. All specimens were stored in 75 - 80% ethanol (EtOH). For scanning, the specimen's valves were slightly opened with a needle and transferred gradually to distilled water. The scanning was performed dry to prevent soft tissue from moving, especially of the lophophore and the tentacles. For scanning the calcareous shell, it made no difference whether the prefixation was performed in alcohol, formaldehyde or Bouin's solution.

The specimens were X-rayed using a Phoenix Nanotom XS 180 NF (max. operating data 180kv, 1,0 mA, 15W – GE Sensing & Inspection Technologies GmbH, Wunstorf, Germany). The samples were placed in custom built carriers made of centrifuge tubes of 1,5 ml, 25 ml and 50 ml or Kautex bottles with a foam plastic inlay to prevent movement artefacts. Once the whole geometry of the specimen was scanned the X-ray images were reconstructed with "datos|x – reconstruction 1.5.0.22" (GE Inspection Technologies) using 4 cluster PCs to create the actual working file. Minimal movement of the specimen within the carrier (along distinct axes – not rotation) could be subtracted prior to reconstruction by aligning the first and an additionally taken image (number of images +1) after a full 360° rotation (tool name: scan optimisation with first and last image). For scanning parameters used in this study see Table 3 and Additional file 19: Table S3.

Image processing of the initial 3D reconstruction with earmarking of the "regions of interests", cleaning of the volume, specimen analyses and morphometries was carried out on a Dell Workstation (Dell Precision T7500, Intel Xeon E5530, 2.4 GHz; graphics card: NVIDIA GeForce GTX 285; operating system: Microsoft Windows XP Professional x64 Edition) using the software "Volume Graphics Studio Max 2.1" (Volume Graphics GmbH, Heidelberg, Germany, www.volumegraphics.com). The reconstructions were virtually sectioned to illustrate key features. Images (also 360° full HD (1920×1080) movie clips, data not shown, for details please contact the authors) were obtained using an isosurface or volume (PHONG) rendering algorithm. To illustrate the natural position of the spiculation and internal shell features the isosurface render algorithm was used to simulate transparency of the shell.

For subsequent data processing mostly open source software was used. As for the 3D interactive models, the open source software "MeshLab 1.3.2" (MeshLab, Visual Computing Lab – ISTI – CNR; http://meshlab.sourceforge.net/) for cleaning and down sampling the meshes and "DAZ Studio 4.5.0.114 Standart"(DAZ Productions, Inc., dba DAZ 3D; 224 South 200 West #250, Salt Lake City, UT 84101 USA; http://www.daz3d.com/) for creating the u3d file format were used. The final 3D interactive pdf file was created using the commercial software Adobe Acrobat 8 Professional 8.1.0. (Adobe® Systems Incorporated, San José, CA).

Additional files

Additional file 1: Table S1. Species used in the initial study. Species of a subgroup are listed in alphabetical order.

Additional file 2: Table S2. Volume analyses and morphometric measurements of all species of the initial study using VG Studio Max 2.1. Abbr.: DV = dorsal valve, VV = ventral valve, Spic Loph = spiculation within lophophore, Spic DV/VV spiculation within mantle of dorsal/ventral valve, Vol = Volume, Vol. total = Vol. VV + DV + Loph + Spic, × = no spiculation, na = not applicable.

Additional file 3: 3D model. *Terebratulina retusa.*

Additional file 4: 3D model. *Rectocalathis schemmgregoryi* n.gen., n. sp.

Additional file 5: 3D model. *Eucalathis* sp.

Additional file 6: 3D model. *Gryphus vitreus.*

Additional file 7: 3D model. *Liothyrella neozelanica.*

Additional file 8: Figure S1. *Laqueus rubellus.*

Additional file 9: 3D model. *Laqueus rubellus.*

Additional file 10: Figure S2. *Megerlia truncata.*

Additional file 11: 3D model. *Megerlia truncata.*

Additional file 12: Figure S3. *Platidia anomioides.*

Additional file 13: 3D model. *Platidia animioides.*

Additional file 14: Figure S4. *Pumilus antiquatus.*

Additional file 15: 3D model. *Pumilus aquaticus.*

Additional file 16: Figure S5. *Pajaudina atlantica.*

Additional file 17: 3D model. *Pajaudina atlantica.*

Additional file 18: Figure S6. Phylogeny.

Additional file 19: Table S3. Scanning parameters of all species of the initial study. Specimens are listed in alphabetical order. *P. atlantica* 3 was scanned twice using different parameters.

Competing interests

The authors declare that they have no competing interests.

Authors' contribution

RS: conception and design of the study; acquisition, analysis and interpretation of data, writing of the manuscript. CL: conception of the study; acquisition of samples; revision of the manuscript. Both authors read and approved the final manuscript.

Acknowledgments

Thanks are due to Johannes Müller and Christy Hipsley (both Museum für Naturkunde, Berlin) for their support and invaluable help with setting up the µCT to optimize the results. We are indebted to the crews, the scientific leaders (namely Pedro Martinez Arbizu, Kaj Hörnle, Reinhard Werner, and Gert Wörheide) and the shipboard scientific parties of several deep-sea expeditions (SO 205 "Mangan", SO 208 "Plumeflux", MSM 19-3, LU 839/2 and WO 896/7 "DeepDownunder") for providing material for this study. We also thank Jana Hoffmann, Bernard L. Cohen, Georg Maghon, Claudia Wolter, Paul Meredith, Sally Carson, Jim Fyfe, Nina Furchheim, and Peter Wirtz for collecting and providing additional specimens of different species. Collection campaigns and expeditions were financially supported within several projects funded by the German Research Foundation (DFG) and the German Ministry of Education and Research (BMBF).

References

1. Griesshaber E, Schmahl W, Neuser R, Job R, Blüm M, Brand U: **Microstructure of Brachiopod Shells - An Inorganic/Organic Fibre Composite with Nanocrystalline Protective Layer.** In *Materials Research Society Symposium Proceedings 2005, 844.* Warrendale, Pa: Materials Research Society; 1999:2005.

2. Williams A, Wright AD: **The origin of the loop in articulate brachiopods.** *Palaeontology* 1961, **3:**149–176.

3. Williams A, James MA, Emig CC, MacKay S, Rhodes MC: **Anatomy.** In *Treatise on Invertebrate Paleontology, Part H: Brachiopoda, Revised, Volume 1.* Edited by Kaesler RL. Boulder C, Lawrence K: The Geological Society of America and University of Kansas; 1997:7–188.

4. Williams A: **A history of skeletal secretion among articulate brachiopods.** *Lethaia* 1968, **1:**268–287.

5. Prenant M: **Notes histologiques sur *Terebratulina caput-serpentis* Linnaeus.** *Bull Soc Zool France* 1928, **53:**113–125.

6. Schumann D: **Mesodermale Endoskelette terebratulider Brachiopoden.** *Paläontol Z* 1973, **47:**77–103.

7. Eudes-Deslongchamps ME: **Recherches sur l'organisation du manteau chez les brachiopodes articulés et principalement sur les spicules calcaires contenus dans son intérieur.** In *Dissertation, Préparateur de Géologie à la Faculté des Sciences de Paris.* Edited by Savy F. Caen: Le Blanc-Hardel F; 1864:1–35.

8. Blochmann F: **Zur Systematik und geographischen Verbreitung der Brachiopoden.** *Z Wiss Zool* 1908, **90:**596–644.

9. Hancock A: **On the organization of the Brachiopoda.** *Phil Trans R Soc Lond* 1858, **184:**791–869.

10. Blochmann F: **Neue Brachiopoden der Valdivia- und Gaußexpedition.** *Zool Anz* 1906, **30:**690–720.

11. Lacaze-Duthiers H: **Histoire naturelle des Brachiopodes vivants de la Mediterranée.** *Ann Sci Nat Zool (serie 4)* 1861, **15:**259–330.

12. Thomson A: **Brachiopoda of the Australasian Antarctic Expedition.** *Scientific reports, ser C* 1918, **4**(3):1–76.

13. James MA: **Brachiopoda: internal anatomy, embryology, and development.** In *Microscopic Anatomy of Invertebrates, Volume 13: Lophophorata, Entoprocta, and Cycliophora.* Edited by Harrison FW, Woollacott RM. New York: Wiley-Liss; 1997:297–407.

14. Neall VE: **Systematics of the endemic New Zealand brachiopod *Neothyris*.** *J Royal Soc New Zealand* 1972, **2:**229–247.

15. Laurin B, Gaspard D: **Variations morphologiques et croissance du brachiopode abyssal *Macandrevia africana* Cooper.** *Oceanol Acta* 1987, **10:**445–454.

16. Südkamp W: **Discovery of soft parts of a fossil brachiopod in the "Hunsrückschiefer" (Lower Devonian, Germany).** *Paläontol Z* 1997, **71:**91–95.

17. Hagadorn JW: **Imaging of pyritized soft tissues in Paleozoic Konservatlagerstätten.** *GSA Ann Meet* 2001, **33**(1770):s430–s431.

18. Pakhnevich AV: **The Use of a Tomograph for the Analysis of Brachiopods.** In *Modern Paleontology: Classical and New Methods: 3rd AllRussia Scientific School for Young Scientists–Paleontologists.* Moscow: Paleontological Institute RAS; 2006. (Paleontol. Inst. Ross. Akad. Nauk, Moscow, 2006):48–50 [in Russian].

19. Pakhnevich AV: **Microtomography of Paleontological Objects: Advantages and Problems.** In *Modern Paleontology: Classical and New Methods: 5th AllRussia Scientific School for Young Scientists–Paleontologists.* Moscow: Borissiak Paleontological Institute RAS; 2008. (Paleontol. Inst. Ross. Akad. Nauk, Moscow, 2008):40–41 [in Russian].

20. Pakhnevich AV: **Study of fossil and recent brachiopods, using a skyscan 1172 X-ray microtomograph.** *Paleontol J* 2010, **44:**1217–1230.

21. Pakhnevich AV: **The type specimens of the Holocene brachiopod *Diestothyris frontalis* (Middendorf, 1849).** In *Brachiopods: extant and extinct. Proceedings of the Sixth International Brachiopod Congress.* Edited by Shi GR, Weldon EA, Percival IG, Pierson RR, Laurie JR: Mem Assoc Australas Palaeontol; 2011, 41:269–272.

22. Błażejowski B, Binkowski M, Bitner MA, Gieszcz P: **X – ray microtomography (XMT) of fossil brachiopod shell interiors for taxonomy.** *Acta Palaeontol Pol* 2011, **56:**439–440.

23. Pérez-Huerta A, Cusack M, McDonald S, Marone F, Stampanoni M, MacKay S: **Brachiopod punctae - A complexity in shell biomineralisation.** *J Struct Biol* 2009, **167:**62–67.

24. Motchurova – Dekova N, Harper DAT: **Synchrotron radiation X-ray tomographic microscopy (SRXTM) of brachiopod shell interiors for taxonomy: preliminary report.** *Annales Géologiques De La Péninsule Balkanique* 2010, **71:**109–117.

25. Logan A: **Geographic distribution of extant articulated brachiopods.** In *Treatise on Invertebrate Paleontology, part H: Brachiopoda, Revised,* Volume 6. Edited by Selden PA. Boulder C, Lawrence K: The Geological Society of America and University of Kansas; 2007:3082–3115.

26. Harper EM: ***Discinisca stella* (Gould, 1860): an intertidal inarticulate brachiopod from the Cape d' Aguilar Marine Reserve, Hong Kong.** In *The Marine Flora and Fauna of Hong Kong and Southern China IV. Proceedings of the Eighth International Marine Biological Workshop: the Marine Flora and Fauna of Hong Kong and Southern China.* Edited by Morton B. Hong Kong: Hong Kong University Press; 1995:235–247.

27. Thayer CW: **Brachiopods versus mussels: competition, predation, and palatability.** *Science* 1985, **228:**1527–1528.

28. Thayer CW, Allmon R: **Unpalatable thecideid brachiopods from Palau: ecological and evolutionary implications.** In *Brachiopods Through Time. Proceedings of the 2nd International Brachiopod Congress.* Edited by MacKinnon DI, Lee DE, Campbell JD. Rotterdam: Balkema; 1991:253–260.

29. Witman JD, Cooper RA: **Disturbance and contrasting patterns of population structure in the brachiopod *Terebratulina septentrionalis* (Couthouy) from two subtidal habitats.** *J Exp Mar Biol Ecol* 1983, **73:**57–79.

30. Harper EM: **What do we really know about predation on modern rhynchonelliforms?** In *Brachiopods: extant and extinct. Proceedings of the Sixth International Brachiopod Congress.* Edited by Shi GR, Weldon EA, Percival IG, Pierson RR, Laurie JR: Mem Assoc Australas Palaeontol; 2011, 41:45–57.

31. Alexander RR: **Mechanical strength of shells of selected extant articulated brachiopods: implications for Paleozoic morphologic trends.** *Hist Biol* 1990, **3:**169–188.

32. Peck L: **The tissues of articulate brachiopods and their value to predators.** *Philos TransR Soc Lond* 1993, **39:**17–32.

33. Fouke BW: **The functional significance of spicule-reinforced connective tissues in *Terebratulina unguicula* (Carpenter).** In *Les Brachiopodes fossiles et actuels. Actes du 1er Congres international sur les brachiopodes, Biostratigraphie du Paléozoique,* Volume 4. Edited by Rachebeuf PR, Emig CC. Brest: Université de Bretagne Occidentale; 1986:271–279.

9

Fine taxonomic sampling of nervous systems within Naididae (Annelida: Clitellata) reveals evolutionary lability and revised homologies of annelid neural components

Eduardo E Zattara[1,2]* and Alexandra E Bely[1]

Abstract

Introduction: An important goal for understanding how animals have evolved is to reconstruct the ancestral features and evolution of the nervous system. Many inferences about nervous system evolution are weak because of sparse taxonomic sampling and deep phylogenetic distances among species compared. Increasing sampling within clades can strengthen inferences by revealing which features are conserved and which are variable within them. Among the Annelida, the segmented worms, the Clitellata are typically considered as having a largely conserved neural architecture, though this view is based on limited sampling.

Results: To gain better understanding of nervous system evolution within Clitellata, we used immunohistochemistry and confocal laser scanning microscopy to describe the nervous system architecture of 12 species of the basally branching family Naididae. Although we found considerable similarity in the nervous system architecture of naidids and that of other clitellate groups, our study identified a number of features that are variable within this family, including some that are variable even among relatively closely related species. Variable features include the position of the brain, the number of ciliary sense organs, the presence of septate ventral nerve cord ganglia, the distribution of serotonergic cells in the brain and ventral ganglia, and the number of peripheral segmental nerves.

Conclusions: Our analysis of patterns of serotonin immunoreactive perikarya in the central nervous system indicates that segmental units are not structurally homogeneous, and preliminary homology assessments suggest that whole sets of serotonin immunoreactive cells have been gained and lost across the Clitellata. We also found that the relative position of neuroectodermal and mesodermal segmental components is surprisingly evolutionarily labile; in turn, this revealed that scoring segmental nerves by their position relative to segmental ganglia rather than to segmental septa clarifies their homologies across Annelida. We conclude that fine taxonomic sampling in comparative studies aimed at elucidating the evolution of morphological diversity is fundamental for proper assessment of trait variability.

Keywords: Ancestral character estimation, Annelida, Clitellata, Comparative morphology, Evolution, Homology, Naididae, Nervous systems, Neurophylogeny

* Correspondence: ezattara@gmail.com
[1]Department of Biology, University of Maryland, College Park, MD 20740, USA
[2]Current address: Department of Biology, Indiana University, 915 E. Third Street, Myers Hall 150, Bloomington, IN 47405-7107, USA

Introduction

Complex nervous systems are characteristic of eumetazoan taxa and, because their study can help to understand organismal function and evolution, they have been of particular interest to zoologists for several centuries [1-4]. Nervous systems play crucial roles integrating internal and external information into physiological and behavioral responses [2]. While incredibly diverse across major animal groups, nervous system architectures tend to be, by comparison, relatively well conserved within phyla [1,2]. As a result, many studies aimed at understanding the evolution of animal nervous systems have drawn conclusions from comparisons of only a few representatives from widely distant groups (e.g., flies and mice) [3-6]. Inferences from such studies are typically based on the similarities identified across these distantly related species, but these inferences hinge on the assumption that the traits in question are invariable at lower taxonomic levels. In order to make strong inferences about the evolution of animal nervous systems, their structure needs to be investigated in a broad array of taxa and with fine taxonomic sampling.

The nervous system of the phylum Annelida (segmented worms) comprises a central nervous system (CNS), composed of an anterior dorsal brain linked via circumesophageal connectives to a ventral nerve cord that is segmentally ganglionated, and a peripheral nervous system (PNS) composed of nerves branching off of the CNS components (Figure 1). Based on descriptions from a limited number of primarily polychaete species (summarized by Bullock and Horridge [2]), the annelid

nervous system was originally inferred to have a highly conserved ground plan. However, more recent studies on a broader range of annelids have revealed enormous variation of the annelid nervous system, especially regarding the morphology of the ventral nerve cord and the number and pattern of peripheral nerves, raising new questions about the ancestral architecture and evolution of the annelid nervous system [7].

The Clitellata are a large annelid subclade to which most freshwater and terrestrial annelids belong. The nervous system of clitellates has often been considered to be a simpler and less variable version of the nervous system typical of the primarily marine polychaetes; however, this inference is based on studies of a few clitellate species, mostly earthworms and leeches, with rather specialized morphology [8-10] and which may not closely reflect the ancestral clitellate condition. Clitellates comprise Naididae (water nymph worms), Crassiclitellata (earthworms), Enchytraeidae (pot worms), Lumbriculidae (blackworms) and Hirudina (leeches). The Naididae (sensu Erséus et al. [11]) are the sister clade to most other clitellates [9,11,12] and knowledge of naidid nervous system architecture is thus of particular importance for inferring how the nervous system has evolved within the clitellates, what the ancestral clitellate nervous system was like, and how it relates to the nervous system of closely related polychaetes.

Available studies of nervous system structure in naidids are few and are difficult to analyze comparatively. Older descriptions based on direct observation, light microscopy, and histological sectioning [13-15] provide different kinds

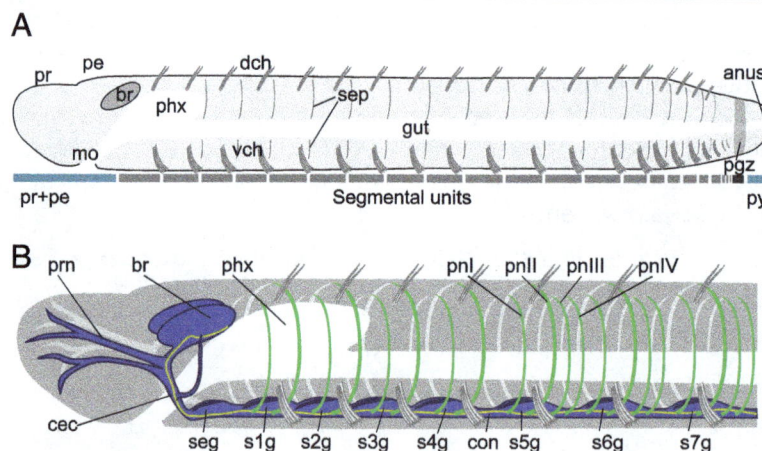

Figure 1 Overview of the naidid ground plan. **A)** Basic annelid body plan. The annelid body consists of an anterior non-segmental region composed of the prostomium (pr) and peristomium (pe), followed by a variable number of segments (grey bars), and a posterior non-segmental region, the pygidium (py). In front of the pygidium is the posterior growth zone (pgz), where new segments are made. **B)** Generalized structure of the nervous system in naidids. This schematic shows the anterior central nervous system (blue), ventral nerve cord neuropil (yellow) and peripheral nervous system (green). Anterior is to the left in this and all figures unless otherwise indicated. Labels: br: brain; cec: circumesophageal connective; con: interganglion connective; dch: dorsal chaetae; gut: ciliated gut; mo: mouth; pe: peristomium; pgz: posterior growth zone; phx: pharynx; pnI-IV: peripheral segmental nerve I-IV; pr: prostomium; prn: prostomial nerves; py: pygidium; sXg: segment x ganglion; seg: subesophageal ganglion; sep: intersegmental septum; vch: ventral chaetae.

of information than newer studies using immunohisto-chemistry and whole-mount confocal laser scanning microscopy [16-19]. Studies using consistent techniques, sampling at a fine taxonomic scale, and analyzing data in a phylogenetic framework are needed in order to reconstruct the ancestral naidid nervous system architecture and how it has evolved. Such studies can identify conservative and variable elements of the nervous system and should be particularly useful in identifying possible homologies between neural elements (e.g., nerves, cell types) across species, a task usually made challenging by the high degree of serial duplication characteristic of nervous system evolution.

In this paper, we describe and compare the nervous system architecture of 12 species of Naididae Ehrenberg, 1828 (*sensu* Erseus *et al.* [11]), representing four out of seven naidid subfamilies: Tubificinae - *Tubifex tubifex*; Pristininae - *Pristina leidyi* and *Pristina æquiseta*; Rhyacodrilinae - *Monopylephorus rubroniveus;* and Naidinae - *Dero digitata, Dero furcata, Allonais paraguayensis, Paranais litoralis, Amphichaeta* sp., *Chaetogaster diaphanus, Nais stolci* and *Stylaria lacustris*. We base our descriptions on adult individuals immunostained for acetylated-alpha-tubulin and serotonin, known to label a significant fraction of the neurites and some perikarya [16-21], along with labeled phalloidin to visualize muscular F-actin and DAPI as a nuclear counterstain, and imaged using confocal laser scanning microscopy. We focus in particular on the location and organization of immunoreactive elements of the brain and ventral nerve cord, the topological relationship between the ventral ganglia and the mesodermal septa, and the number and branching architecture of peripheral nerves. Based on our new descriptions and available published data, we identify conserved and variable elements of the naidid nervous system and propose possible homologies for some of these elements. We discuss our findings in the context of current knowledge about the phylogenetic relationships within this family, as well as relationships within the Clitellata and Annelida more broadly, providing insight into the evolution of the nervous system within these groups.

Results and discussion

An important goal for understanding how animals have evolved is to reconstruct the ancestral features and evolution of the nervous system. Many inferences about nervous system evolution are weak, though, because taxonomic sampling is sparse and phylogenetic distances between species compared are deep. Increasing sampling within specific clades can strengthen such inferences by revealing which features are conserved and which are variable within these groups. In the Annelida, the segmented worms, considerable variation in nervous system architecture has been reported for marine polychaete

families [7] but the terrestrial and freshwater Clitellata are typically viewed as having a simple and conserved nervous system [7,15]. However, this view is based on information from a limited number of species spanning this clade and, importantly, no detailed comparative studies within subgroups, such as within families, are available to provide insight into variability and conservation of neural architecture.

To address this gap, we characterized the morphology of the nervous system in 12 species of naidids using immunohistochemistry and confocal laser scanning microscopy. In the interest of brevity, we provide detailed descriptions and diagrams as Supplementary Information, including diagrams of the nervous system of 10 species (Additional file 1: Figure S1, Additional file 2: Figure S2, Additional file 3: Figure S3, Additional file 4: Figure S4, Additional file 5: Figure S5, Additional file 6: Figure S6, Additional file 7: Figure S7, Additional file 8: Figure S8, Additional file 9: Figure S9 and Additional file 10: Figure S10), an overview of a generalized naidid body segment (Additional file 11: Figure S11), image panels showing data for all species (Additional file 12: Figure S12, Additional file 13: Figure S13, Additional file 14: Figure S14, Additional file 15: Figure S15, Additional file 16: Figure S16, Additional file 17: Figure S17 and Additional file 18: Figure S18), and morphological descriptions for each species (Additional file 19). In our descriptions, we use whenever possible the terminology defined by Richter *et al.* [22]. A summary of the character states for all neural traits we found to be variable is provided in Table 1.

Below, we first synthesize the results of our comparative analysis of the nervous system morphology of the 12 naidid species we studied, giving an overview of the common patterns of nervous system components, followed by remarks on their variability. We then discuss the implications of our findings for understanding the stability or lability of neural traits, and the consequences of finding the appropriate homology criteria for inferring the naidid, clitellate and annelid ancestor.

Overview of naidid nervous system components

The general body and nervous system morphology of all naidid species examined follows the basic clitellate plan (Figure 1A, B and Additional file 11: Figure S11). The nervous system of naidids has three main components: the anterior brain and associated peripheral nervous system, the ganglionated ventral nerve cord, and the segmental peripheral nerves (Figure 1B). The brain, located dorsal to the mouth, is a paired bilobed structure composed of an outer cell cortex (comprising neuron cell bodies and supporting cells) surrounding an inner neuropil (formed by cell free neurites, or neuronal processes), and is linked to the ventral nerve cord by paired

Table 1 Character state of variable traits in the naidid nervous sytem

Traits		Species	*Tubifex tubifex*	*Pristina aequiseta*	*Pristina leidyi*	*Monopylephorus rubroniveus*	*Dero digitata*	*Dero furcata*	*Allonais paraguayensis*	*Paranais litoralis*	*Amphichaeta sp.*	*Chaetogaster diaphanus*	*Nais stolci*	*Stylaria lacustris*
		Subfamily	T	P	P	R	N	N	N	N	N	N	N	N
Anterior nervous system		brain, anterior edge	pr	pr	pr	s1	pe	pr/pe	pe	s1	pr	s1	pr/pe	pr/pe
		brain, posterior edge	pe/s1	pe/s1	pe/s1	s2	s1	s1	s1	s2	pe	s1	s1	pe/s1
		#brain SIR cells	2/6	2	2	2	4	4/6	2	10	2	0	8	4
		#ciliary sense organs	1	2	2	0*	2	2	2	2	2	4/6	2	2
		position ciliary sense organs	br	br	br	NA*	br	br	br	pr	pr	br/pr	br	br
		prostomium shape	cone	prob.	prob.	blunt	cone	cone	cone	blunt	cone	lips	cone	prob.
		eyes	no	no	no	no	no	no	no	no	no	no	yes	yes
Ventral nerve cord ganglia		#parachaetal	0-3	0-2	1-2	1-4	1-4	3-4	2-6	1	2	1-2	1-2	1-2
		#axillar	0-2	1	1	1(2)	1	1	1-2	1-2	1	1	1	1
		#central	0-1	0-2	1-3	1-2	1-3	2	1-2	0-2	0	1-2	1-2	1-2
		#rear	0-many	0-1	0-1	1-2	1	1	1	1	1	1	1	1
		#segments with ant. SIR pattern	3	4	4	4	4	4	4	4	4	3	4	4
		#medullary ant. segments	4	4	4	?	5	4	5	3	2	2	3 + 2	4
		first septum	2/3	2/3	2/3	3/4	3/4	3/4	3/4	3/4	2/3	3,4/5	3/4	3/4
		ganglion type	non-sept	sept	sept	non-sept*	non-sept	sept	non-sept	sept	non-sept	non-sept	sept	non-sept
PNS		#seg. nerves	4	4	4	4*	4	4	4	4	4	5	4	4
		#segments with ant. PNS	2	4	4	0	4	4	4	1	2	4	4	2

Summary of main nervous system traits found to be variable across the twelve species of Naididae presented in this study. Character states with an asterisk (*) are based on observations of poor quality images and should not be considered as confirmed. See Main Text and Additional file 19 for explanation of traits. Abbreviations by row: Subfamily - T: Tubificinae; P: Pristininae; R: Rhyacodrilinae; N: Naidinae; brain - pr: prostomium, pe: peristomium, s1: chaetigerous segment 1; s2: chaetigerous segment 2; x/y: boundary between x and y; prostomium shape - prob: proboscis; ganglion type – non-sep: non-septate, sep: septate.

circumesophageal connectives, which also connect to paired sets of prostomial peripheral nerves. The prostomium is usually cone-shaped, but may be blunt (*Monopylephorus, Paranais*), elongated into a proboscis (*Pristina, Stylaria*), or very reduced (*Chaetogaster*). *Stylaria* and *Nais* have a pair of lateral pigment-cup eyes located near the posterior edge of the prostomium; other species we studied are eyeless. The ventral nerve cord runs longitudinally down the length of the animal, between the ventral blood vessel and the ventral body wall (Additional file 11: Figure S11). It is composed of clusters of cell bodies (ganglia) linked by short connectives. There is one ganglion per segment, plus a subesophageal ganglion at the anterior end of the cord (Figure 1B); the cell cortex is trough-shaped and a neuropil runs through the trough (Additional file 11: Figure S11). In each segment, a number of peripheral segmental nerves (variously referred to in the literature as ring, circular, peripheral or segmental nerves) branch off perpendicular to the nerve cord (Figure 1B, Additional file 11: Figure S11A). These nerves, designated nerves I to IV based on the antero-posterior order of their roots along the ganglion, pass through the body wall's muscle layers and run between the muscle and outer epidermis to the dorsal side of the animal.

The following sections describe and compare the anterior nervous system, ventral nerve cord and peripheral nervous system across our study species; within each section, common patterns and conserved elements are described first, followed by a description of the variable elements.

Anterior nervous system: brain and prostomial nerves

The anterior nervous system is composed of common elements connected to each other in a similar manner across all 12 species studied (Figures 2 and 3, Additional file 12: Figure S12 and Additional file 13: Figure S13). The brain is formed by a lobulated cell cortex connected by a dense neuropil, and in most species is located behind the prostomium, spanning the peristomium and segment 1 (exemplified in Figure 3B by *Allonais*). The anterior part of the brain neuropil has a network of serotonin immunoreactive (SIR) neurites, while acetyl-tubulin immunoreactive (acTIR) neurites form an arc across the posterior part (Figure 2A, B). Behind the neuropil, one or more pairs of SIR perikarya are found in the posterior lobe. At the posterior edge of the brain is a set of acTIR hairs that are often associated with muscle fibers attached to that point and to the dorsal body wall. A pair of acTIR ciliary sense organs, evidenced by a coiled neurite structure, are located in front of the brain (Figure 2C-H, Additional file 12: Figure S12, Additional file 13: Figure S13, and Additional file 14: Figure S14). The acTIR tracts in the brain neuropil and the connectives project forward and arborize into a series of prostomial nerves that innervate a dense collection of epidermal acTIR sensory hair cells on the prostomium surface (Figure 4 and Additional file 14: Figure S14). Based on the relative location of nerve roots branching off the circumesophageal connectives, the general patterns of nerve arborization, and the prostomial regions innervated, we tentatively identify six

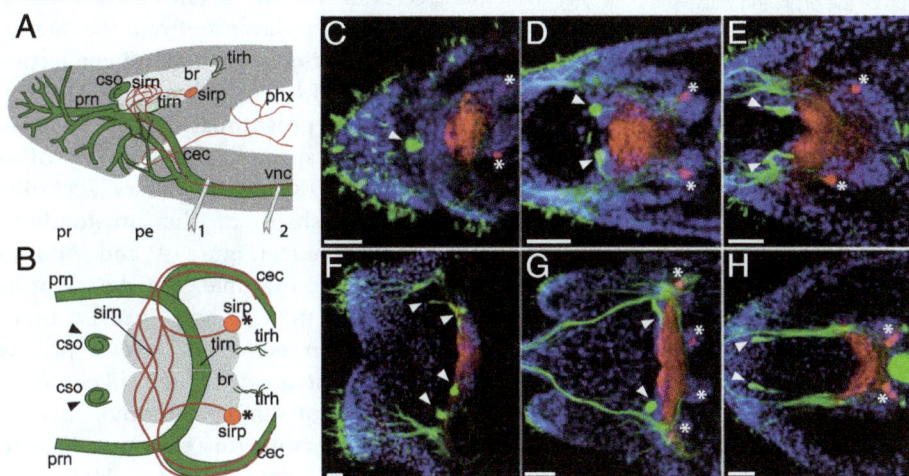

Figure 2 Naidid anterior nervous system. **A-B)** General morphology of the anterior nervous system in naidids in lateral (**A**) and dorsal (**B**) views. Brain cell bodies are represented in light grey, acetyl-tubulin immunoreactive (acTIR) structures in green, and serotonin immunoreactive (SIR) structures in red. **C-H)** Diversity in naidid anterior nervous system morphology, as represented by 6 out of the 12 species studied; see also Additional file 12: Figure S12 for data on the full set. Note the variation in the position and number of ciliary sense organs (arrowheads) relative to the SIR brain neuropil, visible as a red mass of SIR neurites; note also the location and number of SIR perikarya (asterisks). Images are intensity sum projection of dorsal Z-stacks of *Tubifex tubifex* (**C**), *Pristina leidyi* (**D**), *Dero furcata* (**E**), *Chaetogaster diaphanus* (**F**), *Stylaria lacustris* (**G**) and *Paranais litoralis* (**H**). Specimens were stained for DNA (blue), serotonin (red) and acetyl-tubulin (green). br: brain; cec: circumesophageal connectives; cso: ciliary sense organs (arrowheads); phx: pharynx; prn: prostomial nerves; sirn: serotonin immunoreactive neuropil; sirp: serotonin immunoreactive perikarya (asterisks); tirh: acetyl-tubulin immunoreactive hairs; tirn: acetyl-tubulin immunoreactive neurites, vnc: ventral nerve cord. Scale bars: 25 μm.

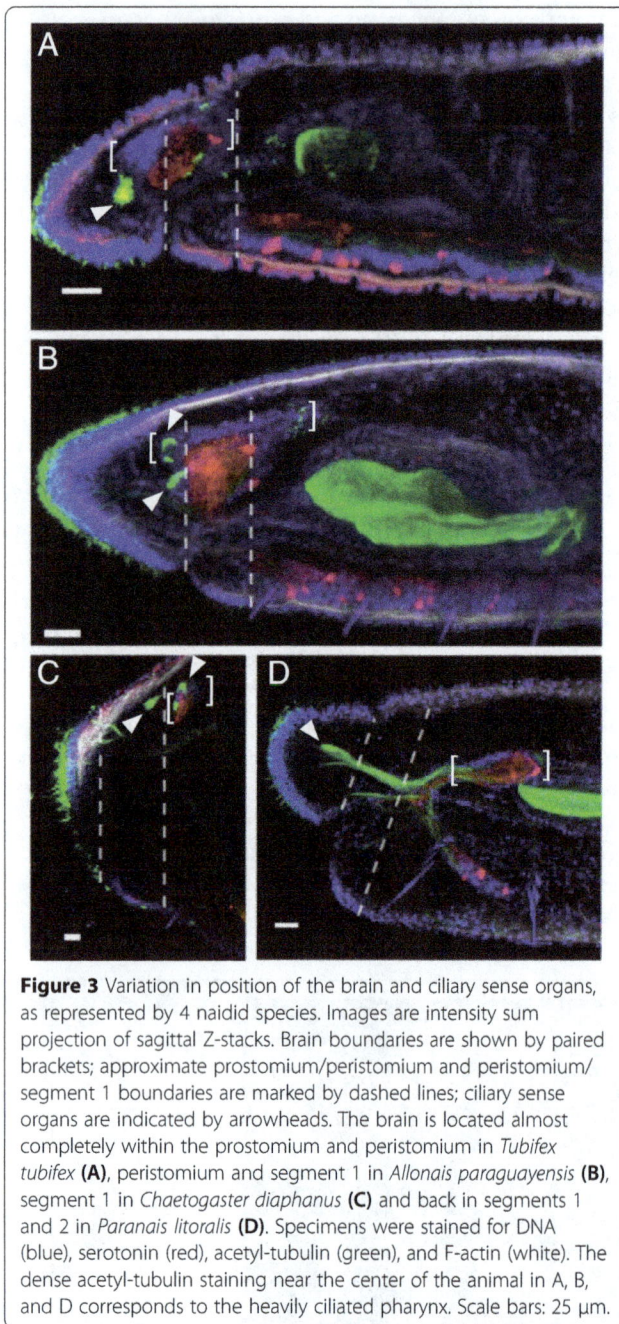

Figure 3 Variation in position of the brain and ciliary sense organs, as represented by 4 naidid species. Images are intensity sum projection of sagittal Z-stacks. Brain boundaries are shown by paired brackets; approximate prostomium/peristomium and peristomium/ segment 1 boundaries are marked by dashed lines; ciliary sense organs are indicated by arrowheads. The brain is located almost completely within the prostomium and peristomium in *Tubifex tubifex* (**A**), peristomium and segment 1 in *Allonais paraguayensis* (**B**), segment 1 in *Chaetogaster diaphanus* (**C**) and back in segments 1 and 2 in *Paranais litoralis* (**D**). Specimens were stained for DNA (blue), serotonin (red), acetyl-tubulin (green), and F-actin (white). The dense acetyl-tubulin staining near the center of the animal in A, B, and D corresponds to the heavily ciliated pharynx. Scale bars: 25 μm.

With respect to the position of the brain (Figure 3 and Additional file 13: Figure S13), we found that *Tubifex* has a brain that is displaced anteriorly, straddling the prostomium/peristomium (Figure 3A and Additional file 13: Figure S13A); *Chaetogaster*, which has a reduced prostomium, has a relatively small brain located level with the chaetae of segment 1 (Figure 3C and Additional file 13: Figure S13J); and *Monopylephorus* and *Paranais* have a brain that is displaced posteriorly into segments 1 and 2 (Figure 3D, Additional file 13: Figure S13D and S13H). The number and location of SIR perikarya in the brain also varies among species, even between close relatives (Figure 4A; see also Table 1). We detected no SIR perikarya in *Chaetogaster*, one pair in both *Pristina* species, *Monopylephorus*, *Amphichaeta* and *Allonais*, two pairs in *Stylaria* and *Dero digitata*, three pairs in *Tubifex* and *Dero furcata*, four pairs in *Nais* and five pairs in *Paranais*. Interestingly, we only detected a single pair of SIR perikarya in the brain of a recently hatched *Tubifex* (data not shown), instead of the three pairs scored in older worms, suggesting that the number of SIR perikarya may also vary with developmental stage. All species examined have a pair of acTIR ciliary sense organs located at the anterior edge of the brain with the exception of *Tubifex*, which has a single, medial organ which may represent fusion of the original pair, and *Chaetogaster*, which has two pairs of ciliary sense organs (Figures 2, 3, 4A, Additional file 12: Figure S12, Additional file 13: Figure S13, and Additional file 14: Figure S14). *Paranais*, *Amphichaeta* and *Chaetogaster* also differ from the rest of the species in that the acTIR ciliary sense organs are located at a short distance from the brain, rather than against the anterior edge, and are connected to the main neuropil by acTIR neurite bundles. Although the number of major prostomial nerve branches is largely similar across species, the location of roots of the prostomial acTIR nerves differs according to differences in the shape of the prostomium, and some nerves are absent (Figure 4 and Additional file 14: Figure S14): for example, the dorsal projecting nerve F is not found in *Nais* and *Stylaria*. In species where the prostomium elongates into a proboscis, namely *Pristina* and *Stylaria* (which likely evolved prostomium elongation independently), nerve D projects forward to innervate this structure (Figure 4B). While nerve A is closest to the eye in *Nais* and *Stylaria*, we could not verify whether these were actually connected due to signal masking by the eye's pigment.

Ventral nerve cord

The ventral nerve cord has a similar architecture in all studied species (Figures 5, 6, 7, Additional file 15: Figure S15, Additional file 16: Figure S16 and Additional file 17: Figure S17). It is formed by a continuous neuropil running

major prostomial nerve branches in the naidids (branches A to F, color coded in Figure 4B). Given the limited number of specimens and range of ages examined here, these homology assignments are necessarily preliminary, but should prove useful as a guide for future studies.

While this general pattern of anterior nervous system elements is shared by most species, we found considerable variation in the location of the brain, number of SIR perikarya, number and location of acTIR ciliary sense organs, and structure and origin of acTIR prostomial nerves across the species examined in this study.

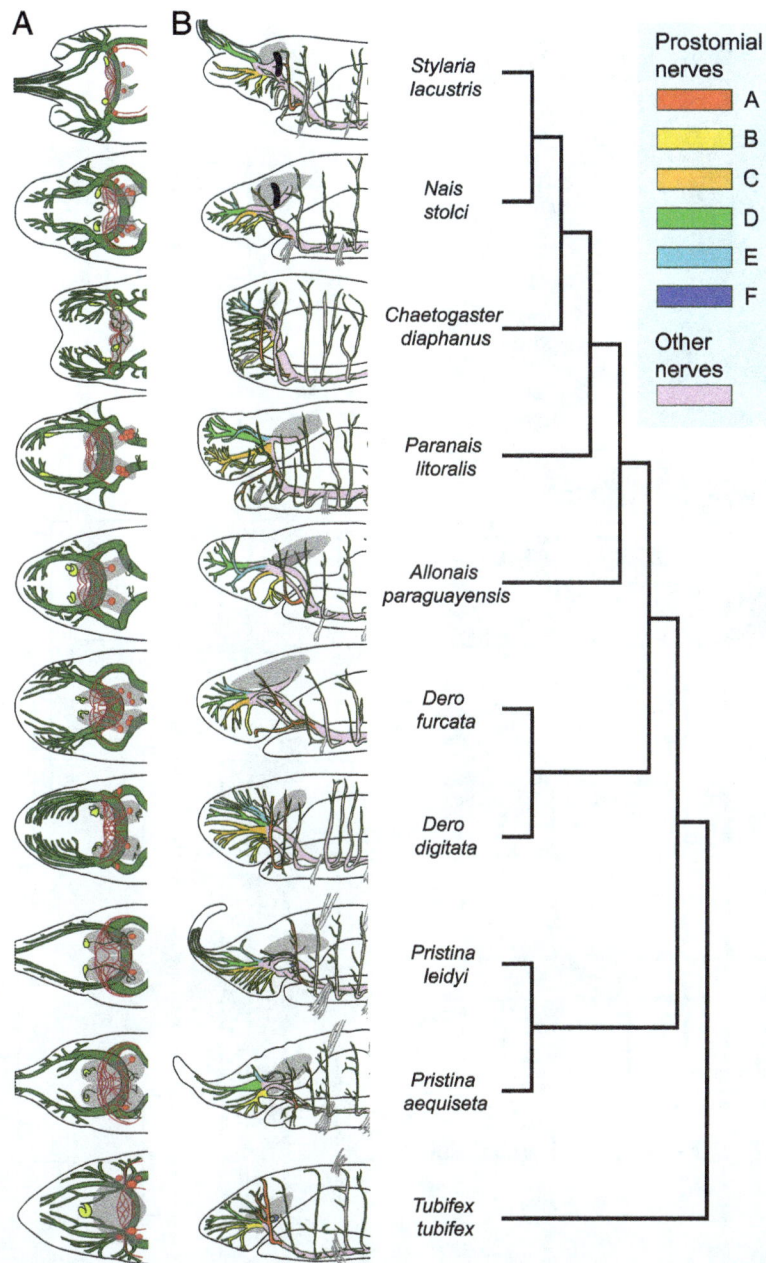

Figure 4 Variation in the architecture of the anterior nervous system across 10 naidid species. **A)** Schematic drawings of the anterior nervous system in dorsal view. Brain lobes are shown in grey, serotonin immunoreactive perikarya and neurites in red, ciliary sense organs in light green, and other acetyl-tubulin immunoreactive nerves in dark green. **B)** Schematic drawings of the prostomial/peristomial nerves in lateral view, color-coded to highlight putative homology assignments. The phylogenetic relationships among the species are shown to the right and are based on recent molecular analyses, as described in the Methods section. Brain is shown in grey; black patches in *Nais stolci* and *Stylaria lacustris* are lateral pigmented eyespots.

through segmentally iterated ganglia, with the neuropil containing acetyl-tubulin immunoreactive (acTIR) and serotonin immunoreactive (SIR) neurites that form longitudinal nerve tracts. The acTIR neurites are found latero-ventrally while SIR neurites tend to be medial and dorsal (Figure 5A, A1 and Additional file 15: Figure S15). These nerve tracts are linked by segmentally iterated transverse

commissures of variable acTIR neurite density. Phalloidin staining indicates that the ventral cord is sheathed by a thin muscular tunic (Figure 5A1). In all ventral nerve cord ganglia a number of SIR perikarya are found connected to the longitudinal SIR nerve tracts. Based on their location in the ganglion relative to the peripheral nerve roots, we recognize four positional types of SIR perikarya

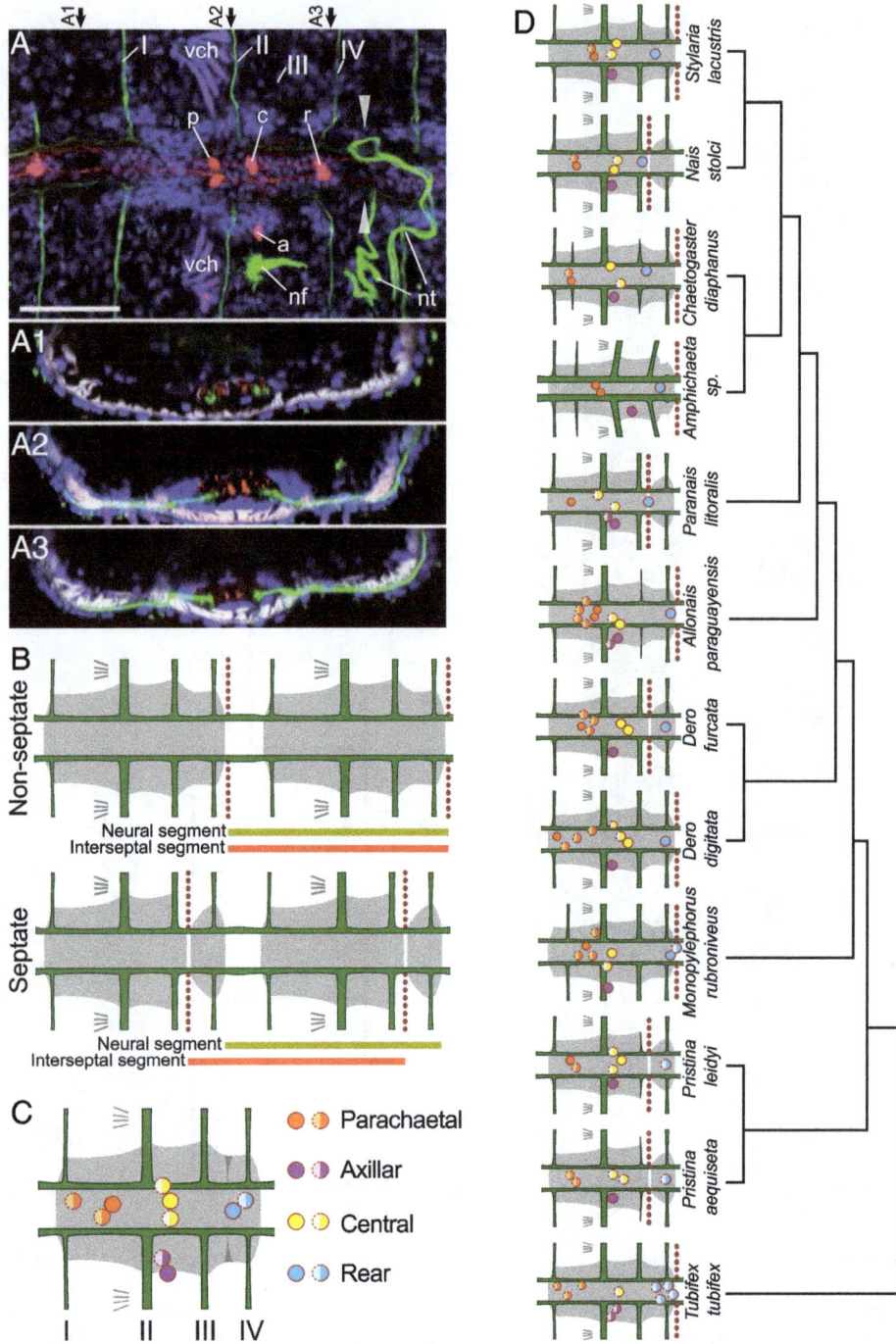

Figure 5 (See legend on next page.)

(Figure 5C): 1) parachaetal cells, located within the anterior third of the ganglion, approximately medial to the neuropil and level with the ventral chaetae, between nerve roots I and II; 2) axillar cells, located within the middle third of the ganglion, lateroventrally outside of the neuropil, either to the left or right, and always behind nerve root II; 3) central cells, located more medially within the middle third of the ganglion and closer to the neuropil but level with the axillar cells; and 4) rear cells, located within the posterior third of the ganglion and medial to the neuropil, between nerve roots III and IV. Notably, in all species examined, the anterior-most ganglia have a pattern of SIR perikarya that is clearly different from more posterior segments (Figure 6A-C and Additional file 16: Figure S16). Within a single trunk ganglion, SIR cells are asymmetrically distributed, mostly evidenced by the presence of axillar cells at only one side; however, consecutive ganglia show alternating mirror symmetry with respect to the mid-sagittal plane. In contrast, anterior ganglia have a larger number of SIR perikarya and these are arranged symmetrically with respect to the mid-sagittal plane. Based on counts from 1–4 individuals per species (~30 specimens across all species), the following general patterns emerge: a) on average, there are approximately 80% more SIR perikarya in anterior ganglia than in more posterior ganglia; b) the subesophageal ganglion has fewer SIR perikarya than the ganglia of the anterior segments, but more than more posterior ganglia; c) ganglia of segments 1 and 2 have about the same number of SIR perikarya per segment, and this number is the highest in the body; d) ganglia of segments 3 and 4 have fewer perikarya than segments 1 and 2, but more than more posterior ganglia. In all cases, SIR perikarya of the anterior ganglia are arranged in a bilaterally symmetric pattern resembling what would be obtained by the superposition of perikarya from two consecutive more posterior ganglia. In all species, the ganglia of the peristomium and anterior-most segments adjoin one another, forming a superganglion that adopts a medullary configuration (Figure 6D-G and Additional file 17: Figure S17).

These ganglia, especially the first three, are displaced posteriorly relative to other segmental organs, as compared to their position in more posterior segments.

Despite the common architecture of the ventral nerve cord, we found considerable variation among species in the specific location of SIR perikarya in the ganglia, the location along the body where perikarya switch from the above described anterior pattern to the general distribution pattern found in the rest of the segments, and the number of anterior segments forming a medullary superganglion. We also noticed unexpected inter-species variability in the position of the mesodermal segmental septa relative to the ventral nerve cord ganglia. A general pattern of the ganglion SIR cells for each species is shown in Figure 5D; however, these patterns should be considered as approximate, since there is considerable variation from segment to segment even within the same worm. Interpretation is further complicated by the fact that some perikarya show weaker serotonin signal in some segments, and their detection may depend both on their developmental state and the quality of immunostaining. Despite these caveats, we found that the asymmetric trunk pattern of SIR perikarya begins at segment 5 in all species studied (Figure 6B-C and Additional file 16: Figure S16), with the exception of *Tubifex* and *Amphichaeta*, where it starts at segment 4 (Figure 6A, Additional file 16: Figure S16F). The extent of the anterior medullary superganglion is more variable: it includes two ganglia in *Chaetogaster* (Additional file 17: Figure S17E) and *Amphichaeta* (Figure 6E), three ganglia in *Paranais* (Additional file 17: Figure S17D), four ganglia in *Tubifex* (Figure 6D), *Pristina* (Figure 6F and Additional file 17: Figure S17A), *Dero furcata* (Additional file 17: Figure S17C), and *Stylaria* (Additional file 17: Figure S17G), and five ganglia in *Dero digitata* (Additional file 17: Figure S17B) and *Allonais* (Figure 6G). In *Nais*, we observed a gap between ganglion 3 and 4, and another between 5 and 6, but not between 4 and 5 (Additional file 17: Figure S17F). *Nais* also was the only species in which we detected SIR

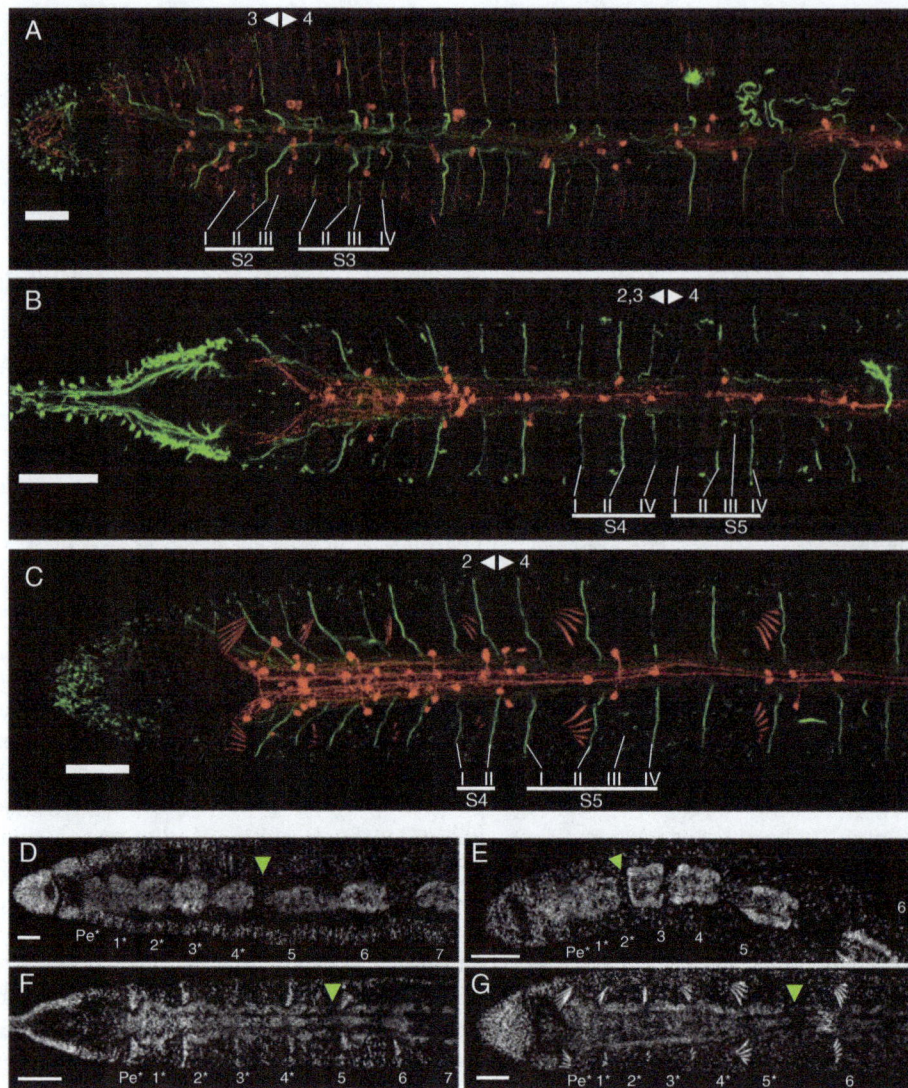

Figure 6 Central and peripheral nervous system of anterior segments of naidids. **A-C)** Ventral maximum intensity projections of the anterior end of *Tubifex tubifex* (**A**), *Pristina aequiseta* (**B**) and *Allonais paraguayensis* (**C**), showing the ventral nerve cord neuropil and segmental peripheral nerves. Labeled acetyl-tubulin immunoreactive nerves (green) and serotonin immunoreactive nerves and perikarya (red) consistently show a different pattern in anterior-most segments as compared to more posterior segments, but the level at which this transition occurs varies across species, as well as across the elements of the nervous system. The transition between the anterior and posterior pattern of segmental nerves is indicated by the back-to-back arrowheads (number of segmental nerves per segment indicated beside arrowheads). Nerve identity is shown below for segments flanking the boundary. **D-G)** Ventral views of DAPI stained specimens of *Tubifex tubifex* (**D**), *Amphichaeta* sp. (**E**), *Pristina leidyi* (**F**), and *Allonais paraguayensis* (**G**). The green arrowhead marks the location of the anterior-most connective; anterior segments forming a medullary superganglion are labeled by asterisks. Scale bars: 25 μm.

perikarya outside of the central nervous system, specifically as segmentally repeated single cells embedded in the body wall, over or close to the lateral line, a subepidermal, lateral cord of cells of unknown function (Additional file 18: Figure S18).

Although a ganglionated ventral nerve cord is present in all species, we found that the position of the ganglia relative to the segmental septa is variable, and specifically takes one of two configurations: non-septate or septate (Figure 5B). Each ganglion is composed of three

cell clusters, one anterior, one central and one posterior (Figure 5A). In non-septate ganglia, the intersegmental septum is located behind the posterior cluster, so the whole ganglion is located within the segment; in septate ganglia, the septum is located between the central and the posterior cluster, so the ganglion straddles two segments (Figure 5B). *Tubifex, Monopylephorus, Dero digitata, Allonais, Amphichaeta, Chaetogaster* and *Stylaria* have non-septate ganglia, while both *Pristina* species, *Dero furcata, Paranais* and *Nais* have septate ganglia (Figure 5D); note that this trait

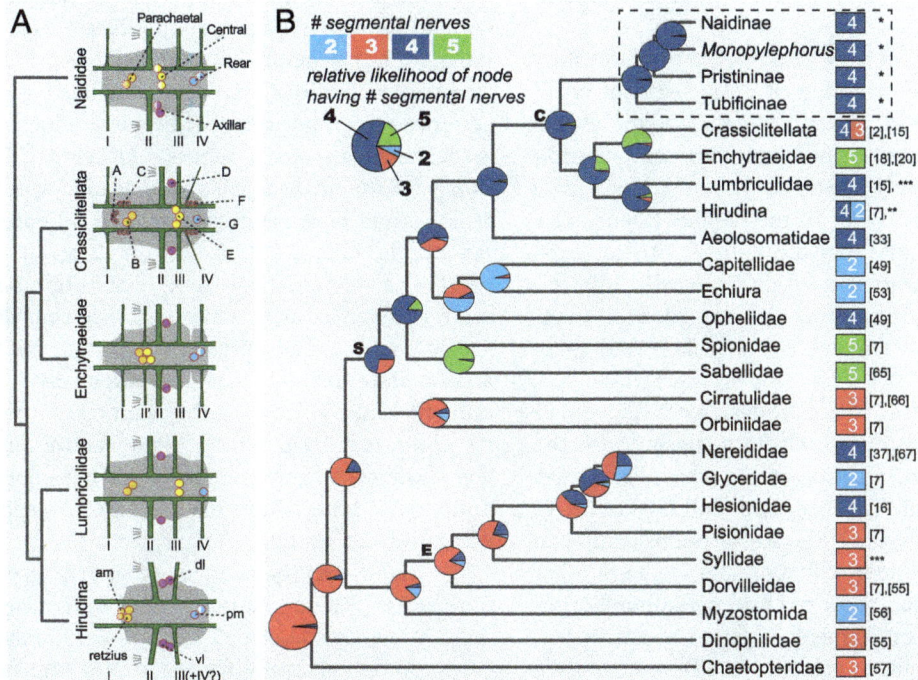

Figure 7 Phylogenetic distribution of serotonin immunoreactive perikarya and peripheral nerve patterns. **A)** Reconstruction of the basal architecture of a trunk ventral ganglion across clitellate groups, based on this study and previous reports [20,27,29,32,35,36]. Colored circles represent serotonin immunoreactive perikarya, with color representing putative homologous groups; A to G cells in Crassiclitellata refer to cell groups defined for *Lumbricus terrestris* [35]; anteromedial (am), dorsolateral (dl), ventrolateral (vl), posteromedial (pm) and Retzius cells in Hirudina refer to cell groups defined for leech species [27,29,32]. Segmental peripheral nerves are shown aligned according to their position along the ganglion. **B)** Phylogenetic mapping of the number of segmental peripheral nerves across Annelida, and maximum likelihood estimations of the ancestral state at each node. Pie charts illustrate the relative likelihood of a given node having had each character state, and were calculated in R [63] using the *ace* function from the *ape* package [64]. Annelid relationships are based on recent phylogenetic studies [50,52]. Labeled ancestral nodes are for Errantia (E) polychaetes, Sedentaria (S) polychaetes and Clitellata (C); the dashed box highlights data from this paper. Nerve number for each group is based on information from one or a few species, except for Naididae, where ancestral condition was estimated from our 12 study species. References are indicated on the right, in brackets for published works; *: data from this study; **R. Hessling, personal communications in [7]; ***E.E. Zattara, unpublished observations.

varies even between phylogenetically close species pairs (e.g. *Stylaria* and *Nais*; *Dero furcata* and *Dero digitata*).

Peripheral nervous system

The main component of the naidid peripheral nervous system is a system of segmentally iterated acTIR nerves that originate from the neuropil of the ventral nerve cord, exit through the ganglia, pass ventrally through the body wall muscle layer and run subepidermally towards the segment's dorsum (Figures 1B, 5A and Additional file 11: Figure S11A). Each nerve innervates a number of epidermal sensory hairs. We found that certain nerves can be found in all species examined, while others appear to have been recently gained or lost; we also found that the number of segmental nerves is reduced in the anterior-most segments of most species.

All examined species have four segmental nerves in each trunk segment, except for *Chaetogaster diaphanus* which has five (Figure 5D, Additional file 15: Figure S15). We designated the segmental nerves as nerves I to

IV, according to the antero-posterior order in which they branch from the ganglion. The largest nerves are nerve I, located in the anterior half of the segment, and nerve II, which is always located just posterior to the chaetal plane (defined by the ventral and dorsal chaetal bundles) and which sends branches anteriorly to a set of epidermal sensory structures closely associated with the chaetae. Nerves III and IV are usually smaller, and one of the two is generally associated with the intersegmental septum; which nerve this is depends on whether the ganglion is septate or not (Figure 5B). Some species diverge from this general pattern, however (Figure 5D): in *Pristina leidyi* (Additional file 15: Figure S15C), *Pristina aequiseta* (Additional file 15: Figure S15B), and *Allonais paraguayensis* (Figure 5A) nerve III is very small or absent; in *Amphichaeta* sp. (Additional file 15: Figure S15G) and *Chaetogaster diaphanus* (Additional file 15: Figure S15H) there is a small fifth nerve located between nerves I and II (labeled as "2" in arabic numerals in Additional file 15: Figure S15G, H); in *Amphichaeta* sp.

we could not detect nerve IV and nerve III is displaced backwards.

We performed an ancestral character estimation analysis to reconstruct the evolution of trunk segment nerve number across annelids, using our data and published information from across the phylum. This analysis indicates that four segmental nerves is the most likely state for the last common ancestor of the naidids (Figure 7B). Our analysis also suggests that the number of segmental nerves has been evolutionarily labile, increasing and decreasing at several points during the evolution of annelids, and that basal stem annelids most likely had three segmental nerves.

The segmental nerve patterns described above are consistent across all segments within a species with the exception of the anterior-most segments, which have fewer nerves the closer they are to the anterior end (Figure 6A-C, Additional file 16: Figure S16). Species vary considerably in the degree of nerve number reduction and in the number of anterior segments that show reduced nerve number. In most species, the full complement of nerves begins in segment 5. However, it begins in segment 3 in *Tubifex, Amphichaeta* and *Stylaria*, segment 2 in *Paranais* and in segment 1 (no reduction) in *Monopylephorus.*

Fine taxonomic sample of Naididae reveals conservation and lability of neural traits

To gain better understanding of nervous system evolution within Clitellata, we described and made a comparative analysis of the nervous system of 12 species of Naididae, a basally branching clitellate family [9,23-25]. Although we found many similarities between nervous system architecture in naidids and that of other clitellate groups [15,20,23], our study identified many features that are variable within this family of clitellates, including some that are variable even among relatively closely related species (Table 1). Variable features of the nervous system include the location of the brain, the number of ciliary sense organs, the presence of non-septate or septate ganglia along the ventral nerve cord, the distribution of serotonergic cells in the brain and ventral ganglia, and the number of peripheral nerves. Below, we discuss interspecific and intra-individual variation in the distribution of serotonin immunoreactive perikarya in the central nervous system and potential homologies between Naididae and other clitellates; we address the unexpected interspecific variability in the position of segmental septa relative to ventral ganglia; and we show how using ganglia rather than body segments to identify peripheral nerves can help reveal nerve homologies across Annelida. We end by highlighting the importance of fine taxonomic sampling in comparative studies aimed at elucidating the evolution of morphological diversity.

Patterns of serotonin immunoreactive perikarya in the annelid central nervous system indicate that segmental units are not structurally homogeneous

The central nervous system of annelids typically includes serotonin immunoreactive (SIR) cells, which are putatively associated with motor neurons [17,26-31]. The distribution patterns of these cells vary among species and body regions, and have been suggested to be potentially useful taxonomic traits [17]. Our data on 12 species of naidid annelids show that SIR perikarya distribution patterns in the brain and ventral nerve cord can vary considerably across species and even within individuals, both along the antero-posterior body axis and potentially between developmental stages. Despite this variability, the positions of serotonin-positive perikarya in the ventral nerve cord ganglia show consistent enough patterns to suggest putative homologies both within naidids and between naidids and other clitellate groups.

We found that the number of paired serotonin immunoreactive (SIR) perikarya in the brain varies across naidids. While a single pair of SIR perikarya is the most common arrangement for the group, the range is quite broad, from no SIR perikarya in *Chaetogaster* to 5 pairs in *Paranais*. The number of perikarya does not appear to be related in any obvious manner to species attributes such as size, habitat or preferred movement type. While in most species the number of serotonin-positive perikarya is lower in the brain than in the body ganglia, we found that the reverse is true in both *Nais* and *Paranais*. Thus, the number of perikarya in the brain is not always less than that in the ventral ganglia, as had been previously suggested [17]. Our observation that *Tubifex* juveniles had only two serotonin-positive cells in the brain while adult worms have six cells suggests that this number can change post-embryonically; caution is recommended when using the number of serotonin-positive cells in the brain as a diagnostic feature for species identification.

In all but the anterior-most segments, serotonin immunoreactive (SIR) perikarya in ventral ganglia have an asymmetrical distribution, with consecutive segments having patterns that are mirror-images of one another. This alternating pattern has been previously described in three naidid species, all three being naidines [17]; our study extends the distribution of this pattern in naidids and close relatives to seven other naidines, a rhyacodriline, two pristinines, and a tubificine. Such an alternating pattern is also present in leeches [27,29,32], a more distantly related group of clitellates, and even the polychaetous *Aeolosoma* sp. has a single yet alternating serotonergic cell in each ventral ganglion [33]. Thus, the alternating pattern is not an autapomorphy of Naididae, as suggested elsewhere [17], but a feature common to many clitellates and related annelids. The developmental processes that generate this pattern have not been studied in naidids, but during development of the leech species *Helobdella triserialis*, *Theromizon rude* and *Hirudo medicinalis* paired seroto-nergic precursor cells in consecutive segments make con-tact with each other and one member of each pair undergoes cell death in a pattern that alternates across segments, setting up a similar alternating pattern of un-paired cells as seen in the naidids [27,29,32]. Whether a similar mechanism is responsible for the SIR perikarya dis-tribution in naidids remains to be determined.

In contrast to trunk ganglia, serotonin immunoreactive (SIR) perikarya in the anterior-most ganglia of naidids are more numerous and show a symmetrical pattern that roughly matches the overlay of two consecutive trunk gan-glia. In naidines, anterior segments often develop post-embryonically by paratomic fission [34], and it has been suggested that the symmetrical pattern of SIR perikarya in those segments results from such segments being at an earlier developmental stage in which differential cell death in consecutive segments has not yet taken place [17]. However, our study indicates that anterior segments have a symmetrical pattern not only in fissioning spe-cies but in two non-fissioning species as well, *Tubifex* and *Monopylephorus*. Furthermore, in the pristinines, six anterior segments form during fission [19] yet only the four anterior-most segments show symmetrically arranged SIR perikarya, indicating that an asymmetrical pattern can be established during fission as well.

Our finding that naidids have a different distribution of serotonin immunoreactive (SIR) perikarya in anterior segments as compared to more posterior segments is consistent with data from several other annelid groups. In the crassiclitellate earthworm *Lumbricus terrestris*, a subesophageal medullary superganglion formed by two fused anterior ganglia has over three times as many SIR perikarya as trunk ganglia, while segments 4–10 have twice as many [28]. In the leech *Hirudo medicinalis*, eight SIR perikarya per ganglion are found in the six anterior ganglia, with four of these being fused into a subesophageal medul-lary ganglion and two remaining as free ganglia; in contrast, only seven SIR perikarya per ganglion are found in more posterior cells [26]. In the freshwater polychaete *Aeolosoma* sp., the anterior-most two ganglia have four SIR perikarya, while remaining trunk ganglia have just a single SIR cell [33]. Thus, the presence of two clearly different distribution patterns of SIR perikarya along the body, with a clear break at a specific point along the antero-posterior axis, is wide-spread among clitellates and is even found outside of this clade. The location of this transition between the two pat-terns is different for different groups, however. It would be interesting to investigate whether other nervous system components and other organ systems also show a similar antero-posterior break, and whether the presence of such boundaries indicates some degree of "cephalization" of the anterior-most segments.

Preliminary homology assessments suggest gains and losses of serotonin immunoreactive cell sets across the Clitellata

Based on our studies of naidids, serotonin immunoreac-tive (SIR) perikarya within the ventral nerve cord ganglia appear to show a common pattern across species. These cells form four spatially segregated sets of cells, which we termed parachaetal, central, axillar and rear cells (Figure 5C). When the naidid pattern is compared with SIR cell patterns reported for representative crassiclitellates (*Lumbricus terrestris* and *Eisenia foetida*) [35,36], enchy-traeids (*Enchytraeus crypticus*) [20], hirudines (*Helobdella triserialis*, *Theromizon rude* and *Hirudo medicinalis*) [27,29,32] and lumbriculids (*Lumbriculus variegatus*, E.E.Z. unpublished observations), a very preliminary homology assignment across the clitellates is possible, using as criteria the axial position of the cells along the length of the ganglion and the topological relationship of the cells to the peripheral nerve roots (Figure 7A). By these criteria, the naidid parachaetal cells might be homologous to *Lumbri-cus*' B cells, the leech's anteromedial or Retzius cells, and unnamed anterior cells found in *Enchytraeus* and *Lumbri-culus*; the central cells might be homologous to E cells in *Lumbricus* and to unnamed midline cells in *Enchytraeus* and *Lumbriculus*, while being absent in hirudines; the axil-lar cells might be homologous to D cells in *Lumbricus*, ven-tro and dorsolateral cells in hirudines, and unnamed lateral cells in *Enchytraeus* and *Lumbriculus*; and the rear cells might be homologous to G cells in *Lumbricus*, posterome-dial cells in hirudines, and unnamed midline posterior cells in *Enchytraeus* and *Lumbriculus*. Interestingly, *Lumbricus terrestris* seems to have a number of SIR cell sets clearly ab-sent in more distantly related groups (namely, A, C and F cells). The total number of cells per trunk ganglion varies significantly among groups: naidids reported here have 4–11 cells, leeches [26,27,29], enchytraeids [20] and lumbricu-lids (E.E.Z. unpublished observations) have 7–9 cells, while earthworms have 30–80 cells [35,36]. The homology as-signments we propose here are necessarily tentative, and future studies including developmental and neuronal con-nectivity studies are needed to confirm them; nonetheless, even this preliminary comparison across clitellates high-lights that whole sets of serotonin immunoreactive cells have been gained or lost throughout the evolution of this group of annelids.

The relative position of neuroectodermal and mesodermal segmental components is evolutionarily labile

An unexpected finding from our study is that the position of the segmental ganglia relative to the septa (which are used to define morphological segment boundaries) varies across species, and even among close relatives. We found that some naidids have non-septate ganglia (in which the entire ganglion falls between consecutive septa) while others have septate ganglia (in which the ganglion straddles two consecutive segments, with a septum falling across the ganglion). Among our study species, we found examples of ganglion type being variable among closely related genera (*Nais* and *Stylaria*), and even within a single genus (*Dero digitata* and *Dero furcata*). Among other clitellates, ganglion type is also variable. Currently available descriptions indicate that *Lumbricus* and *Eisenia* (both in Crassiclitellata) as well as *Lumbriculus* (Lumbriculidae) have non-septate ganglia [15,23], while *Enchytraeus* (Enchytraeidae) has septate ganglia [20]; leeches cannot be categorized since they have no septa at all [23]. Among polychaetes, septate ganglia have been described in the nereids *Neanthes*, *Platynereis* and *Hediste* [37] and this ganglion type is not uncommon in other polychaete groups as well [7]. Although variability in ganglion type can be inferred from the literature, sampling density remains sparse outside of the naidids and none of these previously published observations have been made within a comparative framework; to our knowledge, this is the first report showing that ganglion type can vary even between closely related species. Given our observations, we conclude that the relative position of neuroectodermal elements (i.e., neural ganglia) and mesodermal elements (i.e., septa) of each segment can experience frequent shifts over evolutionary time, and these elements should not be assumed to be in the same register within a group.

We hypothesize that the evolutionary lability of the relative position of neural ganglia and mesodermal septa in clitellates likely reflects the considerable developmental independence of ectodermal and mesodermal teloblastic bandlets during embryogenesis [38,39]. As indicated by work done in *Tubifex hattai* [38–40] and several leech species [41–44], in clitellates much of a segment's ectodermal and mesodermal tissues arises through teloblastic growth, in which a small set of large stem cells (four ectodermal and one mesodermal teloblast pairs) divide asymmetrically leaving behind bandlets of founder cells that will each form components of one or two consecutive segments. Ablation experiments in *Tubifex* suggest that ectodermal segmentation comprises an initial autonomous morphogenetic stage, followed by mesoderm-dependent alignment of segmental elements [39]. Mesodermal segmentation, on the other hand, does not require segmented ectoderm [38]. Similar results have been reported for leeches [45–47]. Given this degree of independence in the development of these two tissue layers, evolutionary changes in ectoderm/mesoderm alignment may arise relatively easily within clitellates.

Interestingly, even though the relative position of septa and ganglia varies across species, we found that all septate ganglia are septate in a similar manner, with the septum consistently located at two-thirds of the ganglion length. This uniformity in the configuration of septate ganglia

suggests that there may be developmental constraints restricting the possible locations of the septa relative to the ganglia. Furthermore, the septate/non-septate condition appears to be fixed within a species, suggesting a strong genetic control. The developmental mechanisms that keep ectoderm and mesoderm in the same register intraspecifically while allowing for interspecific shifts in Naididae are unknown and warrant further investigation.

The homologies of segmental nerves are clarified by scoring their position relative to segmental ganglia rather than to segmental septa

Among our study species, the most common configuration of the peripheral nervous system was four segmental nerves per trunk ganglion. We did find variation in this pattern, however, with *Chaetogaster* having a fifth small nerve located between nerves I and II, *Pristina* spp. and *Allonais* having nerve III very reduced, and *Amphichaeta* having no detectable nerve IV but a posteriorly displaced nerve III.

Given the variation in ganglion septation we found in our study, we determined that the common approach of naming nerves based on their position within the segment [19,20,37], that is, relative to segmental septa, might not reflect underlying nerve homologies. Instead, we based our naming scheme for segmental nerves on the order of nerve roots along the segmental ganglion (rather than position within the segment), such that nerve I is the anterior-most nerve emanating from the ganglion, nerve II is the next most anterior nerve, etc. In addition to the order of nerve roots along the ganglion, we also considered in our naming scheme the relative sizes and innervation patterns of these nerves, to account for the possibility that certain nerves may be gained or lost over evolutionary time. Using this approach, the arrangement, relative size and innervated structures of nerve roots along the ganglion are almost identical in *Dero furcata* and *Dero digitata* (which, respectively, have septate and non-septate ganglia), whereas a septum-based naming scheme would entail nerves with quite different sizes and innervation patterns being assigned the same nerve number, and nerves associated with the same structures being assigned different nerve numbers (e.g., the nerve innervating the chaetae would be the third nerve in *D. furcata* and the second nerve in *D. digitata*). The homologies of segmental nerves across species that are implied by our naming scheme follow from our expectation that relative slippage of the mesodermal/ectodermal boundaries (see above) is more likely than concerted change in size and innervation patterns of all segmental nerves; thus, naming nerves based on their position along the ganglion rather than relative to mesodermal septa is more likely to reflect nerve homologies. We would discourage others from using septum boundaries to identify segmental nerves in annelids, and encourage the use of ganglion boundaries instead; such an approach should facilitate efforts to trace how segmental nerves evolved in clitellates and other annelids.

Based on our data and ancestral character estimation (*ace*) analysis, four peripheral nerves per trunk ganglion is the most likely ancestral state for Naididae. Outside of the naidids but still among the clitellates, the number of nerves described is four for *Lumbriculus* [48], five for *Enchytraeus* [7,18,20], three for *Lumbricus* [23], and two for adult leeches [23,26,27] (although four nerves are described in embryos of *Erpobdella octoculata*; R. Hessling, unpublished observations in [7]). Under the assumption that these species are representative of their respective groups, our ancestral character estimation analysis supports four peripheral nerves as the ancestral state for Clitellata as a whole (Figure 7B). Out of the four nerves, nerve II is larger and associates with the chaetae in members of Naididae, in *Lumbriculus variegatus* [48] and in *Lumbricus terrestris* [15]. Nerve II is also associated with chaetae in *Enchytraeus crypticus* [20], but it is smaller than the following nerve (Figure 7A). Since five rather than four nerves are present in this species, we speculate that enchytraeids must have either evolved a novel nerve (II' in Figure 7A), intercalated between nerves I and II, or duplicated nerve II (II' and II in Figure 7A).

How does this basal clitellate state for segmental nerve number relate to the rest of Annelida? Within polychaetous annelids, the number of nerves reported ranges from none (*Trilobodrilus hermaphroditus*) to eight or more (*Protodrilus* sp.) [7,49]. Putative nerve homologies among taxa have been difficult to establish, in part due to a poor understanding of deep phylogenetic relationships among annelid taxa. However, recent progress in understanding annelid phylogeny [50-52] now provides a stronger framework for mapping of variation in segmental nerves across the phylum, including for ancestral character estimation to reconstruct the most likely number of nerves at the main nodes of the annelid tree, as we have done here (Figure 7B) [7,18,20,23,37,48,49,53-57]. Our analysis supports a previous claim that three segmental nerves represents the ancestral condition for the phylum [2,7]. Interestingly, the analysis suggests that a fourth nerve evolved either at the base of the Sedentaria, the clade of mostly less motile worms within which the clitellates are nested, or shortly thereafter, after the cirratulids + orbinids lineage branched off. According to this reconstruction, the two-nerve state of capitellids and echiurids and the five-nerve state of sabellids and spionids would both be derived from a four-nerve condition. Even if our specific results regarding ancestral character estimation for nerve number at each node are later revised, it is clear that segmental nerves unquestionably have been gained and lost several times during annelid evolution. Finer taxonomic sampling would allow a more precise mapping of novel origins and losses of segmental nerves, which in turn could reveal useful groups in

which to study the evolution of neuronal elements within annelids.

Conclusions

Our comparative description of the nervous system of several species of Naididae and the resulting identification of common patterns and differences in nervous system architecture in this group highlight the potential insights that can be gleaned from comparative morphological studies made at a fine taxonomic scale. Such studies have the power to confirm the deep conservation of certain widespread traits, whose evolution may be strongly constrained by functional or developmental constraints, but also to reveal highly labile traits that can change readily during evolution. Furthermore, such fine-scale comparative studies can contradict prior inferences about which characters are unlikely to vary, as we have shown here, for example, for the septate/non-septate ganglion condition. Obtaining strong data for the conservation or lability of characters is particularly important since traits thought to be well conserved tend to be used as landmarks for homology assignments; thus we recommend ensuring that such conservation has been evaluated at a relatively fine taxonomic scale before using such traits more broadly as a basis for determining homologies.

Methods
Animal samples
Specimens of *Tubifex tubifex*, *Pristina leidyi*, *Pristina aequiseta*, *Dero digitata*, *Dero furcata*, *Allonais paraguayensis*, *Paranais litoralis*, *Chaetogaster diaphanus*, *Nais stolci* and *Stylaria lacustris* were obtained from established laboratory cultures [58,59] and specimens of *Monopylephorus rubroniveus* and *Amphichaeta* sp. were field collected. Table 2 provides worm sources, culture conditions, and NCBI accession numbers for available COI and 16S sequence barcodes for the strains used. New COI or 16S sequences were obtained for *Paranais litoralis*, *Dero furcata*, *Pristina leidyi* and *Pristina aequiseta*; primers, PCR parameters, and sequencing methods are as described elsewhere [58].

Immunocytochemistry
Samples were relaxed 10 min in cold (4°C) relaxant solution (10 mM $MgCl_2$/5 mM NaCl/1 mM KCl/8% ethanol), fixed 30 min in 4% formaldehyde in 0.75x PBS, and rinsed in 1x PBS. Then they were permeabilized with 0.1% Triton-X in PBS (PBTx), blocked 1 h in 10% normal goat serum (NGS) in PBTx, and incubated 15–20 h at 4°C with mouse anti-acetylated-α-tubulin monoclonal antibody (T6793, Sigma, St. Louis, MO, USA) and rabbit anti-serotonin polyclonal antibodies (S5545, Sigma), both diluted 1:100 in blocking solution. Specimens were then washed in PBTx and incubated 15–20 h at 4°C in

Table 2 Classification, source and culture conditions for study species

Subfamily	Species	Source	Culture conditions	Acc. number
Tubificinae Vejdovsky, 1876	*Tubifex tubifex*	Western Fisheries Research Center, USGS, Sand Point, Lake Washington, WA, USA (supplied by C. Rasmussen)	Sand substrate in artificial spring water aerated flasks, 15C. Fed spirulina pellets.	GenBank: AF534866 (COI)
Pristininae Lastočkin, 1921	*Pristina leidyi*	Pond, Terrapin Softball Complex, University of Maryland, College Park, MD	Paper substrate in artificial spring water bowls, room temperature. Fed spirulina powder.	Genbank: KR296707 (16S rRNA)
Pristininae Lastočkin, 1921	*Pristina œquiseta*	Cichlid fish tanks, Biology/Psychology Building, University of Maryland, College Park, MD	Paper substrate in artificial spring water bowls, room temperature. Fed powdered fish food flakes.	Genbank: KR296708 (16S rRNA)
Rhyacodrilinae Hrabě, 1963	*Monopylephorus rubroniveus*	Charleston, SC, USA	N.A.	GenBank: GQ355379 (COI)
Naidinae Ehrenberg, 1828	*Amphichaeta sp.*	Rhode River, Smithsonian Environmental Research Center, Edgewater, MD, USA	N.A.	GenBank: AF534829 (COI)
Naidinae Ehrenberg, 1828	*Chaetogaster diaphanus*	Edwards Lake, University of Maryland, College Park, MD, USA	Paper substrate in artificial spring water bowls, room temperature. Fed chopped *Allonais*.	GenBank: GQ355366 (COI)
Naidinae Ehrenberg, 1828	*Nais stolci*	Paint Branch, University of Maryland, College Park, MD, USA	Paper substrate in artificial spring water bowls, room temperature. Fed spirulina powder.	GenBank: GQ355369 (COI)
Naidinae Ehrenberg, 1828	*Stylaria lacustris*	Paint Branch, University of Maryland, College Park, MD, USA	Paper substrate in artificial spring water bowls, room temperature. Fed spirulina powder.	GenBank: AF534861 (COI)
Naidinae Ehrenberg, 1828	*Paranais litoralis*	Herring Bay, Fairhaven, MD, USA	Mud substrate in artificial brackish water bowls, 15C. Fed mud supplemented with fish food flakes.	GenBank: KP204261 (COI)
Naidinae Ehrenberg, 1828	*Allonais paraguayensis*	Ward's Natural Science (sold as *Stylaria*). USA.	Paper substrate in artificial spring water bowls, room temperature. Fed rolled oats.	GenBank: AF534828 (COI)
Naidinae Ehrenberg, 1828	*Dero furcata*	Carolina Biological Supply (found inside a *Daphnia magna* flask)	Paper substrate in artificial spring water bowls, room temperature. Fed rolled oats.	Genbank: KP204260 (COI)
Naidinae Ehrenberg, 1828	*Dero digitata*	Edwards Lake, University of Maryland, College Park, MD	Paper substrate in artificial spring water bowls, room temperature. Fed rolled oats.	GenBank: GQ355368 (COI)

Classification, source and culture conditions for the 12 study species (Annelida: Clitellata: Naididae Ehrenberg, 1828, *sensu* Erseus *et al.* [11]). NCBI accession numbers for 16S rRNA (*Pristina* species) or cytochrome oxidase I (all other species) partial sequences are provided as barcoding reference.

blocking solution containing Alexa-Fluor-647-conjugated goat anti-mouse IgG(H + L) antibodies (1:200, A21236, Invitrogen, Carlsbad, CA, USA) and Cy3-conjugated goat anti-rabbit IgG antibodies (1:200, 111-166-003, Jackson Immunoresearch,West Grove, PA, USA), 60 nM Alexa-Fluor-488 phalloidin (A12379, Invitrogen), and 10 µg/mL DAPI. After washing with PBTx and PBS, specimens were transferred through a graded glycerol series and mounted in 25 mM *n*-propyl-gallate in 75% glycerol/25% PBS.

Image acquisition and analysis
Labeled specimens were mounted on glass slides and imaged using a Leica SP5X confocal laser scanning microscope (Leica, Wetzlar, Germany) under 20x or 40x oil immersion lenses. Z-stacks with 0.5-1.0 µm steps were acquired using the Leica LAS AF software. For each species, the anterior region (prostomium and first seven segments) and several mid-body segments were imaged from a total of at least 6 individuals (two each for lateral, ventral and dorsal anterior views, plus mid-body views). Z-stack images from different specimens and views were analyzed using ImageJ [60] (Bethesda, MD, USA) and Zen 2009 LE (Zeiss, Oberkochen, Germany) to infer the characteristic acetyl-tubulin immunoreactive (acTIR) and serotonin immunoreactive (SIR) elements of the nervous system, as well as other morphological landmarks. We

used a combination of maximum intensity projections, depth-color-coded projections and 3D volume reconstructions to guide our interpretation. Based on these image data, we generated hand-drawn representative diagrams (lateral and ventro-dorsal views) of the morphology of anterior and mid-body regions of each species. Representative drawings of the nervous system and associated structures for each species were traced and colored using Adobe Illustrator CS3. We used these summary drawings along with the actual Z-stacks to compare the morphology of all twelve species.

Phylogenetic analyses and ancestral character estimation

Phylogenetic relationships among the 12 study species were established primarily based on previously published studies [10,12,58,61,62]; where prior studies conflict, phylogenetic positions were resolved according to the results of a recent analysis using the largest dataset of naidid sequences yet analyzed (C. Erséus, personal communications). Relationships among clitellate and polychaete groups were also established based on previous studies [50-52]. Ancestral character estimation of peripheral nerve numbers was made using R [63] with the *ace* function from the *ape* package [64]; the trait value for each group was based on results from either this study or from existing reports for representatives of the group [2,7,15,16,20,37,49,55-57,65-67]. We used maximum likelihood estimation with a custom symmetrical transition rate matrix allowing single-step changes in character state.

Availability of supporting data

The data supporting the results of this article are included within the article and its additional files.

Additional files

Additional file 1: Figure S1. Nervous system of *Tubifex tubifex* (Clitellata: Naididae: Tubificinae). Drawings based on observations of individuals stained using anti-acetylated-alpha-tubulin and anti-serotonin antibodies, fluorescently labeled phalloidin and DAPI, imaged as Z-stacks under a confocal laser scanning microscope. Acetylated-tubulin immunoreactive (acTIR) neuropil shown in green, serotonin immunoreactive (SIR) neurites and perikarya shown in dark gray in A and B, brain shown in dark grey, ventral nerve cord ganglia shown in dark grey in C and D, and intersegmental septa shown as dashed dark red lines. A) Lateral view of the anterior end showing ventral nerve cord, segmental peripheral nerves, prostomial nerves, circumesophageal connective and brain. SIR structures shown only for brain. B) Dorsal view of the anterior end, showing same structures as A, plus SIR elements in ventral nerve cord and pharyngeal plexus. C) Lateral view of a typical trunk body segment showing localization of peripheral nerve roots relative to ganglia, septa and chaetae; SIR elements not shown. D) Schematic of the structure of the ventral nerve cord showing localization of peripheral nerve roots relative to ganglia, septa and chaetae. Abbreviations: br: brain; cec: circumesophageal connectives; cso: ciliary sense organ; dch: dorsal chaetae (notochaetae); eye: pigment cup eye (not present in all species); gut: digestive tract; mo: mouth; php: pharyngeal plexus; phx: pharynx; prb:

proboscis (not present in all species); prn: prostomial nerve; psn: peripheral segmental nerve; sep: mesodermal septum; sirl: serotonin immunoreactive lateral neuron (not present in all species); sirn: serotonin immunoreactive neuropil; sirp: serotonin immunoreactive perikaryon; tirn: acetyl-tubulin immunoreactive neuropil; vch: ventral chaetae (neurochaetae); vg: ventral ganglion; vnc: ventral nerve cord.

Additional file 2: Figure S2. Nervous system of *Pristina aequiseta* (Clitellata: Naididae: Pristininae). Drawings based on specimens prepared and labeled as in Additional file 1: Figure S1. A) Lateral view of the anterior end showing ventral nerve cord, segmental peripheral nerves, prostomial nerves, circumesophageal connective and brain. Serotonin immunoreactive (SIR) structures shown only for brain and pharyngeal plexus. B) Dorsal view of the anterior end, showing same structures as A, plus SIR elements in ventral nerve cord and pharyngeal plexus. C) Lateral view of a typical trunk body segment showing localization of peripheral nerve roots relative to ganglia, septa and chaetae; SIR elements not shown. D) Schematic of the structure of the ventral nerve cord showing localization of peripheral nerve roots relative to ganglia, septa and chaetae. Abbreviations as in Additional file 1: Figure S1.

Additional file 3: Figure S3. Nervous system of *Pristina leidyi* (Clitellata: Naididae: Pristininae). Drawings based on specimens prepared and labeled as in Additional file 1: Figure S1. A) Lateral view of the anterior end showing ventral nerve cord, segmental peripheral nerves, prostomial nerves, circumesophageal connective and brain. Serotonin immunoreactive (SIR) structures shown only for brain and pharyngeal plexus. B) Dorsal view of the anterior end, showing same structures as A, plus SIR elements in ventral nerve cord and pharyngeal plexus. C) Lateral view of a typical trunk body segment showing localization of peripheral nerve roots relative to ganglia, septa and chaetae; SIR elements not shown. D) Schematic of the structure of the ventral nerve cord showing localization of peripheral nerve roots relative to ganglia, septa and chaetae. Abbreviations as in Additional file 1: Figure S1.

Additional file 4: Figure S4. Nervous system of *Dero digitata* (Clitellata: Naididae: Naidinae). Drawings based on specimens prepared and labeled as in Additional file 1: Figure S1. A) Lateral view of the anterior end showing ventral nerve cord, segmental peripheral nerves, prostomial nerves, circumesophageal connective and brain. Serotonin immunoreactive (SIR) structures shown only for brain and pharyngeal plexus. B) Dorsal view of the anterior end, showing same structures as A, plus SIR elements in ventral nerve cord and pharyngeal plexus. C) Lateral view of a typical trunk body segment showing localization of peripheral nerve roots relative to ganglia, septa and chaetae; SIR elements not shown except for lateral subepidermal perikarya. D) Schematic of the structure of the ventral nerve cord showing localization of peripheral nerve roots relative to ganglia, septa and chaetae. Abbreviations as in Additional file 1: Figure S1.

Additional file 5: Figure S5. Nervous system of *Dero furcata* (Clitellata: Naididae: Naidinae). Drawings based on specimens prepared and labeled as in Additional file 1: Figure S1. A) Lateral view of the anterior end showing ventral nerve cord, segmental peripheral nerves, prostomial nerves, circumesophageal connective and brain. Serotonin immunoreactive (SIR) structures shown only for brain and pharyngeal plexus. B) Dorsal view of the anterior end, showing same structures as A, plus SIR elements in ventral nerve cord and pharyngeal plexus. C) Lateral view of a typical trunk body segment showing localization of peripheral nerve roots relative to ganglia, septa and chaetae; SIR elements not shown except for lateral subepidermal perikarya. D) Schematic of the structure of the ventral nerve cord showing localization of peripheral nerve roots relative to ganglia, septa and chaetae. Abbreviations as in Additional file 1: Figure S1.

Additional file 6: Figure S6. Nervous system of *Allonais paraguayensis* (Clitellata: Naididae: Naidinae). Drawings based on specimens prepared and labeled as in Additional file 1: Figure S1. A) Lateral view of the anterior end showing ventral nerve cord, segmental peripheral nerves, prostomial nerves, circumesophageal connective and brain. Serotonin immunoreactive (SIR) structures shown only for brain and pharyngeal plexus. B) Dorsal view of the anterior end, showing same structures as A, plus SIR elements in ventral nerve cord and pharyngeal plexus. C) Lateral view of a typical trunk body segment showing localization of peripheral nerve roots relative to ganglia, septa and chaetae. D) Schematic of the structure of the ventral nerve cord showing localization of peripheral nerve roots relative to ganglia, septa and chaetae. Abbreviations as in Additional file 1: Figure S1.

Additional file 7: Figure S7. Nervous system of *Paranais litoralis* (Clitellata: Naididae: Naidinae). Drawings based on specimens prepared and labeled as in Additional file 1: Figure S1. A) Lateral view of the anterior end showing ventral nerve cord, segmental peripheral nerves, prostomial nerves, circumesophageal connective and brain. Serotonin immunoreactive (SIR) structures shown only for brain and pharyngeal plexus. B) Dorsal view of the anterior end, showing same structures as A, plus SIR elements in ventral nerve cord and pharyngeal plexus. C) Lateral view of a typical trunk body segment showing localization of peripheral nerve roots relative to ganglia, septa and chaetae. D) Schematic of the structure of the ventral nerve cord showing localization of peripheral nerve roots relative to ganglia, septa and chaetae. Abbreviations as in Additional file 1: Figure S1.

Additional file 8: Figure S8. Nervous system of *Chaetogaster diaphanus* (Clitellata: Naididae: Naidinae). Drawings based on specimens prepared and labeled as in Additional file 1: Figure S1. A) Lateral view of the anterior end showing ventral nerve cord, segmental peripheral nerves, prostomial nerves, circumesophageal connective and brain. Serotonin immunoreactive (SIR) structures shown only for brain. B) Dorsal view of the anterior end, showing same structures as A, plus SIR elements in ventral nerve cord and pharyngeal plexus. C) Lateral view of a typical trunk body segment showing localization of peripheral nerve roots relative to ganglia, septa and chaetae; SIR elements not shown. D) Schematic of the structure of the ventral nerve cord showing localization of peripheral nerve roots relative to ganglia, septa and chaetae. Abbreviations as in Additional file 1: Figure S1.

Additional file 9: Figure S9. Nervous system of *Nais stolci* (Clitellata: Naididae: Naidinae). Drawings based on specimens prepared and labeled as in Additional file 1: Figure S1. A) Lateral view of the anterior end showing ventral nerve cord, segmental peripheral nerves, prostomial nerves, circumesophageal connective and brain. Serotonin immunoreactive (SIR) structures shown only for brain and pharyngeal plexus. B) Dorsal view of the anterior end, showing same structures as A, plus SIR elements in ventral nerve cord and pharyngeal plexus. C) Lateral view of a typical trunk body segment showing localization of peripheral nerve roots relative to ganglia, septa and chaetae; SIR elements not shown except for lateral subepidermal perikarya. D) Schematic of the structure of the ventral nerve cord showing localization of peripheral nerve roots relative to ganglia, septa and chaetae. Abbreviations as in Additional file 1: Figure S1.

Additional file 10: Figure S10. Nervous system of *Stylaria lacustris* (Clitellata: Naididae: Naidinae). Drawings based on specimens prepared and labeled as in Additional file 1: Figure S1. A) Lateral view of the anterior end showing ventral nerve cord, segmental peripheral nerves, prostomial nerves, circumesophageal connective and brain. Serotonin immunoreactive (SIR) structures shown only for brain and pharyngeal plexus. B) Dorsal view of the anterior end, showing same structures as A, plus SIR elements in ventral nerve cord and pharyngeal plexus. C) Lateral view of a typical trunk body segment showing localization of peripheral nerve roots relative to ganglia, septa and chaetae. D) Schematic of the structure of the ventral nerve cord showing localization of peripheral nerve roots relative to ganglia, septa and chaetae. Abbreviations as in Additional file 1: Figure S1.

Additional file 11: Figure S11. Naidid segment ground plan. A) Stereogram of a generic naidid segmental unit. B) Internal structure of a body segment. Parasagittal maximum intensity projection of a lateral Z-stack of *Stylaria lacustris* stained for DNA (blue - nuclei), F-actin (red - muscle), acetyl-tubulin (green - primarily peripheral nervous system and ciliated structures (e.g., nephridia and gut)) and serotonin (yellow - primarily ventral nerve cord). Labels: bw: body wall; chl: chloragogen cells; cm: circular muscle; dbv: dorsal blood vessel; dbw: dorsal body wall; dch: dorsal chaetae; ep: epidermis; gut: ciliated gut; ll: lateral line; lm: longitudinal muscle; nf: nephridial funnel; nph: nephridium; nt: nephrotubule; pnI-IV: peripheral segmental nerve I-IV; sep: intersegmental septum; vbv: ventral blood vessel; vbw: ventral body wall; vch: ventral chaetae; vg: ventral ganglion; vnc: ventral nerve cord.

Additional file 12: Figure S12. Diversity in anterior nervous system morphology of 12 species of Naididae. Note the variation in the position and number of ciliary sense organs (arrowheads) relative to the SIR brain neuropil, visible as a red mass of SIR neurites; note also the location and number of SIR perikarya (asterisks). Images are intensity sum projections of dorsal Z-stacks of *Tubifex tubifex* (A), *Pristina aequiseta* (B), *Pristina leidyi* (C), *Monopylephorus rubroniveus* (D), *Dero digitata* (E), *Dero furcata* (F), *Allonais paraguayensis* (G),

Paranais litoralis (H), *Amphichaeta* sp. (I), *Chaetogaster diaphanus* (J), *Nais stolci* (K) and *Stylaria lacustris* (L). Specimens were stained for DNA (blue), serotonin (red) and acetyl-tubulin (green), except for *Monopylephorus* (D), where DNA was degraded and failed to stain. Scale bars: 25 μm.

Additional file 13: Figure S13. Variation in position of the brain and ciliary sense organs of 12 species of Naididae. Images are intensity sum projection of sagittal Z-stacks. Brain boundaries are shown by paired brackets; approximate prostomium/peristomium and peristomium/segment 1 boundaries are marked by dashed lines; ciliary sense organs are indicated by arrowheads. *Tubifex tubifex* (A), *Pristina aequiseta* (B), *Pristina leidyi* (C), *Monopylephorus rubroniveus* (D), *Dero digitata* (E), *Dero furcata* (F), *Allonais paraguayensis* (G), *Paranais litoralis* (H), *Amphichaeta* sp. (I), *Chaetogaster diaphanus* (J), *Nais stolci* (K) and *Stylaria lacustris* (L). Specimens were stained for DNA (blue), serotonin (red) and acetyl-tubulin (green), except for *Monopylephorus* (D), where DNA was degraded and failed to stain. Scale bars: 25 μm.

Additional file 14: Figure S14. Diversity in prostomial innervation in 12 species of Naididae. Images are maximum intensity projections of sagittal Z-stacks, color coded by depth along the stack (relative color reference scale at bottom right). All panels show staining for acetylated-tubulin. Ciliary sense organs are indicated by arrowheads. Images are of *Tubifex tubifex* (A), *Pristina aequiseta* (B), *Pristina leidyi* (C), *Monopylephorus rubroniveus* (D), *Dero digitata* (E), *Dero furcata* (F), *Allonais paraguayensis* (G), *Paranais litoralis* (H), *Amphichaeta* sp. (I), *Chaetogaster diaphanus* (J), *Nais stolci* (K) and *Stylaria lacustris* (L). Scale bars: 25 μm.

Additional file 15: Figure S15. Diversity of ventral nerve cord ganglion architecture in 10 species of Naididae. Images are intensity sum projections of ventral Z-stacks. Specimens were stained for DNA (blue), serotonin (red) and acetyl-tubulin (green). Segmental nerves are labeled I-IV; alternative arabic numerals are shown in G and H. Ventral chaetae (ch) are visible due to birefringence. The paired arrowheads mark the position of the mesodermal septum. The looping, acetyl-tubulin positive structures in the image corresponds to ciliated nephridia (nf: nephridial funnel; nt: nephrotubule). *Tubifex tubifex* (A), *Pristina aequiseta* (B), *Pristina leidyi* (C), *Dero digitata* (D), *Dero furcata* (E), *Paranais litoralis* (F), *Amphichaeta* sp. (G), *Chaetogaster diaphanus* (H), *Nais stolci* (I) and *Stylaria lacustris* (J). *Monopylephorus* is not included here due to failure of DAPI staining and low quality of acITR detection. Scale bars: 25 μm.

Additional file 16: Figure S16. Central and peripheral nervous system of anterior segments of 9 species of Naididae. Images are intensity sum projections of ventral Z-stacks showing the ventral nerve cord neuropil and segmental peripheral nerves. Labeled acetyl-tubulin immunoreactive nerves (green), and serotonin immunoreactive nerves and perikarya (red) consistently show a different pattern in anterior-most segments relative to more posterior segments, but the level at which this transition occurs varies across species, as well as across the elements of the nervous system. The transition between the anterior and posterior pattern of segmental nerves is indicated by occurs the opposed arrowheads (number of segmental nerves per segment indicated beside arrowheads). Nerve identity is shown below for segments flanking the boundary. Images are from *Pristina leidyi* (A), *Monopylephorus rubroniveus* (B), *Dero digitata* (C), *Dero furcata* (D), *Paranais litoralis* (E), *Amphichaeta* sp. (F), *Chaetogaster diaphanus* (G), *Nais stolci* (H) and *Stylaria lacustris* (I). Scale bars: 25 μm.

Additional file 17: Figure S17. Diversity in architecture of anterior ventral nerve cord ganglia. Images are intensity sum projections of ventral Z-stacks showing cell nuclei. The green arrowhead marks the location of the anterior-most connective. Images are from *Pristina aequiseta* (A), *Dero digitata* (B), *Dero furcata* (C), *Paranais litoralis* (D), *Chaetogaster diaphanus* (E), *Nais stolci* (F) and *Stylaria lacustris* (G). Scale bars: 25 μm.

Additional file 18: Figure S18. Single serotonin-immunoreactive cells in the lateral body wall of *Nais stolci*. A,B) Representative images of two different segments from different individuals; maximum intensity projections of sagittal Z-stacks of representing thick optical sections level to the lateral body wall of mid-trunk segments. Specimens were stained for DNA (blue), serotonin (red) and acetyl-tubulin (green). ll: lateral line; psn: peripheral segmental nerve; sirl: serotonin-immunoreactive cell; vch: ventral chaetae. Scale bars: 25 μm.

Additional file 19: Supplementary results.

Abbreviations
acTIR: acetyl-tubulin immunoreactive; CNS: Central nervous system; PNS: Peripheral nervous system; SIR: Serotonin immunoreactive.

Competing interests
The authors declare that they have no competing interests.

Authors' contributions
EEZ conceived the study, conducted the fluorescent stainings, the CLSM imaging and 3D reconstruction, made interpretative hand and digital drawings, and wrote the first draft of the manuscript and assembled the figures. AEB contributed substantially to the writing of the manuscript. Both authors read and approved the final manuscript.

Acknowledgments
We are thankful to Amy E. Beaven for her continual support during confocal imaging at the UMCP-CMNS-CBMG Imaging Core Facility. E.E.Z. would like to warmly thank Kyomu Magoya for diligently taking care of so many tasks no one else was willing to do. The authors also thank two anonymous reviewers whose comments and suggestions greatly enhanced our original manuscript. This research was supported in part by National Science Foundation grant IOS-0920502 to A.E.B. and by scholarships from the University of Maryland Graduate School and the Behavior, Ecology, Evolution, and Systematics Graduate Program to E.E.Z.

References
1. Brusca RC, Brusca GJ. Invertebrates. Sinauer: Sunderland, MA; 1990.
2. Bullock TH, Horridge GA. Structure and Function in the Nervous System of Invertebrates. London: W.H. Freeman and Company; 1965.
3. Arendt D, Denes AS, Jékely G, Tessmar-Raible K. The evolution of nervous system centralization. Philos Trans R Soc B Biol Sci. 2008;363:1523–8.
4. Holland LZ, Carvalho JE, Escriva H, Laudet V, Schubert M, Shimeld SM, et al. Evolution of bilaterian central nervous systems: a single origin? EvoDevo. 2013;4:27.
5. Engelbrecht van Z D. The annelid ancestry of the chordates and the origin of the chordate central nervous system and the notochord. J Zool Syst Evol Res. 1969;7:18–30.
6. Northcutt RG. Evolution of centralized nervous systems: Two schools of evolutionary thought. Proc Natl Acad Sci. 2012;109:10626–33.
7. Orrhage L, Müller MCM. Morphology of the nervous system of Polychaeta (Annelida). Hydrobiologia. 2005;535/536:79–111.
8. McHugh D. Molecular phylogeny of the Annelida. Can J Zool. 2000;78:1873–84.
9. Erséus C. Phylogeny of oligochaetous Clitellata. Hydrobiologia. 2005;535–536:357–72.
10. Erséus C, Envall I, Marchese M, Gustavsson L. The systematic position of Opistocystidae (Annelida, Clitellata) revealed by DNA data. Mol Phylogenet Evol. 2010;54:309–13.
11. Erséus C, Wetzel MJ, Gustavsson LN. IZCD rules - a farewell to Tubificidae (Annelida, Clitellata). Zootaxa. 2008;1744:66–8.
12. Envall I, Källersjö M, Erséus C. Molecular evidence for the non-monophyletic status of Naidinae (Annelida, Clitellata, Tubificidae). Mol Phylogenet Evol. 2006;40:570–84.
13. Dehorne L. Les naidimorphes et leur reproduction asexuée. Arch Zool Exp Générale. 1916;56:25–157.
14. Brode HS. A contribution to the morphology of Dero vaga. J Morphol. 1898;14:141–80.
15. Stephenson J. The Oligochaeta. Oxford: Oxford University Press; 1930.
16. Muller MCM. Polychaete nervous systems: Ground pattern and variations–cLS microscopy and the importance of novel characteristics in phylogenetic analysis. Integr Comp Biol. 2006;46:125–33.
17. Hessling R, Müller MC, Westheide W. CLSM analysis of serotonin-immunoreactive neurons in the central nervous system of Nais variabilis, Slavina appendiculata and Stylaria lacustris (Oligochaeta: Naididae). Hydrobiologia. 1999;406:223–33.
18. Müller M. Nerve development, growth and differentiation during regeneration in Enchytraeus fragmentosus and Stylaria lacustris (Oligochaeta). Dev Growth Differ. 2004;46:471–8.
19. Zattara EE, Bely AE. Evolution of a novel developmental trajectory: fission is distinct from regeneration in the annelid Pristina leidyi. Evol Dev. 2011;13:80–95.
20. Hessling R, Westheide W. CLSM analysis of development and structure of the central nervous system of Enchytraeus crypticus ("Oligochaeta", Enchytraeidae). Zoomorphology. 1999;119:37–47.
21. Müller M, Berenzen A, Westheide W. Experiments on anterior regeneration in Eurythoe complanata ("Polychaeta", Amphinomidae): reconfiguration of the nervous system and its function for regeneration. Zoomorphology. 2003;122:95–103.
22. Richter S, Loesel R, Purschke G, Schmidt-Rhaesa A, Scholtz G, Stach T, et al. Invertebrate neurophylogeny: suggested terms and definitions for a neuroanatomical glossary. Front Zool. 2010;7:29.
23. Bullock TH. Annelida. In: Bullock TH, Horridge GH, editors. Structure and Function in the Nervous System of Invertebrates. Volume Vol. 1. San Francisco: Freeman; 1965. p. 661–789.
24. Erséus C, Kallersjo M. 18S rDNA phylogeny of Clitellata (Annelida). Zool Scr. 2004;33:187–96.
25. Marotta R, Ferraguti M, Erseus C, Gustavsson LM. Combined-data phylogenetics and character evolution of Clitellata (Annelida) using 18S rDNA and morphology. Zool J Linn Soc. 2008;154:1–26.
26. Marsden CA, Kerkut GA. Fluorescent microscopy of the 5HT- and catecholamine-containing cells in the central nervous system of the leech Hirudo medicinalis. Comp Biochem Physiol. 1969;31:851–62.
27. Stuart DK, Blair SS, Weisblat DA. Cell lineage, cell death, and the developmental origin of identified serotonin- and dopamine-containing neurons in the leech. J Neurosci. 1987;7:1107–22.
28. Spörhase-Eichmann U, Gras H, Schürmann F-W. Patterns of serotonin-immunoreactive neurons in the central nervous system of the earthworm Lumbricus terrestris L. II. Rostral and caudal ganglia. Cell Tissue Res. 1987;249:625–32.
29. Macagno ER, Stewart RR. Cell death during gangliogenesis in the leech: competition leading to the death of PMS neurons has both random and nonrandom components. J Neurosci. 1987;7:1911–8.
30. Anctil M, Waele J-PD, Miron M-J, Pani AK. Monoamines in the nervous system of the tube-worm Chaetopterus variopedatus (Polychaeta): Biochemical detection and serotonin immunoreactivity. Cell Tissue Res. 1990;259:81–92.
31. Reglödi D, Lubics A, Szelier M, Lengvári I. Serotonin immunoreactivity in the peripheral nervous system of oligochaeta. Acta Biol Hung. 1997;48:439–51.
32. Stewart RR, Gao WQ, Peinado A, Zipser B, Macagno ER. Cell death during gangliogenesis in the leech: bipolar cells appear and then degenerate in all ganglia. J Neurosci. 1987;7:1919–27.
33. Hessling R, Purschke G. Immunohistochemical (cLSM) and ultrastructural analysis of the central nervous system and sense organs in Aeolosoma hemprichi (Annelida, Aeolosomatidae). Zoomorphology. 2000;120:65–78.
34. Zattara EE, Bely AE. Investment choices in post-embryonic development: Quantifying interactions among growth, regeneration, and asexual reproduction in the annelid Pristina leidyi. J Exp Zoolog B Mol Dev Evol. 2013;320:471–88.
35. Spörhase-Eichmann U, Gras H, Schürmann F-W. Patterns of serotonin-immunoreactive neurons in the central nervous system of the earthworm Lumbricus terrestris L. Cell Tissue Res. 1987;249:601–14.
36. Koza A, Wilhelm M, Hiripi L, Elekes K, Csoknya M. Embryogenesis of the serotonergic system in the earthworm Eisenia fetida (Annelida, Oligochaeta): Immunohistochemical and biochemical studies. J Comp Neurol. 2006;497:451–67.
37. Smith JE. The Nervous Anatomy of the Body Segments of Nereid Polychaetes. Philos Trans R Soc Lond B Biol Sci. 1957;240:135–96.
38. Goto A, Kitamura K, Shimizu T. Cell lineage analysis of pattern formation in the Tubifex embryo. I. Segmentation in the mesoderm. Int J Dev Biol. 1999;43:317–27.
39. Nakamoto A, Arai A, Shimizu T. Cell lineage analysis of pattern formation in the Tubifex embryo. II. Segmentation in the ectoderm. Int J Dev Biol. 2000;44:797–805.
40. Goto A, Kitamura K, Arai A, Shimizu T. Cell fate analysis of teloblasts in the Tubifex embryo by intracellular injection of HRP. Dev Growth Differ. 1999;41:703–13.
41. Fernández J, Stent GS. Embryonic development of the hirudinid leech Hirudo medicinalis: structure, development and segmentation of the germinal plate. J Embryol Exp Morphol. 1982;72:71–96.
42. Irvine SM, Martindale MQ. Cellular and molecular mechanisms of segmentation in annelids. Semin Cell Dev Biol. 1996;7:593–604.
43. Seaver EC, Shankland M. Leech Segmental Repeats Develop Normally in the Absence of Signals from either Anterior or Posterior Segments. Dev Biol. 2000;224:339–53.
44. Gline SE, Kuo D-H, Stolfi A, Weisblat DA. High resolution cell lineage tracing reveals developmental variability in leech. Dev Dyn. 2009;238:3139–51.
45. Blair SS. Interactions between mesoderm and ectoderm in segment formation in the embryo of a glossiphoniid leech. Dev Biol. 1982;89:389–96.

46. Gleizer L, Stent GS. Developmental origin of segmental identity in the leech mesoderm. Development. 1993;117:177–89.

47. Shain DH, Stuart DK, Huang FZ, Weisblat DA. Segmentation of the central nervous system in leech. Development. 2000;127:735–44.

48. Isossimow VV. Zur Anatomie des Nervensystems der Lumbriculiden. Zool Jb Anat. 1926;48:365–404.

49. Müller, M. C. Das Nervensystem der Polychaeten: Immunhistochemische Untersuchungen an ausgewählten Taxa. Dissertation. Univ. Osnabrück; 1999.

50. Struck TH, Paul C, Hill N, Hartmann S, Hosel C, Kube M, et al. Phylogenomic analyses unravel annelid evolution. Nature. 2011;471:95–8.

51. Zrzavy J, Riha P, Pialek L, Janouskovec J. Phylogeny of Annelida (Lophotrochozoa): total-evidence analysis of morphology and six genes. BMC Evol Biol. 2009;9:189.

52. Weigert A, Helm C, Meyer M, Nickel B, Arendt D, Hausdorf B, et al. Illuminating the base of the annelid tree using transcriptomics. Mol Biol Evol. 2014;31(6):1391–401.

53. Hessling R. Novel aspects of the nervous system of *Bonellia viridis* (Echiura) revealed by the combination of immunohistochemistry, confocal laser-scanning microscopy and three-dimensional reconstruction. Hydrobiologia. 2003;496:225–39.

54. Müller MCM, Henning L. Ground plan of the polychaete brain—I. Patterns of nerve development during regeneration in *Dorvillea bermudensis* (Dorvilleidae). J Comp Neurol. 2004;471:49–58.

55. Müller MCM, Westheide W. Comparative analysis of the nervous systems in presumptive progenetic dinophilid and dorvilleid polychaetes (Annelida) by immunohistochemistry and cLSM. Acta Zool. 2002;83:33–48.

56. Müller MC, Westheide W. Structure of the nervous system of *Myzostoma cirriferum* (Annelida) as revealed by immunohistochemistry and cLSM analyses. J Morphol. 2000;245:87–98.

57. Martin N, Anctil M. The nervous system of the tube-worm *Chaetopterus variopedatus* (Polychaeta). J Morphol. 1984;181:161–73.

58. Bely AE, Sikes JM. Latent regeneration abilities persist following recent evolutionary loss in asexual annelids. Proc Natl Acad Sci. 2010;107:1464–9.

59. Zattara EE. Regeneration, fission and the evolution of developmental novelty in naid annelids. In: Ph.D. Thesis. College Park: University of Maryland; 2012.

60. Abramoff MD, Magalhaes PJ, Ram SJ. Image processing with ImageJ. Biophotonics Int. 2004;11:36–42.

61. Erséus C, Prestegaard T, Kallersjo M. Phylogenetic analysis of Tubificidae (Annelida, Clitellata) based on 18S rDNA sequences. Mol Phylogenet Evol. 2000;15:381–9.

62. Bely AE, Wray GA. Molecular phylogeny of naidid worms (Annelida: Clitellata) based on cytochrome oxidase I. Mol Phylogenet Evol. 2004;30:50–63.

63. R Development Core Team. R: A Language and Environment for Statistical Computing. Vienna, Austria: R Foundation for Statistical Computing; 2011.

64. Paradis E, Claude J, Strimmer K. APE: Analyses of Phylogenetics and Evolution in R language. Bioinformatics. 2004;20:289–90.

65. Nicol J a C. The Giant Nerve-Fibres in the Central Nervous System of *Myxicola* (Polychaeta, Sabellidae). Q J Microsc Sci. 1948;89(3):1–45.

66. Filippova AV, Tzetlin AB, Purschke G. Morphology and ultrastructure of the anterior end of *Diplocirrus longisetosus* Marenzeller, 1890 (Flabelligeridae, Polychaeta, Annelida). Hydrobiologia. 2003;496:215–23.

67. Winchell CJ, Valencia JE, Jacobs DK. Confocal analysis of nervous system architecture in direct-developing juveniles of *Neanthes arenaceodentata* (Annelida, Nereididae). Front Zool. 2010;7:17.

10

Myoinhibitory peptide regulates feeding in the marine annelid *Platynereis*

Elizabeth A Williams, Markus Conzelmann and Gáspár Jékely*

Abstract

Background: During larval settlement and metamorphosis, marine invertebrates undergo changes in habitat, morphology, behavior and physiology. This change between life-cycle stages is often associated with a change in diet or a transition between a non-feeding and a feeding form. How larvae regulate changes in feeding during this life-cycle transition is not well understood. Neuropeptides are known to regulate several aspects of feeding, such as food search, ingestion and digestion. The marine annelid *Platynereis dumerilii* has a complex life cycle with a pelagic non-feeding larval stage and a benthic feeding postlarval stage, linked by the process of settlement. The conserved neuropeptide myoinhibitory peptide (MIP) is a key regulator of larval settlement behavior in *Platynereis*. Whether MIP also regulates the initiation of feeding, another aspect of the pelagic-to-benthic transition in *Platynereis*, is currently unknown.

Results: Here, we explore the contribution of MIP to the regulation of feeding behavior in settled *Platynereis* postlarvae. We find that in addition to expression in the brain, MIP is expressed in the gut of developing larvae in sensory neurons that densely innervate the hindgut, the foregut, and the midgut. Activating MIP signaling by synthetic neuropeptide addition causes increased gut peristalsis and more frequent pharynx extensions leading to increased food intake. Conversely, morpholino-mediated knockdown of MIP expression inhibits feeding. In the long-term, treatment of *Platynereis* postlarvae with synthetic MIP increases growth rate and results in earlier cephalic metamorphosis.

Conclusions: Our results show that MIP activates ingestion and gut peristalsis in *Platynereis* postlarvae. MIP is expressed in enteroendocrine cells of the digestive system suggesting that following larval settlement, feeding may be initiated by a direct sensory-neurosecretory mechanism. This is similar to the mechanism by which MIP induces larval settlement. The pleiotropic roles of MIP may thus have evolved by redeploying the same signaling mechanism in different aspects of a life-cycle transition.

Background

Many organisms have a complex life cycle consisting of distinct stages that differ in form, physiology, behavior and habitat. Among benthic marine invertebrates, a common life cycle strategy, the biphasic life cycle, consists of a free-swimming larva that settles to the ocean floor and undergoes metamorphosis to a bottom-dwelling adult [1]. Marine invertebrate larval settlement is often coupled to the initiation of feeding or a change in diet [2]. These behavioral, physiological and morphological changes have to be tightly coordinated for the successful transition to a benthic life style. Knowledge of how this transition is regulated is important for understanding population structure in the ocean, life-history evolution, and how the environment influences animal life cycles [3-8].

The marine annelid *Platynereis dumerilii* has recently proven to be a useful marine invertebrate model for studying the molecular details of marine larval behavior, including settlement [9-12]. *Platynereis* has a biphasic life cycle with free-swimming, non-feeding larval (trochophore and nectochaete) stages and bottom-dwelling, feeding postlarval, juvenile and adult stages [13]. Larval settlement is followed by a period of growth and feeding, during which juvenile *Platynereis* add additional posterior segments. Cephalic metamorphosis, in which the first pair of parapodia are transformed into a second pair of tentacular cirri

---
* Correspondence: gaspar.jekely@tuebingen.mpg.de
Max Planck Institute for Developmental Biology, Spemannstrasse 35,
Tübingen 72076, Germany

on the head, occurs after the juveniles have begun to add their sixth posterior segment [13-16].

Recently, we identified myoinhibitory peptide (MIP) as an inducer of rapid larval settlement behavior in *Platynereis* [11]. MIP is expressed in anterior chemosensory-neurosecretory neurons of the larva. Exogenous application of MIP inhibits the activity of the locomotor cilia, resulting in rapid sinking, and induces sustained contact with the substrate. *Platynereis* MIP belongs to an ancient neuropeptide family of Wamides, which are characterized by their amidated C-terminal tryptophan residue preceded by a small aliphatic residue [11,17]. Wamides are widespread among eumetazoans, except deuterostomes, and recently emerged as conserved regulators of life-cycle transitions [18]. For example, in some insects, MIP (also known as prothoracicostatic peptide (PTSP) or allatostatin-B (AST-B)) regulates ecdysone [19-21] and juvenile hormone levels [22], potentially influencing the timing of larval ecdysis and pupation. In cnidarians, including some corals and hydrozoans, GLWamide (also called metamorphosin) is known to induce larval settlement and metamorphosis [23-25].

How changes in feeding are regulated during marine life-cycle transitions is less well understood. Many neuropeptides are known to have roles in regulating different aspects of feeding [26-28]. MIPs/Wamides are also pleiotropic [29-35] and regulate aspects of feeding and gut muscle activity in some insects and cnidarians. The first MIP described had a myoinhibitory function on adult locust hindgut [36]. In several insects, MIP is expressed in the adult stage and can suppress muscle contractions of the hindgut [36-40]. MIP is also expressed in the stomatogastric nervous system of the adult crab, *Cancer borealis*, where it decreases the frequency of pyloric rhythm [41,42]. In addition, cnidarian GLWamide increases myoactivity in both hydra and sea anemone polyps, potentially influencing feeding [43,44]. Although none of these studies directly quantified feeding in the whole organism, MIP is a strong candidate for the regulation of feeding during marine life-cycle transitions.

Here, we study the expression and function of MIP in *Platynereis* during late larval (3-6 days) and early juvenile development (6-30 days). We found MIP expression in sensory neurons of the gut of 6 days and older *Platynereis*. We used both peptide-soaking and morpholino-mediated knockdown approaches to establish a role for MIP in the regulation of postlarval feeding and gut peristalsis. MIP treatment also resulted in faster juvenile growth, probably as a consequence of increased food ingestion and gut movement. Our results establish MIP as a pleiotropic neuropeptide in *Platynereis* that links behavioral and physiological components of a life-cycle transition.

Results
Platynereis MIP is expressed in the brain and gut of postlarvae, juveniles and adults
Expression profiling of the *MIP* precursor gene by RNA *in situ* hybridization showed that *MIP* is expressed during both larval and postlarval development and continues to be expressed after cephalic metamorphosis, in the early adult stage (Figure 1A-C, G-H; Additional file 1). At 6 days and older, *MIP* is expressed in both the median brain and the trunk nervous system, in paired cells and also in single cells closer to the larval midline. We also found *MIP* expression in the digestive system, in the fore-, mid- and hindgut. The different regions of the gut are delineated by the differential expression patterns of *Platynereis* digestive system marker genes (Figure 1D, I; discussed below). The *MIP*-expressing cells in the gut have sensory dendrites that project toward the lumen of the gut (Figure 1B-C; Additional files 2, 3). In some of these dendrites we could even detect the *MIP* RNA *in situ* signal, allowing the unambiguous assignment of these acetylated tubulin-positive cellular projections to the MIP-expressing cells (Additional file 3).

In addition to *MIP*, we also attempted to characterize the expression of the *MIP receptor* in 6 dpf and older larvae. We previously described the expression of the MIP receptor in the head of 2 dpf *Platynereis* larvae [11], however, in older larvae and postlarvae, the levels of *MIP receptor* expression proved too low to detect reliably with our RNA *in situ* hybridization method. The low expression of the *Platynereis MIP receptor* is typical of most G protein-coupled receptor expression levels [45].

Immunostaining with an antibody against *Platynereis* MIP showed that in addition to the neurosecretory plexus of the brain, MIP peptide is transported throughout the ventral nerve cord. At 6 days post fertilization (dpf), MIP-expressing neurons in the digestive system innervate both the foregut and hindgut (Figure 1E, F). As larvae progress from 3 to 6 dpf, during which time the digestive system develops, MIP expression first emerges in the developing hindgut at 4 dpf, followed by the expression in the foregut at 6 dpf (Additional file 4A-L). By one month, MIP-expressing cells densely innervate the entire length of the gut, forming a nerve-net. Using an antibody against the conserved C-amidated dipeptide VWamide [11], we found similar immunolabeling in the brain, ventral nerve cord and gut of larvae of *Capitella teleta*, a distantly related annelid species [46] (Additional file 4M-P).

By combining phalloidin staining and MIP immunostaining in *Platynereis* 1 month post fertilization (1 mpf), we could assess the location of MIP-expressing neurons in the gut in relation to the digestive system musculature. In the foregut and in the sphincter muscle that separates foregut from hindgut, MIP-expressing neurons are intermingled with the muscle tissue of the pharynx and

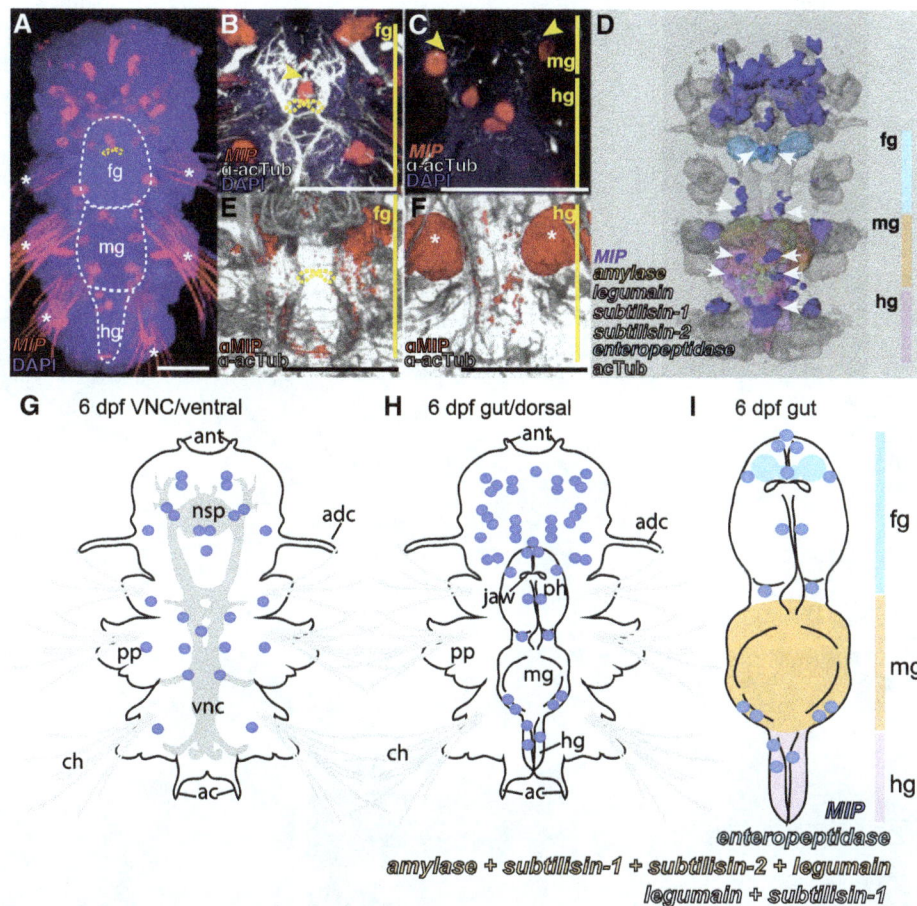

Figure 1 MIP is expressed in the digestive system of 6 dpf *Platynereis*. (A) Whole-mount RNA *in situ* hybridization (WMISH) for the *Platynereis MIP* precursor (red) counterstained with DAPI nuclear stain (blue). Ventral view of 6 dpf *Platynereis* with the image stacks corresponding to the ventral nerve cord (VNC) region not included in the maximal projection (outlined with white dashed line). White asterisks indicate background reflection from chaetae. **(B, C)** Close-up WMISH for the *Platynereis MIP* precursor (red) counterstained for acetylated tubulin (α-acTub) (white) and DAPI nuclear stain (blue). Yellow arrowheads indicate sensory dendrites of *MIP*-expressing cells. **(B)** Ventral view of the foregut. **(C)** Ventral view of the midgut and hindgut. **(D)** Dorsal view of surface representation of average *MIP* precursor expression domains registered to a 6 dpf nuclear stain reference template. Expression domains of *alpha-amylase, legumain protease precursor, subtilisin-1, subtilisin-2* and *enteropeptidase* are included as digestive system markers. White arrows indicate areas of *MIP* expression associated with the digestive system. **(E, F)** Immunostaining with *Platynereis* MIP antibody (red) counterstained with acetylated tubulin (grey) at 6 dpf, with the image stacks corresponding to the VNC region not included in the maximum projection to show only the digestive system. **(E)** Ventral view of foregut. **(F)** Ventral view of hindgut. White asterisks indicate background reflection from spinning glands. **(G-I)** Schematic representation of *MIP* precursor expression in 6 dpf *Platynereis*. **(G)** Ventral side. **(H)** Dorsal side. **(I)** *MIP* expression relative to expression of digestive system marker genes in the gut. In **(A, B, E)**, yellow dashed lines indicate the jaws. Scale bars: 50 μM. Abbreviations: fg, foregut; mg, midgut; hg, hindgut; ant, antenna; nsp, neurosecretory plexus; adc, anterior dorsal cirrus; pp, parapodia; vnc, ventral nerve cord; ch, chaetae; ac, anal cirrus; ph, pharynx.

sphincter (Figure 2A-C). In the mid- and hindgut, MIP-expressing neurons sit in the inner epithelial cell layer underlying the smooth muscles of the gut (Figure 2D-K). The axons of the MIP-expressing cells in the mid- and hind-gut of 1 month post fertilization (mpf) *Platynereis* run parallel to and just beneath the muscle fibers of both circular and longitudinal smooth muscles (Figure 2I-K). The spatial expression patterns of *Platynereis MIP* and MIP peptide suggest a potential role for MIP signaling in feeding and digestion during larval and early juvenile stages of the life cycle.

Characterization of normal gut development and the initiation of feeding in postlarvae

At 6 dpf *Platynereis* postlarvae have a through gut with clearly recognizable fore-, mid- and hindgut regions (Figure 1). The foregut contains the muscular and extendable pharynx with the jaws and salivary glands (Additional file 2). Phyllodocid polychaetes, such as *Platynereis*, have an axial muscular pharynx consisting of circular, longitudinal and radial muscle fibers, which allow for complex sucking and swallowing movements [47,48]. The foregut-midgut boundary is marked by the presence of a sphincter

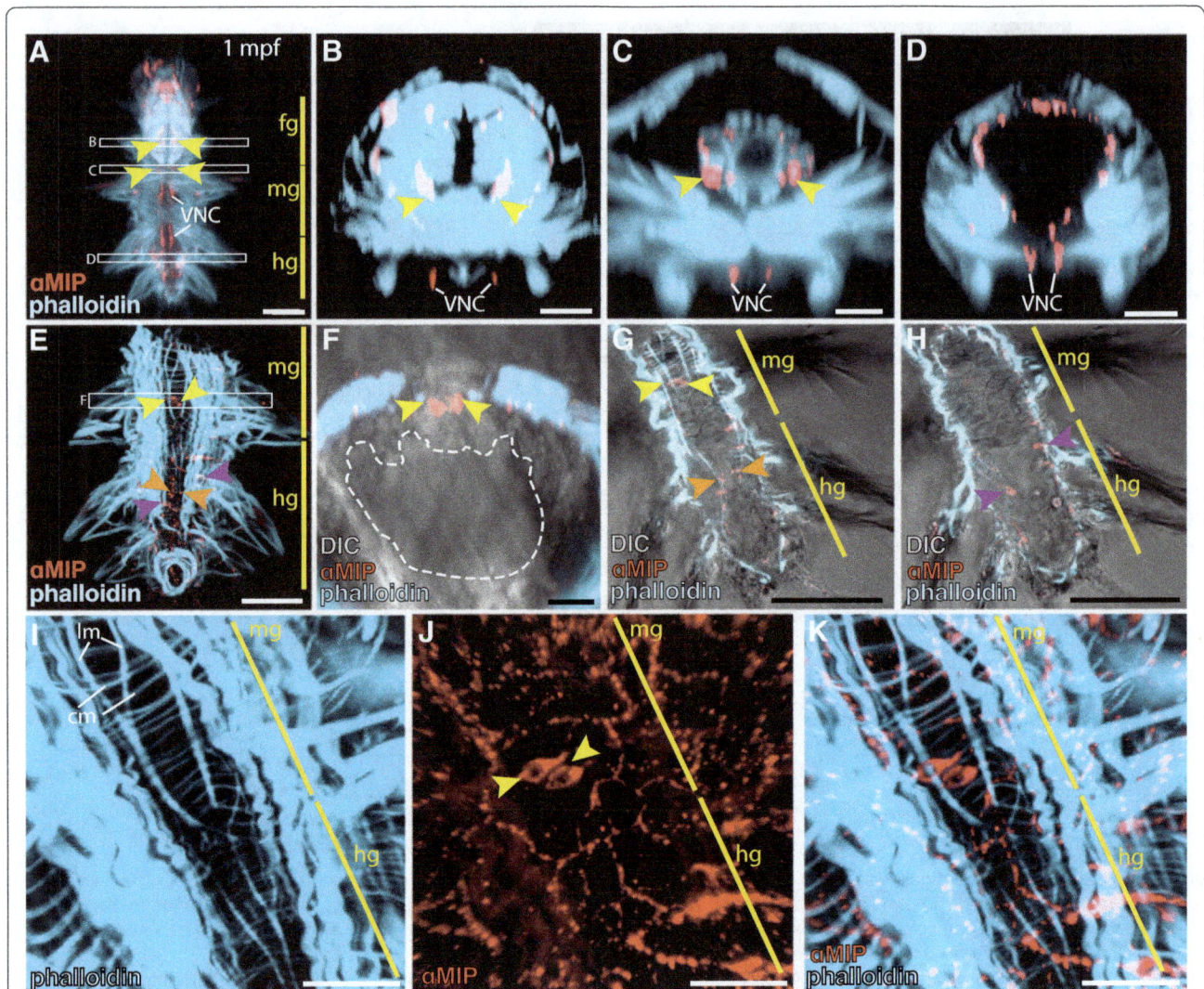

Figure 2 Co-staining of MIP and phalloidin in 1 month post fertilization (mpf) *Platynereis*. (A-K) Immunostaining of 1mpf *Platynereis* with an antibody raised against *Platynereis* MIP7 (red), counterstained with phalloidin. **(A)** Full body ventral view. White boxes indicate areas examined in cross-section in **(B-D)**. Yellow arrowheads in **(A-D)** indicate cell bodies of MIP-expressing neurons. **(B-D)** Apical views of 10 μM cross-sections of the foregut **(B)**, foregut-midgut boundary **(C)** and hindgut **(D)**. **(E)** Ventral view of mid- and hindgut with ventral nerve cord region removed to expose digestive system. White box indicates area examined in cross-section in **(F)**. Yellow arrowheads in **(E-H, J)** indicate cell bodies of MIP-expressing neurons in the midgut. Orange and purple arrowheads in **(E, G, H)** mark cell bodies of MIP-expressing neurons in the hindgut. **(F)** Apical view of 10 μM cross-section in the midgut. Dashed white line marks the boundary between the gut lumen and epithelial cell layer. **(G, H)** Lateral layers of **(E)** indicating the MIP-expressing cell bodies of the midgut and hindgut that sit in the gut epithelium, just underlying the gut musculature. **(I-K)** Close-up ventral view of the mid- and hindgut. **(I)** Phalloidin staining shows the circular and longitudinal smooth muscle fibres of the gut. **(J)** MIP immunostaining. **(K)** Overlay of MIP immunostaining and phalloidin staining. The axons of MIP-expressing neurons run parallel to the muscle fibres of the gut. Scale bars in **(A, E, G-K)**: 50 μM, in **(B-D, F)**: 20 μM. Abbreviations: fg, foregut; mg, midgut; hg, hindgut; VNC, ventral nerve cord; DIC, differential interference contrast; cm, circular muscle fibre; lm, longitudinal muscle fibre.

muscle. This muscle showed regular contractions in larvae expressing a genetically encoded calcium indicator GCaMP6 (Additional file 5). The broad midgut does not show regionalization and is followed by a short and narrow hindgut.

To gain insight into the morphology and maturation of the *Platynereis* digestive system, we carried out whole-mount RNA *in situ* hybridization on 6 dpf, 14 dpf and 1 mpf *Platynereis* with marker genes selected from an

ongoing broad RNA *in situ* hybridization screen, based on their expression in the digestive system at 6 dpf. The digestive system marker genes were identified through domain conservation, reciprocal BLAST and phylogenetic analyses as: extracellular digestive enzymes, peptidases *subtilisin-1 and subtilisin-2* (Peptidase_S8; Pfam domain: PF00082), the protease *enteropeptidase*, responsible for the activation of proteolytic enzymes [49], the polysaccharide-digesting enzyme *alpha-amylase*, and the intracellular digestive enzyme *legumain protease precursor* (Figure 1D, I; Additional files 6, 7, 8). *Alpha-amylase* and *subtilisin-1* expression was restricted to the midgut at 6 dpf, but expanded to mid- and hindgut at 14 dpf and 1 mpf. *Legumain protease precursor* was constantly expressed in both mid- and hindgut from 6 dpf to 1 mpf, while *subtilisin-2* expression was restricted to the midgut from 6 dpf to 1 mpf. *Enteropeptidase* was the only gene with expression in the foregut, including the salivary glands, at 6 and 14 dpf. At 1 mpf, *enteropeptidase* remained strongly expressed in the foregut, but expression also extended to the mid- and hindgut. Registration of these marker gene expression patterns [50] at 6 dpf to a common nuclear stain reference scaffold, along with the average 6 dpf *MIP* expression, highlighted the close association of *MIP*-expressing cells with the digestive system at this stage (Figure 1D; Additional file 9).

We also looked at the change in expression of these digestive system marker genes and *MIP* across the *Platynereis* life cycle in stage-specific RNA-seq datasets [51].

With the exception of *legumain protease precursor*, the expression of all digestive system marker genes was undetectable in non-feeding larval stages but sharply increased between 4 and 10 dpf (Additional file 10). In accordance with a digestive function, these genes were strongly down-regulated in the adult non-feeding epitokes. *MIP* expression also increased sharply between 4 and 10 dpf, although it continued to be expressed in the non-feeding epitokes, suggesting further functional roles beyond feeding in *Platynereis*.

Following settlement, *Platynereis* larvae have been reported to begin feeding between 5 – 8 dpf, with considerable variation between individuals [13,16]. Due to this variability, we decided to document feeding initiation in our own laboratory culture (Figure 3A). We added *Tetraselmis marina* microalgae to the larval cultures and documented feeding based on chlorophyll fluorescence in the gut (Figure 4C). Most larvae initiated feeding between 6 and 7 dpf; by 8 dpf, nearly all larvae had started feeding (Figure 3A).

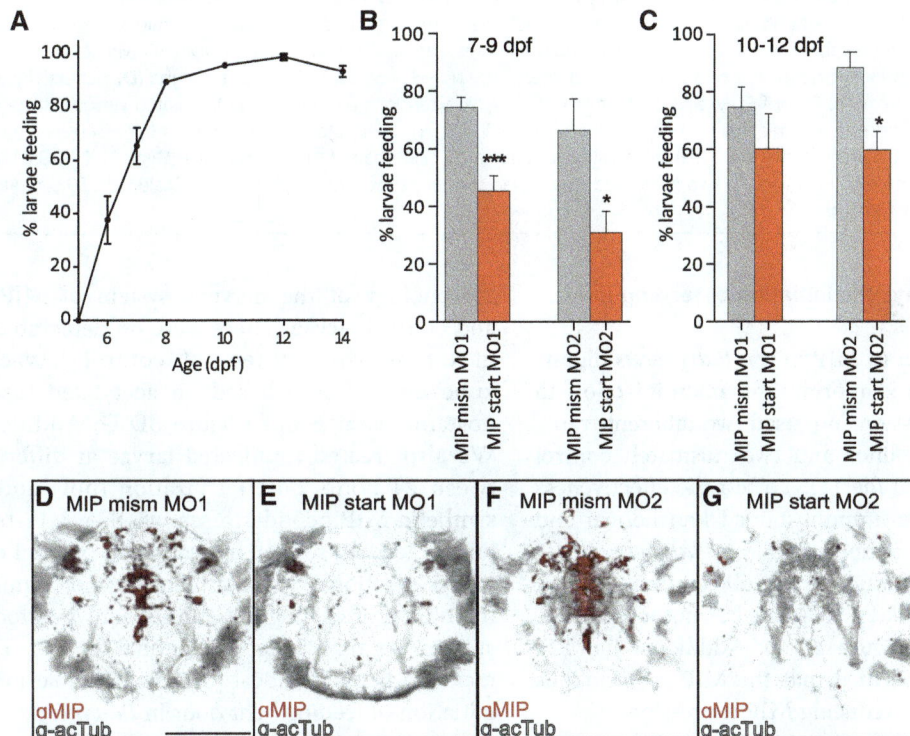

Figure 3 Feeding in control and MIP knockdown larvae. (A) Timecourse of initiation of feeding during *Platynereis* larval development. % larvae feeding on *Tetraselmis marina* algae between 5 – 14 dpf, n = 3 x 30 larvae. **(B)** % larvae feeding at 7 - 9 dpf following injection of MIP mismatch (n = 210 larvae mismatch MO1, 118 larvae mismatch MO2) or start (n = 424 larvae start MO1, 79 larvae start MO2) morpholinos **(C)** % larvae feeding at 10 - 12 dpf following injection of MIP mismatch (n = 115 larvae mismatch MO1, 72 larvae mismatch MO2) and start (n = 185 larvae start MO1, 62 larvae start MO2) morpholinos. Data in **(A-C)** are shown as mean +/- s.e.m. p-value cut-offs based on unpaired *t*-test: *** <0.001; ** < 0.01; * <0.05. **(D–G)** Anterior view of 6 dpf *Platynereis* injected with MIP mismatch or start morpholinos and immunostained with *Platynereis* MIP antibody (red) counterstained with acetylated tubulin (grey). Identical confocal microscopy and image processing parameters were applied to all images. Scale bar: 100 μm. Abbreviations: mism, mismatch; MO, morpholino; a-acTub, anti acetylated tubulin.

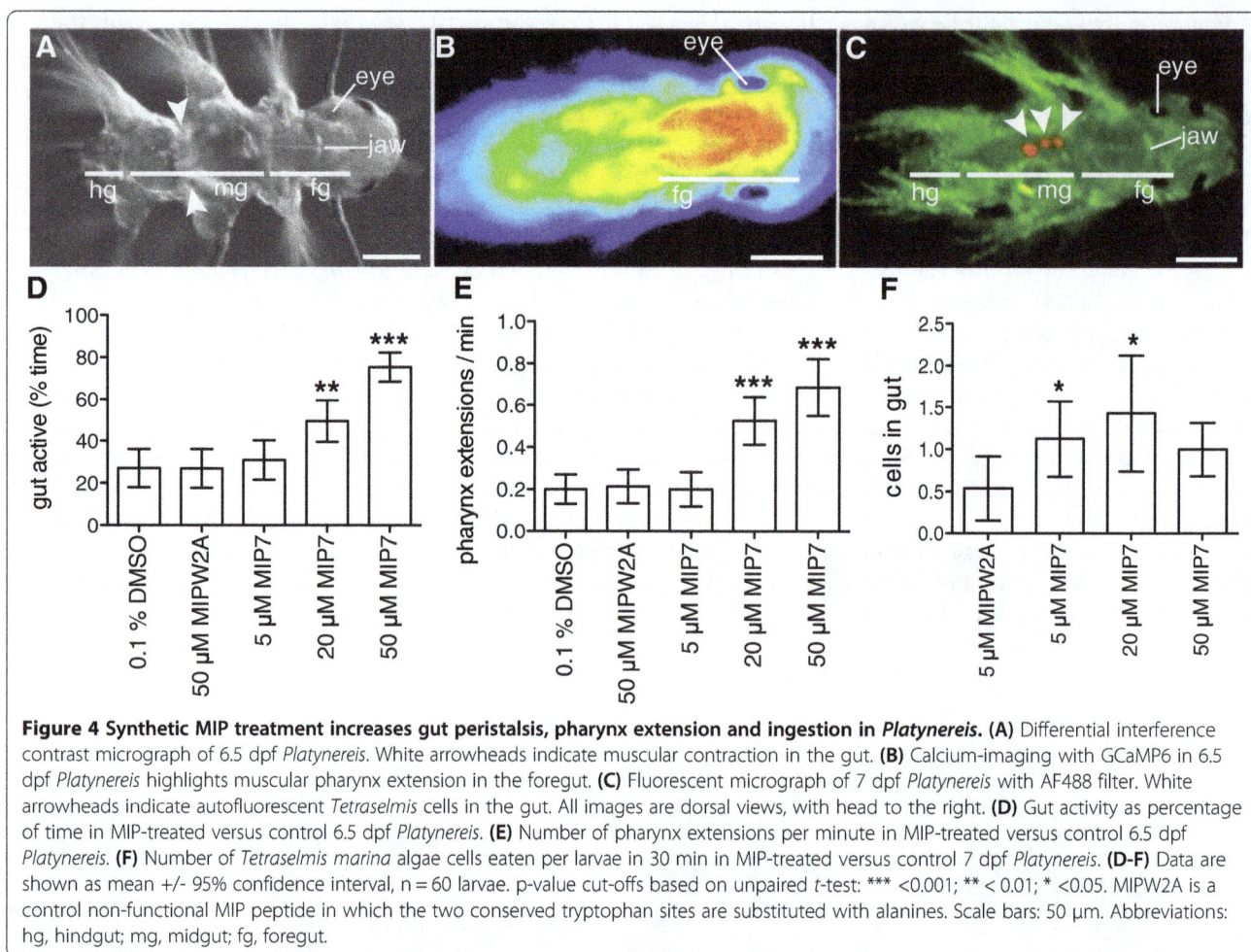

Figure 4 Synthetic MIP treatment increases gut peristalsis, pharynx extension and ingestion in *Platynereis*. (A) Differential interference contrast micrograph of 6.5 dpf *Platynereis*. White arrowheads indicate muscular contraction in the gut. (B) Calcium-imaging with GCaMP6 in 6.5 dpf *Platynereis* highlights muscular pharynx extension in the foregut. (C) Fluorescent micrograph of 7 dpf *Platynereis* with AF488 filter. White arrowheads indicate autofluorescent *Tetraselmis* cells in the gut. All images are dorsal views, with head to the right. (D) Gut activity as percentage of time in MIP-treated versus control 6.5 dpf *Platynereis*. (E) Number of pharynx extensions per minute in MIP-treated versus control 6.5 dpf *Platynereis*. (F) Number of *Tetraselmis marina* algae cells eaten per larvae in 30 min in MIP-treated versus control 7 dpf *Platynereis*. (D-F) Data are shown as mean +/- 95% confidence interval, n = 60 larvae. p-value cut-offs based on unpaired t-test: *** <0.001; ** < 0.01; * <0.05. MIPW2A is a control non-functional MIP peptide in which the two conserved tryptophan sites are substituted with alanines. Scale bars: 50 μm. Abbreviations: hg, hindgut; mg, midgut; fg, foregut.

Knockdown of MIP delays the initiation of feeding in *Platynereis* larvae

To explore the function of MIP in the *Platynereis* digestive system, we employed morpholino microinjection to knockdown MIP expression. We used two different translation blocking morpholinos and two mismatch control morpholinos (Additional file 11). To test the effectiveness of MIP-knockdown, we immunostained knockdown and control larvae with an antibody against *Platynereis* MIP. We observed a strong reduction in MIP immunostaining in *Platynereis* MIP-knockdown larvae, but not in controls, up to at least 6 dpf (Figure 3D-G, Additional file 12). These experiments confirmed that the MIP morpholinos were capable of strongly reducing MIP expression.

Next, we documented feeding in MIP-knockdown and control larvae. Similar to untreated larvae, most larvae injected with a control morpholino had initiated feeding between 7-9 dpf, whereas a significantly lower number of MIP-knockdown larvae had food in the gut at 7-9 dpf. This effect was still observed between 10-12 dpf (Figure 3B-C).

To rule out that the reduced feeding in MIP-knockdown larvae is due to a developmental delay, we compared the morphology of the nervous system of MIP-knockdown and control larvae. There were no detectable differences in the nervous system of control larvae and MIP-knockdown larvae based on acetylated tubulin immunostainings at 6 dpf (Figure 3D-G; Additional file 13). We also treated uninjected larvae at different ages between 24 hours post fertilization (hpf) and 5 dpf with synthetic MIP peptide to see whether MIP-treated larvae initiate feeding sooner, indicating a potential developmental acceleration. MIP treatment did not significantly alter the timing of feeding initiation, even when food was available earlier than 5 dpf (Additional file 14). These experiments indicate a critical physiological role for MIP in the initiation of feeding behaviour in *Platynereis* postlarvae.

MIP treatment has a myostimulatory effect on the digestive system of *Platynereis* postlarvae

In order to understand how the morpholino knockdown of MIP resulted in reduced larval feeding, we examined the effect of synthetic MIP treatment on postlarvae, focusing on the digestive system. Treatment of 6.5 dpf postlarvae with synthetic MIP caused a significant increase in gut peristalsis (Figure 4A, D; Additional file 15). MIP-treated postlarvae also displayed increased rates of pharynx extension (Figure 4B, E; Additional file 16).

To determine whether these effects on gut and pharynx movement resulted in increased ingestion of algal cells in MIP-treated postlarvae, we then scored the number of algal cells consumed by MIP-treated versus control 7 dpf postlarvae. Treatment with 5 µM and 20 µM MIP significantly increased postlarval algal cell consumption (Figure 4C, F; Additional file 17). Additionally, MIP-treated postlarvae have decreased locomotion, indicating a switch in the nervous system from a locomotory to a feeding program (Additional file 18). These results show that MIP up-regulates feeding activity and gut peristalsis.

Long-term MIP treatment enhances growth in *Platynereis* postlarvae

Given the effect of MIP on the digestive system and feeding in *Platynereis* postlarvae, we next investigated the long-term effects of MIP treatment on postlarval growth. At approximately two weeks of age, feeding *Platynereis* begin to add new posterior segments [13]. After the development of the 5th segment, juveniles undergo cephalic metamorphosis, a morphogenetic process in which the first chaetigerous segment loses its chaetae, develops a pair of tentacular cirri and fuses with the head (Figure 5A-D). The timing of cephalic metamorphosis and the addition of new segments vary between individuals. Even juveniles cultured individually showed variation in the timing of posterior segment addition, with segment number varying between 4 and 8 segments at 34 dpf (Additional file 19D). On a diet of *Tetraselmis*, the shortest interval for an individually-raised juvenile to develop an additional posterior segment was 4 days. The addition of new segments required that larvae begin to feed. Unfed larvae never develop beyond the 3-segmented stage (Additional file 19E). Growth in other nereid species depends on culture density [52-55]. We documented the growth of *Platynereis* juveniles cultured at different densities with excess food and determined the maximal density that still allowed optimal growth (3 larvae/ml) (Additional file 19 A-C). Under these conditions, juvenile *Platynereis* begin to develop the 5th segment at 16 dpf, and start to undergo cephalic metamorphosis at 24 dpf. Morpholino knockdown methods are not applicable to such late stage animals, therefore we tested the effects of MIP treatment on errant juvenile growth. We found that the time to the addition of new posterior segments, and to cephalic metamorphosis, was reduced by sustained exposure to five different versions of mature MIP encoded by the *Platynereis MIP* preproneuropeptide gene (Figure 5E, F). At 25 dpf, some MIP-treated individuals had completed cephalic metamorphosis, while control individuals were yet to undergo cephalic metamorphosis. Comparing the body length of MIP-treated and control *Platynereis* at 25 dpf revealed that MIP-treated individuals were on average approximately 100 µM longer than control individuals (Figure 5G). The effect of MIP treatment on growth was only seen in the presence of food. In the absence of food, MIP treatment could not induce the addition of any new posterior segments, and postlarvae remained at the 3-segmented stage (Additional file 19E). Additionally, MIP-treated larvae, both fed and unfed, exhibited altered pigmentation of the gut and the body (Figure 5H, Additional file 19F, G).

Discussion

Our results established the MIP neuropeptide as a regulator of postlarval feeding and gut activity in *Platynereis*. At 6 dpf, MIP is expressed in both the pharynx and the hindgut in neurons with a sensory morphology with dendrites projecting to the lumen. The MIP-expressing neurons of the *Platynereis* gut possess several hallmark features of mammalian enteroendocrine cells, including a scattered distribution, dendrites extending towards the gut lumen and long branching axons in the gut epithelial layer underlying the gut musculature [56]. These results are consistent with a model where MIP cells receive sensory signals from inside the mouth and the gut and respond by releasing MIP in a neurosecretory manner in the vicinity of the pharynx and hindgut muscles. However, given the use of bath-application and whole-body morpholino knock-down, we could not analyze the function of individual MIP-expressing cells. In principle, MIP-expressing neurons in other parts of the body may also affect gut activity by hormonal action.

Contrary to its name, MIP plays a myostimulatory role in the *Platynereis* digestive system. This may be the result of a direct effect whereby MIP directly acts on the digestive system musculature to increase the rate of pharynx extensions and peristaltic movements. Alternatively, the myostimulatory action of MIP may be caused indirectly through the regulation by MIP of other neurons in the gut, for example, in an as yet unidentified central pattern generator circuit responsible for regular gut contractions, as seen in crustaceans [57]. Knowledge of the spatial expression pattern of the MIP receptor in 6 dpf and older *Platynereis* larvae could help to resolve this.

Our results show that the increase in pharynx extensions in *Platynereis* postlarvae has a direct effect on the amount of food ingested. Increased gut peristalsis could promote the passage of food within the gut or the mixing of food with digestive enzymes, speeding up digestion. The fact that the highest concentration of MIP treatment, 50 µM, did not increase the amount of food ingested compared to control postlarvae is likely a result of the simultaneous reduction in locomotor activity caused by MIP treatment. At the highest concentrations of MIP (20 – 50

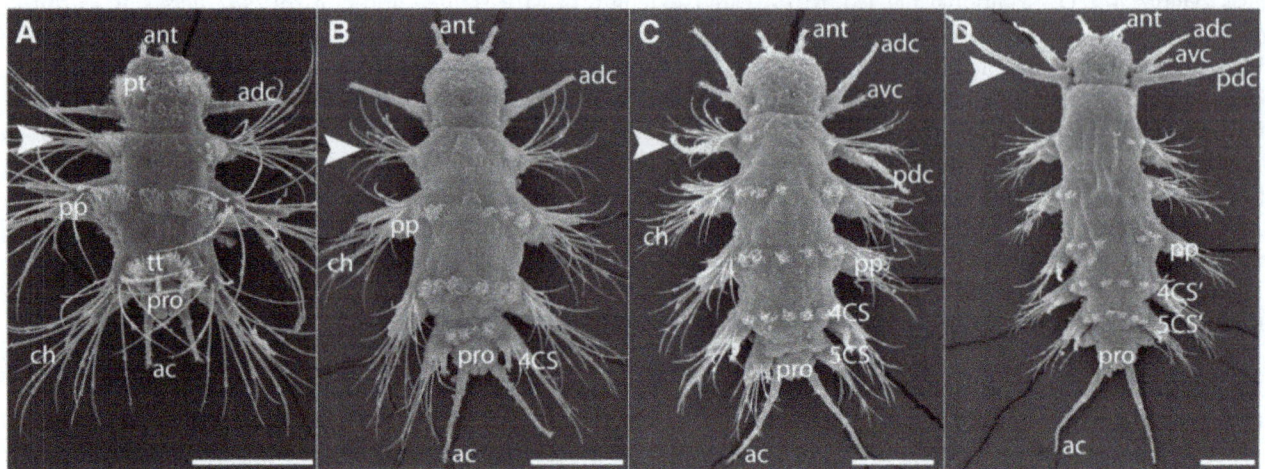

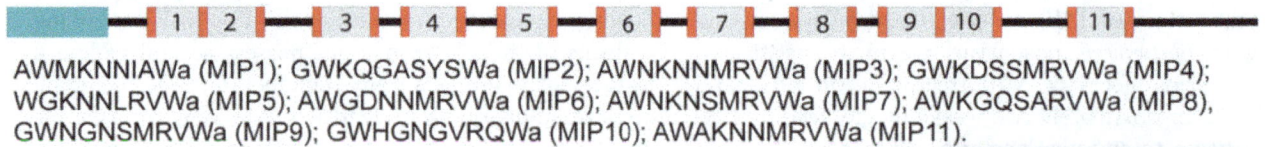

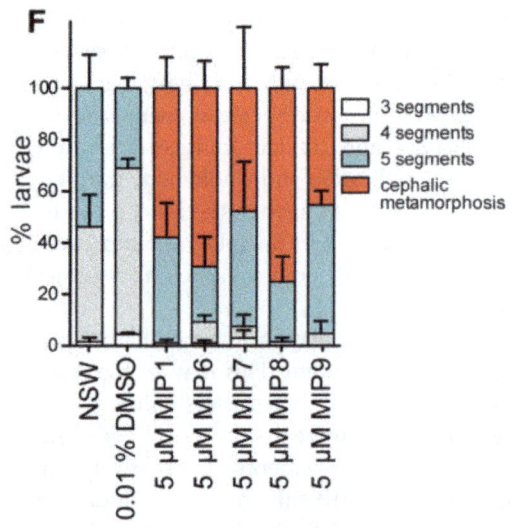

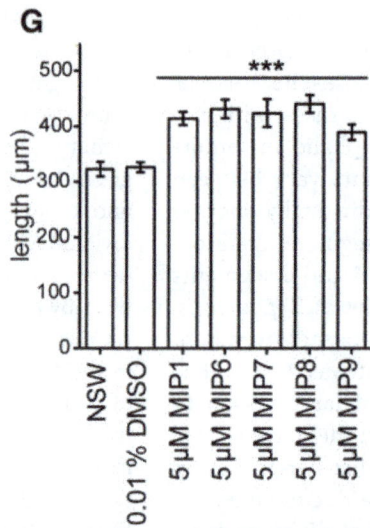

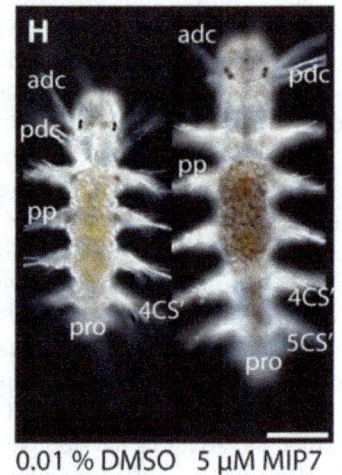

Figure 5 Long-term treatment of *Platynereis* with synthetic MIP enhances growth and decreases time to cephalic metamorphosis.
(A-D) SEM images depicting posterior segment addition followed by cephalic metamorphosis in *Platynereis dumerilii*, dorsal views. White arrowhead indicates parapodia of the 1st chaetigerous segment, which are transformed into the posterior tentacular cirri of the head during cephalic metamorphosis. **(A)** 6 dpf 3-segmented nectochaete larva. **(B)** 4-segmented errant juvenile. **(C)** 5-segmented errant juvenile. Cephalic metamorphosis has begun. **(D)** Cephalic metamorphosis is complete. **(E)** Schematic of the *Platynereis* MIP precursor protein. The N-terminal signal peptide (teal) and the predicted peptides (grey) flanked by basic cleavage sites (red) are shown. The predicted mature MIP peptide sequences are indicated below. **(F)** Percentage of 25 dpf *Platynereis* larvae/juveniles with 3, 4, 5 segments, or complete cephalic metamorphosis following exposure to 5 µM synthetic MIP peptides, from 4 dpf onwards. Data are shown as mean +/- s.e.m, n = 3 x 30 larvae. NSW, natural seawater control. **(G)** Total length of 25 dpf *Platynereis* larvae/juveniles exposed to 5 µM synthetic MIP peptides, from 4 dpf onwards. Data are shown as mean +/- 95% confidence interval, n = 90 larvae. **(F, G)** p-value cut-offs based on unpaired *t*-test: *** <0.001; ** < 0.01; * <0.05. **(H)** Differential interference contrast light micrographs of example control and MIP7-treated individuals at 1 month post fertilization. Scale bars: 100 µM. Abbreviations: ac, anal cirrus; adc, anterior dorsal cirrus; ant, antenna; avc, anterior ventral cirrus; ch, chaetae; pdc, posterior dorsal cirrus; pp, parapodia; pro, proctodeum; pt, prototroch; tt, telotroch; 4CS, 4th chaetigerous segment; 5CS, 5th chaetigerous segment; 4CS', 4th chaetigerous segment after cephalic metamorphosis; 5CS', 5th chaetigerous segment after cephalic metamorphosis.

μM), increased gut peristalsis and pharynx extension activity may be offset by a decrease in locomotion, resulting in treated individuals encountering fewer algal cells.

The sustained expression of MIP in the gut and the long-term effects of MIP on juvenile growth indicate that MIP also has an important physiological role later in the life cycle. We interpret the enhancement of juvenile growth in long-term MIP treatment experiments to be a consequence of a sustained increase in feeding caused by MIP. Given its effect on both settlement and growth, MIP treatment may be a useful means of enhancing both larval settlement and juvenile growth in polychaete aquaculture [58].

Interestingly, MIP has a myostimulatory role in cnidarians and in the *Platynereis* digestive system, but a myoinhibitory role in the arthropod digestive system [36,41,44]. This could mean that either MIP was independently recruited to regulate gut activity in different phyletic lineages, or that the sign of the regulation switched during evolution. In the latter case, MIP would represent a conserved bilaterian gut peptide influencing feeding. Further comparative morphological and molecular studies of MIP cells and signaling pathways in a broader range of taxa will be needed to resolve this.

MIP regulates both settlement behavior and feeding, two aspects of the pelagic-to-benthic transition of the non-feeding *Platynereis* larvae. What could be the reason for the redeployment of the same peptidergic signal at different times during development and in different contexts? One possibility is that the anterior MIP-expressing sensory-neurosceretory cells of the larva and the MIP cells in the gut of the postlarva sense the same chemical cues released by potential food sources. Some marine larvae are induced to settle by their future juvenile food source [2]. Testing this hypothesis will require the identification of naturally occurring settlement cues and their corresponding receptors in *Platynereis*.

In *Platynereis*, juvenile feeding is an essential requirement for the completion of cephalic metamorphosis. In other polychaete species, where feeding often begins in the pelagic larval stage before settlement, feeding is also an essential component for settlement and metamorphosis. Starved larvae of *Capitella sp.*, *Polydora ligia*, *Hydroides elegans* and *Phragmatopoma lapidosa* all lose or have decreased ability to complete settlement and metamorphosis [59-62]. Exploration of the roles of MIP in polychaete species with feeding larvae would increase our understanding of the links between MIP signaling, larval settlement and feeding.

Conclusions

We have described a role for MIP in *Platynereis* postlarval feeding and established methods for studying the neuroendocrine regulation of feeding, providing the basis for future studies in this area. The amenability of *Platynereis* larvae to peptide treatments by soaking, their transparent body wall, and a neuropeptide complement that overlaps with that of both vertebrates and arthropods, make *Platynereis* an ideal model with which to study the neuroendocrine regulation of feeding in an evolutionary context.

Methods

Platynereis culture

Platynereis larvae were obtained from an in-house culture as previously described [15]. After fertilization of eggs, developing embryos and larvae were kept in an incubator at a constant temperature of 18°C with a regular light-dark cycle.

Platynereis digestive system marker genes

Five genes with spatial expression domains restricted to the digestive system were identified from an ongoing RNA *in situ* hybridization screen of 48 hpf, 72 hpf and 6 dpf *Platynereis*. These genes were identified as *subtilisin-1* and *subtilisin-2* (Genbank Accession KM577672, KM577673), *alpha amylase* (Genbank Accession KM577675), *legumain-protease-precursor* (Genbank Accession KM77676) and *enteropeptidase* (Genbank Accession KM577674).

Genes were named according to their common conserved domains, reciprocal BLAST to the *Homo sapiens* peptidome, and neighbor-joining and maximum likelihood phylogenetic analyses (Additional files 7, 8). Genes were analyzed for the presence of a signal peptide assigned using the SignalP 4.1 server (http://www.cbs.dtu.dk/services/SignalP/) and conserved domains assigned by searches in the Pfam database with an e-value cutoff 1e-06 (http://pfam.xfam.org/search). To find additional sequences for use in phylogenetic analyses, the candidate gene sequences were used as queries in BLAST searches against the NCBI nr and Swissprot databases, taking the top 50 hits from each BLAST search. To diversify the range of phyla represented, BLAST searches were also performed with different restrictions, including 'non-mammal', 'non-*Drosophila*' and 'Lophotrochozoans'. Sequence redundancy was reduced to 90% identity using CD-HIT [63]. Genes were aligned to candidate orthologues from other taxa with MUSCLE (http://www.ebi.ac.uk/Tools/msa/muscle/). Conserved regions of sequence alignment were select for phylogenetic analysis using Gblocks with minimal stringency settings [64]. Phylogenetic trees were constructed from trimmed sequence alignments using the neighbor-joining methods with 1000 bootstrap replicates in CLC Genomics Workbench 5.5.1 (CLC Bio, Qiagen), with a Gap Open Cost of 10 and a Gap Extension Cost of 1. Maximum likelihood trees with 100 bootstrap replicates were constructed using PhyML 3.0 using an LG substitution model and SPR and NNI tree searching methods [65]. Trees were inspected and taxa with long branches were removed to avoid longbranch attraction bias. The phylogenetic analyses were then re-run with the remaining taxa. We then went on to examine the expression of the digestive system marker genes in 6 dpf, 15 dpf and 1 mpf *Platynereis* by RNA *in situ* hybridization methods as described below.

RNA *In situ* hybridization

Different developmental stages of *Platynereis* were collected for fixation for use in wholemount RNA *in situ* hybridization and immunostaining techniques. Individuals 6 days and older were relaxed using 1 M MgCl$_2$ [66] prior to fixation. Postlarvae and juveniles that had begun feeding were starved for a few days prior to fixation to avoid the presence of autofluorescent algae cells in the gut, which interfere with fluorescent signals from immunostaining. All animals were fixed in 4% paraformaldehyde (PFA) in 0.1 M MOPS (pH 7.5), 2 mM MgSO$_4$, 1 mM EGTA, 0.5 M NaCl for 1 h at room temperature. Fixed larvae were dehydrated through a MeOH series and stored in 100% MeOH at -20°C.

DIG-labelled antisense RNA probes for the *Platynereis MIP precursor* (JX513877), *MIP receptor* (JX513876), *alpha-amylase, subtilisin-1, subtilisin-2, legumain-protease precursor*, and *enteropeptidase* were synthesized from purified PCR products of clones sourced from a *Platynereis* cDNA library [51]. RNA *in situ* hybridization using nitroblue tetrazolium (NBT)/5-bromo-4-chloro-3-indolyl phosphate (BCIP) staining combined with mouse anti-acetylated-tubulin staining to highlight cilia and nervous system, followed by imaging with a Zeiss LSM 780 NLO confocal system and Zeiss ZEN2011 Grey software on an AxioObserver inverted microscope, was performed as previously described [50], with the following modification: fluorescence (instead of reflection) from the RNA *in situ* hybridization signal was detected using excitation at 633 nm in combination with a Long Pass 757 filter. Animals were imaged with a 40X oil objective.

Image registration of RNA *in situ* hybridization patterns

We projected average *MIP* expression pattern of four 6 dpf individuals onto a common 6 dpf whole-body nuclear reference template generated from DAPI signal of 40 individuals as described previously for 72 hpf larvae [50]. Acetylated tubulin and expression patterns of digestive system marker genes of select individuals were also projected onto the reference template. Snapshots and video of the projected genes were generated in Blender 2.7.1 (http://www.blender.org/).

Immunostaining

Immunohistochemistry with 1 µg/ml rabbit anti-MIP (AWNKNSMRVWamide) or cross-species anti-VWa, 0.5 µg/ml mouse anti-acetylated tubulin (Sigma) primary antibodies, and 1 µg/ml anti-rabbit Alexa Fluor® 647

(Invitrogen) and 0.5 µg/ml anti-mouse FITC (Jackson Immuno Research) secondary antibodies, was performed as previously described [67]. After the staining procedure, samples were mounted in 97% 2,2'-thiodiethanol (TDE).

For phalloidin stainings, we used freshly-PFA-fixed larvae dehydrated in 100% acetone for 5 minutes. Staining with rhodamine phalloidin (Molecular Probes) 1:100 in combination with the rabbit anti-MIP antibody was performed with the standard protocol adapted from [67]. After the staining procedure, samples were transferred to the mounting medium 87% glycerol containing 2.5 mg/mL of anti-photobleaching reagent DABCO (Sigma, St. Louis, MO, USA), as phalloidin-conjugated rhodamine destabilizes in TDE [68].

Confocal images were processed with Imaris 6.4 (Bitplane Inc., Saint Paul, USA) software. Raw image stacks and the *Platynereis* 6 dpf nuclear stained reference are available at the Dryad data repository.

Calcium imaging

Fertilized eggs were injected with 500 ng/µl capped and polyA-tailed GCaMP6 [69] RNA generated from a vector (pUC57-T7-RPP2-GCaMP6) containing the GCaMP6 ORF fused to a 169 base pair 5' UTR from the *Platynereis* 60S acidic ribosomal protein P2, as in [70]. Injection protocol is described in more detail in the 'Morpholino Knockdown of *Platynereis* MIP' section. Larvae were imaged with a 488 nm laser and transmission imaging with DIC optics on a Zeiss LSM 780 NLO confocal system on an AxioObserver inverted microscope (Additional file 5), or using a Zeiss AxioZoom V16 microscope with Hamamatsu Orca-Flash 4.0 digital camera (Additional file 13).

RNA-Seq

RNA-Seq analysis of digestive enzyme and *MIP* precursor gene expression was performed on an existing dataset of 13 different stages spanning the *Platynereis* life cycle, from egg to mature adults. Methods used were described in [51].

Documentation of normal feeding in *Platynereis*

To document variation in the commencement of feeding in *Platynereis* larvae from our laboratory culture, larvae were kept in Nunclon 6-well tissue culture dishes, with 10 ml sterile filtered seawater (FSW) per well. Each well contained 30 larvae. Larvae from 6 different batches, with different parents, were used in our analysis. Larvae were fed 5 µl *Tetraselmis marina* algae culture at 5 dpf. Larvae were then tested for feeding by checking for the presence of fluoresent *Tetraselmis* algae in the gut using a Zeiss Axioimager Z1 microscope with an AF488 fluorescent filter and a 20X objective. Larvae were checked for signs of feeding at 5.5, 6, 7, 8, 10, 12 and 14 dpf. After ingestion, algal cells can remain in the gut for up to 48 h before digestion causes a loss of fluorescence. Although larvae with a full gut can also be identified with normal light microscopy due to the transparent body wall, fluorescent microscopy enables the detection of even a single alga cell in the gut, due to the strong chlorophyll fluorescence of the *Tetraselmis* cells.

Morpholino knockdown of *Platynereis* MIP

Two translation blocking morpholinos (MOs) and two corresponding 5 base pair mismatch control morpholinos were designed to target the *Platynereis-MIP-precursor* (GeneTools, LLC):

Pdu-MIP-start MO1 TGATAGTGACGCGATCC*A*TTG GACT

Pdu-MIP-mism MO1 TG*T*TAGTGAC*C*CG*T*TCGA*A*TG GACT

Pdu-MIP-start MO2 CTAGTTCCTTCTCTCCCTCTT ATCT

Pdu-MIP-mism MO2 CTAC*T*TG*C*TTG*T*CTCC*G*TGT TATCT

Nucleotides altered in mismatch control morpholinos are in italics. Information on the position of the morpholinos in relation to the *MIP* start codon can be found in Additional file 11.

MOs were diluted in water with 12 μg/μl fluorescein dextran (*Mr* 10,000, Invitrogen) as a fluorescent tracer. 0.6 mM MOs were injected with an injection pressure of 600 hPa for 0.1 s and a compensation pressure of 35 hPa using Eppendorf Femtotip II needles with a Femtojet microinjector (Eppendorf) on a Zeiss Axiovert 40 CL inverted microscope equipped with a Luigs and Neumann micromanipulator. The temperature of developing zygotes was maintained at 16°C throughout injection using a Luigs and Neumann Badcontroller V cooling system and a Roth Cyclo 2 water pump.

For microinjection, fertilized *Platynereis* eggs developing at 16°C were rinsed 1 h after fertilization with sterile 0.2 μm filtered seawater (FSW) in a 100 μM sieve to remove the egg jelly, followed by a treatment with 70 μg/ml proteinase K for 1 min to soften the vitellin envelope. Following injection, embryos were raised in Nunclon 6-well plates in 10 ml FSW and their development was monitored daily.

Larvae were fed 5 μl *Tetraselmis marina* algal culture at 6 dpf. Feeding in 7 – 14 dpf injected larvae was assessed by checking for the presence of fluoresent *Tetraselmis marina* algae in the gut using a Zeiss Axioimager Z1 microscope with an AF488 fluorescent filter and a 20X objective. Larvae were checked for signs of feeding as described above every 24 h from 7 dpf on. We scored a minimum of 62 larvae (maximum 424 larvae) from a minimum of 3 separate microinjection sessions (with 3 different batches of larvae) for each translation-blocking and control

morpholino. Photomicrographs of morpholino-injected larvae were also taken and larval body length was measured from these pictures using Image J 64 software. Some morpholino-injected larvae were also fixed at 6 days for immunostaining with the anti-MIP antibody (as described above) in order to assess morpholino specificity and effectiveness.

Effect of synthetic MIP on *Platynereis* feeding behaviour

Peptide functions can be investigated in *Platynereis* larvae by bath application of synthetic neuropeptides [10,11]. To test whether synthetic MIP treatment increased developmental speed, leading to early initiation of feeding in *Platynereis* larvae, experiments were performed in Nunclon 6-well plates, with 10 ml FSW per well. Each control and peptide treatment was replicated across three wells, with 30 larvae per well. Larvae were treated with 5 μM MIP7 or controls at 24 hpf, 60 hpf, 4 dpf or 5 dpf, then fed at 4 or 5 dpf (depending on the age at which MIP treatment occurred) with 5 μl *Tetraselmis marina* algal culture. Larvae were fed at an earlier age due to the possibility of MIP treatment causing an earlier initiation of feeding. Larvae were checked for feeding by monitoring algal cell fluorescence in the gut as described above. Larvae were monitored from 5 or 5.5 dpf (depending on age at which larvae were first fed) until 7 or 8 dpf. A control non-functional MIP peptide (MIPW2A, AANKNSMRVAamide), in which the two conserved tryptophan sites were replaced with alanines (this prevents MIP from activating its receptor, see [11]) was also tested. A further control of larvae treated with DMSO alone was also included, as MIP peptides require DMSO to be dissolved in solution.

To test the effects of synthetic MIP peptide treatment on the digestive system of *Platynereis* larvae, we recorded videos of groups of 60 larvae at 6.5 dpf in a square glass cuvette 1.5 cm x 1.5 cm x 0.3 cm in 500 μl of FSW using a Zeiss AxioZoom .V16 microscope with Hamamatsu Orca-Flash 4.0 digital camera. For each treatment and control, 3 biological replicates (larval batches with different parentage, fertilized on different days) were carried out. We tested three concentrations of synthetic MIP: 5, 20 and 50 μM, plus 50 μM control non-functional MIP peptide MIPW2A and a 0.1 % DMSO control. A 2.5 min video at 10 frames per second was recorded 10 min after peptide or DMSO addition. Videos were analyzed manually in Fiji (Image J 1.48s, Wayne Rasband, http://imagej.nih.gov/ij). For each video, 20 larvae that remained within the frame of the video for the entire 2.5 min were scored for gut peristalsis and pharynx extension activity. Distance traveled and speed of the larvae was also measured using the MTrack2 plugin [71]. Significant differences in gut peristalsis, pharynx extension activity and locomotion in MIP-treated versus control larvae were tested in an unpaired t test.

To test the effects of synthetic MIP treatment on short-term ingestion of algal cells in *Platynereis* larvae, experiments were performed in Nunclon 24-well plates, with 2 ml FSW per well. Each control and peptide treatment was replicated across three wells, with 20 larvae per well. 7 dpf postlarvae were treated with 5 µM MIPW2A control peptide, 5 µM MIP, 20 µM MIP or 50 µM MIP for 10 min. Following this, 20 µl *Tetraselmis marina* algal culture was added to each well and larvae were left to feed for 30 min. All larvae were then immediately fixed in 0.5 mL 4% paraformaldehyde in 1X PBS with 0.01% Tween (PTw) for 1 hour. Following 4 washes in 1 ml PTw, larvae were mounted on glass slides and the number of algal cells in the digestive system of each larva was counted using a Zeiss Axioimager Z1 microscope with an AF488 fluorescent filter and a 20X objective. Significant differences in MIP-treated versus control larvae were tested in an unpaired *t* test.

Scanning electron microscopy (SEM)

Platynereis larvae and juveniles of different developmental stages were fixed with 3% glutaraldehyde in 0.1 M phosphate buffer pH 7.2, rinsed in phosphate buffer, further fixed with 1 % osmium tetroxide in water and dehydrated in an ascending EtOH series over several days. Critical point drying with carbon dioxide was performed in a Polaron E 3000. The samples were coated with gold-palladium in a Balzers MED 010. Images were taken on a Hitachi S-800 Scanning electron microscope.

Calculation of optimal culture density

The assessment of growth in larvae cultured individually was performed in a Nunclon 24-well tissue culture dish with 1 larva per well in 2 ml FSW. Larvae were fed from 6 dpf with 3 µl *Tetraselmis marina* algae culture. Larvae were scored under a dissection microscope for number of segments and cephalic metamorphosis every 48 h from 14 dpf to 34 dpf.

Documentation of growth in larvae cultured at different densities was carried out in Nunclon 6-well plates with 10 mL FSW/well and 30, 50 or 100 larvae per well. Three replicate wells were included for each culture density. Larvae were fed with surplus *Tetraselmis marina* algae throughout the experiment. Larvae were scored for segment number and cephalic metamorphosis every 4 days from 16 to 32 dpf.

Long term treatment of *Platynereis* with synthetic MIP

To test the effect of synthetic MIP treatment on growth in *Platynereis* larvae, we again carried out experiments in Nunclon 6-well plates as described above, with 30 larvae per well and 3 replicate wells per treatment and control. 5 µM synthetic peptides were added at 4 dpf. Different versions of mature MIP peptide tested were:

MIP1 – AWNKNNIAWamide, MIP6 – AWGDNNMRVWamide, MIP7 – AWNKNSMRVWamide, MIP8 – AWKGQSARVWamide, and MIP9 – GWNGNSMRVWamide. Larvae were also fed at 4 dpf with 5 µl *Tetraselmis sp.* algal culture. At 25 dpf (21 days after peptide addition), errant juveniles were scored for number of segments and cephalic metamorphosis (as above). Juveniles were also photodocumented using a Zeiss Axioimager Z1 microscope with differential interference contrast (DIC) and size of control and treated larvae (end of head to end of pygidium, excluding cirri) was measured in Fiji (Image J 1.48s, Wayne Rasband, http://imagej.nih.gov/ij).

We also tested the effects of synthetic MIP treatment on growth of unfed *Platynereis* larvae. Experiments were carried out as above, 5 µM MIP7 was added at 4 dpf. 5 µM MIPW2A and 0.1% DMSO were included as negative controls. At 24 dpf, errant juveniles were scored for number of segments and cephalic metamorphosis. Most unfed juveniles died between 24 and 28 dpf.

Additional files

Additional file 1: MIP expression in 14 dpf and 1 mpf *Platynereis*. Whole-mount RNA *in situ* hybridization (WMISH) for the *Platynereis* MIP precursor (red) counterstained for acetylated tubulin (white) (A, C, D, E, G, H), or DAPI nuclear stain (blue) (A, C, D, E, G, H). All images shown in ventral view, with head to top. (A-D) 14 dpf, (E-H) 1 mpf. In (B-D) and (F-H), the ventral nerve cord region has been cut away to reveal digestive system. (C, G) close-up foregut (D, H) close-up mid- and hindgut. (I-K) Schematic of *MIP* precursor expression in 1 mpf *Platynereis*. (I) Ventral side. (J) Dorsal side. (K) MIP expression relative to expression of digestive system marker genes. In (B) and (F), white dashed lines indicate digestive system. In (B), (C), (F) and (G), yellow dashed lines mark jaws. Scale bars: 50 µm. Abbreviations: fg, foregut; mg, midgut; hg, hindgut; ant, antenna; nsp, neurosecretory plexus; adc, anterior dorsal cirrus; pp, parapodia; vnc, ventral nerve cord; ch, chaetae; ac, anal cirrus; ph, pharynx.

Additional file 2: Movie of MIP expression in foregut sensory cell of 6 dpf *Platynereis*. Whole-mount RNA *in situ* hybridization (WMISH) for the *Platynereis* MIP precursor (red) counterstained for acetylated tubulin (white) and DAPI nuclear stain (blue). Ventral view of foregut. Yellow arrowhead marks sensory dendrite of MIP-expressing cell. Green arrows indicate salivary glands.

Additional file 3: Movie of MIP expression in mid- and hindgut sensory cells of 6 dpf *Platynereis*. Whole-mount RNA *in situ* hybridization (WMISH) for the *Platynereis* MIP precursor (red) counterstained for acetylated tubulin (white) and DAPI nuclear stain (blue). Ventral view of mid- and hindgut. Yellow arrowheads mark sensory dendrites of MIP-expressing cells.

Additional file 4: MIP peptide expression timecourse in *Platynereis*, MIP peptide expression in *Capitella* larvae. (A-L) Immunostaining of *Platynereis* larvae, postlarvae and juveniles with an antibody raised against *Platynereis* MIP7 (red), counterstained for acetylated tubulin (white) or DAPI nuclear stain (blue). All images ventral view with head to top. (A, B) 3 dpf, (C, D) 4 dpf, (E, F) 5 dpf, (G, H) 6 dpf, (I, J) 10 dpf, (K, L) 1 mpf. (M-P) Immunostaining of *Capitella teleta* Stage 9 larvae with an antibody raised against VVamide (red), counterstained for acetylated tubulin (white) or DAPI nuclear stain (blue). (M, N) ventral view, (O, P) lateral view. In (B, D, F, H, J, L, N), ventral nerve cord area has been removed to expose the underlying digestive system. Scale bars: 50 µm. White asterisks mark background fluorescence of parapodia, chaetae or spinning glands. White dashed lines indicate the developing digestive system. Abbreviations: fg, foregut; mg, midgut; hg, hindgut; pro, proctodeum; sto, stomodeum; dpf, days post fertilization; mpf, month post fertilization.

Additional file 5: Video of *Platynereis* 6 dpf digestive system. Calcium imaging with GCaMP6 shows muscular and neuronal activity in the digestive system of 6 dpf *Platynereis*. A sphincter muscle (indicated by yellow arrowheads) marks the boundary between foregut and midgut. Dorsal view with head to right.

Additional file 6: *Platynereis* digestive system marker genes. Schematic representation of *Platynereis* digestive enzyme genes *alpha-amylase, enteropeptidase, legumain protease precursor, subtilisin-1* and *subtilisin-2*. Signal peptide sequence and conserved domains are marked by coloured boxes, with e-value of an HMM search against the Pfam database (http://pfam.xfam.org) indicated in box. Length of amino acid sequence obtained is shown to the right of each gene.

Additional file 7: Phylogenetic trees of *Platynereis* digestive system marker genes. Neighbour-joining trees with 1000 bootstrap repetitions (NJ-1000) and maximum likelihood trees with 100 bootstrap repetitions (ML-100) for digestive system markers *alpha amylase* (A, B), *enteropeptidase* (B, C), *legumain protease precursor* (D, E), and *subtilisin-1* and *subtilisin-2* (F, G). The positions of the *Platynereis* candidate genes are highlighted by green boxes. Bootstrap values are indicated at branch nodes. *Platynereis alpha amylase* and *legumain protease precursor* cluster with their counterparts in a fellow polychaete, *Capitella teleta*, within invertebrate-specific clades. The orthology of these genes was confirmed by reciprocal BLAST to the *Homo sapiens* peptidome. *Platynereis enteropeptidase* clusters in a weakly-supported group of annelid enteropeptidases, however the identity of this gene is confirmed by the presence of conserved MAM and trypsin domains (Additional file 6) [49]. *Platynereis subtilisin-1* and *subtilisin-2* are intermingled with several bacterial, fungal, annelid and echinoderm sequences in a poorly resolved tree, suggesting a possible horizontal gene transfer event. *Subtilisin-1* and *–2* contain conserved peptidase domains (Additional file 6), including the presence of a catalytic triad, suggesting that they maintain enzymatic function in *Platynereis*.

Additional file 8: Expression timecourse of *Platynereis* digestive system marker genes. Whole-mount RNA *in situ* hybridization (WMISH) for *Platynereis* digestive system marker genes (red) counterstained for DAPI nuclear stain (blue). Ventral view with the head to the top, ventral nerve cord region removed to show the digestive system. (A-C) *alpha amylase*, (D-F) *enteropeptidase*, (G-I) *legumain protease precursor*, (J-L) *subtilisin-2*, (M-O) *subtilisin-2*. (A, D, G, J, M) 6 dpf, (B, E, H, K, N) 14 dpf, (C, F, I, L, O) 1 mpf. White asterisks mark background fluorescence from jaws or chaetae. White dashed lines indicate the digestive system. Abbreviations: fg, foregut; mg, midgut; hg, hindgut. Scale bars: 50 μm.

Additional file 9: Movie of average MIP expression in 6 dpf *Platynereis* in relation to digestive system marker genes. Surface representation of the average expression domains of MIP (blue) at 6 dpf relative to the axonal scaffold (grey). Surface representations of digestive enzyme genes are added sequentially to mark the digestive system. In order of appearance, genes are: *alpha amylase* (orange), *enteropeptidase* (cyan), *subtilisin-2* (green), *legumain protease precursor* (purple), and *subtilisin-1* (pink). Movie starts in dorsal view with head to the top and rotates around anterior-posterior axis.

Additional file 10: Expression of *MIP* and digestive system marker genes throughout *Platynereis* life cycle. Histograms of normalized gene expression generated from RNA-Seq libraries of 13 different life-cycle stages, from egg to sexually mature epitokes. (A) *alpha amylase* (B) *enteropeptidase*, (C) *legumain protease precursor*, (D) *subtilisin-1* (E) *subtilisin-2* and (F) *MIP*. Life-cycle stages during which *Platynereis* feeds are 10 dpf, 15 dpf, 1 mpf pre-cephalic metamorphosis, 1 mpf post-cephalic metamorphosis, and 3 mpf atokous adult. Habitat transitions are also indicated under the histograms: pelagic, free-swimming; benthic, crawling on bottom; pelagobenthic, may switch between swimming and crawling.

Additional file 11: Target sites of *Platynereis* MIP start morpholinos. The sequence of the 5′ region of the *Platynereis* MIP precursor gene is shown. Binding sites of MIP start morpholinos (MOs) 1 and 2 are outlined in red. The start codon for the MIP precursor peptide is underlined and in bold. The 5′ predicted peptide sequence is indicated in grey. MIP start

MO1 binds to the start codon region of the gene, while MIP start MO2 binds in the upstream 5′-UTR region.

Additional file 12: MIP knockdown in *Platynereis* is confirmed by anti-MIP immunostaining. Immunostaining of 6 dpf *Platynereis* with an antibody raised against *Platynereis* MIP7 (red). All images in apical view. (A-C) Larvae injected with mismatch control morpholino 1. (D-K) Larvae injected with MIP start morpholino 1. (L-N) Larvae injected with mismatch control morpholino 2. (O-Q) Larvae injected with MIP start morpholino 2. Injection of morpholinos targeting the start site of MIP results in reduced expression of MIP peptide compared to injection of control morpholinos. Black asterisks in (A-K) indicate background fluorescence caused by the oxidation of eye pigment proteins in larvae exposed to strong light. Scale bars: 20 μM.

Additional file 13: MIP knockdown *Platynereis* are morphologically similar to control *Platynereis*. (A,B) Ventral view of 6 dpf *Platynereis* injected with MIP mismatch morpholino 1 (A) or MIP start morpholino 1 (B) and immunostained with *Platynereis* MIP antibody (red) counterstained with acetylated tubulin (grey). Identical confocal microscopy and image processing parameters were applied to all images. Scale bar: 100 μm. Abbreviations: mism, mismatch; MO, morpholino; α-acTub, anti acetylated tubulin.

Additional file 14: Early treatment with MIP does not induce early onset feeding. Graphs of % larvae with food in gut over time after treatment with 5 μM MIP, 5 μM control MIPW2A or 0.1% DMSO from (A) 24 hpf, (B) 60 hpf, (C) 4 dpf, or (D) 5 dpf. Data are shown as mean +/- s.e. m., n = 3 x 30 larvae. p-value cut-offs based on unpaired *t*-tests indicated no significant difference in initiation of feeding between MIP-treated and control larvae. MIPW2A is a control non-functional MIP peptide in which the two conserved tryptophan sites are substituted with alanines.

Additional file 15: Movie of gut peristalsis in 6.5 dpf *Platynereis*. Dorsal view with head to the right. Extension of pharynx can also be seen at 10 sec and 15 sec.

Additional file 16: Movie of pharynx extension in 6.5 dpf *Platynereis*. Dorsal view with head to the right. Calcium imaging with GCaMP6 shows muscular extension of the pharynx in the foregut. Top panel is differential interference contrast (DIC), bottom panel is colour-coded with Jet-LUT colour map in Image J, most intense signal in red.

Additional file 17: Movie of feeding in 7 dpf *Platynereis*. *Platynereis* (green) feeds on autofluorescent *Tetraselmis* algae (red). Movie was filmed on a Zeiss AxioZoom microscope with AF488 fluorescent filter.

Additional file 18: MIP treatment decreases distance traveled and speed of 6 dpf *Platynereis*. (A) Distance travelled by MIP-treated versus control 6.5 dpf *Platynereis*. (B) Speed of MIP-treated versus control 6.5 dpf *Platynereis*. Data are shown as mean +/- 95% confidence interval, n = 60 larvae. p-value cut-offs based on unpaired *t*-test: *** <0.001; ** < 0.01; * <0.05. MIPW2A = control non-functional MIP peptide in which the two conserved tryptophan sites are substituted with alanines.

Additional file 19: Long-term growth of *Platynereis*. Addition of posterior segments in errant juveniles raised at a density of (A) 3 larvae/ mL (30 larvae in 10mL NSW), n = 3 x 30, (B) 5 larvae/mL (50 larvae in 10 mL NSW), n = 3 x 50, or (C) 10 larvae/mL (100 larvae in 10 mL NSW), n = 3 x 100. (A-C) Data are shown as mean + s.e.m. All larvae were fed from 6 dpf onwards. 'Cephalic metamorphosis' encompasses all worms that have completed cephalic metamorphosis and have 5 or more chaetigerous segments. (D) Addition of posterior segments in errant juveniles raised individually. 24 larvae were raised individually in a 24-well tissue culture dish with 2 mL NSW per well. Note: Larva #13 died prior to 12 dpf. (E) No addition of posterior segments in unfed 5 μM MIP7-treated and control errant juveniles. Data shown are mean + s.e.m, n = 3 x 30 larvae. NSW, 0.22 μM filtered natural seawater. MIPW2A is a control non-functional MIP peptide in which the two conserved tryptophan sites are substituted with alanines. (F, G) Differential interference contrast (DIC) light micrographs of example unfed (F) control and (G) MIP-treated individuals at 24 dpf. Scale bar: 50 μm.

Abbreviations

4CS: 4th chaetigerous segment; 4CS′: 4th chaetigerous segment after cephalic metamorphosis; 5CS: 5th chaetigerous segment; 5CS′: 5th chaetigerous segment after cephalic metamorphosis; α-acTub: anti acetylated tubulin; ac: anal cirrus; adc: anterior dorsal cirrus; ant: antenna; AST-B: allatostatin-B; avc: anterior ventral cirrus; ch: chaetae; cm: circular muscle fibre; DAPI: 4',6-Diamidino-2-Phenylindole, Dihydrochloride; DIC: differential interference contrast; DMSO: dimethyl sulfoxide; dpf: days post fertilization; fg: foregut; FSW: 0.2 μm filtered seawater; hg: hindgut; hpf: hours post fertilization; lm: longitudinal muscle fibre; mg: midgut; MIP: myoinhibitory peptide; mism: mismatch; MO: morpholino; mpf: months post fertilization; nsp: neurosecretory plexus; pdc: posterior dorsal cirrus; ph: pharynx; pp: parapodia; pro: proctodeum; pt: prototroch; PTSP: prothoracicostatic peptide; PTw: 1X PBS with 0.01% Tween; SEM: scanning electron microscopy; sto: stomodeum TDE, 2,2'-thiodiethanol; tt: telotroch; vnc: ventral nerve cord.

Competing interests

A patent application for the potential use of MIP/allatostatin B in lophotrochozoan larval culture has been submitted.

Authors' contributions

EAW and GJ conceived and designed the study. EAW performed expression analysis, morpholino knockdown and behavioral/growth assays. MC performed expression analysis. EAW, MC and GJ analyzed the data. EAW drafted the manuscript. EAW, GJ and MC revised the manuscript. All authors read and approved the final manuscript.

Acknowledgements

The authors thank Dorothee Hildebrandt for her assistance in caring for our *Platynereis* laboratory culture, Jürgen Berger for contributions to SEM, and Günter Purshke for sharing his knowledge of polychaete digestive systems. Aurora Panzera, Christian Liebig and Csaba Verasztó provided assistance with calcium imaging and microscopy. Reza Shahidi and Albina Asadulina assisted with image and video construction in Blender. Heiko Müller provided *in situ* probes for digestive system marker genes. Nadine Randel provided fixed 6 dpf *Platynereis* larvae for SEM. The research leading to these results received funding from the European Research Council under the European Union's Seventh Framework Programme (FP7/2007-2013)/European Research Council Grant Agreement 260821.

References

1. Rieger RM. The Biphasic Life-Cycle - a Central Theme of Metazoan Evolution. Am Zool. 1994;34:484–91.
2. Pawlik JR. Chemical Ecology of the Settlement of Benthic Marine-Invertebrates. Oceanography and Marine Biology. 1992;30:273–335.
3. Rodriguez SR, Ojeda FP, Inestrosa NC. Settlement of Benthic Marine-Invertebrates. Mar Ecol Prog Ser. 1993;97:193–207.
4. Underwood AJ, Keough MJ. Supply side ecology: the nature and consequences of variations in recruitment of intertidal organisms. In: Bertness MD, Gaines SD, Hay MD, editors. Marine Community Ecology. Sunderland, Massachusetts: Sinauer and Associates; 2000. p. 183–200.
5. Jackson D, Leys SP, Hinman VF, Woods R, Lavin MF, Degnan BM. Ecological regulation of development: induction of marine invertebrate metamorphosis. Int J Dev Biol. 2002;46:679–86.
6. Marshall DJ, Morgan SG. Ecological and evolutionary consequences of linked life-history stages in the sea (vol 21, pg R718, 2011). Curr Biol. 2011;21:1771–1771.
7. Gilbert SF. Ecological developmental biology: environmental signals for normal animal development. Evol Dev. 2012;14:20–8.
8. Nielsen C. Life cycle evolution: was the eumetazoan ancestor a holopelagic, planktotrophic gastraea? BMC Evol Biol. 2013;13:171.
9. Tessmar-Raible K, Arendt D. Emerging systems: between vertebrates and arthropods, the Lophotrochozoa. Curr Opin Genet Dev. 2003;13:331–40.
10. Conzelmann M, Offenburger SL, Asadulina A, Keller T, Munch TA, Jekely G. Neuropeptides regulate swimming depth of Platynereis larvae. Proc Natl Acad Sci U S A. 2011;108:E1174–83.
11. Conzelmann M, Williams EA, Tunaru S, Randel N, Shahidi R, Asadulina A, et al. Conserved MIP receptor-ligand pair regulates Platynereis larval settlement. Proc Natl Acad Sci U S A. 2013;110:8224–9.
12. Zantke J, Bannister S, Rajan VB, Raible F, Tessmar-Raible K. Genetic and genomic tools for the marine annelid Platynereis dumerilii. Genetics. 2014;197:19–31.
13. Fischer AH, Henrich T, Arendt D. The normal development of Platynereis dumerilii (Nereididae, Annelida). Front Zool. 2010;7:31.
14. Hempelmann F. Zur Naturgeschichte von Nereis dumerilii Aud. et Edw. Zoologica. 1911;25:1–135.
15. Hauenschild C, Fischer A. Platynereis dumerilii. Mikroscopische Anatomie, Fortpflanzung, Entwicklung. Stuttgart: Gustav Fischer; 1969.
16. Fischer A, Dorresteijn A. The polychaete Platynereis dumerilii (Annelida): a laboratory animal with spiralian cleavage, lifelong segment proliferation and a mixed benthic/pelagic life cycle. Bioessays. 2004;26:314–25.
17. Jekely G. Global view of the evolution and diversity of metazoan neuropeptide signaling. Proc Natl Acad Sci U S A. 2013;110:8702–7.
18. Schoofs L, Beets I. Neuropeptides control life-phase transitions. Proc Natl Acad Sci U S A. 2013;110:7973–4.
19. Hua YJ, Tanaka Y, Nakamura K, Sakakibara M, Nagata S, Kataoka H. Identification of a prothoracicostatic peptide in the larval brain of the silkworm, Bombyx mori. J Biol Chem. 1999;274:31169–73.
20. Davis NT, Blackburn MB, Golubeva EG, Hildebrand JG. Localization of myoinhibitory peptide immunoreactivity in Manduca sexta and Bombyx mori, with indications that the peptide has a role in molting and ecdysis. J Exp Biol. 2003;206:1449–60.
21. Liu X, Tanaka Y, Song Q, Xu B, Hua Y. Bombyx mori prothoracicostatic peptide inhibits ecdysteroidogenesis in vivo. Arch Insect Biochem Physiol. 2004;56:155–61.
22. Lorenz MW, Kellner R, Hoffmann KH. A family of neuropeptides that inhibit juvenile hormone biosynthesis in the cricket, Gryllus bimaculatus. J Biol Chem. 1995;270:21103–8.
23. Leitz T. Metamorphosin A and related compounds - A novel family of neuropeptides with morphogenic activity. Trends in Comparative Endocrinology and Neurobiology. 1998;839:105–10.
24. Iwao K, Fujisawa T, Hatta M. A cnidarian neuropeptide of the GLWamide family induces metamorphosis of reef-building corals in the genus Acropora. Coral Reefs. 2002;21:127–9.
25. Erwin PM, Szmant AM. Settlement induction of Acropora palmata planulae by a GLW-amide neuropeptide. Coral Reefs. 2010;29:929–39.
26. Audsley N, Weaver RJ. Neuropeptides associated with the regulation of feeding in insects. Gen Comp Endocrinol. 2009;162:93–104.
27. Matsuda K, Kang KS, Sakashita A, Yahashi S, Vaudry H. Behavioral effect of neuropeptides related to feeding regulation in fish. Ann N Y Acad Sci. 2011;1220:117–26.
28. Valassi E, Scacchi M, Cavagnini F. Neuroendocrine control of food intake. Nutr Metab Cardiovasc Dis. 2008;18:158–68.
29. Simo L, Koci J, Park Y. Receptors for the neuropeptides, myoinhibitory peptide and SIFamide, in control of the salivary glands of the blacklegged tick Ixodes scapularis. Insect Biochem Mol Biol. 2013;43:376–87.
30. Simo L, Zitnan D, Park Y. Two novel neuropeptides in innervation of the salivary glands of the black-legged tick, Ixodes scapularis: myoinhibitory peptide and SIFamide. J Comp Neurol. 2009;517:551–63.
31. Carlsson MA, Diesner M, Schachtner J, Nassel DR. Multiple neuropeptides in the Drosophila antennal lobe suggest complex modulatory circuits. J Comp Neurol. 2010;518:3359–80.
32. Schulze J, Neupert S, Schmidt L, Predel R, Lamkemeyer T, Homberg U, et al. Myoinhibitory peptides in the brain of the cockroach Leucophaea maderae and colocalization with pigment-dispersing factor in circadian pacemaker cells. J Comp Neurol. 2012;520:1078–97.
33. Kolodziejczyk A, Nassel DR. A novel wide-field neuron with branches in the lamina of the Drosophila visual system expresses myoinhibitory peptide and may be associated with the clock. Cell Tissue Res. 2011;343:357–69.
34. Kolodziejczyk A, Nassel DR. Myoinhibitory peptide (MIP) immunoreactivity in the visual system of the blowfly Calliphora vomitoria in relation to putative clock neurons and serotonergic neurons. Cell Tissue Res. 2011;345:125–35.
35. Moroz LL, Edwards JR, Puthanveettil SV, Kohn AB, Ha T, Heyland A, et al. Neuronal transcriptome of Aplysia: neuronal compartments and circuitry. Cell. 2006;127:1453–67.
36. Schoofs L, Holman GM, Hayes TK, Nachman RJ, De Loof A. Isolation, identification and synthesis of locustamyoinhibiting peptide (LOM-MIP), a novel biologically active neuropeptide from Locusta migratoria. Regul Pept. 1991;36:111–9.
37. Blackburn MB, Wagner RM, Kochansky JP, Harrison DJ, Thomas-Laemont P, Raina AK. The identification of two myoinhibitory peptides, with sequence similarities to the galanins, isolated from the ventral nerve cord of Manduca sexta. Regul Pept. 1995;57:213–9.

38. Blackburn MB, Jaffe H, Kochansky J, Raina AK. Identification of four additional myoinhibitory peptides (MIPs) from the ventral nerve cord of Manduca sexta. Arch Insect Biochem Physiol. 2001;48:121–8.

39. Lange AB, Alim U, Vandersmissen HP, Mizoguchi A, Vanden Broeck J, Orchard I. The distribution and physiological effects of the myoinhibiting peptides in the kissing bug, rhodnius prolixus. Front Neurosci. 2012;6:98.

40. Simo L, Park Y. Neuropeptidergic control of the hindgut in the black-legged tick Ixodes scapularis. Int J Parasitol. 2014;44:819–26.

41. Szabo TM, Chen R, Goeritz ML, Maloney RT, Tang LS, Li L, et al. Distribution and physiological effects of B-type allatostatins (myoinhibitory peptides, MIPs) in the stomatogastric nervous system of the crab Cancer borealis. J Comp Neurol. 2011;519:2658–76.

42. Fu Q, Tang LS, Marder E, Li L. Mass spectrometric characterization and physiological actions of VPNDWAHFRGSWamide, a novel B type allatostatin in the crab, Cancer borealis. J Neurochem. 2007;101:1099–107.

43. Takahashi T, Muneoka Y, Lohmann J, Lopez De Haro MS, Solleder G, Bosch TC, et al. Systematic isolation of peptide signal molecules regulating development in hydra: LWamide and PW families. Proc Natl Acad Sci U S A. 1997;94:1241–6.

44. Takahashi T, Kobayakawa Y, Muneoka Y, Fujisawa Y, Mohri S, Hatta M, et al. Identification of a new member of the GLWamide peptide family: physiological activity and cellular localization in cnidarian polyps. Comp Biochem Physiol B Biochem Mol Biol. 2003;135:309–24.

45. Fredriksson R, Schioth HB. The repertoire of G-protein-coupled receptors in fully sequenced genomes. Mol Pharmacol. 2005;67:1414–25.

46. Struck TH, Paul C, Hill N, Hartmann S, Hosel C, Kube M, et al. Phylogenomic analyses unravel annelid evolution. Nature. 2011;471:95–8.

47. Saulnier-Michel C. Polychaeta: Digestive System. In: Harrison FW, editor. Microscopic Anatomy of Invertebrates. Volume 7. New York, USA: Wiley-Liss, Inc; 1992. p. 53–69. [Harrison FW, Gardiner SL (Series Editor).

48. Tzetlin A, Purschke G. Pharynx and intestine. In: Bartolomaeus T, Purschke G, editors. Morphology, molecules, evolution and phylogeny in polychaeta and related taxa. Dordrecht, The Netherlands: Springer; 2005. p. 199–225.

49. Kitamoto Y, Yuan X, Wu Q, McCourt DW, Sadler JE. Enterokinase, the initiator of intestinal digestion, is a mosaic protease composed of a distinctive assortment of domains. Proc Natl Acad Sci U S A. 1994;91:7588–92.

50. Asadulina A, Panzera A, Veraszto C, Liebig C, Jekely G. Whole-body gene expression pattern registration in Platynereis larvae. Evodevo. 2012;3:27.

51. Conzelmann M, Williams EA, Krug K, Franz-Wachtel M, Macek B, Jekely G. The neuropeptide complement of the marine annelid Platynereis dumerilii. BMC Genomics. 2013;14:906.

52. Nesto N, Simonini R, Prevedelli D, Da Ros L. Effects of diet and density on growth, survival and gametogenesis of Hediste diversicolor (OF Muller, 1776) (Nereididae, Polychaeta). Aquaculture. 2012;362:1–9.

53. Bridges TS, Farrar JD, Gamble EV, Dillon TM. Intraspecific density effects in Nereis (Neanthes) arenaceodentata Moore (Polychaeta: Nereidae). J Exp Mar Biol Ecol. 1996;195:221–35.

54. Scaps P. Intraspecific agonistic behaviour in the polychaete Perinereis cultrifera (Grübe). Vie et milieu. 1995;45:123–8.

55. Miron G, Desrosiers G, Retiére C, Lambert R. Dispersion and prospecting behaviour of the polychaete Nereis virens (Sars) as a function of density. J Exp Mar Biol Ecol. 1991;145:65–77.

56. Gunawardene AR, Corfe BM, Staton CA. Classification and functions of enteroendocrine cells of the lower gastrointestinal tract. Int J Exp Pathol. 2011;92:219–31.

57. Marder E, Bucher D, Schulz DJ, Taylor AL. Invertebrate central pattern generation moves along. Curr Biol. 2005;15:R685–99.

58. Olive PJW. Polychaete aquaculture and polychaete science: a mutual synergism. Hydrobiologia. 1999;402:175–83.

59. Qian PY, Chia FS. Larval Development as Influenced by Food Limitation in 2 Polychaetes - Capitella Sp and Polydora-Ligni Webster. J Exp Mar Biol Ecol. 1993;166:93–105.

60. Pawlik JR, Mense DJ. Larval transport, food limitation, ontogenetic plasticity, and the recruitment of sabellariid polychaetes. In: Wilson JWH, Stricker SA, Shinn GL, editors. Reproduction and Development of Marine Invertebrates. Baltimore, Maryland: John Hopkins University Press; 1994. p. 275–86.

61. Qian PY, Pechenik JA. Effects of larval starvation and delayed metamorphosis on juvenile survival and growth of the tube-dwelling polychaete Hydroides elegans (Haswell). J Exp Mar Biol Ecol. 1998;227:169–85.

62. McEdward LR, Qian PY. Effects of the duration and timing of starvation during larval life on the metamorphosis and initial juvenile size of the polychaete Hydroides elegans (Haswell). J Exp Mar Biol Ecol. 2001;261:185–97.

63. Li W, Godzik A. Cd-hit: a fast program for clustering and comparing large sets of protein or nucleotide sequences. Bioinformatics. 2006;22:1658–9.

64. Castresana J. Selection of conserved blocks from multiple alignments for their use in phylogenetic analysis. Mol Biol Evol. 2000;17:540–52.

65. Guindon S, Dufayard JF, Lefort V, Anisimova M, Hordijk W, Gascuel O. New algorithms and methods to estimate maximum-likelihood phylogenies: assessing the performance of PhyML 3.0. Syst Biol. 2010;59:307–21.

66. Degnan BM, Groppe JC, Morse DE. Chymotrypsin Messenger-Rna Expression in Digestive Gland Amebocytes - Cell Specification Occurs Prior to Metamorphosis and Gut Morphogenesis in the Gastropod, Haliotis-Rufescens. Rouxs Archives of Developmental Biology. 1995;205:97–101.

67. Conzelmann M, Jekely G. Antibodies against conserved amidated neuropeptide epitopes enrich the comparative neurobiology toolbox. Evodevo. 2012;3:23.

68. Staudt T, Lang MC, Medda R, Engelhardt J, Hell SW. 2,2'-thiodiethanol: a new water soluble mounting medium for high resolution optical microscopy. Microsc Res Tech. 2007;70:1–9.

69. Chen TW, Wardill TJ, Sun Y, Pulver SR, Renninger SL, Baohan A, et al. Ultrasensitive fluorescent proteins for imaging neuronal activity. Nature. 2013;499:295–300.

70. Randel N, Asadulina A, Bezares-Calderon LA, Veraszto C, Williams EA, Conzelmann M, et al. Neuronal connectome of a sensory-motor circuit for visual navigation. Elife. 2014;e02730

71. Klopfenstein DR, Vale RD. The lipid binding pleckstrin homology domain in UNC-104 kinesin is necessary for synaptic vesicle transport in Caenorhabditis elegans. Mol Biol Cell. 2004;15:3729–39.

11

Brain functioning under acute hypothermic stress supported by dynamic monocarboxylate utilization and transport in ectothermic fish

Yung-Che Tseng[1][†], Sian-Tai Liu[1][†], Marian Y Hu[2][†], Ruo-Dong Chen[2], Jay-Ron Lee[2] and Pung-Pung Hwang[2]*

Abstract

Background: The vertebrate brain is a highly energy consuming organ that requires continuous energy provision. Energy metabolism of ectothermic organisms is directly affected by environmental temperature changes and has been demonstrated to affect brain energy balance in fish. Fish were hypothesized to metabolize lactate as an additional energy substrate during acute exposure to energy demanding environmental abiotic fluctuations to support brain functionality. However, to date the pathways of lactate mobilization and transport in the fish brain are not well understood, and may represent a critical physiological feature in ectotherms during acclimation to low temperature.

Results: We found depressed routine metabolic rates in zebrafish during acute exposure to hypothermic (18°C) conditions accompanied by decreased lactate concentrations in brain tissues. No changes in brain glucose content were observed. Acute cold stress increased protein concentrations of lactate dehydrogenase 1 (LDH1) and citrate synthase (CS) in brain by 1.8- and- 2.5-fold, paralleled by an increased pyruvate to acetyl-CoA transformation. To test the involvement of monocarboxylate transporters (MCTs) under acute cold stress in zebrafish, we cloned and sequenced seven MCT1-4 homologues in zebrafish. All drMCT1-4 are expressed in brain tissues and in response to cold stress *drmct2a* and *drmct4a* transcripts were up-regulated 5- and 3-fold, respectively. On the contrary, mRNA levels of *drmct1a*, *-1b* and *-4b* in zebrafish brain responded with a down regulation in response to cold stress. By expressing drMCTs in *Xenopus* oocytes we could provide functional evidence that hypothermic stress leads to a 2-fold increase in lactate transport in drMCT4b expressing oocytes. Lactate transport of other paralogues expressed in oocytes was unaffected, or even decreased during cold stress.

Conclusion: The present work provides evidence that lactate utilization and transport pathways represent an important energy homeostatic feature to maintain vital functions of brain cells during acute cold stress in ectotherms.

Keywords: Monocarboxylate transporter, Cold stress, Energy homeostasis, Brain, Ectotherms

Background

Fluctuating environmental temperatures are a major stressor for ectothermic animals which can severely affect vital biochemical and physiological processes [1,2]. Ectothermic organisms have evolved a range of mechanisms to maintain physiological functions during environmental temperature fluctuations [1,3,4]. In the context of hypothermic tolerance special attention has been dedicated

* Correspondence: pphwang@gate.sinica.edu.tw
[†]Equal contributors
[2]Institute of Cellular and Organismic Biology, Academia Sinica, Nankang, Taipei City, Taiwan
Full list of author information is available at the end of the article

to energy metabolism related processes. For example, in blood and liver of common eelpout (*Zoarces viviparus*), the levels of adenosine were elevated under cold stress via adenosine-triphosphate (ATP) hydrolysis indicating that immediate energy supply and the activation of associated enzymes are essential for ectothermic animals undergoing hypothermic acclimation [5]. Furthermore the utilization of lactate to generate energy equivalents has been hypothesized to represent an essential adaptive mechanism to support cellular and organismic functionality during hypothermic stress. For example, Antarctic fish brains were demonstrated to have higher activities of lactate dehydrogenases (LDH) and citrate synthases (CS) when compared to tropical and subtropical species [6,7]. Neuronal functions may thus remain regular and well-coordinated by maintaining metabolic rates in brain tissues during cold acclimation. For example, in isolated brain slices of teleost fishes, high oxygen consumption rates during cold acclimation have been reported [7,8]. These results suggest that at least a partial compensation to cold stress may exist, which is fueled by aerobic ATP producing pathways in fish brain. Moreover, earlier studies using zebrafish could demonstrate a mild mitochondrial uncoupling resulting in enhanced heat production in brain during acute cold stress [1]. Additionally increased activation of glucose transporter (GLUT) and ATP production via glycolysis has been suggested to allow for a metabolic shift to maintain physiological energy balance in brains of zebrafish under acute cold stress [1]. Therefore it was concluded that despite reductions in metabolic rate on the whole animal level, cold acclimated/adapted fish need to maintain brain energy homeostasis by temperature-compensatory mechanism to support proper functionality. This suggests that brain tissues have evolved compensatory metabolic features that allow fish to maintain brain functionality in a hypothermic environment [6-8].

The brain is a highly energy consuming organ and very sensitive to fluctuations in energy supply [9,10]. Most energy consumed by brain cells is devoted to message signal transduction, systemic endocrine regulation, ionic gradient maintenance and restoration after depolarization [11-13]. The brain belongs to the central nervous system and is encased in fluid-filled meninges which consist of neurons, glial cells and capillaries separated from other tissues by a blood-brain barrier (BBB) [10,14]. High concentrations of GLUTs, monocarboxylate transporters (MCTs) and excitatory amino acid transporters (EAATs), facilitate the transport of energy equivalents across membranes of diverse cell types that build up the central nervous system (CNS) [9,15,16]. Although glucose represents the most important energy substrate [9,17-19] a growing number of studies suggested that besides glucose oxidation lactate may also represent an important substrate to provide energy to brain cells [18,20,21]. The transport and utilization of lactate is based on the "astrocyte-neuron lactate shuttle hypothesis" and has already been proposed as a central energy providing pathway in the mammalian CNS [9,22]. Lactate is transported from astrocytes to neurons through MCT1 and 4 [23] and is converted to pyruvate by LDH1. Subsequently pyruvate dehydrogenase (E1) catalyzes the transformation of pyruvate to acetyl-CoA and enters the Krebs cycle to produce ATPs to fuel the energetic demands of neuron cells [24-27]. In this context special attention has been dedicated to monocarboxylate

transporters (MCTs) that enable the transport of lactate in the mammalian brain [9,22,28,29]. MCTs belong to the solute carrier family (SLC) 16, which consists of 14 members in mammals [30-33]. Only MCT1, MCT2, MCT3 and MCT4 have been demonstrated to be responsible for the trans-membrane transport of relevant monocarboxylic acids, such as lactate, pyruvate and ketone bodies [31,33,34]. In mammals, MCT1 can be found in membranes of neurons and astrocytes of the CNS [31]. MCT1 was also detected in the sarcolemma of oxidative skeletal muscle fibers and cardiomyocytes, but never in fast-twitch glycolytic skeletal muscle fibers [35-37]. The lactate transport mechanism of MCT1 has been thoroughly studied [38,39]. MCT1 binds one proton and undergoes a conformational change, which transports the lactate molecule across the membrane [40-42]. Another paralogue, MCT2 has been demonstrated to be the major player for monocarboxylate transport in rodent brain neurons [17,43-45]. Overexpression of MCT2 in *Xenopus* oocytes revealed that this protein is also driven by proton fluxes and has a higher affinity for monocarboxylates than MCT1. The high affinity of MCT2 for monocarboxylates makes it especially suitable for transporting lactate into neurons for oxidation as an important energy source [46]. Moreover, MCT3 has a unique distribution in the basal membrane of retinal pigment epithelium and choroid plexus epithelia of mammals, but the detailed characterization of the transport kinetics have not been studied so far [47,48]. MCT4 has been reported to be localized in astrocytes, fast-twitch oxidative glycolytic skeletal muscle fibers and cardiomyocytes [35,46]. The lactate affinity of MCT4 is lower than that of MCT1 and MCT2, but MCT4 has a high capacity for lactate transport [46,49,50]. In summary, MCTs play an important role in the trafficking of lactate into energy-consuming cells to meet their energetic demands. Regarding lactate utilization in brains of teleosts, Polakof and colleagues [51,52] have demonstrated that lactate can act as an alternative energy fuel in glucose-sensing brain regions such as hypothalamus and hindbrain. Thus, in ectothermic fish brain, lactate may have a similar metabolic role in maintaining energy homeostasis as observed in mammals [52,53]. However, the cellular mechanistic basis for lactate utilization and transport in brain of ectothermic animals are largely unexplored and may significantly differ from those described for mammalian systems. Zebrafish (*Danio rerio*) has been recently explored as a suitable model organism to study the effects of environmental stressors including hypothermic stress on various physiological processes [1,54-56]. Therefore, we use this ectothermic model species to study lactate utilization strategies and further demonstrate the role of MCTs in energy metabolism of brain under acute cold stress. First, we investigated the effect of cold acclimation on several metabolic indexes (O_2 consumption and NH_4^+ excretion), carbohydrate contents and relevant metabolites enzymes to clarify if cold stress induced changes in metabolism are associated to enhanced lactate utilization. Based on our previous study that identified drMCT2 as an indispensable monocarboxylate-transporting route for brain development and function in zebrafish [57]. In order to extend our knowledge regarding differential functions of MCT isoforms in fish during cold-acclimation we identified and characterized additional teleost-specific isoforms. We hypothesize an evolutionary trend for diversified functions and dynamic expression patterns of MCT isoforms in the brain of ectothermic fish (Tseng et al. [57]). We further asked if drMCTs are involved in lactate utilization during cold acclimation and if drMCT isoforms play differential roles in this process. Accordingly, temperature-dependent lactate transport of MCT homologues' was characterized via expression and overexpression in *Xenopus* oocytes. These findings contribute to a better understanding for energy provision by monocarboxylate transport pathways during cold acclimation in brains of ectothermic vertebrates.

Results
Routine metabolic rates under acute cold exposure
Oxygen consumption rates were examined in zebrafish (body length: 4.0-4.5 cm) acclimated to control temperatures (28°C) and those directly transferred to 18°C for 1 and 24 h. As shown in Figure 1A, oxygen consumption rates of adult zebrafish exposed to 18°C for 1 and 24 h were about 2- and 6-fold lower than those determined for control animals kept at 28°C. NH_4^+ excretion rates significantly decreased in animals exposed to hypothermic stress (Figure 1B). After 24 h experimental period, NH_4^+ excretion rates in 28°C acclimated fish was about 4 folds higher than in the hypothermic group.

Glucose and lactate concentrations in zebrafish brain under acute hypothermic stress
At 28°C glucose content in zebrafish brain was 0.43 ± 0.08 nmole/mg. No significant differences in brain glucose contents were observed between the animals from the 28°C group and the acute cold exposure group after 1 h (0.33 ± 0.15 nmole/mg) as well as the 24 h (0.68 ± 0.45 nmole/mg) group (Figure 2A).

Lactate levels in brain of zebrafish kept at 28°C was 48.33 ± 7.48 nmole/mg. In comparison to control (28°C) animals, lactate contents in zebrafish brains which were exposed to 18°C for 1 and 24 h were decreased to 26.66 ± 2.48 and 25.36 ± 1.95 nmole/mg, respectively (Figure 2B). This indicates a 45% to 49% decrease in brain lactate contents during 1 and 24 h 18°C exposure compared to control animals.

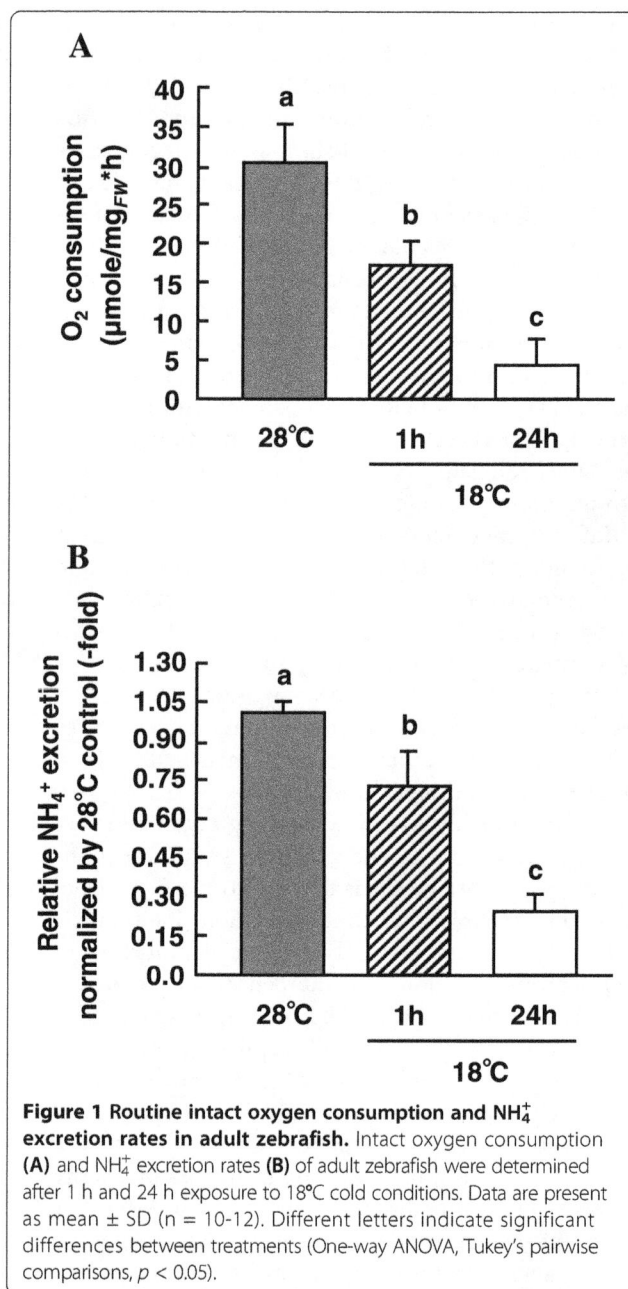

Figure 1 Routine intact oxygen consumption and NH_4^+ excretion rates in adult zebrafish. Intact oxygen consumption (**A**) and NH_4^+ excretion rates (**B**) of adult zebrafish were determined after 1 h and 24 h exposure to 18°C cold conditions. Data are present as mean ± SD (n = 10-12). Different letters indicate significant differences between treatments (One-way ANOVA, Tukey's pairwise comparisons, $p < 0.05$).

Protein concentration and transcript abundance of metabolic genes during hypothermic stress
LDH1 protein concentrations were found to be increased 1.8-fold at the 24 h time point after exposure to 18°C (Figure 3A). In response to cold exposure for 24 h, the relative protein level of citrate synthase (CS) was 2.5-fold higher compared to control conditions (Figure 3B). Gene expression analyses demonstrated that among the three pyruvate dehydrogenase phosphatase (PDP) paralogues expressed in zebrafish brain, *drpdp1b* is expressed 1 to 2-fold higher than the other two isoforms (Figure 4A-C).

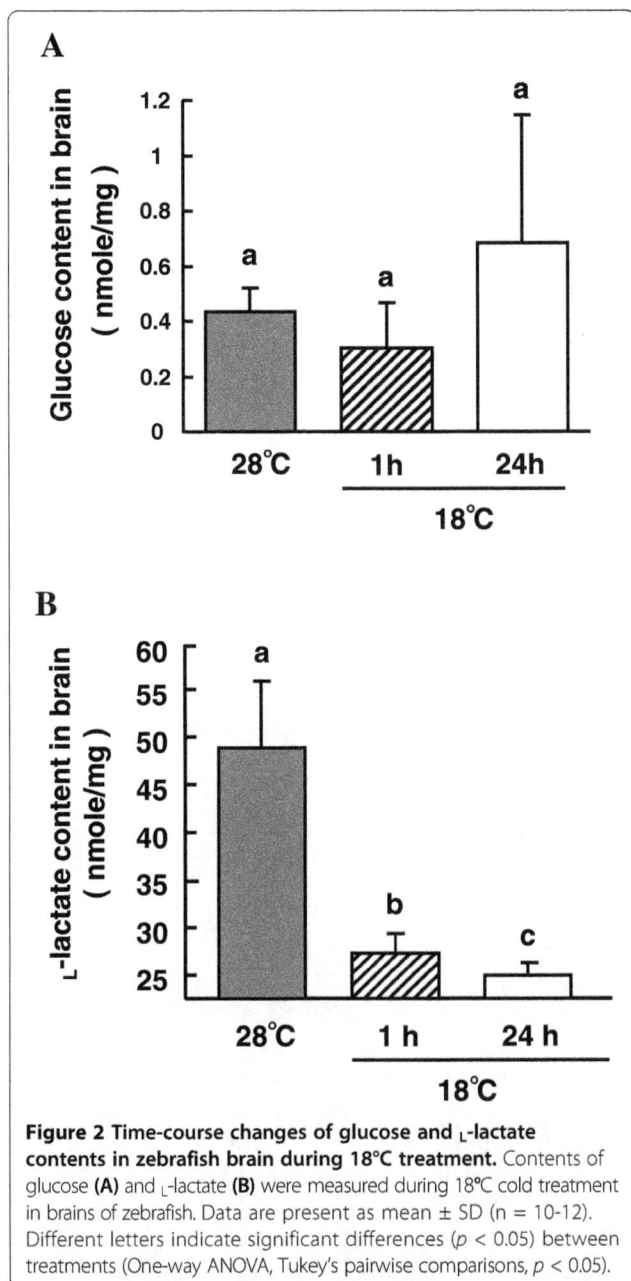

Figure 2 Time-course changes of glucose and ʟ-lactate contents in zebrafish brain during 18°C treatment. Contents of glucose **(A)** and ʟ-lactate **(B)** were measured during 18°C cold treatment in brains of zebrafish. Data are present as mean ± SD (n = 10-12). Different letters indicate significant differences ($p < 0.05$) between treatments (One-way ANOVA, Tukey's pairwise comparisons, $p < 0.05$).

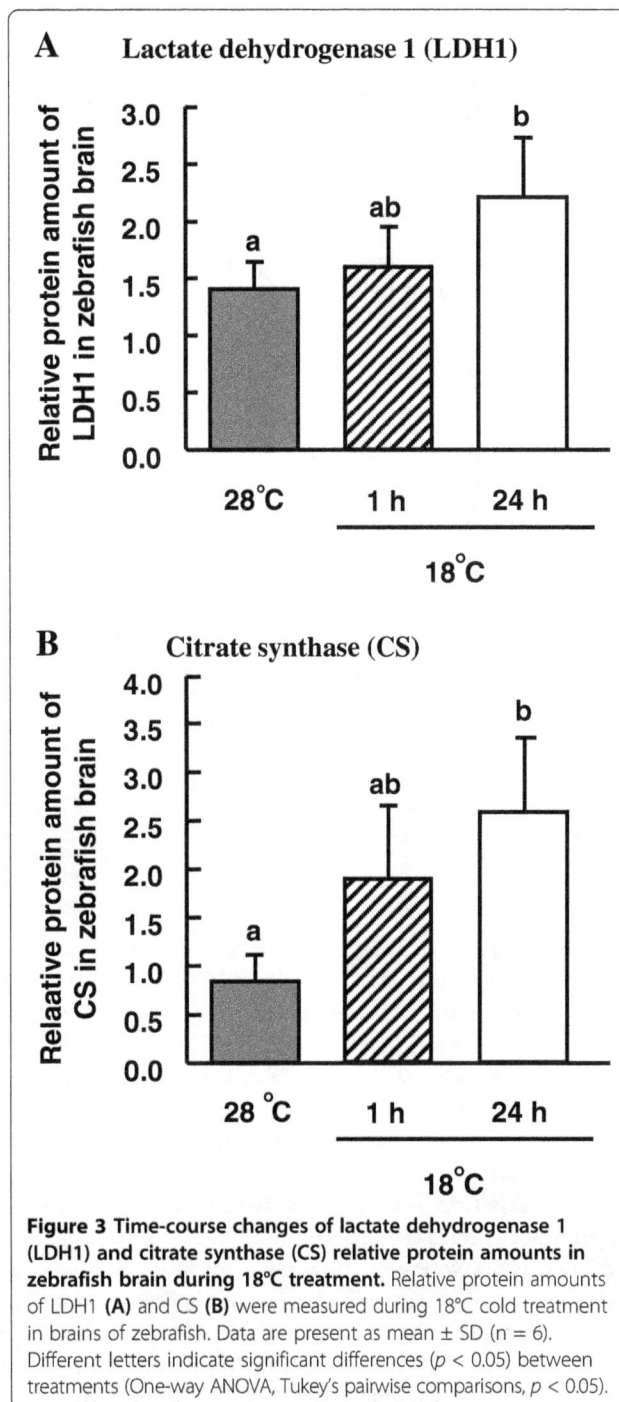

Figure 3 Time-course changes of lactate dehydrogenase 1 (LDH1) and citrate synthase (CS) relative protein amounts in zebrafish brain during 18°C treatment. Relative protein amounts of LDH1 **(A)** and CS **(B)** were measured during 18°C cold treatment in brains of zebrafish. Data are present as mean ± SD (n = 6). Different letters indicate significant differences ($p < 0.05$) between treatments (One-way ANOVA, Tukey's pairwise comparisons, $p < 0.05$).

Within the first hour of acute 18°C cold shock only the *drpdp1b* increased significantly by about 27% ($p < 0.05$) (Figure 4B). After 18°C cold exposure for 24 h, *drpdp1b* and *drpdp2* were up-regulated by about 80% and 52%, respectively (Figure 4B and C). In addition, pyruvate kinase (PK) was also found to be expressed in zebrafish brain and transcript abundance of both, *drpkma* and *drpkmb* were not affected by ambient hypothermic exposure after 24 h (Figure 4D and E). However, transcript abundance of *drpkma* was decreased by about 39% in response to acute (1 h) cold exposure compared to control groups (Figure 4E).

Identification of drMCTs in zebrafish

In the present study, seven isoforms of MCTs1 ~ 4 were cloned and sequenced from zebrafish. There were two drMCT1 isoforms including drMCT1a (annotated as zMCT1 in the previous study) and drMCT1b; two drMCT2 isoforms including drMCT2a (annotated as zMCT2 in the previous study) and drMCT2b; one drMCT3 (annotated as zMCT3 in the previous study);

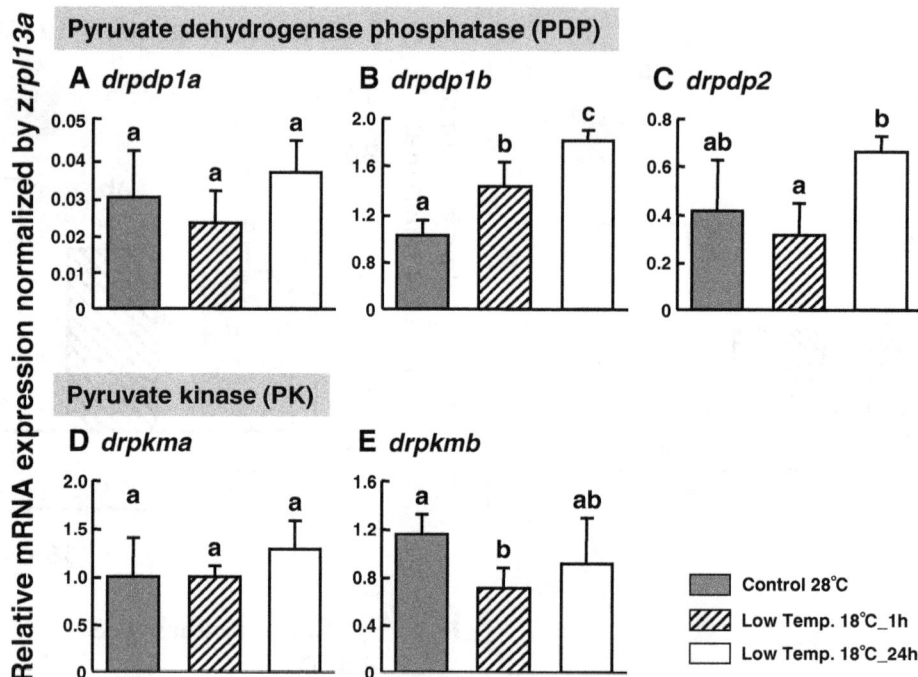

Figure 4 Time-course changes of pyruvate dehydrogenase phosphatase (PDP) and pyruvate kinase (PK) transcripts levels in zebrafish brain during 18°C treatment. Transcripts expression levels of PDP **(A-C)** and PK **(D, E)** were examined during 18°C cold treatment in brains of zebrafish. Expressions of the gene candidates were normalized to ribosomal protein L13A (*drrpl13a*) and presented as relative change. Data are present as mean ± SD (n = 6). Different letters indicate significant differences ($p < 0.05$) between treatments (One-way ANOVA, Tukey's pairwise comparisons, $p < 0.05$). Gray panels: 28°C control; striped panels: 18°C treatment for 1 h; white panels: 18°C treatment for 24 h.

and two drMCT4 including drMCT4a (annotated as zMCT4 in the previous study) and drMCT4b. Neighbor joining (NJ) analysis was used to generate the phylogenetic tree of full-length amino acid sequences of MCT1 (SLC16A1), MCT2 (SLC16A7), MCT3 (SLC16A8) and MCT4 (SLC16A3) from human, mouse, chicken, frog, zebrafish and other teleosts. The phylogenetic analysis demonstrates that seven drMCT isoforms were clearly classified into different groups and clustered with orthologues from other species (Figure 5).

In order to characterize the physicochemical properties of drMCTs, protein trans-membrane predictions were generated (Figure 6). The analysis demonstrates that drMCT1a, -1b and -2a have 11 putative trans-membrane domains (TMDs) with extracellular N-termini and cytoplasmic C-termini (Figure 6A-C) whereas drMCT2b, -3, -4a and -4b have 12 putative TMDs with cytoplasmic C- and N-termini (Figure 6D-G). Based on this analysis, a large cytoplasmic loop (30-53 AA) between TMDs 6 and 7 was found in drMCT2b, -3, -4a and -4b (Figure 7). Moreover, drMCT1a, -1b and -2a can be characterized by a large cytoplasmic loop (41-44 AA) between TMDs 5 and 6 and an additional extracellular loop consisting of about 20-30 AA between TMDs 2 and 3 (Figure 6).

Expression of drMCTs in brains of zebrafish

35 cycles amplification of RT-PCR analysis for various MCTs was conducted in different tissues including brain, gill, eye, spleen and liver from adult zebrafish. Except from *drmct1b* and *-4b* which were not detected in liver, *drmct1a*, *-2a*, *-2b*, *-3* and *-4a* were ubiquitously expressed in various tissues of zebrafish. All the transcripts of *drmct1-4* paralogues were detected in brain tissues (Figure 7A).

Based on our previous studies, spatial expressions of *drmct1a*, *-2a*, *-3* and *-4a* mRNAs in brains of adult zebrafish were already proved to be expressed in both neurons and astrocytes [57]. To further identify cell types that express three novel drMCT1b, -2b and -4b, *in vitro* synthesized RNA probes were used to detect mRNA of these paralogues in transverse sections of zebrafish brains hypothalamus. Subsequent immunocytochemical labeling with ZN12 and glial fibrillary acidic protein (GFAP) antibodies was carried out to specifically identify neurons and astrocytes, respectively [58,59]. In our preliminary test, ZN12 and GFAP were shown to specifically characterize neurons and astrocytes, respectively, in the midbrain of zebrafish [57]. On one hand as shown in Figure 7B and C, *drmct1b* and *-2b* (fluorescence *in situ* hybridization) were both detected in specified groups of

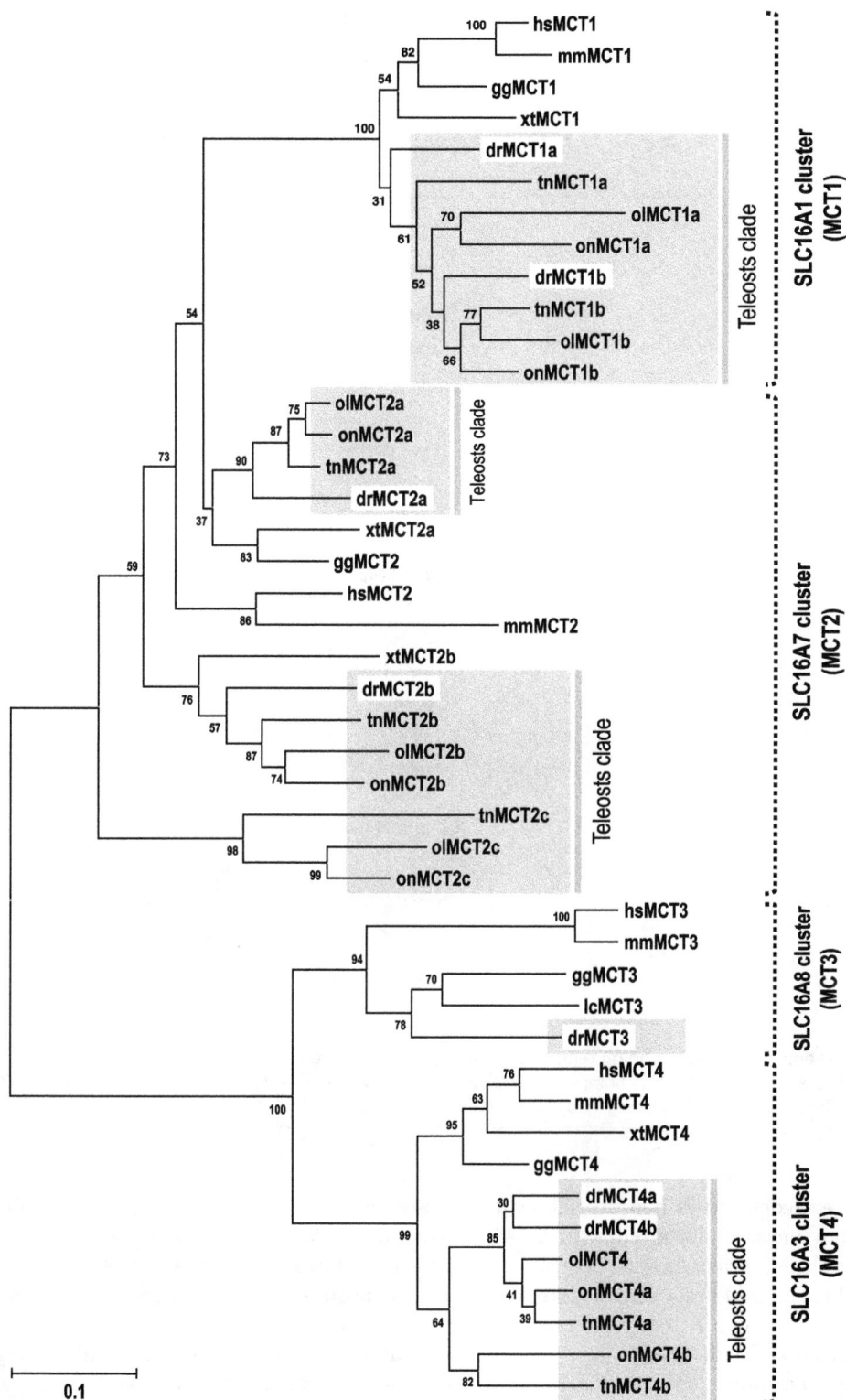

Figure 5 Routed phylogenetic analysis of the MCT1-4 amino acid sequences. The putative MCT sequences of other species were obtained from the Ensembl database, as shown in Table 2. Consensus trees were generated using the Neighbor-joining method with the pairwise deletion gap calculating option. The results were confirmed by 10,000 bootstraps. Numbers indicate bootstrap values and the scale bar units are the number of amino acid substitutions per site.

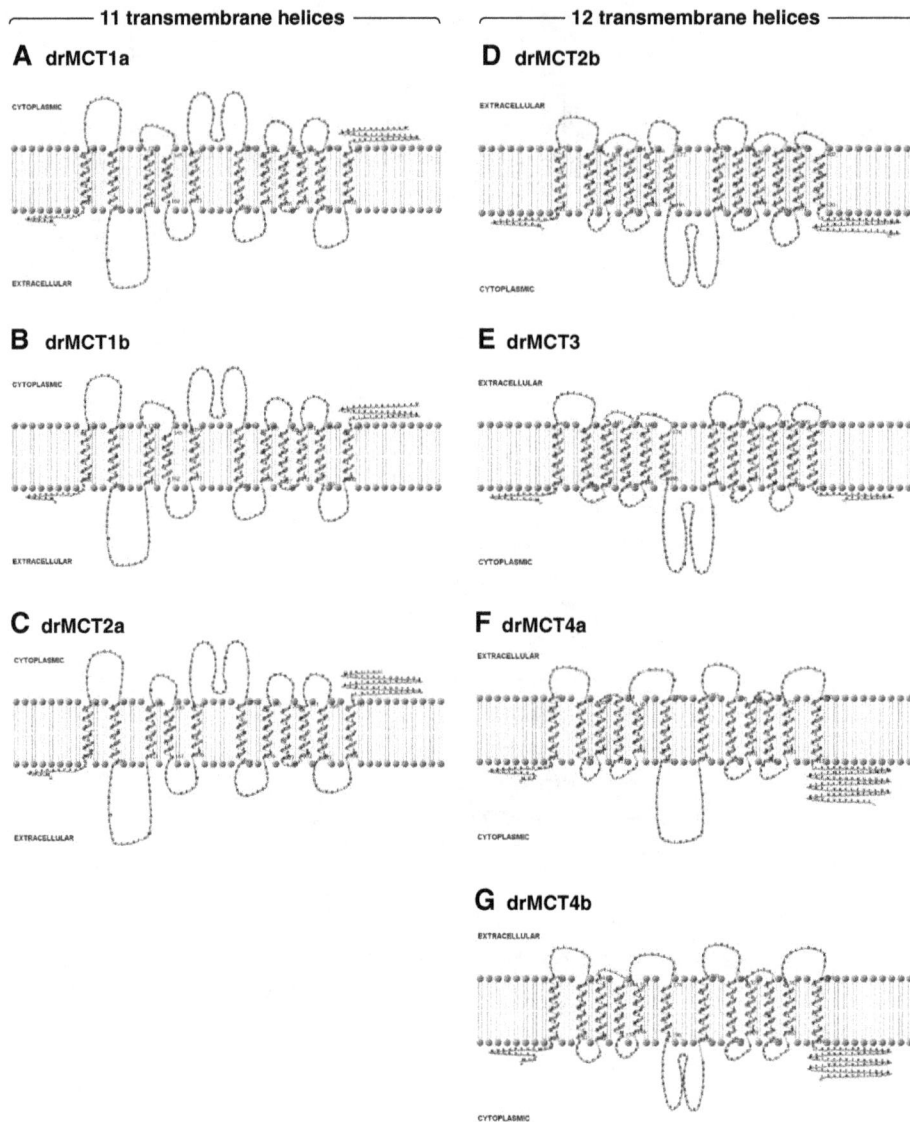

Figure 6 Topological plots and membrane-spanning regions of zebrafish MCT. Deduced amino acid topological plots and putative membrane-spanning regions of seven zebrafish MCT homologues including drMCT1 **(A, B)**, drMCT2 **(C, D)**, drMCT3 **(E)**, drMCT4 **(F, G)** were analyzed by "HMMTOP". The topological hydropathicity plots were draw by TMRres2D.

brain cells, which were recognized as neuron cells with the ZN12 antibody but never co-localized with GFAP signals. On the other hand, *drmct4b* was detected in both ZN12 and GFAP recognized brain cells (Figure 7D). Therefore, these images indicated that *drmct4b* was expressed in both neurons and astrocytes.

Figure 8 shows the time-course changes of *drmct1-4* mRNA expression levels in zebrafish brain in response to acute transfer from 28°C to 18°C with *drrpl13a* serving as an internal control. Furthermore, the results are presented relative to *drmct2b*. In zebrafish brain, on one hand, the mRNA expression of *drmct1b*, *drmct4a* and *drmct4b* were 60 to 400-fold higher compared to

expression levels of other paralogues. On the other hand, *drmct2b* and *drmct3* expression levels were comparatively lower than those of other paralogues. After 18°C exposure for 24 h, mRNA expression of *drmct2a* and *drmct4a* transcripts in brain tissues were up-regulated by 5- and 3-fold, respectively (Figure 8C and F). On the contrary, transcript levels of *drmct1a*, *drmct1b* and *drmct4b* in zebrafish brain were decreased after 1 h exposure to cold stress by about 33%, 81% and 62%, respectively (Figure 8A,B and G). However, after exposure to 18°C for 24 h, *drmct1b* and *drmct4b* mRNA levels in brain tissues were still decreased by about 90% and 58%, respectively, compared to the

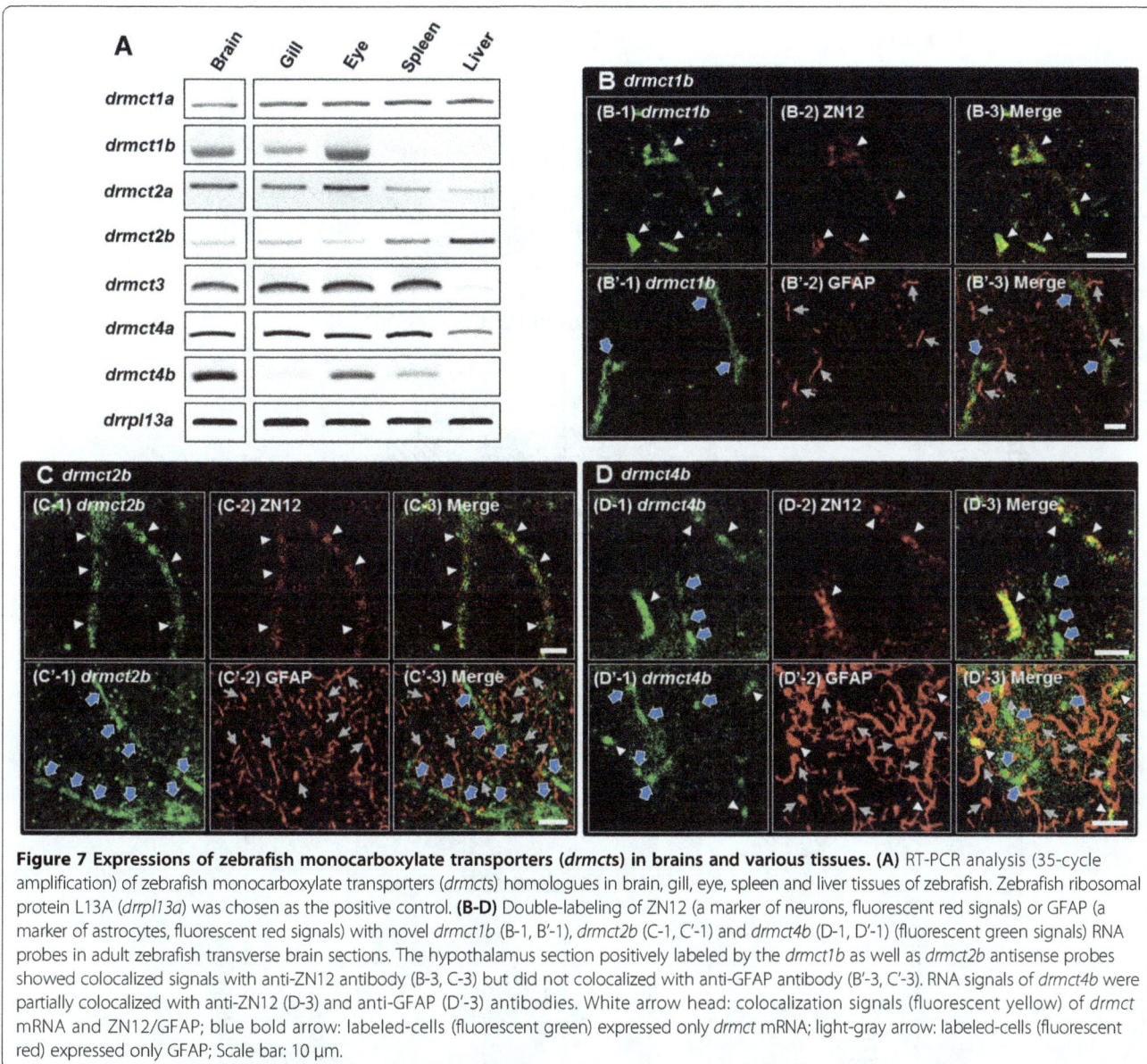

Figure 7 Expressions of zebrafish monocarboxylate transporters (*drmcts*) in brains and various tissues. (A) RT-PCR analysis (35-cycle amplification) of zebrafish monocarboxylate transporters (*drmcts*) homologues in brain, gill, eye, spleen and liver tissues of zebrafish. Zebrafish ribosomal protein L13A (*drrpl13a*) was chosen as the positive control. **(B-D)** Double-labeling of ZN12 (a marker of neurons, fluorescent red signals) or GFAP (a marker of astrocytes, fluorescent red signals) with novel *drmct1b* (B-1, B'-1), *drmct2b* (C-1, C'-1) and *drmct4b* (D-1, D'-1) (fluorescent green signals) RNA probes in adult zebrafish transverse brain sections. The hypothalamus section positively labeled by the *drmct1b* as well as *drmct2b* antisense probes showed colocalized signals with anti-ZN12 antibody (B-3, C-3) but did not colocalized with anti-GFAP antibody (B'-3, C'-3). RNA signals of *drmct4b* were partially colocalized with anti-ZN12 (D-3) and anti-GFAP (D'-3) antibodies. White arrow head: colocalization signals (fluorescent yellow) of *drmct* mRNA and ZN12/GFAP; blue bold arrow: labeled-cells (fluorescent green) expressed only *drmct* mRNA; light-gray arrow: labeled-cells (fluorescent red) expressed only GFAP; Scale bar: 10 μm.

control group (Figure 8B and G). Compared to other paralogues, *drmct2b* and *drmct3* mRNA levels maintained stable during the entire cold-shock period (Figure 8D and E).

In vitro functional analysis of drMCTs in *Xenopus laevis* oocytes

Using functional expression of drMCTs in oocytes we determined $_L$-lactate uptake by measuring the accumulation of ^{14}C-labeled $_L$-lactate during exposure to 28°C and 18°C (Figure 9). Two days after injection, the rate of lactate uptake was 1.3 ~ 4.1-fold higher in oocytes injected with drMCTs' cRNA, compared to control oocytes (RNase-free MBS injection). The rate of lactate uptake by drMCT1b expressing oocytes was highest

(~1200 nmole/hr*oocyte) compared to oocytes expressing other paralogues at 28°C, while the highest uptake rates (~1700 nmole/hr*oocyte) were measured in drMCT4b expressing oocytes at 18°C (Figure 9). In addition, the amount of lactate transported by oocytes expressing drMCT2a was significantly decreased by approximately 73% at 18°C compared to those measured at 28°C; in contrast, lactate transport abilities of drMCT4b expressing oocytes was increased up to 2-fold at 18°C. Besides drMCT2a and drMCT4b, *in vitro* examination in *Xenopus* oocytes revealed that $_L$-lactate transport characteristics of other drMCT paralogues were not significantly affected by ambient temperature.

Kinetic analyses of $_L$-lactate transport at 28°C and 18°C in drMCT2a- and drMCT4b-injected oocytes are shown

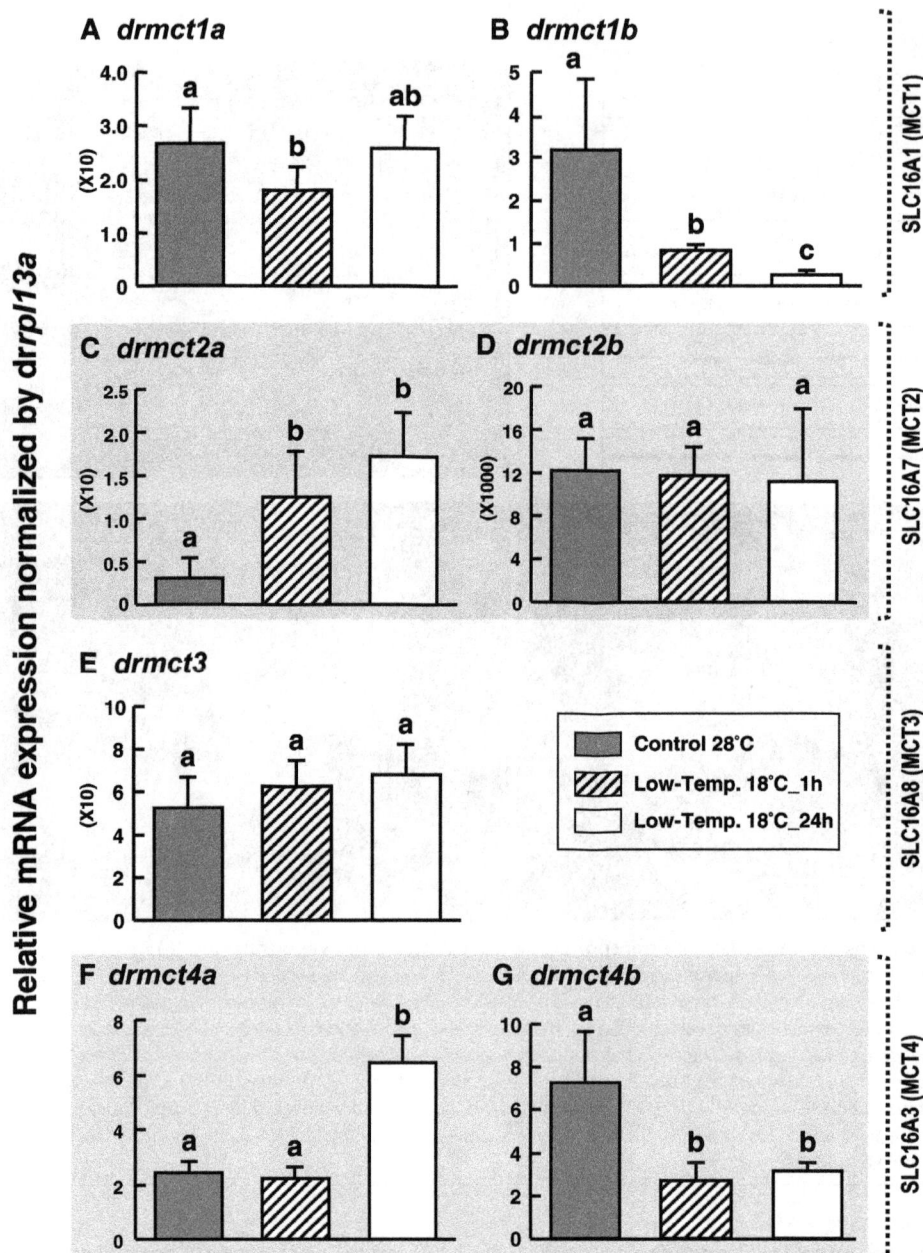

Figure 8 Time-course changes of zebrafish monocarboxylate transporters (*drmct*s) transcripts levels in brain during 18°C treatment. Quantitative real-time PCR (qRT-PCR) were utilized to compare *drmct*s mRNA expressions in brains between 28°C control group and 18°C cold treatment counterparts. Zebrafish ribosomal protein L13A (*drrpl13a*) was used as an internal control. Values are presented as the mean ± SD (n = 6). Different letters indicate significant differences between treatments. (One-way ANOVA, Tukey's pairwise comparisons, $p < 0.05$). Gray panels: 28°C control; striped panels: 18°C treatment for 1 h; white panels: 18°C treatment for 24 h.

in Figure 10. After exposure to 18°C for 0.5 h, lactate transport activity by drMCT2a was 8-fold less than in the 28°C treatment group. After 2 h lactate transport ability of drMCT2a-injected oocytes at 18°C was still about 6-fold lower than at 28°C (Figure 10A). Nevertheless, during 1 h exposure to 18°C L-lactate transport ability of drMCT4b-injected oocytes was observed to be maintained at about 2.3-fold higher levels compared to the 28°C control counterpart.

Discussion

The present work demonstrated that lactate utilization in zebrafish brain represents an advantageous feature to adequately and timely provide energy in brain tissues during acute (1- and 24 h) cold exposure. Furthermore, drMCT isoforms expressed in oocytes reveal differential characteristics in terms of temperature-dependent lactate transport. The lactate uptake of drMCT2a expressing

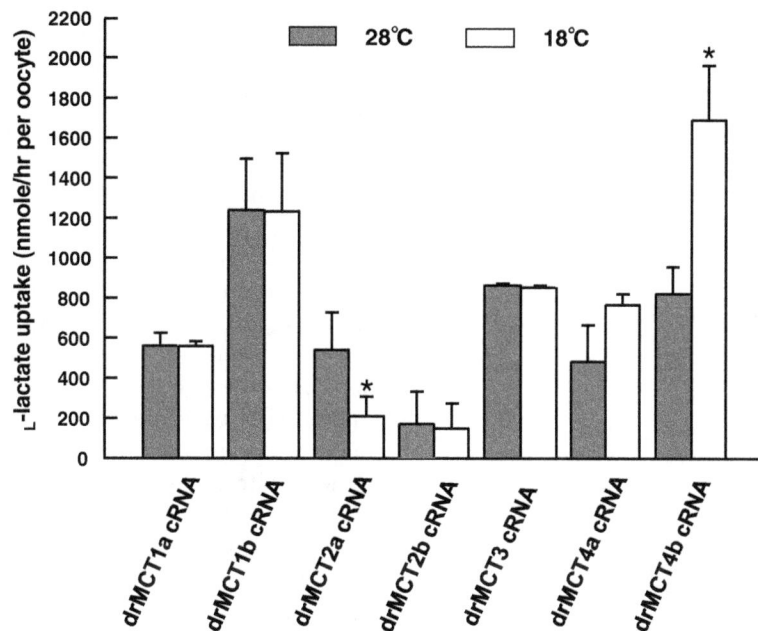

Figure 9 Functional analysis of *drmct* homologues in *Xenopus* oocytes. After the injection of 5 ng *drmcts* cRNA into Stage V-VI Xenopus oocytes for 2 days, the uptake experiment was initiated by transferring the oocytes to 28°C or 18°C for 6 h and then L-[^{14}C]-lactate (5 kBq) was added into the incubation medium. Data are present as mean ± SD (n = 20). Different letters indicate significant differences between treatments (One-way ANOVA, Tukey's pairwise comparisons). *, Significantly different between 28°C and 18°C (Student's t-test, $p < 0.05$).

oocytes was decreased while those expressing drMCT4b were stimulated by low temperature (18°C). These findings, relevant for brain functionality during cold stress are discussed in the following and embedded in the existing body of knowledge.

In the present study, oxygen consumption rates of adult zebrafish transferred from 28°C to 18°C for 1 and 24 h were 2- to 6-fold decreased, indicating that acute hypothermic stress directly translates into decreased metabolic rates on the whole animal level. On one hand, hypothermic challenges may directly affect metabolic enzyme activities and muscle contraction; on the other hand, cold-induced activation of the neuroendocrine system (e.g. TRH release) may be beneficial to counteract physiological compromises during hypothermic stress [60]. Similar observations were made for a range of other fish species with Q_{10} values between 2 and 3 although Q_{10} may show strong differences between species and ontogenetic stages [61-64]. The fact that Q_{10} increased further in response to 24 h cold acclimation is in accordance to Q_{10} values determined for adult zebrafish which are typically in the range between 4-6 [65]. Therefore it can be expected that the highly energy consuming vertebrate brain [1,9,10] is exposed to acute energy fluctuations as well. Thus it is interesting and important to understand how brain energy metabolism of ectotherms is regulated under hypothermic stress. Previous studies demonstrated that Antarctic fish have higher lactate dehydrogenase (LDH) and citrate synthase (CS) activity in brain tissues

compared to tropical and subtropical fishes [6,7]. This suggests that despite comparatively lower metabolic rates and depressed enzymatic activities in locomotory muscle of cold adapted fish, the energy providing pathways that fuel metabolic demands of the brain have to compensate for cold-dependent activity losses [6,8]. This ability is essential for brain functionality under hypothermic stress and allows fish to survive at low temperatures. On one hand, earlier studies indicated that fish can use lactate as an additional energy substrate during acclimation to severe environmental conditions including salinity and temperature changes [8,66,67]. On the other hand, lactate utilization is involved in glucose-sensing capacity and further allowed the maintenance of energy homeostasis in fish brain [68,69]. In the present study, a significant 2-fold decrease, of brain lactate content was found in zebrafish brain transferred from 28°C to 18°C for 1 and 24 h. This decrease of lactate concentrations may indicate enhanced lactate utilization from the lactate pool in fish brain under hypothermic conditions. Together with increased LDH protein concentrations we suggest that lactate is metabolized at higher rates in response to acute cold stress in brain tissues. Kawall and colleagues [6] also denoted that LDH activities in Antarctic fish brains were significant higher compared to those in tropical/subtropical fish species. Therefore lactate utilization and LDH activation for ATP production may represent a potential benefit for energy mobilization

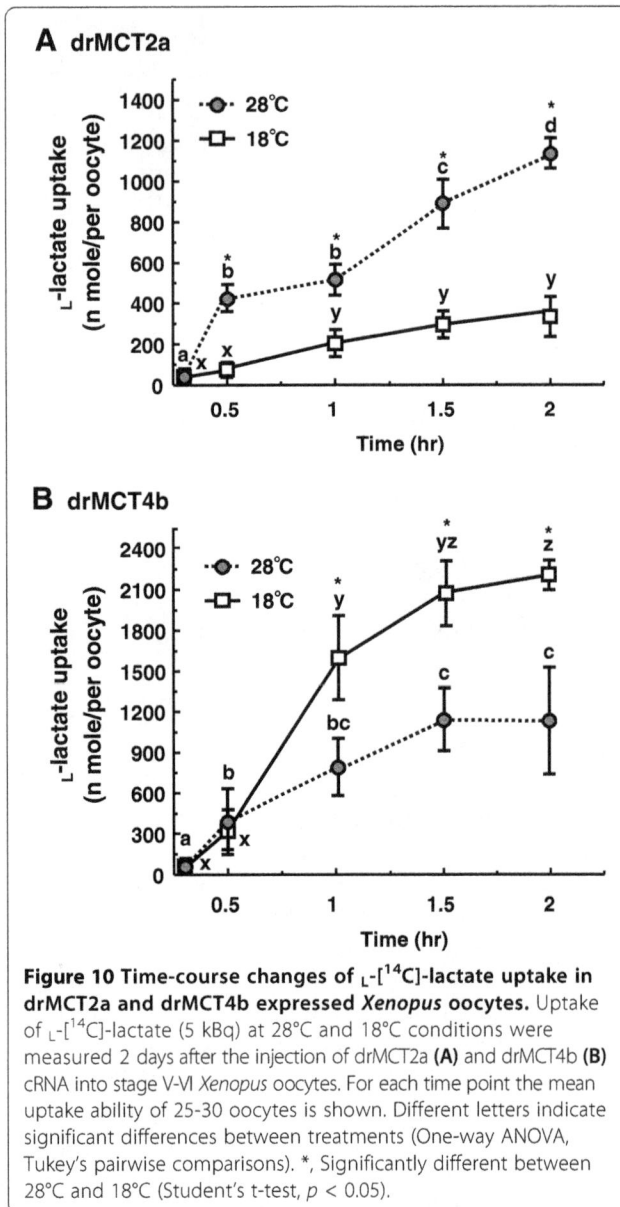

Figure 10 Time-course changes of $_L$-[^{14}C]-lactate uptake in drMCT2a and drMCT4b expressed *Xenopus* oocytes. Uptake of $_L$-[^{14}C]-lactate (5 kBq) at 28°C and 18°C conditions were measured 2 days after the injection of drMCT2a (A) and drMCT4b (B) cRNA into stage V-VI *Xenopus* oocytes. For each time point the mean uptake ability of 25-30 oocytes is shown. Different letters indicate significant differences between treatments (One-way ANOVA, Tukey's pairwise comparisons). *, Significantly different between 28°C and 18°C (Student's t-test, $p < 0.05$).

context of acclimation to cold stress, lactate may represent an important metabolite to support energy metabolism in fish brain. Ivanov and colleagues [70] demonstrated that lactate is an efficient energy substrate to fuel brain aerobic energy metabolism in case of insufficient glucose supply under intense synaptic activity. In this example LDH1 converts lactate to pyruvate and generates cellular NADH pools which represent an important physiological response to neural activation. This lactate metabolizing pathway has been suggested to be more beneficial for rapid energy supply compared to pyruvate formation via glycolysis [71]. Although lactate has long been considered as a potentially toxic metabolic waste product recent findings could demonstrate that neurons preferentially consume lactate as energy substrate using rat neuronal culture experiments [28,72]. In addition, lactate also evokes neuroprotective effects via transcriptional activation of brain-derived neurotrophic factor (BDNF), an essential factor for nerve cells survival [73]. Based on our acute hypothermic challenge in zebrafish, we provided convincing molecular evidences to prove that the astrocyte-neuron lactate shuttle hypothesis (ANLSH) that has been proposed in mammals and other teleosts is an important pathway in fish brain, as well. The existence of the lactate shuttle in fish brain mediated by MCTs further suggests that lactate may serve as an alternative metabolite in brain tissues [66]. Furthermore, transport of lactate between astrocytes and neurons via MCTs in ectothermic fish brain not only plays a critical role for glucosensing function [67], but is also beneficial for fish to cope with acute hypothermic stress.

In contrast to mammals, where only four MCT isoforms have been described, more additional paralogues have been found in teleosts. Some of these novel isoforms were demonstrated to mediate lactate cycling between cells in fish swim bladder [74]. In this study, three novel MCT isoforms, drMCT1b, -2b and -4b, were explored and sequenced from zebrafish and most of them specifically clustered with those from other teleosts. Phylogenetic analysis inferred that diverse functionalities of MCT homologues could have evolved among teleosts and other higher vertebrates. Moreover according to the analysis of physicochemical properties, amino acid residues and membrane-spanning domains, all the drMCT paralogues were predicted to have a long C-terminus, and drMCT2b, -3, -4a and -4b have conventional features of 12 putative TMDs similar to MCT homologues of mammals [30,49]. Interestingly only 11 putative TMDs were predicted for drMCT1a, -1b and -2a which differs from other homologues [75]. According to findings in mammalian MCT structural characterizations, the greatest sequence variation between different MCT

in fish brain under hypothermic conditions. Moreover, increased protein levels of pace-determining metabolic enzymes involved in lactate metabolism and energy equivalent synthesis within the krebs cycle also support our hypothesis that lactate utilization is increased to control energy homeostasis in zebrafish brain during acute (1 and 24 h) cold stress. Our results further suggest that the "lactate-shuttle hypothesis" model established in mammalian systems has also evolved in the teleost brain [57]. This hypothesis denotes that lactate formed in astrocytes is shuttled into neurons via MCTs as an alternative energy source. Thus, increased mRNA levels of PDP and PK particularly support this pathway because lactate transported into neurons needs to be converted to pyruvate and NADH. In the isoforms has been observed in the large cytoplasmic loop between TMDs 6 and 7 [30]; therefore it would be interesting to investigate whether structural differences in TMD numbers or interval loop between TMDs can be related to differential lactate transport abilities of MCTs from ectothermic animals. To offer more definite conclusions regarding structural and functional characteristics of MCT isoforms deeper protein structural analysis such as protein chimera studies, nuclear magnetic resonance spectroscopic (NMRS) or X-ray spectroscopic investigations in combination with determinations of lactate transport rates are needed.

Based on our semi-quantitative PCR results, each organ tested expressed more than one MCT isoforms in zebrafish with all seven MCT isoforms detected in brain tissues. In mammals, MCT1 was expressed in most of tissues and sometimes in combination with other MCT isoforms [29-31]. However, compared to MCT1, a higher affinity for lactate and pyruvate was found for MCT2 in mammalian systems [31]. This paralogue is primarily expressed in liver, kidney, brain, sperm tail, skeletal muscle, and heart and has been described to transport significant amounts of lactate that can be used as energy source [76,77]. In zebrafish, drMCT2a (annotated as zMCT2 in the previous study) has been previously described as an essential component for the development of the central nervous system (CNS) by mediating monocarboxylate transport [57]. Compared with our previous study, the novel drmct2b orthologues were found to be specifically expressed in neurons while drmct2a was localized both in neuron and astrocyte [57] and expressed at higher rates than drmct2b. MCT3 has a unique distribution in the basal membrane of the mouse choroid plexus epithelia [47]. Nevertheless in this study, drmct3 was highly expressed in most examined tissues except for liver, different from the mammalian MCT3 which was expressed in retinal cells only [30,47,48]. However, the physiological role of MCT3 remains unclear in the mammalian homologue [31]. The fact that all the drMCT isoforms were expressed in different cell types of the zebrafish brain suggest that lactate transport between neuron and glial cells and its utilization are probably even more complex in ectothermic fish than in warm-blooded vertebrates.

In zebrafish brain, mRNA expressions of drmct1b and drmct4a were about 10-folds higher than that of drmct3, while drmct2a and drmct2b expression levels were comparatively lower than those of drmct3. This is different from findings in mammals where the expression level of MCT2 was higher than that of MCT1 in brain tissues [78]. Compared to other MCT homologues, expression of drmct2a transcripts in fish brain was up-regulated both at 1- and 24-h cold shock, while the transcript of drmct4a was only increased after 24 h hypothermic

stress. In contrast drmct1b/-2b (specifically expressed in neurons) and drmct3/-4b (expressed both in neurons and astrocytes) mRNA levels in brain were down-regulated or remained unchanged during the acute cold-shock period. According to these differential and partly opposing expression patterns of drMCT isoforms in brain we hypothesize that during acute cold stress an increased lactate demand is mediated by a shift between different MCT isoforms. In order to provide a deeper insight regarding the differential functionality of MCT isoforms we conducted lactate uptake experiments using MCT expressing oocytes.

The present study demonstrated positive lactate uptake rates in all drMCT expressing oocytes with highest uptake rates for drMCT1b compared to all other paralogues. These findings are slightly differing from those obtained for mammals where only MCT1, -2 and -4 have been clearly demonstrated to transport lactate with different capacities [29,30]. Also in chicken, MCT3 has been demonstrated to transport lactate when expressed in a yeast cells [47]. Interestingly, an opposing response of drMCT2a and drMCT4b were observed under 18°C treatment. While lactate uptake of drMCT2a expressing oocytes was decreased (~66%), oocytes expressing drMCT4b increased lactate uptake 1.8-fold during exposure to 18°C. This observation was further supported in vitro by higher lactate transport rates for drMCT4b and lower lactate uptake rates for drMCT2a under hypothermic conditions. The other drMCT paralogues remained unaffected by hypothermic conditions suggesting that transport characteristics of drMCT1a, -1b, -2b, -3 and -4a have a low thermal sensitivity. These observations could suggest that i) cold induced mRNA activation of drmct2a may serve compensatory processes for the decreased expression of drmct4b; ii) and/or increased transcript abundance of drmct2a is necessary to compensate for decreased transport rates under hypothermic conditions; iii) and/or a compensatory mechanism may also apply for drmct4b (the highest expression transcript in brain) where increased cold-induced lactate transport rates are balanced via down regulated mRNA synthesis for metabolic savings. The calculation of a theoretical total lactate transport by combining relative changes of in-vivo expression levels and in-vitro lactate transport between 28° and 18°C indicates a stronger relative response for drMCT2a (2.20 fold-change) than for drMCT4b (0.86 fold-change) (Table 1). These findings corroborate with our previous abrogation test demonstrating an essential role of drMCT2a in zebrafish CNS development [57]. However, drMCT4b has much higher lactate transport capacities and mRNA levels under control conditions compared to drMCT2a. This suggests a significant contribution to lactate transport in fish brain by drMCT4b, as well. The fact that this novel isoform is absent in mammals suggests that drMCT4b is teleost specific, and evolved

Table 1 Comparison of the expression and function of drMCT2aand-4bunder acute cold exposure in zebrafish

Average value	drMCT2a			drMCT4b		
	28°C (Control)	18°C (for 24 h)	Fold of change	28°C (Control)	18°C (for 24 h)	Fold of change
(A)						
Relative expression (normalized by *drrpl13a*)	0.03	0.17	5.67	7.4	3.1	0.41
(B)						
Lactate uptake rate (n mole/h/oocyte)	525.73	202.42	0.39	806.90	1688.62	2.09
(A x B)						
Theoretical lactate uptake capacity	15.75	34.34	2.20	5964.40	5232.80	0.86

to support brain energy homeostasis in this group of ectotherms. Moreover, the results on mRNA transcript and spatial expressions in brain together with functional characterization in oocytes could demonstrate that the timely activations of drMCT2a mRNA as well as drMCT4b kinetic properties are probably more essential for nutrient supply in neurons compared to other isoforms under hypothermic stress.

Conclusion
To our best knowledge, the present work is the first report on the differential *in vivo* and *in vitro* responses of MCT isoforms expressed in fish brain during cold shock. The present study identified relevant MCT isoforms in zebrafish and analyzed their differential relevance for lactate transport in zebrafish brain *in vivo and in vitro* under hypothermic stress. In accordance with our previous study demonstrating different MCT isoforms' spatial localization and functional characterization in zebrafish [57], the present study further indicates a lactate utilization pathway between astrocytes and neurons summarized in Figure 11. Our findings corroborate earlier studies in rainbow trout showing conserved features of a ANLSH in the teleost brain which is similar to that described in mammalian systems [68]. Future studies will investigate the lactate transport kinetics of all seven drMCT isoforms in combination with specific inhibitors in order broaden our understanding regarding isoform-specific lactate transport characteristics. Although lactate is deemed as a toxic metabolite for cells this work highlights its important role in fast energy supply during cold stress to maintain brain functionality in fish. Therefore lactate utilization is not only beneficial for fish brain to maintain physiological status by mediating glucosensing system, but also plays a substantial nourishment role under hypothermic stress. In this context the transport of lactate via MCT paralogues has been demonstrated to represent a novel, highly dynamic process that can be differentially expressed even under non-stressful conditions, and can additionally be timely regulated in response to ambient temperature as well. These unique characteristics of MCTs may represent

an important feature in ectothermic animals that enables utilization of lactate and thus, supports energy homeostasis in fish brain under acute cold stress.

Materials and methods
Experimental animals
Zebrafish
The wild type AB strain of adult zebrafish (*Danio rerio*) were obtained from the stock of the Institute of Cellular and Organismic Biology, Academia Sinica. Animals were kept in a circulating system at 28°C under 14 hour/10 hour of light/dark photoperiod. Fish were fed with dry food (Hai Feng, Nantou, Taiwan) twice a day.

Xenopus laevis
Imported mature female African *X. laevis* frogs were purchased from the African Xenopus facility c.c., Noordhoek, South Africa. They were housed in cages in 18°C carbon-filtered water under standard conditions, and were fed small-sized live goldfish. Through an abdominal small (<1 cm) incision, oocytes were isolated by a partial ovariectomy from female *Xenopus* anesthetized with 0.1% tricaine (3-aminobenzoic acid ethyl ester) on ice. The incision was sutured, and the animal was monitored during the recovery period before it was returned to its tank. Oocytes were maintained at 18°C in Barth's solution containing 88 mM NaCl, 1 mM KCl, 2.4 mM $NaHCO_3$, 0.3 mM CaN_2O_6, 0.41 mM $CaCl_2$, 0.82 mM $MgSO_4$, and 15 mM HEPES (pH 7.6), with gentamycin (20 µg/mL). Oocytes were utilized within 2 days after sampling.

Hypothermic experiments
Zebrafish acclimated in 28°C for over than six months and were directly transferred to 18°C circulating tanks. 8 males and 8 females (body length: 4.0~4.5 cm) were collected and both incubated in 5 L aquaria which were placed in a 300 L water bath at a constant temperature of 18 ± 0.5°C. The four replicate tanks were randomly distributed within the water bath and connected to a flow through system providing freshwater. After 1 h and 24 h cold transferred, fish were anesthetized with

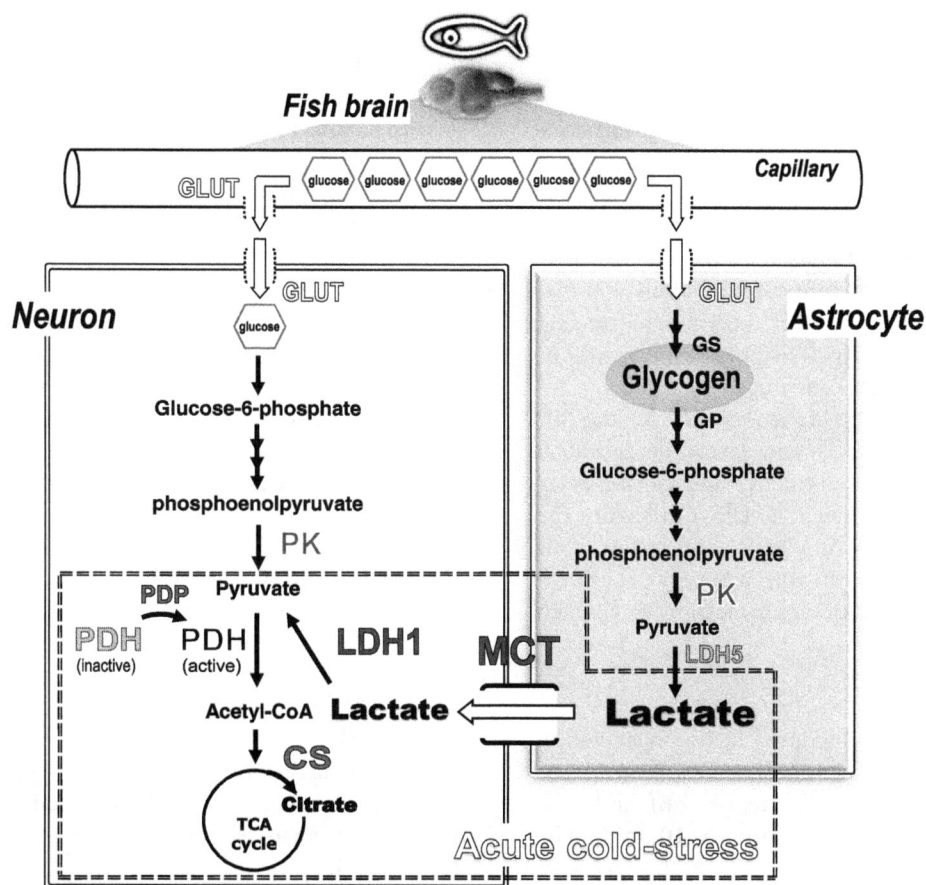

Figure 11 Proposed model for glucose and lactate metabolism in ectothermic fish brain neurons and astrocytes. Lactate is metabolized from glycogenolysis via LDH5 and accumulated in astrocytes. In case of environmental cold stress, lactate is rapidly transport to adjacent neurons through MCT and then converted to pyruvate for immediate ATP production and supply. Different expression patterns of MCT homologues can mediate compensatory function for monocarboxylate transport between neurons and astrocytes in fish brain. Confirmed transporter and enzymes (MCT, LDH1, PK, CS and PDP) are colored as black or dark grey whereas the LDH5 proteins is not yet confirmed and thus hypothetical (light grey).

buffered MS222 (Sigma, St. Louis, MO, USA) before sampling and brains were sampled for total RNA, protein extraction and glucose/lactate content analysis. They were always sacrificed at the same time between 11:00 AM to 12:00 AM, in order to minimize the effect of circadian rhythm. In addition, our preliminary test could demonstrate that oxygen consumption rates were not affected by food deprivation for 24 h at 28°C. Therefore during the acclimation experiments, fish were not fed.

Oxygen consumption and NH$_4^+$ excretion rates

Oxygen consumption rates were determined by closed respirometry in 2 L respiration chambers at atmospheric pressure. Three chambers were filled with filtered freshwater (0.2 μm) and equipped with three animals each. An additional chamber without fish served as a control. Fish were pre acclimated in the respiration chamber for 2-h while the water was aerated to reach full oxygen saturation. After closing the respiration chambers animals

were incubated for 1 h. In order to keep the incubation temperature constant, test chambers were placed in a temperature controlled water bath. A decrease of oxygen concentrations below 70% was avoided. Before and after incubation, water samples (3 × 50 ml plus overflow) were carefully siphoned off from each respiration chamber, and the amount of dissolved oxygen was determined following the Winkler method [79]. The amount of oxygen consumed by bacteria (control chamber) was subtracted from the overall oxygen consumption of the fish. Oxygen consumption rates were expressed as μmole/mg$_{FW}$*h.

To assess the effect of hypothermic treatment on ammonia excretion, one fish was placed in a 3 L tank, and 400 μL of water samples were be collected from the incubation water at the start of the experiment and after 1- and 24-h incubation respectively. Water samples were deproteinized in 4 volumes of ice-cold 6% perchloric acid, and centrifuged at 16000 g for 15 min at 4°C. The supernatant was further neutralized with ice-cold 2 M

K_2CO_3 and spun again at 16000 g for 10 min (4°C). The final supernatant was directly analyzed for NH_4^+ concentrations using a tissue ammonia kit (AA0100; Sigma, St Louis, MO, USA). Ammonia contents in the incubation water were determined colorimetrically at an absorbance of 390 nm with a Synergy HT spectrophotometer (BIO-TEK, Winooski, Vermont, VT, USA).

Glucose and L-lactate content analysis

Isolated brain tissues were homogenized by polytron disruption with 7.5 volume of ice-cooled 6% perchloric acid (PCA), and neutralized (using 1 M potassium bicarbonate). The homogenate was centrifuged at $1000 \times g$ for 15 min in 4°C, and the supernatant was used for the following assay. Glucose and lactate levels in zebrafish brain were assessed using the glucose and lactate assay kit (BioVision, Mountain View, CA, USA) following the manufacturer's protocols. Both, glucose and lactate contents were measured at 570 nm with a Synergy HT spectrophotometer (BIO-TEK) for colorimetric assay. Each sample was assayed in triplicate.

Western blotting

Isolated brain tissues were disrupted in homogenization buffer (100 mM imidazole, 5 mM EDTA, 200 mM sucrose, and 0.1% sodium deoxycholate; pH 7.6), and then centrifuged at 4°C and 10,000 rpm for 10 min. The supernatant (a volume equivalent to 20 μg protein) was supplemented with electrophoresis sample buffer (250 mM Tris-base, 2 mM Na_2EDTA, 2% SDS, and 5% dithiothreitol), and then incubated at 95°C for 10 min. The denatured samples were subjected to 10% sodium dodecylsulfate (SDS)-polyacrylamide gel electrophoresis then transferred to polyvinylidene difluoride membranes (Millipore, Billerica, CA, USA). After blocking in 5% nonfat milk, the blots were incubated with a mouse anti-human lactate dehydrogenase 1 (LDH1) monoclonal antibody (Acris Antidodies GmbH, Germany, diluted 1:1000), a mouse anti-porcine citrate synthase (CS) monoclonal antibody (US Biological, Massachusetts, MA, USA, diluted 1:1000) and a mouse anti-chicken α-tubulin (Sigma), respectively. Blots were then incubated with alkaline phosphatase (AP)-conjugated goat anti-mouse immunoglobulin G (Pierce, Rockford, IL, USA, diluted 1:1000) for one hour. The immunoreactive proteins were visualized with a 5-bromo-4-chloro-3-indolyl phosphate (BCIP)/nitro blue tetrazolium (NBP) substrate kit for AP (Zymed Laboratories, San Francisco, CA, USA). Immunoblots were scanned and exported to TIFF files, and the differences between the band intensities of control fishes and cold acclimation fishes were compared using a commercial software package (Image-Pro Plus 7.0, Media Cybernetics, Silver Spring, MD, USA).

Preparation of total RNA

250-300 mg brain tissue were collected and homogenized in 4 mL of Trizol Reagent (Invitrogen, Carlsbad, CA, USA). Genomic DNA was removed by treating total RNA with DNase I (Promega, Madison, WI, USA) at 37°C for 15 min, and then total RNA was purified by a RNA purification kit (MasterPureTM, EPICENTRE Biotechnologies, Madison, WI, USA). The amount and quality of total RNA were determined by measuring the absorbance at 260 nm and 280 nm with NanoDrop spectrophotometer (ND-1000, NadroDrop Technologies, Wilmington, Delaware, USA). Then the completeness of total RNA was checked by RNA denatured gels. The total RNA pellets were stored at -20°C.

Cloning, phylogenetic analysis and transmembrane prediction

In-silico predicted full-length of zebrafish (*D. rerio*) MCT (*drmct*) homologues obtained from the genome were carefully confirmed by the NCBI database. Specific primers (as listed in Additional file 1: Table S1) were designed for the reverse-transcriptase polymerase chain reaction (RT-PCR) analysis. PCR products were subcloned into a pGEM-T Easy vector (Promega, Madison, WI, USA), and the nucleotide sequences were determined with an ABI 377 sequencer (Applied Biosystems, Warrington, UK). Sequence analysis was conducted with a BLASTx program (NCBI).

To verify the membership of identified candidates in the core MCT protein family, the deduced amino-acid sequences of cloned zebrafish MCTs were aligned with ClustalX together with all known MCT protein sequences available from public databases (referred to Table 2) and subjected to phylogenetic inferences using the Neighbor-joining (NJ) method. 10,000 bootstrap replicate analyses were carried out with MEGA 5.

Protein transmembrane predictions were generated using the "HMMTOP" program (http://www.enzim.hu/hmmtop/). On the basis of physicochemical properties (i.e. hydrophobicity, charges, and distribution) of drMCT paralogues' amino acid residues and membrane-spanning features, the two-dimensional model images were created by a TMRPres2D software (http://biophysics.biol.uoa.gr/TMRPres2D/).

Reverse transcription-PCR analysis

Total RNA extracted from brain, eye, gill, heart, stomach, intestine, spleen, liver, kidney, testis, ovary and muscle of zebrafish were diluted to equal concentration as template for reverse transcription (5 μg total RNA/reaction). The final volume of 20 μL contained 0.5 mM dNTPs, 2.5 μM oligo$_{(dT)}$20 primer, 5 mM dithiothreitol, and 200 units PowerScript reverse transcriptase III (Invitrogen, Carlsbad, CA, USA) and was incubated for

Table 2 Summary of the known and predicted homologs of solute carrier 16A (SLC16A) protein family

SLC16A1 orthologues

Gene name	Protein	Species	Gene loci	Accession/Prediction Numbers
hsSLC16A1	hsMCT1	*Homo sapiens*	Ch.1: 113.4m	NP_003042
mmSLC16A1	mmMCT1	*Mus musculus*	Ch.3: 104.6m	NP_033222
ggSLC16A1	ggMCT1	*Gallus gallus*	Ch.26: 3.55m	NP_001006323
xtSLC16A1	xtMCT1	*Xenopus_tropicalis*	Scaffold GL172654.1:2.6m	NP_001015931
drSLC16A1a	**drMCT1a**	***Danio rerio***	**Ch.8: 27.7m**	**NP_956379**
drSLC16A1b	**drMCT1b**	***Danio rerio***	**Ch.6: 48.9m**	**XP_005172152**
tnSLC16A1a	tnMCT1a	*Tetraodon_nigroviridis*	Ch.Un_random:50.4m	ENSTNIP00000004779
tnSLC16A1b	tnMCT1b	*Tetraodon_nigroviridis*	Ch.11:8.5m	ENSTNIP00000021447
olSLC16A1a	olMCT1a	*Oryzias latipes*	Ch.7: 26.4m	ENSORLP00000020855
olSLC16A1b	olMCT1a	*Oryzias latipes*	Ch.5: 11.9m	ENSORLP00000008214
onSLC16A1a	onMCT1a	*Oreochromis_niloticus*	Scaffold GL831365.1: 0.77m	ENSONIP00000020865
onSLC16A1b	onMCT1b	*Oreochromis_niloticus*	Scaffold GL831272.1: 0.54m	ENSONIP00000018476

SLC16A7 orthologues

Gene name	Protein	Species	Gene loci	Accession/Prediction Numbers
hsSLC16A7	hsMCT2	*Homo sapiens*	Ch.12: 59.9m	NP_004722
mmSLC16A7	mmMCT2	*Mus musculus*	Ch.10: 125.2m	NP_035521
ggSLC16A7	ggMCT2	*Gallus gallus*	Ch.1: 31.9m	XP_416057
xtSLC16A7a	xtMCT2a	*Xenopus_tropicalis*	Scaffold GL173247.1:0.46m	NP_001106392
xtSLC16A7b	xtMCT2b	*Xenopus_tropicalis*	Scaffold GL173000.1: 0.36m	ENSXETP00000005367
drSLC16A7a	**drMCT2a**	***Danio rerio***	**Ch.23: 9.9m**	**NP_001092889**
drSLC16A7b	**drMCT2b**	***Danio rerio***	**Scaffold Zv9_NA123: 35,158**	**ENSDARP00000097781**
olSLC16A7a	olMCT2a	*Oryzias latipes*	Ch.23: 1.1m	ENSORLP00000011457
olSLC16A7b	olMCT2b	*Oryzias latipes*	Ch.7: 20.8m	ENSORLP00000017081
olSLC16A7c	olMCT2c	*Oryzias latipes*	Ch.5: 15.1m	ENSORLP00000010065

SLC16A7 orthologues

Gene name	Protein	Species	Gene loci	Accession/Prediction Numbers
tnSLC16A7a	tnMCT2a	*Tetraodon_nigroviridis*	Ch.19: 6.4m	ENSTNIP00000009708
tnSLC16A7b	tnMCT2b	*Tetraodon_nigroviridis*	Ch.9:8.2m	ENSTNIP00000015526
tnSLC16A7c	tnMCT2c	*Tetraodon_nigroviridis*	Ch.11: 2.3m	ENSTNIP00000014066
onSLC16A7a	onMCT2a	*Oreochromis_niloticus*	Scaffold GL831168.1:2.6m	ENSONIP00000012517
onSLC16A7b	onMCT2b	*Oreochromis_niloticus*	Scaffold GL831137.1: 4.3m	ENSONIP00000021355
onSLC16A7c	onMCT2c	*Oreochromis_niloticus*	Scaffold GL831152.1: 4.9m	ENSONIP00000025647

SLC16A8 orthologues

Gene name	Protein	Species	Gene loci	Accession/Prediction Numbers
hsSLC16A8	hsMCT3	*Homo sapiens*	Ch.22: 38.4m	NP_037488
mmSLC16A8	mmMCT3	*Mus musculus*	Ch.15: 79.2m	NP_065262
ggSLC16A8	ggMCT3	*Gallus gallus*	Ch.1: 50.8m	NP_990471
lcSLC16A8	lcMCT3	*Latimeria_chalumnae*	ScaffoldJH127309.1: 0.25m	ENSLACP00000007957
drSLC16A8	**drMCT3**	***Danio rerio***	**Ch.12: 20.2m**	**XP_686130**

SLC16A3 orthologues

Gene name	Protein	Species	Gene loci	Accession/Prediction Numbers
hsSLC16A3	hsMCT4	*Homo sapiens*	Ch.17: 80.1m	NP_001035887
mmSLC16A3	mmMCT4	*Musmusculus*	Ch.11: 120.9m	NP_001033743
ggSLC16A3	ggMCT4	*Gallus gallus*	Ch.18: 4.8m	NP_989994

Table 2 Summary of the known and predicted homologs of solute carrier 16A (SLC16A) protein family *(Continued)*

xtSLC16A3	xtMCT4	*Xenopus_tropicalis*	Scaffold GL172918.1: 1.0m	NP_001007912
drSLC16A3a	**drMCT4a**	***Danio rerio***	**Ch.12: 34.8m**	**NP_997873**
drSLC16A3b	**drMCT4b**	***Danio rerio***	**Ch.3: 36.1m**	**ENSDARP00000032530**
olSLC16A3	olMCT4	*Oryzias latipes*	Ch.8: 25.2m	ENSORLP00000021568
tnSLC16A3a	tnMCT4a	*Tetraodon_nigroviridis*	Ch.Un_random:5.2m	ENSTNIP00000017000
tnSLC16A3b	tnMCT4b	*Tetraodon_nigroviridis*	Ch.Un_random:38.6m	ENSTNIP00000006385
onSLC16A3a	onMCT4a	*Oreochromis_niloticus*	Scaffold GL831179.1: 0.76m	ENSONIP00000007853
onSLC16A3b	onMCT4b	*Oreochromis_niloticus*	Scaffold GL831279.1: 1.7m	ENSONIP00000009194

Gene loci organization in the genomic region are collected from the Release 75 (February2014) of ENSEMBL Genome Browser.

90 min at 50°C followed by 15 min incubation at 70°C. Then 20 units Escherichia coli RNase H (Invitrogen, Carlsbad, CA, USA) was added to remove the remnant RNA. For PCR amplification, 2 µL total cDNA was used as template in a 50 µL final reaction volume containing 0.25 mM dNTPs, 2 units ExTaq polymerase (Takara, Shiga, Japan), and 0.2 µM of each primer. Primers were designed against specific regions of each isoform according to the drMCT alignment results (Figure 6). The primer sets for the PCR are shown in Additional file 1: Table S1. Zebrafish ribosomal protein L13A (zrpl13a) was used to evaluate the relative amounts of cDNAs as an internal control. This gene has been demonstrated to serve as a suitable reference gene in zebrafish under diverse experimental treatments and different developmental periods [80]. All amplicons were sequenced to ensure that PCR products corresponded to the desired gene fragments.

Quantitative real-time PCR (qRT-PCR) analysis of gene expressions

Total RNA was extracted and reverse-transcribed from zebrafish brain as described above. Real-time PCR was performed with Roche LightCycler® 480 System (Roche Applied Science, Mannheim, Germany). The primers for all genes were designed (Additional file 1: Table S1) using Primer Premier 5.0 software (PREMIER Biosoft Int., Palo Alto, CA, USA). The final volume of 10 µL containing 40 ng of cDNA, 50 nM of each primer and the LightCycler® 480 SYBR Green I Master (Roche). The standard curve of each gene was checked in linear range with *drrpl13a* as an internal control. All qRT-PCR reactions were administered as follows: 1 cycle of 50°C for 2 min and 95°C for 10 min, followed by 45 cycles of 95°C for 15 sec and 60°C for 1 min. PCR products were subjected to melting-curve analysis, and representative samples were electrophoresed to verify that only a single product was present. All primer pairs used in this PCR had efficiencies >94%. Control reactions were conducted with sterile water to determine the levels of the background and genomic DNA contamination.

Fluorescent *in situ* hybridization and immunocytochemical staining

Digoxigenin (DIG)-labeled RNA probes of *drmct1b*, *drmct2b* and *drmct4b* sense and antisense strands were synthesized by *in vitro* transcription using DIG-labeling mix (Roche, Grenzach-Wyhlen, Germany) according to the manufacturer's instructions. RNA probes were examined on RNA gels.

The brain was dissected out of the adult zebrafish head capsule and then sliced into transverse sections of 50 µm using a vibrating blade microtome (VT–1200S, Leica, Wetzlar, Germany). Sliced brain samples were attached to poly-$_L$-lysine-coated slides (Erie, Portsmouth, NH, USA). Prepared slides were washed with PBST and then hybridized with prepared probes at 65°C for overnight incubation. On the next day, slides were washed in a hybridization buffer series in 2× SSC at 65°C. After another serial washing with 0.2× SSC and PBST, samples were incubated with 5% sheep serum in 2 mg/mL bovine serum albumin (BSA), and then the anti-DIG-POD antibody (Roche) was added at a 1:500 dilution. Slides were then washed with PBST and stained with the Alexa 488-tyramide substrate (1:100 dilution). After being washed with PBST and 100% methanol, slides were incubated in an H_2O_2/methanol solution to inactivate the POD. The validity of antisense RNA probes' signals was further examined by the same experiments utilizing sense probes (data not shown). After *in situ* hybridization, zebrafish brain slides were rinsed in PBS, and blocked with 3% BSA. Afterwards, slides first incubated with the ZN12 monoclonal antibody (Institute of Neuroscience, University of Oregon, Eugene, OR, USA; diluted 1:100) to label neurons, or with the glial fibrillary acidic protein (GFAP) mAb (DAKO, Glostrup, Denmark; diluted 1:100) to label astrocytes. After washing with PBS, slides were incubated in anti-mouse IgG conjugated with Alexa Fluor 568 (1:300; Molecular Probes, Eugene, OR, USA). Then slides were washed with PBS. Images were acquired with a Leica TCS-SP5 confocal laser scanning microscope (Leica Lasertechnik, Heidelberg, Germany).

Synthesis of *drmct* homologue capped mRNA (cRNA)

The open-reading frame of *drmct* homologues and the designed anchor sites (XhoI and BamHI) were PCR-amplified using primers based on gene coding sequence, including its stop codon. They were cloned into a pGEM-T easy vector (Promega, Madison, WI, USA) and sequenced. The PCR products were then digested with XhoI and BamHI and subcloned in-frame into the pCS2$_{XLT}$ vector, in which the signal sequence was not fused with the N-terminus of the green fluorescent protein GFP. Capped mRNA (cRNA) encoding *drmct* paralogues were transcribed using the SP6 mMessage mMachine (Ambion, Austin, TX, USA) from plasmid template linearized with NotI (Promega, Madison, WI, USA).

Expression of *drmct* homologues in *X. laevis* oocytes

5 ng of *drmct* homologue cRNA was microinjected into stage V-VI *Xenopus* oocytes. An equal volume of RNase-free modified Barth's saline (MBS, containing 85 mM NaCl, 1 mM KCl, 2.4 mM NaHCO$_3$, 0.82 mM MgSO$_4$, 0.33 mM Ca(NO$_3$)$_2$, 0.41 mM CaCl$_2$, 10 mM HEPES, 3 mM NaOH, pH 7.4, supplemented with 10 units/ml penicillin and 10 μg/ml streptomycin) was also injected to serve as the control group. After culturing for 2 days at 18°C MBS, the uptake experiment was initiated by transferring and incubating the oocytes at 18°C or 28°C in 500 μL MBS medium for 6 h and then 5 kBq/ml of ʟ-[^{14}C]-lactate (Amersham, Piscataway, NJ, USA) was added into the MBS medium. After incubation in L-[14C]-lactate containing MBS medium for 2 h, the uptake reaction was terminated by adding 2 mL of ice-cold MBS, followed by washing the oocytes five times with 2 mL MBS. After the wash, each oocyte was dissolved with 300 μL of 10% SDS. Radioactivity was determined by adding 2 mL of ACSII (Amersham) to each solubilized oocyte in a liquid scintillation counter and then total L-[14C]-lactate concentration in experimental oocytes was determined in a beta counter (LS6500 Liquid Scintillation Counter, Beckman Coulter, Fullerton, CA, USA). The lactate uptake rate of MBS injected oocytes is subtracted from the uptake rate of MCT expressing oocytes and presented as nmole/hr per oocyte.

Statistical analysis

Values are presented as mean ± standard deviation (SD). The level of significance was set to $p < 0.05$ in a one-way ANOVA (Tukey's pairwise comparison) for calculating intact O$_2$ consumption and ammonium excretions of zebrafish; glucose and ʟ-lactate contents in brain; LDH1 protein relative abundance; homologues of *drpk*, *drpdp* and *drmct* mRNA expression levels; ʟ-[^{14}C]-lactate transport characterizations of drMCT2a and drMCT4b for each time point along cold treatment. Student's t-test was used for analyzing ammonium concentration, CS protein expression and ʟ-[^{14}C]-lactate uptake between control (28°C) and hypothermic (18°C) groups. Different letters indicate significant differences between treatments ($p < 0.05$), whereas same letters indicate no significant differences between treatments ($p > 0.05$). Asterisks indicate significant different between control and hypothermic group (Student's t-test, $p < 0.05$).

Additional file

Additional file 1: Table S1. Primer used for RT-PCR and real-time PCR.

Competing interests
The authors declare that they have no competing interests.

Authors' contributions
YCT, STL and MYH designed and conducted experiment, analyzed the data and compiled the manuscript. STL, MYH and JRL conducted oxygen consumption and ammonium excretion experiments and analyzed the data. RDC carried out the molecular cloning studies and gene expression analysis. All authors read and approved the final manuscript.

Acknowledgements
This study was financially supported by the grants to Y. C. Tseng from the National Science Council, Taiwan, Republic of China (NSC 102-2321-B-003 -002) and an Alexander von Humbold/National Science Council (Taiwan) grant awarded to M. H. (NSC 102-2911-I-001-002-2).

Author details
[1]Department of Life Science, National Taiwan Normal University, Taipei City, Taiwan. [2]Institute of Cellular and Organismic Biology, Academia Sinica, Nankang, Taipei City, Taiwan.

Reference
1. Tseng YC, Chen RD, Lucassen M, Schmidt MM, Dringen R, Abele D, Hwang PP: **Exploring uncoupling proteins and antioxidant mechanisms under acute cold exposure in brains of fish.** *PLoS One* 2011, **6**:e18180.
2. Windisch HS, Kathöver R, Pörtner H-O, Frickenhaus S, Lucassen M: **Thermal acclimation in Antarctic fish: Transcriptomic profiling of metabolic pathways.** *Am J Physiol Regul Integr Comp Physiol* 2011, **301**:R1453–R1466.
3. Shabtay A, Arad Z: **Ectothermy and endothermy: evolutionary perspectives of thermoprotection by HSPs.** *J Exp Biol* 2005, **208**:2773–2781.
4. Pörtner HO, Hardewig I, Sartoris FJ, Van Dijk PLM: *Energetic Aspects of Cold Adaptation: Critical Temperatures in Metabolic, Ionic and Acid-Base Regulation.* Cambridge: Cambridge University Press; 1998.
5. Eckerle L, Lucassen M, Hirse T, Pörtner H-O: **Cold induced changes of adenosine levels in common eelpout (***Zoarces viviparus***): a role in modulating cytochrome c oxidase expression.** *J Exp Biol* 2008, **211**:1262–1269.
6. Kawall H, Torres J, Sidell B, Somero G: **Metabolic cold adaptation in Antarctic fishes: evidence from enzymatic activities of brain.** *Mar Biol* 2002, **140**:279–286.
7. Somero GN, Fields PA, Hofmann GE, Weinstein RB, Kawall H: **Cold adaptation and stenothermy in Antarctic notothenioid fishes: what has been gained and what has been lost?** In *Fishes of Antarctica*. Milan: Springer; 1998:97–109.
8. Soengas JL, Aldegunde M: **Energy metabolism of fish brain.** *Comp Biochem Physiol B Biochem Mol Biol* 2002, **131**:271–296.
9. Magistretti PJ, Allaman I: **Brain energy Metabolism.** In *Neuroscience in the 21st Century*. New York: Springer; 2013:1591–1620.
10. Magistretti PJ, Pellerin L: **Astrocytes couple synaptic activity to glucose utilization in the brain.** *News Physiol Sci* 1999, **14**:177–182.
11. Buzsáki G, Kaila K, Raichle M: **Inhibition and brain work.** *Neuron* 2007, **56**:771–783.

12. Hylland P, Milton S, Pek M, Nilsson GE, Lutz PL: **Brain Na⁺/K⁺-ATPase activity in two anoxia tolerant vertebrates: crucian carp and freshwater turtle.** *Neurosci Lett* 1997, **235**:89–92.

13. Purdon A, Rapoport S: **Energy requirements for two aspects of phospholipid metabolism in mammalian brain.** *Biochem J* 1998, **335**:313–318.

14. Alvarez JI, Katayama T, Prat A: **Glial influence on the blood brain barrier.** *Glia* 2013, **61**:1939–1958.

15. Simpson IA, Carruthers A, Vannucci SJ: **Supply and demand in cerebral energy metabolism: the role of nutrient transporters.** *J Cereb Blood Flow Metab* 2007, **27**:1766–1791.

16. Tanaka K: **Role of glutamate transporters in brain development.** *Nippon Yakurigaku Zasshi* 2007, **130**:455.

17. Bélanger M, Allaman I, Magistretti PJ: **Brain energy metabolism: focus on astrocyte-neuron metabolic cooperation.** *Cell Metab* 2011, **14**:724–738.

18. Mergenthaler P, Lindauer U, Dienel GA, Meisel A: **Sugar for the brain: the role of glucose in physiological and pathological brain function.** *Trends Neurosci* 2013, **36**:587–597.

19. Hertz L, Peng L: **Energy metabolism at the cellular level of the CNS.** *Can J Physiol Pharmacol* 1992, **70**:S145–S157.

20. Bolaños JP, Almeida A, Moncada S: **Glycolysis: a bioenergetic or a survival pathway?** *Trends Biochem Sci* 2010, **35**:145–149.

21. Rodriguez-Rodriguez P, Almeida A, Bolaños JP: **Brain energy metabolism in glutamate-receptor activation and excitotoxicity: Role for APC/C-Cdh1 in the balance glycolysis/pentose phosphate pathway.** *Neurochem Int* 2013, **62**:750–756.

22. Pellerin L, Halestrap AP, Pierre K: **Cellular and subcellular distribution of monocarboxylate transporters in cultured brain cells and in the adult brain.** *J Neurosci Res* 2005, **79**:55–64.

23. Aubert A, Costalat R, Magistretti PJ, Pellerin L: **Brain lactate kinetics: modeling evidence for neuronal lactate uptake upon activation.** *Proc Natl Acad Sci U S A* 2005, **102**:16448–16453.

24. Bittar PG, Charnay Y, Pellerin L, Bouras C, Magistretti PJ: **Selective distribution of lactate dehydrogenase isoenzymes in neurons and astrocytes of human brain.** *J Cereb Blood Flow Metab* 1996, **16**:1079–1089.

25. Debernardi R, Pierre K, Lengacher S, Magistretti PJ, Pellerin L: **Cell-specific expression pattern of monocarboxylate transporters in astrocytes and neurons observed in different mouse brain cortical cell cultures.** *J Neurosci Res* 2003, **73**:141–155.

26. Hanu R, McKenna M, O'Neill A, Resneck WG, Bloch RJ: **Monocarboxylic acid transporters, MCT1 and MCT2, in cortical astrocytes in vitro and in vivo.** *Am J Physiol Cell Physiol* 2000, **278**:C921–C930.

27. Laughton J, Charnay Y, Belloir B, Pellerin L, Magistretti P, Bouras C: **Differential messenger RNA distribution of lactate dehydrogenase LDH-1 and LDH-5 isoforms in the rat brain.** *Neuroscience* 2000, **96**:619–625.

28. Bouzier-Sore AK, Voisin P, Bouchaud V, Bezancon E, Franconi JM, Pellerin L: **Competition between glucose and lactate as oxidative energy substrates in both neurons and astrocytes: a comparative NMR study.** *Eur J Neurosci* 2006, **24**:1687–1694.

29. Morris ME, Felmlee MA: **Overview of the proton-coupled MCT (SLC16A) family of transporters: characterization, function and role in the transport of the drug of abuse γ-hydroxybutyric acid.** *AAPS J* 2008, **10**:311–321.

30. Halestrap AP, Meredith D: **The SLC16 gene family-from monocarboxylate transporters (MCTs) to aromatic amino acid transporters and beyond.** *Pflugers Arch* 2004, **447**:619–628.

31. Halestrap AP, Wilson MC: **The monocarboxylate transporter family—role and regulation.** *IUBMB Life* 2012, **64**:109–119.

32. Merezhinskaya N, Fishbein WN: **Monocarboxylate transporters: past, present, and future.** *Histol Histopathol* 2009, **24**:243.

33. Meredith D, Christian H: **The SLC16 monocaboxylate transporter family.** *Xenobiotica* 2008, **38**:1072–1106.

34. Pierre K, Pellerin L: **Monocarboxylate transporters in the central nervous system: distribution, regulation and function.** *J Neurochem* 2005, **94**:1–14.

35. Bergersen L, Thomas M, Johannsson E, Waerhaug O, Halestrap A, Andersen K, Sejersted O, Ottersen O: **Cross-reinnervation changes the expression patterns of the monocarboxylate transporters 1 and 4: an experimental study in slow and fast rat skeletal muscle.** *Neuroscience* 2006, **138**:1105–1113.

36. Chiry O, Pellerin L, Monnet-Tschudi F, Fishbein WN, Merezhinskaya N, Magistretti PJ, Clarke S: **Expression of the monocarboxylate transporter MCT1 in the adult human brain cortex.** *Brain Res* 2006, **1070**:65–70.

37. Hashimoto T, Hussien R, Oommen S, Gohil K, Brooks GA: **Lactate sensitive transcription factor network in L6 cells: activation of MCT1 and mitochondrial biogenesis.** *FASEB J* 2007, **21**:2602–2612.

38. Bröer S, Rahman B, Pellegri G, Pellerin L, Martin J-L, Verleysdonk S, Hamprecht B, Magistretti PJ: **Comparison of lactate transport in astroglial cells and monocarboxylate transporter 1 (MCT 1) expressing *Xenopus laevis* oocytes. Expression of two different monocarboxylate transporters in astroglial cells and neurons.** *J Biol Chem* 1997, **272**:30096–30102.

39. Broer S, Schneider H, Broer A, Rahman B, Hamprecht B, Deitmer J: **Characterization of the monocarboxylate transporter 1 expressed in *Xenopus laevis* oocytes by changes in cytosolic pH.** *Biochem J* 1998, **333**:167–174.

40. Deuticke B: **Monocarboxylate transport in erythrocytes.** *J Membr Biol* 1982, **70**:89–103.

41. Deuticke B: **Monocarboxylate transportr in red blood cell: kinetics and chemical modification.** *Methods Enzymol* 1989, **173**:300–329.

42. Poole RC, Halestrap AP: **Transport of lactate and other monocarboxylates across mammalian plasma membranes.** *Am J Physiol Cell Physiol* 1993, **264**:C761–C782.

43. Broer S, Broer A, Schneider H, STEGEN C, HALESTRAP A, DEITMER J: **Characterization of the high-affinity monocarboxylate transporter MCT2 in *Xenopus laevis* oocytes.** *Biochem J* 1999, **341**:529–535.

44. Pierre K, Parent A, Jayet PY, Halestrap AP, Scherrer U, Pellerin L: **Enhanced expression of three monocarboxylate transporter isoforms in the brain of obese mice.** *J Physiol* 2007, **583**:469–486.

45. Zhang SX, Searcy TR, Wu Y, Gozal D, Wang Y: **Alternative promoter usage and alternative splicing contribute to mRNA heterogeneity of mouse monocarboxylate transporter 2.** *Physiol Genomics* 2007, **32**:95–104.

46. Bergersen L: **Is lactate food for neurons? Comparison of monocarboxylate transporter subtypes in brain and muscle.** *Neuroscience* 2007, **145**:11–19.

47. Philp NJ, Yoon H, Lombardi L: **Mouse MCT3 gene is expressed preferentially in retinal pigment and choroid plexus epithelia.** *Am J Physiol Cell Physiol* 2001, **280**:C1319–C1326.

48. Yoon H, Fanelli A, Grollman EF, Philp NJ: **Identification of a unique monocarboxylate transporter (MCT3) in retinal pigment epithelium.** *Biochem Biophys Res Commun* 1997, **234**:90–94.

49. Halestrap AP: **The SLC16 gene family–Structure, role and regulation in health and disease.** *Mol Aspects Me* 2013, **34**:337–349.

50. Fox JEM, Meredith D, Halestrap AP: **Characterisation of human monocarboxylate transporter 4 substantiates its role in lactic acid efflux from skeletal muscle.** *J Physiol* 2000, **529**:285–293.

51. Polakof S, Míguez JM, Soengas JL: **In vitro evidences for glucosensing capacity and mechanisms in hypothalamus, hindbrain, and Brockmann bodies of rainbow trout.** *Am J Physiol Regul Integr Comp Physiol* 2007, **293**:R1410–R1420.

52. Polakof S, Soengas JL: **Involvement of lactate in glucose metabolism and glucosensing function in selected tissues of rainbow trout.** *J Exp Biol* 2008, **211**:1075–1086.

53. Marty N, Dallaporta M, Thorens B: **Brain glucose sensing, counterregulation, and energy homeostasis.** *Physiol* 2007, **22**:241–251.

54. Chou MY, Hsiao CD, Chen SC, Chen IW, Liu ST, Hwang PP: **Effects of hypothermia on gene expression in zebrafish gills: upregulation in differentiation and function of ionocytes as compensatory responses.** *J Exp Biol* 2008, **211**:3077–3084.

55. Hwang PP, Lee TH: **New insights into fish ion regulation and mitochondrion-rich cells.** *Comp Biochem Physiol A Mol Integr Physiol* 2007, **148**:479–497.

56. McClelland GB, Craig PM, Dhekney K, Dipardo S: **Temperature-and exercise-induced gene expression and metabolic enzyme changes in skeletal muscle of adult zebrafish (*Danio rerio*).** *J Physiol* 2006, **577**:739–751.

57. Tseng YC, Kao ZJ, Liu ST, Chen RD, Hwang PP: **Spatial expression and functional flexibility of monocarboxylate transporter isoforms in the zebrafish brain.** *Comp Biochem Physiol A Mol Integr Physiol* 2013, **165**:106–118.

58. Nielsen AL, Jorgensen AL: **Structural and functional characterization of the zebrafish gene for glial fibrillary acidic protein, GFAP.** *Gene* 2003, **310**:123–132.

59. Trevarrow B, Marks DL, Kimmel CB: **Organization of hindbrain segments in the zebrafish embryo.** *Neuron* 1990, **4**:669–679.

60. Pacák K, Palkovits M: **Stressor specificity of central neuroendocrine responses: implications for stress-related disorders.** *Endocr Rev* 2001, **22**:502–548.

61. Herbing I: **Effects of temperature on larval fish swimming performance: the importance of physics to physiology.** *J Fish Biol* 2002, **61**:865–876.

62. Jonassen TM, Imsland AK, Stefansson SO: **The interaction of temperature and fish size on growth of juvenile halibut.** *J Fish Biol* 1999, **54**:556–572.

63. Clarke A, Johnston NM: **Scaling of metabolic rate with body mass and temperature in teleost fish.** *J Animal Ecol* 1999, **68**:893–905.

64. White CR, Alton LA, Frappell PB: **Metabolic cold adaptation in fishes occurs at the level of whole animal, mitochondria and enzyme.** *Proc Biol Sci B* 2012, **279**:1740–1747.

65. Barrionuevo WR, Burggren WW: **O2 consumption and heart rate in developing zebrafish (*Danio rerio*): influence of temperature and ambient O_2.** *Am J Physiol Regul Integr Comp Physiol* 1999, **276**:R505–R513.

66. Sangiao-Alvarellos S, Laiz-Carrión R, Guzmán JM, del Río MPM, Miguez JM, Mancera JM, Soengas JL: **Acclimation of *S. aurata* to various salinities alters energy metabolism of osmoregulatory and nonosmoregulatory organs.** *Am J Physiol Regul Integr Comp Physiol* 2003, **285**:R897–R907.

67. Tseng YC, Lee JR, Chang JCH, Kuo CH, Lee SJ, Hwang PP: **Regulation of lactate dehydrogenase in tilapia (*Oreochromis mossambicus*) gills during acclimation to salinity challenge.** *Zool Stud* 2008, **47**:473–480.

68. Polakof S, Mommsen TP, Soengas JL: **Glucosensing and glucose homeostasis: from fish to mammals.** *Comp Biochem Physiol B Biochem Mol Biol* 2011, **160**:123–149.

69. Polakof S, Panserat S, Soengas JL, Moon TW: **Glucose metabolism in fish: a review.** *J Comp Physiol B* 2012, **182**:1015–1045.

70. Ivanov A, Mukhtarov M, Bregestovski P, Zilberter Y: **Lactate effectively covers energy demands during neuronal network activity in neonatal hippocampal slices.** *Front Neuroenergetics* 2011, **3**:2.

71. Gladden LB: **Lactate metabolism: a new paradigm for the third millennium.** *J Physiol* 2004, **558**:5–30.

72. Newington JT, Harris RA, Cumming RC: **Reevaluating metabolism in Alzheimer's disease from the perspective of the astrocyte-neuron lactate shuttle model.** *J Neurodegenr Dis* 2013, **2013**:234572.

73. Coco M, Caggia S, Musumeci G, Perciavalle V, Graziano ACE, Pannuzzo G, Cardile V: **Sodium L-lactate differently affects brain-derived neurothrophic factor, inducible nitric oxide synthase, and heat shock protein 70 kDa production in human astrocytes and SH-SY5Y cultures.** *J Neurosci Res* 2013, **91**:313–320.

74. Umezawa T, Kato A, Ogoshi M, Ookata K, Munakata K, Yamamoto Y, Islam Z, Doi H, Romero MF, Hirose S: **O_2-filled swimbladder employs monocarboxylate transporters for the generation of O_2 by lactate-induced root effect hemoglobin.** *PLoS One* 2012, **7**:e34579.

75. Halestrap AP: **The monocarboxylate transporter family—structure and functional characterization.** *IUBMB Life* 2012, **64**:1–9.

76. Garcia CK, Brown MS, Pathak RK, Goldstein JL: **cDNA cloning of MCT2, a second monocarboxylate transporter expressed in different cells than MCT1.** *J Biol Chem* 1995, **270**:1843–1849.

77. Jackson V, Price N, Carpenter L, Halestrap A: **Cloning of the monocarboxylate transporter isoform MCT2 from rat testis provides evidence that expression in tissues is species-specific and may involve post-transcriptional regulation.** *Biochem J* 1997, **324**:447–453.

78. Pellerin L, Pellegri G, Martin J-L, Magistretti PJ: **Expression of monocarboxylate transporter mRNAs in mouse brain: support for a distinct role of lactate as an energy substrate for the neonatal vs. adult brain.** *Proc Natl Acad Sci U S A* 1998, **95**:3990–3995.

79. Hansen HP: **Determination of oxygen. Methods of Seawater Analysis.** In Edited by Grasshoff K, Kremling K, Ehrhardt M. Weinheim: Verlag Chemie; 1999:75–89.

80. Tang R, Dodd A, Lai D, McNabb WC, Love DR: **Validation of zebrafish (*Danio rerio*) reference genes for quantitative real-time RT-PCR normalization.** *Acta Biochim Biophys Sin (Shanghai)* 2007, **39**:384–390.

12

A new kind of auxiliary heart in insects: functional morphology and neuronal control of the accessory pulsatile organs of the cricket ovipositor

Reinhold Hustert[1], Matthias Frisch[1], Alexander Böhm[2] and Günther Pass[2*]

Abstract

Introduction: In insects, the pumping of the dorsal heart causes circulation of hemolymph throughout the central body cavity, but not within the interior of long body appendages. Hemolymph exchange in these dead-end structures is accomplished by special flow-guiding structures and/or autonomous pulsatile organs ("auxiliary hearts"). In this paper accessory pulsatile organs for an insect ovipositor are described for the first time. We studied these organs in females of the cricket *Acheta domesticus* by analyzing their functional morphology, neuroanatomy and physiological control.

Results: The lumen of the four long ovipositor valves is subdivided by longitudinal septa of connective tissue into efferent and afferent hemolymph sinuses which are confluent distally. The countercurrent flow in these sinuses is effected by pulsatile organs which are located at the bases of the ovipositor valves. Each of the four organs consists of a pumping chamber which is compressed by rhythmically contracting muscles. The morphology of the paired organs is laterally mirrored, and there are differences in some details between the dorsal and ventral organs. The compression of the pumping chambers of each valve pair occurs with a left-right alternating rhythm with a frequency of 0.2 to 0.5 Hz and is synchronized between the dorsal and ventral organs. The more anteriorly located genital chamber shows rhythmical lateral movements simultaneous to those of the ovipositor pulsatile organs and probably supports the hemolymph exchange in the abdominal apex region. The left-right alternating rhythm is produced by a central pattern generator located in the terminal ganglion. It requires no sensory feedback for its output since it persists in the completely isolated ganglion. Rhythm-modulating and rhythm-resetting interneurons are identified in the terminal ganglion.

Conclusion: The circulatory organs of the cricket ovipositor have a unique functional morphology. The pumping apparatus at the base of each ovipositor valve operates like a bellow. It forces hemolymph via sinuses delimited by thin septa of connective tissue in a countercurrent flow through the valve lumen. The pumping activity is based on neurogenic control by a central pattern generator in the terminal ganglion.

Keywords: Orthoptera, Gryllidae, Abdomen, Circulation, Hemolymph, Neuroanatomy, Neurogenic, Terminal ganglion, Central pattern generator, Evolutionary novelty

Introduction

In the open circulatory system of insects, the pumping dorsal heart tube circulates hemolymph in the central body cavity enabling a constant perfusion of the internal organs and tissues. This flow, however, cannot effect circulation in outlying dead-end structures, such as antennae, legs, wings and abdominal appendages. For this task, insects have special hemolymph guiding structures and/or auxiliary hearts [1-3].

In appendages, such as the thoracic legs and some abdominal appendages, a longitudinal septum divides the lumen into two sinuses. Distally the septum is lacking, and the sinuses are confluent. Thereby a countercurrent flow is enabled within these appendages, and we distinguish between an efferent and afferent sinus. How the hemolymph flow is produced remains unclear in most cases [3,4]. In some appendages, pressure changes due to regular volume alterations of tracheae or tracheal

* Correspondence: guenther.pass@univie.ac.at
[2]Department of Integrative Zoology, University of Vienna, Althanstraße 14, 1090 Vienna, Austria
Full list of author information is available at the end of the article

sacs contribute to the hemolymph exchange [5,6]. More elaborate organs for the supply of hemolymph to long body appendages are the so-called accessory pulsatile organs or auxiliary hearts. These muscle-driven pumps can be very diverse in their functional morphology in the various groups of insects. They may be located at the base or within the appendages and are in general autonomous organs which pump rhythmically, but independently, from that of the dorsal heart. The contractions of these auxiliary hearts are based on a myogenic automatism which can be modulated by neuronal and/or neurohormonal control [7-10]. A thoroughly investigated example of such an auxiliary heart is the antenna-heart of the cockroach *Periplaneta americana* in which the functional morphology, neuroanatomy, neurochemistry, pharmacology and the control mechanisms have been analyzed in detail [11-16].

However, the problem of circulation has not yet been investigated in insect ovipositors although some of them reach considerable length. In this paper we describe for the first time accessory pulsatile organs for these body appendages. The organs were discovered in the female cricket *Acheta domesticus* (preliminary notes [1,17]). In live specimens, hemocyte movements can be observed under the microscope through transparent parts of the ovipositor cuticle. The flow occurs in pulses that are clearly correlated with conspicuous compressions of structures at the base of the ovipositor valves which were revealed to be the pumping organs for hemolymph circulation in these appendages. The functional morphology of these ovipositor pulsatile organs was investigated on the basis of serial semi-thin sections and a microCT scan in combination with *in vivo* observations. In addition, neuroanatomical and physiological studies were performed. Several motoneurons and interneurons involved in the control of the ovipositor pulsatile organs could be identified in the terminal ganglion. The electrophysiological recordings revealed a coordinated and rhythmic bilateral motor output from these neurons. Since the rhythm persists even when the terminal ganglion is completely isolated, it could serve as a model for studies of autonomous rhythm generation in a neural network (preliminary reports [17,18]).

Results
Ovipositor and anatomical condition at the abdominal apex
The ovipositor shaft is composed of four long valves: a ventral pair, referred to as the gonapophyses of the abdominal segment 8 (ga8; synonyms: 1st valvulae of Snodgrass [19,20], 1st gonapophyses of Scudder [21]), and a dorsal pair, referred to as the lateral gonapophyses of the abdominal segment 9 (ga9l; 3rd valvulae of Snodgrass [19,20], gonoplacs of Scudder [21]). The median gonapophyses of the 9th segment (ga9m, synonyms: 2nd valvulae

of Snodgrass [19,20], 2nd gonapophyses of Scudder [21]) are very short and inconspicuous structures in crickets. The long ovipositor valves are distinctly widened at their bases and interconnected by several joints to the highly modified and strongly sclerotized coxosternites of the abdominal segment 8 and 9, respectively (cs8 and cs9; synonyms: 1st and 2nd valvifers of Snodgrass [19,20], gonocoxae of Scudder [21], coxosternite of Klass and Ulbricht [22]). The lumen of the four long ovipositor valves is subdivided by delicate septa of connective tissue into sinuses in which efferent or afferent flows of hemolymph can be observed in live specimens. Aside from the epidermis, the ovipositor valves only contain nerves and few tracheal trunks extending to the apical tip. The ovipositor pulsatile organs (opo) are located at the base of the appendages (Figure 1).

Ovipositor pulsatile organs of the abdominal segment 8
Removal of the subgenital plate exposes the bases of the ga8 (Figure 1A, B; Figure 2A). Their lateral sides consist of a strongly sclerotized cuticle while medially the cuticle is thin and flexible forming a conspicuous bulge at each ga8 base ("soft lateral walls" of [23]). The two bulges form the soft-walled margin of the genital opening. In live specimens conspicuous movements of the bulges can be observed: they are strongly compressed and expanded in an alternating left-right rhythm of 0.2-0.5 Hz at room temperature (see Additional file 1: Video). Our study concludes that the compressible bulges constitute the pumping chambers of the ovipositor pulsatile organs of abdominal segment 8 (opo8). They serve as bellow-like pumping devices for hemolymph transport through the ga8. The compression is caused by a muscle that extends across the lumen of the bulges (cm8, Figure 2B). The muscle originates laterally at the sclerotized cs8 and fans out into individual strands which attach medially to the soft wall of the bulge (Figures 3A, B; Figure 4D). Contraction of this muscle pulls the soft parts of the cuticle in a lateral direction. Thereby the pumping chamber is compressed, and hemolymph is forced into the ga8. The individual muscle strands are extended at slightly different angles enabling compression of a wide area of the bulge during contraction. The compression of one bulge supports the dilation of the other bulge and thereby helps to fill it with hemolymph drawn from the abdominal body cavity. In this way the two opo8 operate as two partly interconnected left-right alternating bellows.

After each pumping stroke the hemolymph slightly flows back before it stalls. This observation is considered an indication for the absence of a discrete valve device within the ga8. However, on each body side some of the more anterior and dorsal muscle strands of the cm8 are attached to the lateral margins of the flat cuticle part between the two bulges (Figure 3A, arrow). During bulge

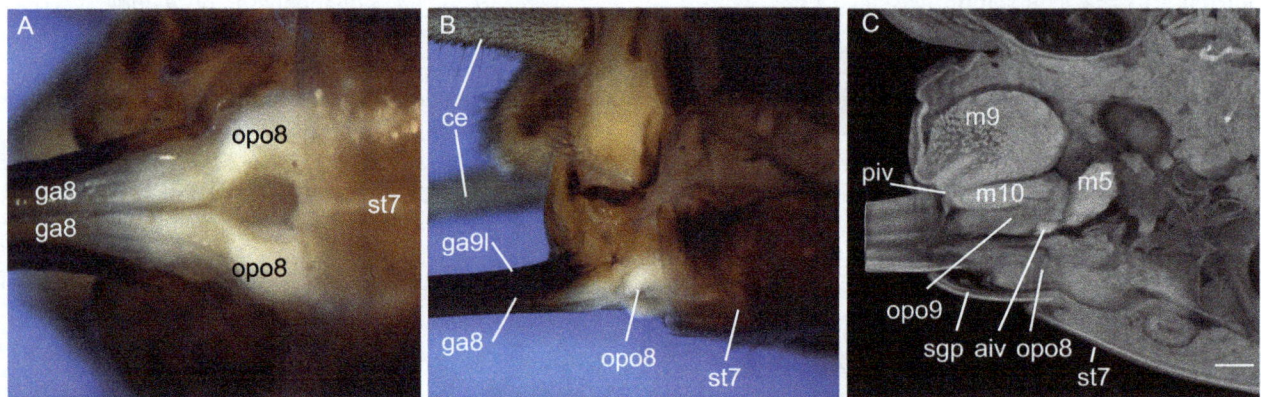

Figure 1 Location of the ovipositor pulsatile organs in the abdominal apex. (A) Ventral view (subgenital plate removed): the distally strongly sclerotized paired gonapophyses 8 (ga8) have proximally an extensive area of thin and flexible cuticle. The distinct bulges at their basal articulation represent the ovipositor pulsatile organs 8 (opo8). **(B)** Lateral view (subgenital plate removed): the ovipositor shaft consists of the ga8 and the lateral gonapophysis 9 (ga9l); opo8 form protruding bulges. **(C)** Volume rendering of microCT data in lateral view showing spatial relationship of opo8 and opo9 to the anterior and posterior intervalvular sclerites (aiv, piv) and some larger muscles (numbering follows [19]). Scale bar: 500 μm. Further abbreviations: ce, cercus; sgp, subgenital plate; st7, sternite of the 7th abdominal segment.

compression this part is pulled strongly into a lateral direction thereby narrowing the base of the bulge. This movement may partly prevent backflow of hemolymph into the abdominal cavity.

The lumen of each bulge is divided by a septum into a wide median sinus (si8m) which contains the cm8 in its proximal portion, and a narrower lateral sinus (si8l) which is devoid of any muscles (Figure 3A; Figure 4C). The septum extends into the ga8 nearly up to the apex and separates the somewhat smaller si8m from the larger si8l (Figure 5A, B). Compressions of the opo8 force hemolymph into a distal direction within the si8m. Observed peak flow velocities of hemocytes range between 5 to 15 mm.s^{-1}, and hemolymph can thereby be moved from the base to the apex of the ga8 with a single pumping stroke. At the lancet-like apex of the ga8, the septum is perforated and the hemolymph can flow into the lateral sinus. There it returns to the base of the ga8 and continues further back dorso-laterally into the body cavity.

Two tracheae are embedded within the septum which separates the si8m and si8l in each ga8 (Figure 4C, D). During each pumping stroke these tracheae become laterally displaced and partly compressed. When the ipsilateral bulge relaxes, the tracheae return to their original shape indicating that during pumping strokes the hemolymph pressure in the si8m is higher than in the si8l.

Ovipositor pulsatile organs of the abdominal segment 9

The pump system of the ovipositor pulsatile organs 9 (opo9) overlies the posterior part of the opo8 (Figure 2C, D). Its functional principle resembles that of the abdominal segment 8, but there are some anatomical differences which

may be explained by the deeper integration of the ga9 bases into the abdominal apex. The soft cuticular parts of the opo9 (Figure 4C, D) extend ventrally between the lateral strongly sclerotized parts of the coxosternite 9 (cs9), as well as the anterior and posterior intervalvular sclerites (aiv and piv after Snodgrass [19], Figure 2D). Medially, the walls of the two ga9 bases are narrowly apposed which appears from outside as slit-like invagination (Figure 4A, B). It consists for the most part of flexible cuticle but in the midline, where the right and left ga9 meet, there is a strongly sclerotized structure (Figure 3A). The small hemocoel spaces lateral of the invagination represent the pump chambers of opo9. Each chamber is continuous with the ventral sinus (si9v) of the ipsilateral ga9 (Figures 2A, 3A, 4 and 5), and dorsally each chamber is covered by a muscle (cm9) that is attached to the upper part of the invagination and laterally to the cs9 (Figure 3A; Figure 4C, D). In live specimens one can observe that alternating contractions of the left and right muscles tilt the median cuticular structure and the flexible median cuticle portion to the corresponding side (see Additional file 1: Video). Thereby the two pump chambers are compressed and widened in alternation and hemolymph is forced into the si9v of each ga9l. The opo9 operates, similar to the opo8, as a pair of interconnected left-right alternating bellows: compression of one pump chamber (systole) simultaneously widens the opposite chamber (diastole) and stretches its compressor muscles (cm9). This leads to contraction of the cm9 thereby completing a full pumping cycle.

Hemolymph flows through the si9v of ga9l to the apex and passes through small gaps in the septa to si9d and si9i (Figure 5C, D). From these two sinuses the hemolymph flows back to the ovipositor base. During each pumping

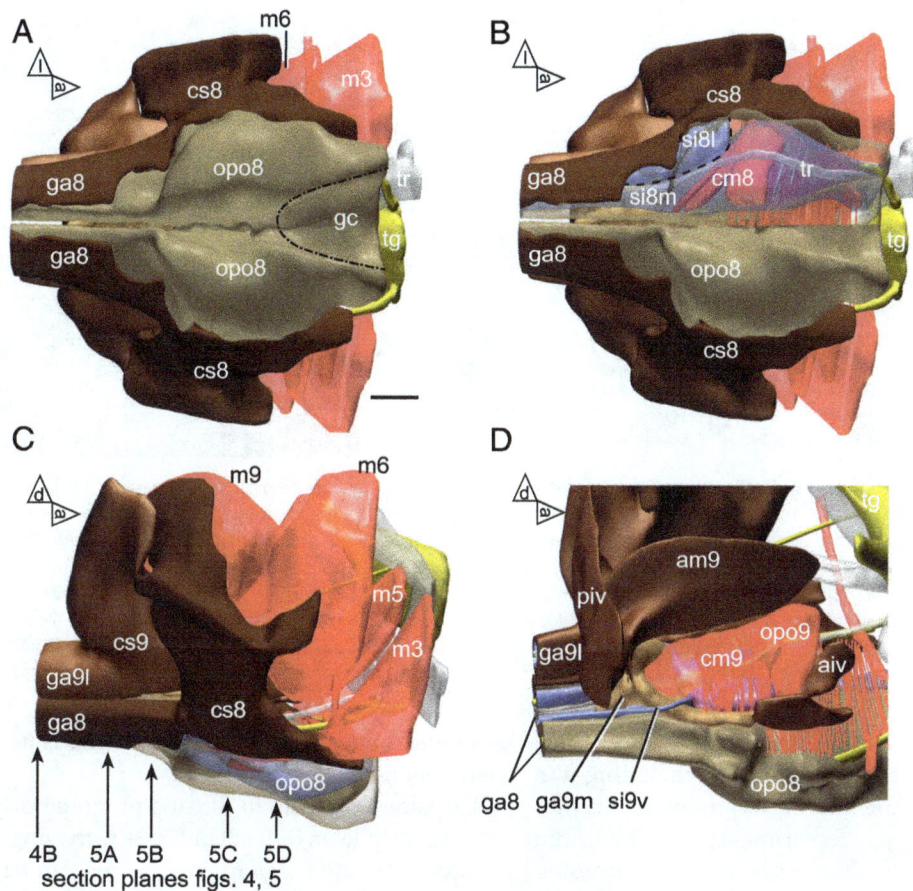

Figure 2 Reconstruction of the ovipositor pulsatile organs and the surrounding structures of the abdominal apex. (A) Ventral view (subgenital plate). Bases of the gonapophyses 8 (ga8) enlarged and with strongly sclerotized lateral parts (*dark brown*), medial parts of flexible cuticle (*light brown*) forming compressible bulges which constitute the ovipositor pulsatile organ 8 (opo8); area of the preceding genital chamber (gc) delimited by dotted line. **(B)** Same view as in A, but right ga8 base is presented transparently; compressor muscle 8 (cm8) (*red*) extends between strongly sclerotized coxosternite 8 (cs8) and medial flexible cuticle of ga8; lumen of ga8 is divided by a delicate septum (indicated by dashed line) into a wide medial sinus (si8m) and a smaller lateral sinus (si8l; both sinuses in *blue*) **(C)** Lateral view. ga8 and lateral gonapophysis 9 (ga9l) are articulately jointed to cs8 and cs9. The bulges of the protruding ovipositor pulsatile organ (opo8) are presented transparently. Arrows refer to section levels of Figures 4 and 5. **(D)** Parasagittal view of the opo, most of left cs9 and many large muscles are removed to expose opo9; cm9 extends between cs9 and a median invagination of flexible cuticle. Scale bar: 250 μm. Further abbreviations (numbers 8 and 9 refer to the respective abdominal segment): aiv, anterior intervalvular sclerite; am, median apodeme (attachment site of muscle 9); m, muscle (numbering follows [19]); piv, posterior intervalvular sclerite; tr, trachea; tg, terminal ganglion. Coloring of structures in italics. Arrowheads outside the diagram indicate reference planes: a, anterior; d, dorsal; l, lateral.

stroke the tracheae in the si9d of the ga9l (Figure 4B) become displaced and partly collapse; in the intervals they return to their original position.

Genital chamber movements

The regular pumping of the left and right opos is well synchronized with a rhythmic bilateral movement of the genital chamber (gc) located in the posterior ventral abdomen (Figure 6). It is caused by the attached bilateral muscles (m2 of Snodgrass [19]) which originate laterally at the sternite of the abdominal segment 7. The tilting of the gc narrows and widens the lateral hemolymph spaces. In intact animals these movements are clearly visible through the transparent parts of the abdominal sternite 7

and were the first and most obvious indication for the presence of the rhythmic pumping apparatus at the base of the ovipositor valves in *Acheta domesticus* [24].

The gc muscles (m2) are always active in synchrony with the ipsilateral opo8/9 contraction muscles as was evident from long-time recording in more than 25 preparations. However, they can halt or remain in tonic contraction when they contribute to other behavior; the opo8/9 muscles however continue their rhythm at the same time.

Innervation and rhythm of the accessory hearts

The terminal ganglion (tg) is a fusion of the abdominal ganglia of the 7th and the more posterior abdominal

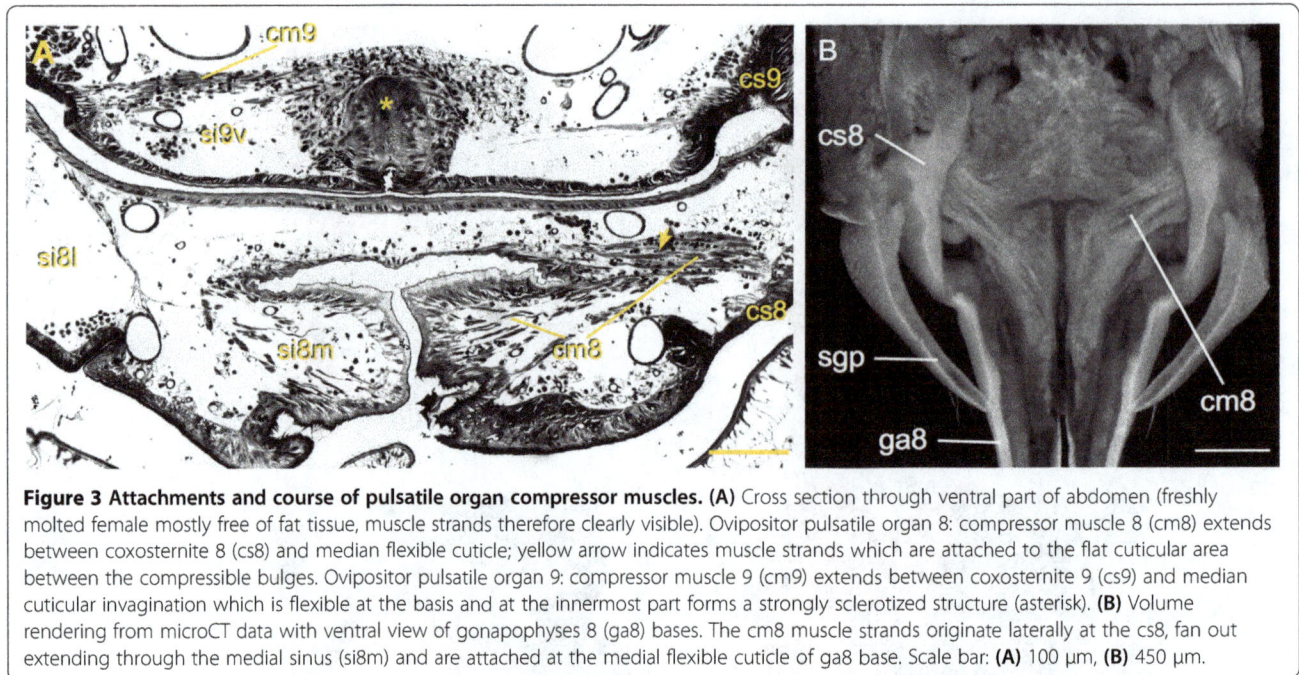

segments. Its ventral nerves innervate the sternal regions of several segments (7v, 8v, 9v) including the motor innervation of the rhythmic pumping muscles of the opos and the gc. Experiments in which the specific nerve branches that innervate the pumping muscle of an opo are cut selectively show that the contraction is based on neural commands. After denervation it stops immediately and permanently. The pumping muscles of the other opos remain active during these experiments as long as their nerve connection to the tg persists.

The rhythmic motor output recorded from the tilting muscles of the gc or the opos typically shows the activity of two excitatory motor units bursting in phase. One unit usually bursts at a higher frequency and longer than the other (Figure 7B, C). The somata of the rhythmic motoneurons for the opos are located ipsilaterally in their respective neuromere. Their neurites exhibit a rich branching, also into the neighboring neuromeres and some even with extensions across the midline (Figure 7B, C).

The opo rhythms of intact animals are occasionally accelerated or decreased respectively terminated during certain activities of the abdomen such as oviposition or strong ventilation. Our intracellular recordings from opo motoneurons reveal that the input, which elicits rhythmic motor bursts, starts with distinct excitatory postsynaptic potentials from premotor interneurons (in) which summate and release efferent spikes when the membrane potential rises over the firing threshold of the motoneuron (Figure 7B, C).

Influences on pattern generation for opos in the terminal ganglion

The pattern generation for opo pumping originates from the tg. While in intact and freely moving animals the motor output to the pumping muscles can be variable, the auxiliary tilting of the gc by the m2 remains consistently regular after severing the neck connectives and persists for several hours (Figure 7A). The common rhythm of the central pattern generator (cpg) even persists after severance of all peripheral nerves and the anterior connectives of the tg. The cpg thereby becomes separated from all sensory inputs and signals from the other ganglia of the cns. In such an isolated tg, the rhythms remain regular at about 0.5-0.2 Hz in aerated saline. Using this preparation we could also substantiate several non-neural influences on the autonomous rhythm generation and bilateral coordination of the cpg in the tg:

(i) Temperature changes applied unilaterally to the tg (Figure 8). By bringing warm or cold Peltier probes close to one side, changes in the left-right alternating rhythm of the opos were provoked. Dramatic inhibitory effects on the rhythm are caused by cooling, which when applied unilaterally decreases and abolishes the ipsilateral motor output of the cpg while the contralateral pattern paces down to a slower rhythm with extended burst durations. Removal of the cold probe gradually restores the initial bilateral rhythm.

(ii) Different concentrations of CO_2. Infusion of air with gradually increasing pCO_2 into the lateral trachea that supplies an isolated tg (Figure 9A) slows the ipsilateral rhythmic motor output to the opos progressively and the

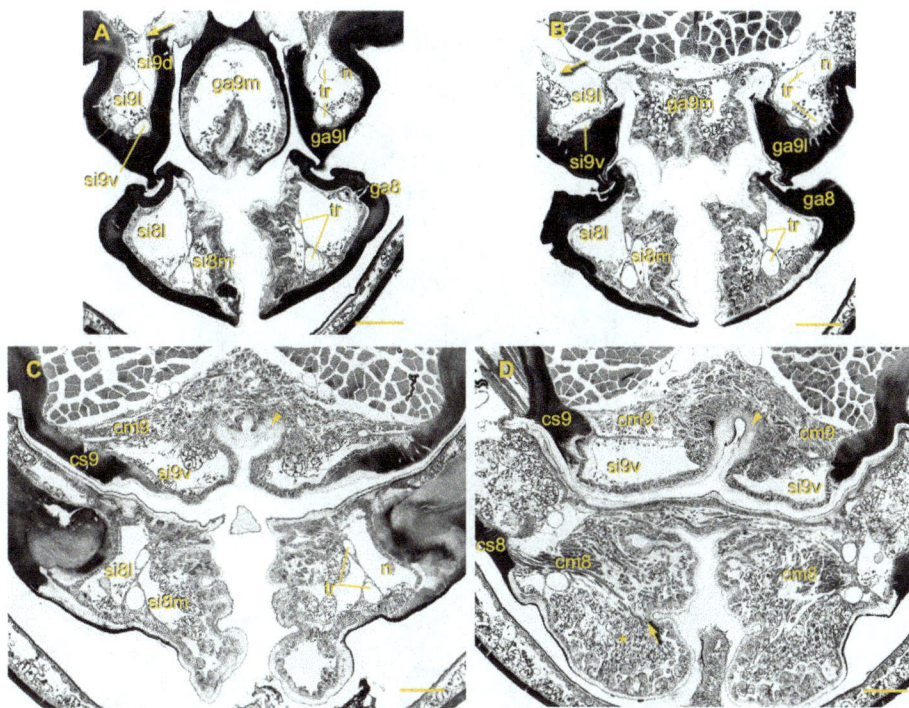

Figure 4 Organization of hemolymph sinuses in the ovipositor and their fusion with the abdominal hemocoel. Cross sections from distal to proximal, section levels indicated in figure 2C by arrows. **(A)** Gonapophysis 8 (ga8): note strong sclerotized lateral and flexible medial cuticle, lumen divided by septum with associated tracheae (tr) into efferent medial sinus (si8m) and afferent lateral sinus (si8l); lateral gonapophysis 9 (ga9l): efferent ventral sinus (si9v), afferent lateral sinus (si9l) and afferent dorsal sinus (si9d); the latter merges on the left side with the abdominal hemocoel (arrow); medial gonapophysis 9 (ga9m): each valve with undivided lumen; **(B)** left afferent si9l merges with abdominal hemocoel. **(C)** Ovipositor pulsatile organ 8: si8m partly filled with fat tissue; ovipositor pulsatile organ 9: the two pumping chambers are separated by invagination of ventral cuticle, compressor muscle 9 (cm9) originates laterally at coxosternite 9 (cs9) and extends to the medial cuticular invagination (arrowhead). **(D)** Ovipositor pulsatile organ 8: cm8 originates laterally at cs8, fans out into individual strands which extend to attachment site at medial flexible cuticle (arrow), in-between fat tissue (asterisk); ovipositor pulsatile organ 9: cm9 dorsal of si9v extending between cs9 and strong sclerotized part of medial cuticular invagination (arrowhead). Scale bars: 100 μm.

amplitudes of action potentials decrease (Figure 9B). Finally the rhythm ceases in one hemiganglion while the regular rhythm of the other side persists. Stopping the CO_2 infusion allows for the rhythm to recover and return to the initial rates. In contrast, when the ganglion surface is superfused with bathing saline in which the pCO_2 is increased (which also lowers the pH of the saline), the cpg rhythm accelerates and finally transits into more tonic activity (Figure 9C).

(iii) Increasing acidity (with drops of HCl) of the saline bathing of the ganglion. This procedure had an accelerating effect on the cpg for the opo rhythms from 0.23 to 0.26 Hz and also lowered the action potential amplitudes (Figure 9D). A similar effect occurs with an increased pH due to the application of CO_2 in saline (lowest trace in Figure 9D).

Interneurons with rhythmic activity for opo muscles

Five interneurons of the tg that burst in synchrony with the rhythm of the opos (opo-in) were identified by recording and staining. Their morphology differs considerably; further

their physiological influence on the opo muscles ranges from transient influences on the rhythm to resetting the rhythm properties. The basic morphological shapes of these interneurons are (Figure 10A-C): (a) local, with branching restricted to the tg and connections between ipsilateral areas of adjacent neuromeres, (b) local, with branches crossing the midline, and (c) intersegmental, with axon collaterals entering anterior connectives and preceding ganglia.

Modifying the activity of the interneurons by electrical stimulation influenced the bilateral motor output to the opos in different ways. Three kinds of affects can be characterized: (i) a transient suppression of the bilateral or only the unilateral motor output (Figure 9A, B). The interneuron opo-in1 has its soma located in the neuromere 9 and branches extensively into all neuromeres but most densely along the median region of the tg. An intersegmental axon collateral ascends in the ipsilateral connective. It exhibits a high tonic spiking activity which can be modulated by irregular bursting. Its effects on the motor output to the opos was most dramatic: when it was released from inhibition the subsequent rebound resulted in

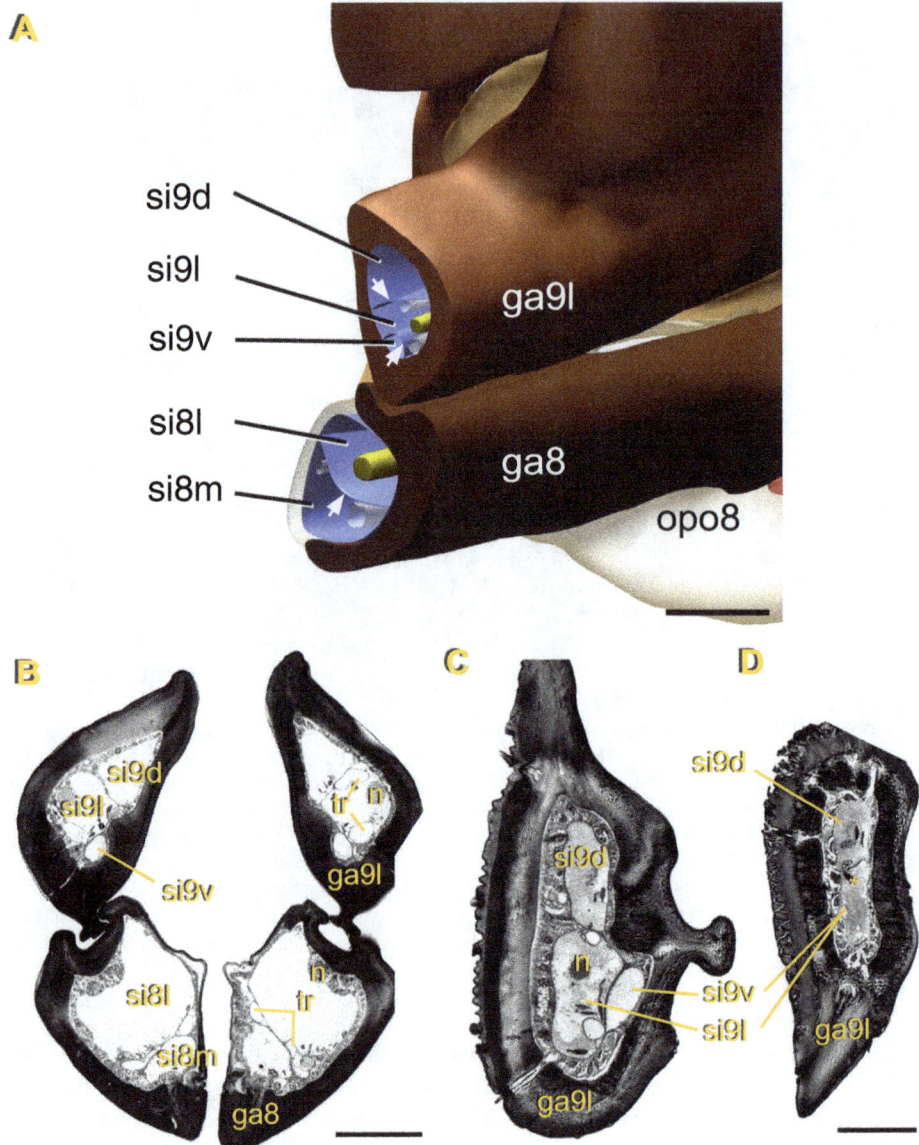

Figure 5 Morphology and inner organization of the gonapophyses. (A) Reconstruction of right half of ovipositor. Aside from epidermis, gonapophyses contain nerves (yellow), tracheae (gray) and hemolymph; their lumen is divided by thin septa of connective tissue (arrows) into hemolymph sinuses guiding the countercurrent flow; gonapophysis 8 (ga8) with efferent medial sinus (si8m) and afferent lateral sinus (si8l), gonapophysis 9 (ga9) with efferent lateral sinus (si9l), afferent dorsal sinus (si9d) and afferent ventral sinus (si9v) **(B)** Cross section through paired ga8 and ga9, same level as in diagram above (n: nerves, tr: tracheae). **(C, D)** Cross sections through most distal region of right ga9, si9v and si9l in **(C)** merge into one sinus **(D)**, asterisk indicates hemocyte moving through septum gap into neighboring sinus. Scale bar: **(A)** 150 µm, **(B)** 100 µm, **(C, D)** 50 µm.

intense spiking that inhibits the bilateral motor output to the opos specifically on the side ipsilateral to the soma. Nevertheless, the basic ongoing rhythm for the opos was maintained and not reset by the opo-in1. The interneuron opo-in2 (Figure 10B) has a large soma located in the neuromere 8, and its neurites extend in the ipsilateral neuromeres 8 and 9 and just one branch into the 7th neuromere. The principal axon crosses to the contralateral side, diverges into a smaller posterior branch and then ascends in the contralateral connective to the anterior abdominal ganglion. The opo-in2 bursts in synchrony with the ipsilateral opo motoneurons and when it is hyperpolarized, the ipsilateral opo motoneuron activity is inhibited. That may also slightly affect the basic opo rhythm. (ii) a resetting of the basic rhythm that is achieved by neurons which may be intrinsic to the cpg (published preliminarily as Figure 2E in [18]). The opo-in3, with a dorsal soma located in the 8th neuromere, extends only in the ipsilateral neuromeres 7 and 8. It bursts rhythmically in synchrony with the ipsilateral opo

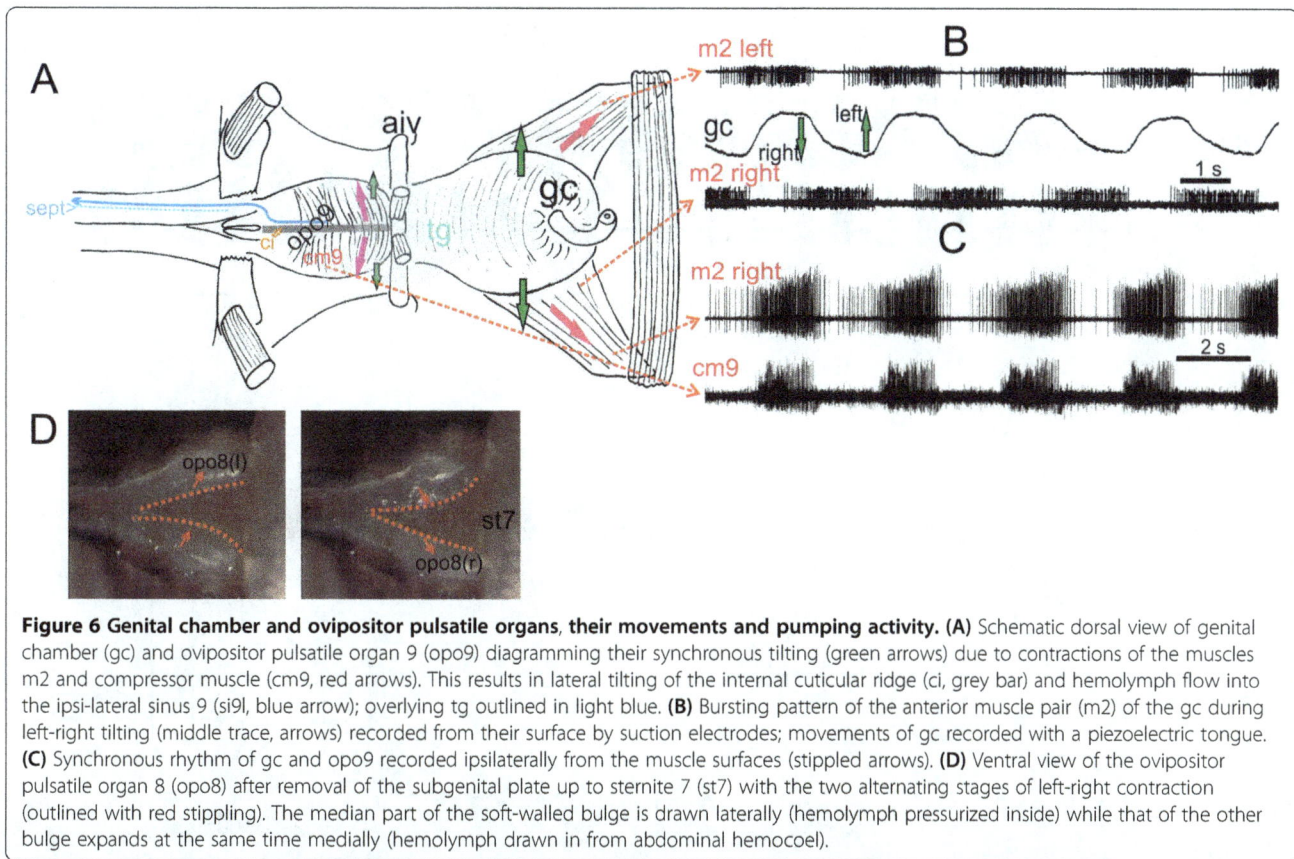

Figure 6 Genital chamber and ovipositor pulsatile organs, their movements and pumping activity. (A) Schematic dorsal view of genital chamber (gc) and ovipositor pulsatile organ 9 (opo9) diagramming their synchronous tilting (green arrows) due to contractions of the muscles m2 and compressor muscle (cm9, red arrows). This results in lateral tilting of the internal cuticular ridge (ci, grey bar) and hemolymph flow into the ipsi-lateral sinus 9 (si9l, blue arrow); overlying tg outlined in light blue. **(B)** Bursting pattern of the anterior muscle pair (m2) of the gc during left-right tilting (middle trace, arrows) recorded from their surface by suction electrodes; movements of gc recorded with a piezoelectric tongue. **(C)** Synchronous rhythm of gc and opo9 recorded ipsilaterally from the muscle surfaces (stippled arrows). **(D)** Ventral view of the ovipositor pulsatile organ 8 (opo8) after removal of the subgenital plate up to sternite 7 (st7) with the two alternating stages of left-right contraction (outlined with red stippling). The median part of the soft-walled bulge is drawn laterally (hemolymph pressurized inside) while that of the other bulge expands at the same time medially (hemolymph drawn in from abdominal hemocoel).

motoneurons. When this rhythm in opo-in3 is abolished by hyperpolarization, the bursting frequency of contralateral opo motoneurons is reduced. Rebounds from inhibition reset the whole opo rhythm starting with ipsilateral excitation and contralateral inhibition. Another interneuron (opo-in4, as APOV-IN4 in [18]) extends ipsilaterally from a particularily posterior and median soma into the 9th, 8th and 7th neuromere with some smaller branches crossing over the midline. It bursts in synchrony with the contralateral motoneurons of the opos and has the strongest driving and resetting properties for the opo rhythm. Depolarizations of the opo-in4 cause immediate rhythm reset which inhibits the ipsilateral motoneurons and excites the contralateral motoneurons. (iii) an unaltered rhythm by current injection which is observable in the rhythmically active opo-in5 (Figure 10C). This local interneuron connects bilaterally the 8th and 9th neuromeres with widespread branches. Its activity pattern corresponds with the motor bursts that move the opo and gc muscles ipsilateral to the soma. The motor output is not altered dramatically when this neuron is de- or hyperpolarized.

Discussion

In the accessory circulatory organs of insects one can distinguish between the pulsatile apparatus and the hemolymph guiding structures which provide for circulation throughout the appendage. In part one of the discussion, we address these two construction elements in the cricket ovipositor circulatory organs with respect to their structure and functional mechanisms and compare them with other accessory pulsatile organs [1-3]. The second part of the discussion is dedicated to the neuroanatomical results and the physiological control of the opos.

Functional morphology and pumping mechanism
The pulsatile part of the opos
The pulsatile part of the opos consists of a pump chamber at the base of each valve (functional scheme Figure 11). The organs of the right and left valve of the same segment have a mirrored but otherwise identical anatomy. The organs of the valves of the 8th and 9th segment function according to the same principle, but have a slightly different anatomy. While the pumps at the base of each ga8 are formed by bulges of flexible cuticle which are compressed by an internal muscle, the pump chambers of the 9th segment are compressed by contractions of an overlying external muscle. Since the compressor muscles of both organs have comparable attachments sites, i.e. laterally at the cs and medially at the flexible cuticle of the base of the ga, they are considered to be

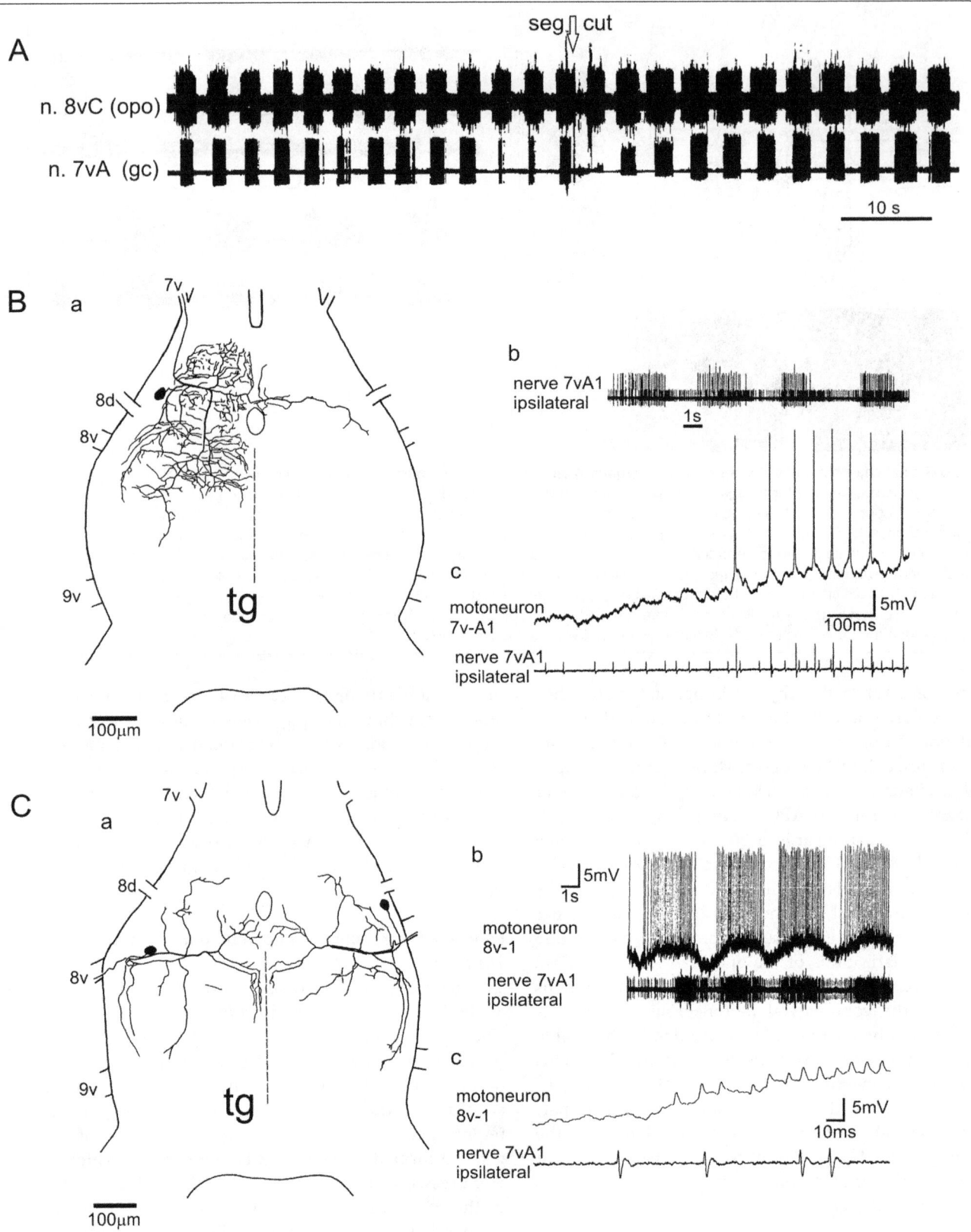

Figure 7 (See legend on next page.)

(See figure on previous page.)

Figure 7 Morphology and activity of motoneurons that drive the ovipositor pulsatile organs. (A) Recording from the efferent nerves (suction electrodes) to a opo8 (upper trace, n.8vC) and to the ipsilateral genital chamber muscle (lower trace, n.7vA, gc) before and after disconnection from cns by severing both connectives (seg cut, arrow) posterior to the subesophageal ganglion. The opo8 and gc rhythms persist almost undisturbed. **(B)** a. Branching of one motoneuron for genital chamber movement (7vA-1) in the terminal ganglion (tg, dorsal view) and b. its rhythmic activity along with the second motoneuron to M2 and, c. onset of a burst due to summation of postynaptic potentials elicited by premotor interneurons. **(C)** a. Branching types of two icM8 motoneurons in the terminal ganglion (tg, bilateral, dorsal view) which supply the opo8. b. Rhythmic activity of motoneuron 8v-1. Summation of excitatory postsynaptic potentials originating from premotor interneurons are the basis for bursting activity of the motoneuron. Lower trace shows synchronous activities of the ipsilateral genital chamber motoneurons. c. Expanded view of the onset of a burst.

serial homologues. Both pumping chambers, as well as the compressor muscles, have not been described in previous anatomical descriptions of the abdominal apex of gryllids (e.g. [19]). The four pumps work continuously whereby the right and left organs of one segment are compressed in alternation. The pumping activities of the 8th and 9th segment are synchronized.

Compared to other accessory pulsatile organs, certain similarities can be found between the functional morphology of the opos and the cercus-hearts in Plecoptera [25]. However, while the cercus-hearts in Plecoptera suck hemolymph out from the cerci into the abdominal cavity, the opos force hemolymph into the valves. Accessory pulsatile organs which likewise force hemolymph into the appendages are the various antenna-hearts; however, they strongly differ in functional morphology and use vessels as hemolymph guiding structures [11,26,27].

Circulation within the valves and tracheal ventilation

The systolic compression of the pumping chambers force hemolymph distally into the efferent sinus of the valves. The presence of non-return valves could neither be demonstrated in any ga nor at their bases. Probably backflow is reduced by the narrowing of the proximal bases of the

pumping chambers during compression. The hemolymph guiding structures are thin septa of connective tissue which extend the whole length of the ovipositor valves up to their apices. There the septa are perforated enabling the passage of hemolymph into the afferent sinuses. Curiously, only one afferent sinus is present in the ga8, while there are two in the ga9. The diameter of the efferent sinuses is much larger in the opo region than that of the afferent sinuses, which may contribute to slowing any backflow when the pump pressure decreases during diastole.

The hemolymph guiding structures in long abdominal appendages of insects are generally vessels [28]. Longitudinal septa which guide the countercurrent hemolymph flow as in the ovipositor valves have been reported from the thoracic legs, the maxillary and labial palps of many insects [3], and the cerci of the cockroach [4]. While in the legs of many Heteroptera, a rhythmically contracting muscle associated with the septum effectuates a countercurrent circulation within the limb [29,30], in most other insects it is not yet fully understood how the observed countercurrent flows are generated [2,3]. In some appendages without specific muscular pumps, the breathing-related collapse and expansion of tracheae

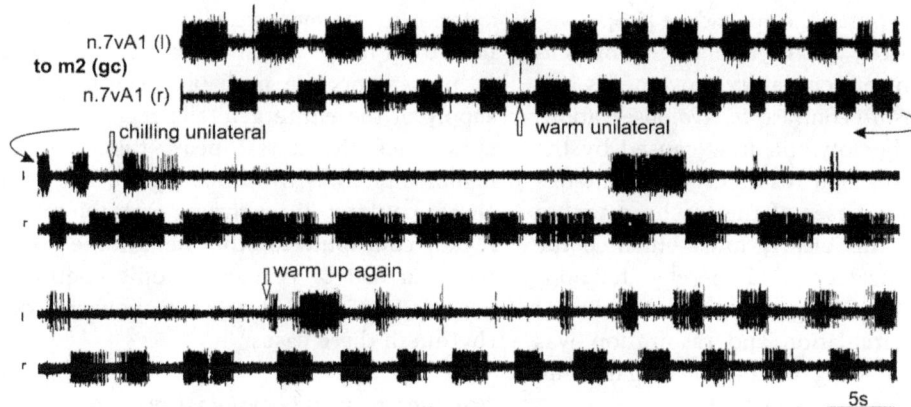

Figure 8 Influences on the rhythmic bilateral genital chamber movements by unilateral warming and cooling. Continuous recording of the efferent output to the genital chamber tilting muscles m2 (n.7vA1 l/r). The left-right rhythm accelerates slightly due to unilateral warming of the terminal ganglion (arrow). Cooling the ganglion with a probe on the other side (arrow, chilling unilateral) abolishes the ipsilateral rhythm and slows the antagonist. Recovery to room temperature (22°C, arrow) occurs after 60 s.

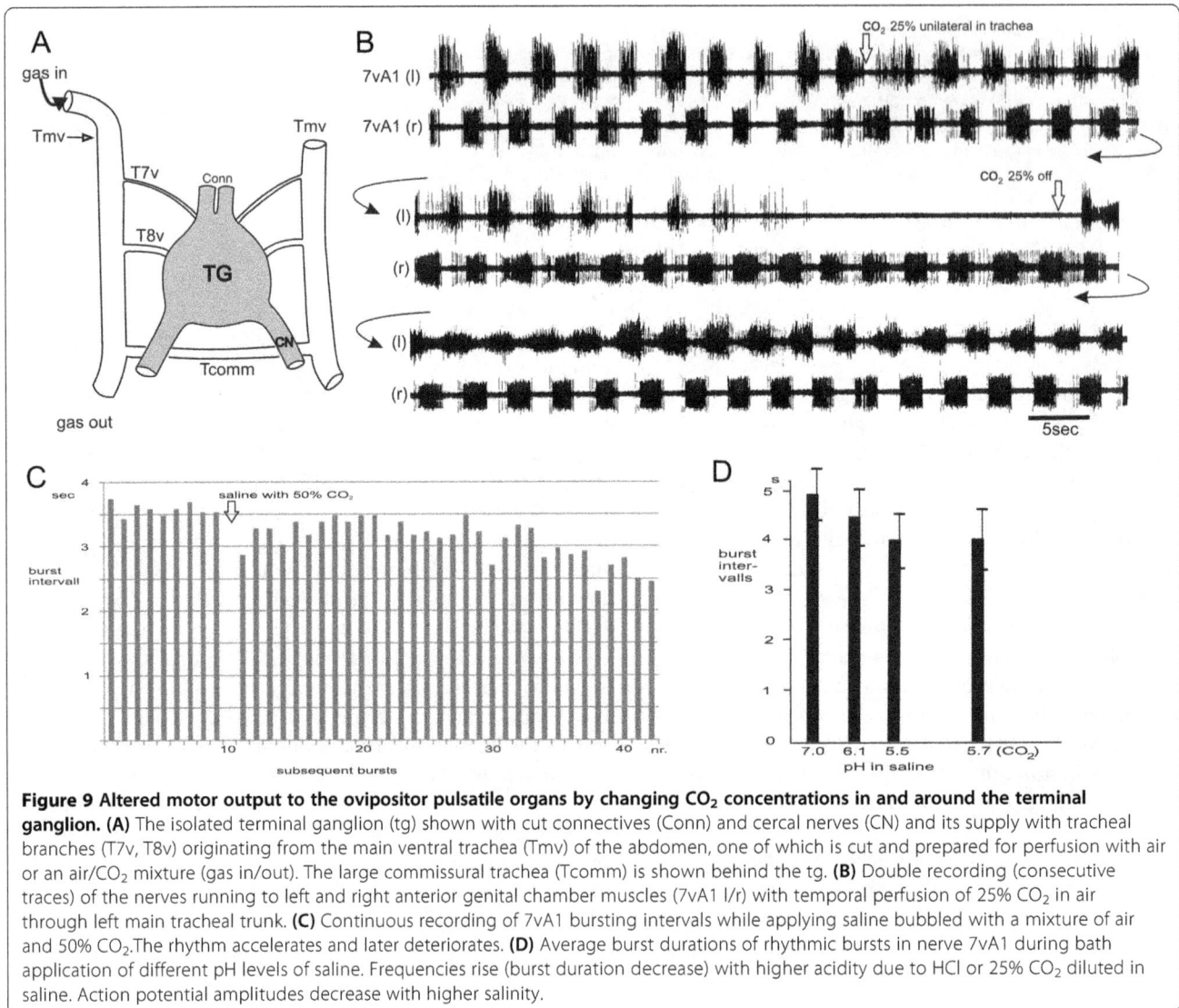

Figure 9 Altered motor output to the ovipositor pulsatile organs by changing CO_2 concentrations in and around the terminal ganglion. (A) The isolated terminal ganglion (tg) shown with cut connectives (Conn) and cercal nerves (CN) and its supply with tracheal branches (T7v, T8v) originating from the main ventral trachea (Tmv) of the abdomen, one of which is cut and prepared for perfusion with air or an air/CO_2 mixture (gas in/out). The large commissural trachea (Tcomm) is shown behind the tg. **(B)** Double recording (consecutive traces) of the nerves running to left and right anterior genital chamber muscles (7vA1 l/r) with temporal perfusion of 25% CO_2 in air through left main tracheal trunk. **(C)** Continuous recording of 7vA1 bursting intervals while applying saline bubbled with a mixture of air and 50% CO_2. The rhythm accelerates and later deteriorates. **(D)** Average burst durations of rhythmic bursts in nerve 7vA1 during bath application of different pH levels of saline. Frequencies rise (burst duration decrease) with higher acidity due to HCl or 25% CO_2 diluted in saline. Action potential amplitudes decrease with higher salinity.

and tracheal sacks cause volume changes that induce hemolymph propagation within the appendage [5,6].

In the cricket ovipositor, the rhythm of the opos is completely independent of ventilatory movements and abdominal compressions. In contrast, *in vivo* observations show that the pulsed hemolymph flow caused by the rhythmic pumping of the opo results in simultaneous collapses of the widened bases of the tracheae within the ovipositor valves. This clearly must enhance the convection of the tracheal gas and thereby the opos also contribute to the O_2-CO_2 gas exchange. A similar relationship between circulation and respiration was also found between the wing circulatory organs and the tracheal tubes in the wing veins [31].

Simultaneous genital chamber movements
In synchrony with the rhythm of the opos, the apex of the gc moves laterally. We conclude that hemolymph is

thereby pressed from the abdomen into the lateral space anterior to the ovipositor base assisting the hemolymph flow toward the ipsilateral ga. Furthermore, the lateral gc movements are probably necessary for hemolymph supply of the entire genitalic region and the abdominal apex since the dorsal heart tube permanently sucks hemolymph away from this region. The gc muscles (m2) always contract in synchrony with the ipsilateral opo8/9. If they contribute to other behavior, e.g. egg laying [32], they can halt or remain in tonic contraction for short periods; the opo8/9 muscles however continue their rhythm in these cases.

Neuroanatomy and physiological control
In the fused tg both motoneurons and interneurons of the opos tend to extend over several neuromeres. This morphological feature may functionally ease the inter-segmental communication between sensory and motor

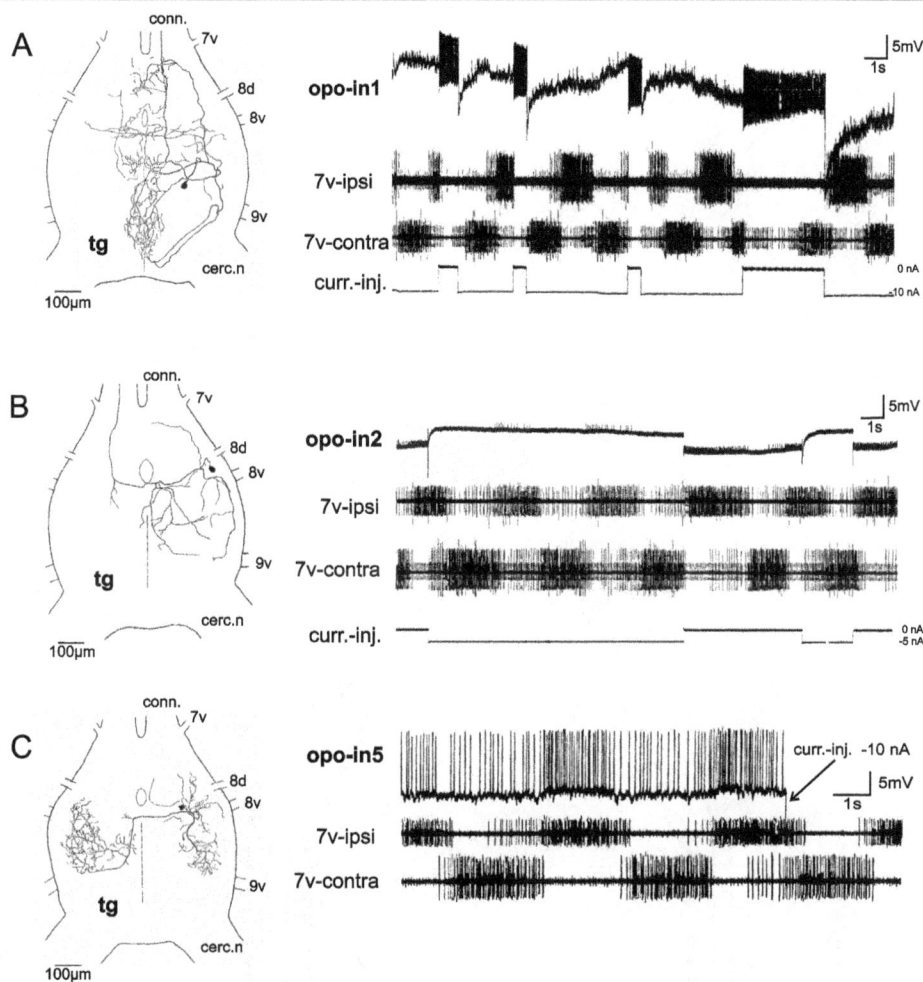

Figure 10 Morphology of terminal ganglion interneurons and their influence on the ovipositor pulsatile organ and genital chamber rhythm. Major nerve roots (7v-9v, cerc. n) from neuromeres 7 to 9 are indicated on the terminal ganglion (tg). **(A)** The opo-in1 is a widespread bilateral interneuron with ipsilateral soma and ascending axon collateral showing little rhythm related activity. Intracellular current injection (curr.-inj., lowest trace) elicits an inhibitory effect on the rhythmic output on both sides that is stronger on the ipsilateral motor output to the genital chamber muscles. **(B)** The opo-in2 is a widespread bilateral interneuron extending over three neuromeres (7–9) with opo motor output, a soma in the neuromere 8 and a contralateral ascending axon collateral. Its own rhythmic activity is enhanced when it is depolarized (bridge balance inverted, current injection in lowest trace) and the efferents to the genital chamber muscles are enhanced ipsilaterally and inhibited contralaterally. **(C)** The opo-in5 is a bilaterally branching interneuron with extensive branching in neuromeres 8 and 9. The interneuron bursts in synchrony with the ipsilateral motoneurons but depolarization causes no major changes in rhythm or intensity of motor output.

activity of the adjacent segments, specifically between the rhythmic neurons influencing the pump muscles of the different opos that originate in different neuromeres. Generally, it is rare in insects that the motoneurons innervating non-tergal muscles, such as in the opos, extend with their branches into two or more neighboring ganglia or neuromeres [33]. Basically, interneurons could achieve motoneuron coordination alone when they branch into several neuromeres.

Influences on the coordination of the opo rhythm

All contractions of opo muscles are coordinated by neuronal control from a common cpg in the tg. The extent of this neuronal network remains unknown but operates continuously and stably when the ganglion is not addressed by descending neuronal commands.

Higher-order descending interneurons are known to originate in the cricket cns in the subesophageal ganglion serving for the control of respiration [34] and oviposition [35]. Influences on the opo rhythms are evident during strong ventilation or the oviposition procedure when an egg enters the gc and the bilateral muscle pair m2 contracts synchronously [32]. Comparable systems with autonomous and spontaneous neuronal rhythms are known from other isolated insect ganglia which coordinate, e.g. locust respiration [18,36], cricket oviposition [35], and feeding

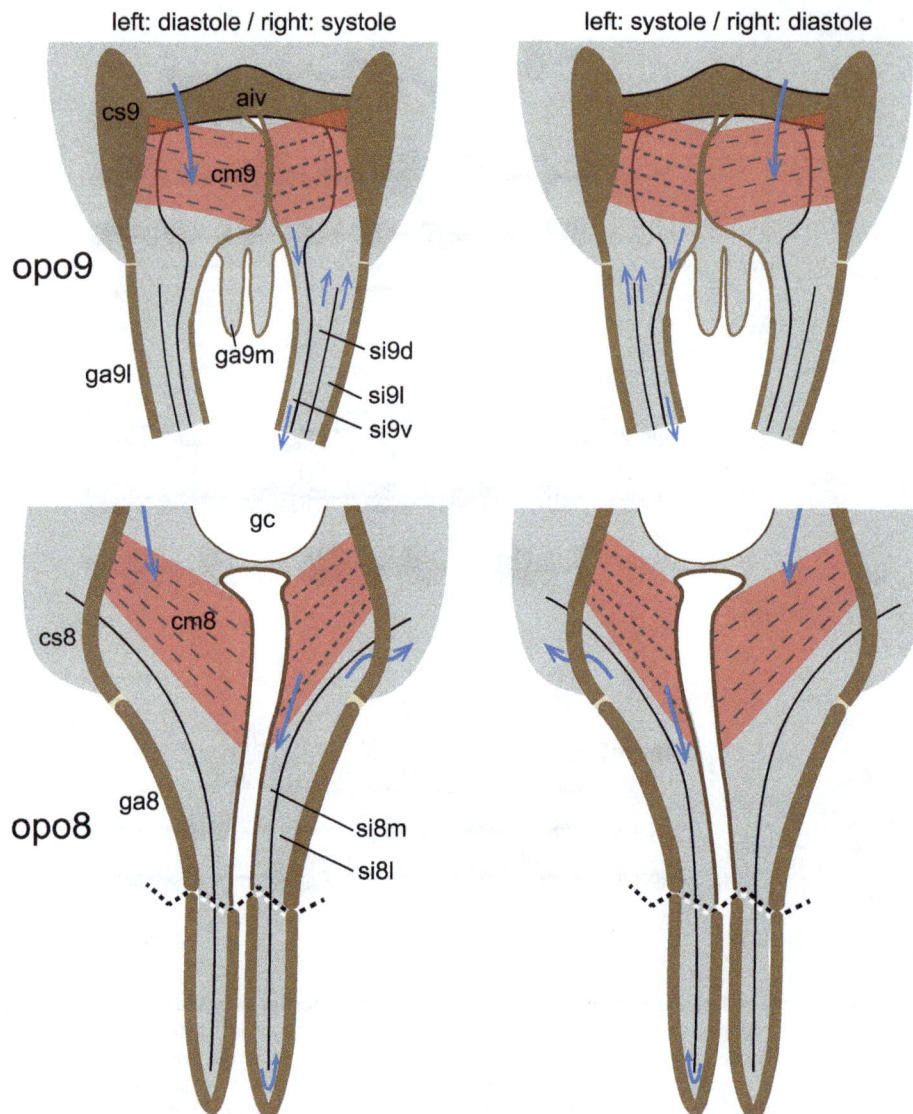

Figure 11 Functional mechanisms of the ovipositor pulsatile organs. Organs of abdominal segment 9 (upper graphs) and segment 8 (lower graphs) in two different phases of action. Ipsilateral compressor muscles of both abdominal segments (cm8, cm9) contract simultaneously, alternating with the other side. Thereby they compress the corresponding pumping chamber, i.e. the systole of the organs, and hemolymph is forced in a countercurrent flow through sinuses in the ovipositor gonapophyses (ga) and further back into the abdominal cavity. At the same time the opposite organs are in the diastole phase during which the pumping chambers dilate and fill with hemolymph from the abdominal hemocoel. Simultaneous lateral movements of the genital chamber probably support aspiration of hemolymph during diastole and hinder backflow at systole. Further abbreviations (numbers 8 and 9 refer to the respective abdominal segment, sometimes followed by an additional letter: d, dorsal; l, lateral; m, medial; v, ventral): aiv, anterior intervalvular sclerite; cs, coxosternite; opo, ovipositor pulsatile organ; gc, genital chamber; si, sinus.

patterns of *Drosophila* larvae [37]. The autonomous cpg rhythms of these systems appear more "natural" than those which require pharmacological or permanent sensory stimulation such as insect walking [38,39], flying [40], and feeding [41,42].

The autonomous and spontaneous cpg for the opos in the cricket tg can be modulated by the following non-neural factors: (i) temperature changes that induce activity changes of the cpg and (ii) lowered pH in the bathing fluid provided by an increased pCO_2 causing rhythm acceleration. In contrast, when higher levels of pCO_2 are introduced into the tg via its tracheal supply, the effect is not rhythm acceleration but rather that of an anesthetic. These contrasting CO_2-effects may reach the cpg in the neuropil by different mechanisms. The rapid effect of pH changes in the bathing fluid may be transferred inward by the glial cells which are interconnected with numerous gap junctions [43]. They may transmit the effect to the cpg neurons for the opos. As an alternative explanation, specific sensory neurons with endings near the surface of a ganglion may monitor pH changes and influence the neurons inside the ganglion – but sensors of this type are so far not known from any insect cns.

The contrasting (non-pH-like) effect of CO_2 after intra-tracheal infusion inhibits the rhythmic motor output of only the ipsilateral hemiganglion of the tg. This speaks against a pH-effect via the ganglion surface and agrees with the notion that there is no tracheal junction over the midline to the contralateral side of the tg [44]. Apparently the gaseous intratracheal CO_2 has a low effect on the pH levels in the environment of the cpg neurons of the tg. That seems to indicate a neuronal tolerance to self-produced metabolic CO_2 in the cns, as was found for single neurons of crickets [45].

The hemolymph that returns from the ovipositor partly overflows the tg with metabolically loaded and more acidic hemolymph caused by a high metabolic rate of the many cuticular sensilla located on the surface of the ovipositor [46]. That may contribute to the regulation of the cpg rhythm as indicated by experimental superfusion of the isolated tg. In this way metabolic requirements may indirectly control the velocity of the hemolymph flow through the ovipositor valves.

Coupling of the left-right opo rhythm

Unilateral changes of external influences on the tg, such as cooling, and unilateral application of CO_2, affect the rhythmic output mainly on the ipsilateral side, at least, for the first minutes of application (Figure 9); the rhythm on the other side remains nearly unchanged. This strong ipsilateral suppression of the motor, and possibly also of premotor neurons, raises the question whether the total ipsilateral cpg is affected. That leads one to assume that the cpg for the opos consists of two (left and right) half centers producing their own – but normally coupled – rhythms.

Interneurons and the cpg

All interneurons exhibiting the rhythm of the opos could belong to the cpg itself or are influenced by it. They extend over, at least, two or more neuromeres of the tg. A similarly extensive wiring is required to connect the cpg to the different motoneurons of the segmental neuromeres 7–9 which has efferents to the opos and gc. Yet the exact location of the rhythm-generating neuronal network and the extent of the essential network remain unclear. The "core" of the cpg may be located in the neuromere 8 where all the motoneurons for the rhythmic muscles have branches. At the level of interneurons, only one potentially rhythm resetting interneuron (opo-in3) was found that branches unilaterally in neuromeres 8 and 9, whereas the opo-in4 reaches all neuromeres mainly on one side and the contralateral neuromere 8 [18]. In contrast, the opo-in5 exhibits a morphology that appears well suited for a left-right coordination of all opo-rhythms. However, the physiology of this interneuron, with its ideal left-right connection and rich bilateral arborizations in the neuromeres 8 and 9, is not sufficiently elaborated to substantiate the proposed function.

Conclusions

Most arthropods have a complex vascular system in which the limbs are supplied with hemolymph by arteries. In insects, this system is greatly reduced and a ventral longitudinal vessel from which such arteries could emanate is lacking [47]. Their thoracic limbs are supplied by sinuses delimited by thin septa of connective tissue which are perforated in the tip region of the appendage enabling a countercurrent hemolymph flow [3]. A comparable condition can also be found in the gonapophyseal appendages in *Acheta*. However, while in most thoracic limbs and cerci it is unclear how the hemolymph flow is generated, a pumping apparatus exists for each of the ovipositor valves. These organs represent evolutionary novelties having a functional morphology which has not been reported from any other auxiliary heart in insects [2]. The origin of the associated pumping muscle must remain unclear since no unambiguous homologization with any of the serial homologues of the abdominal musculature is possible.

With respect to physiological control, it must be emphasized that the neurogenic automatism of the opo is unique among insects. All other known circulatory organs are based on a myogenic automatism which may be neuronally or hormonally modulated [6-10]. The great autonomy of opo rhythm generation is surprising. The only noticeable influence on the cpg interneurons is – apart from general temperature effects and inhibitory cns commands – the pH of the fluid surrounding the tg. This may be linked with the metabolic requirements of the numerous sensilla which are located especially at the ovipositor apex. An additional task of the opos may be the convection of the extensive tracheal system within the ovipositor valves.

From an evolutionary point of view it will be a rewarding task to investigate if corresponding pump organs are associated with the ovipositors in other insects. Future research in this direction could reveal remarkable insights to the evolution of the female ovipositor in insects, a classical topic of comparative morphology in these animals [19,21,22,48-50].

Material and methods
Animals

Females of *Acheta domesticus* used in this study originated from breeding stocks in our laboratories. For immobilization the specimens were cooled to 0-4°C previous to and during preparations. All experiments were carried out respecting the relevant ethical guidelines for experimentation with live animals.

Observation of the pumping organs *in vivo*

The speed and direction of hemolymph flow inside the ovipositor valves is readily recognizable through the transparent regions of the ovipositor cuticle via movement of the hemocytes. Experiments with introducing various vital stains into the hemolymph failed due to immediate clotting that slowed or stopped fluid propagation in the small sinuses of the ovipositor. Observations were made with incident or translucent light under a stereomicroscope. In addition, the pumping action of the opos was video-recorded (camera: Kappa C15) in intact animals (in a small glass chamber from below), as well as from prepared specimens (ventral side up). The range of peak velocities during pumping strokes was calculated (n = 8 preparations) from tracking individual large hemocytes frame by frame in high-speed video sequences (300 fps, Casio Exilim F1) recorded through a dissection microscope in translucent light.

Correlation of the hemolymph pulses in the ovipositor valves to the pumping activity of the opo8 was studied from the ventral side after removal of the subgenital plate that covers the ovipositor base ventrally. The pumping movements of the opo9 system were observed dorsally in semi-intact preparations after removal of overlying muscles and other tissue.

Experimentally induced influences on the opo/gc rhythm were measured in 5–8 animals per parameter, relating the undisturbed burst frequency of the individual preparation with the altered frequency after introducing an influence to the same preparation.

Morphological methods

Chemical fixation: freshly cut last abdominal segments of female crickets were fixed in alcoholic Bouin ("Dubosq-Brasil" mixture) and subsequently washed in ethanol.

Histological sections: the fixed specimens were embedded after dehydration with acetone in low viscosity resin (Agar Scientific). Serial semithin sections (1 μm thickness) were cut with a diamond knife on an ultramicrotome and stained with a mixture of 1% azure II and 1% methylene blue in a 1% aqueous borax solution for approximately 40 s at 80°C.

MicroCT: a female abdomen fixed in alcoholic Bouin was stained in a solution of 1% iodine in 96% ethanol overnight. After this treatment it was imaged with an Xradia MicroXCT x-ray microtomography system (University of Vienna, Department of Theoretical Biology) with a tungsten source at 60 kVp and 66 μA.

3D reconstruction and visualization: the software Amira 5.4.2 was used for 3D reconstruction of the microCT dataset. Blender (www.blender.org) was used to postprocess the meshes exported from Amira and to remodel certain parts using the Amira data as a guide. Images of semi-thin sections were postprocessed with Fiji (www.fiji.sc) using the CLAHE plugin to enhance contrast.

Recording from nerves and muscles

To make preparations of the dorsal side, the median part of the tergites, the gut and the ovaries were removed carefully. That gave access to the tg, peripheral nerves, several muscles of the opos and the gc. The easiest access for recording is to the opo muscles (m2) and its motor nerve 7vA whose bursting activities are always in synchrony with the ipsilateral opo muscles cm8/9. The internal organs were flushed regularly with saline [51]. Care was taken not to block the abdominal spiracles by saline from outside. Extracellular recording was performed with suction electrodes on cut nerve stumps, laterally on intact nerves, or by gently sucking the surface of active muscles near their attachments where movement amplitudes of the fibers were low. The time intervals from the start of a burst to the next burst (myogram or nerve recording) were measured continuously for several hours in more than 25 specimens. In none of these or any of the other 250–300 experiments we found rhythms below 0.2 Hz or above 0.5 Hz at room temperature.

Intracellular recording required a supporting silver platform for the tg. The electrodes for intracellular recording were made of borosilicate glass with 50–80 MΩ tip resistances and had their shaft filled with 1 M LiCl while their tip contained about 1-2% Lucifer yellow in LiCl for iontophoretic staining. Intracellular recording focused on rhythmically active or rhythm-influencing interneurons and motoneurons; the data were stored on magnetic tape (Racal Store 7) or on a PC after digitalization (Datapac K2).

Temperature application (n = 6 preparations): Short metal studs connected to a regulated Peltier element (Peltron, Nürnberg) were brought close to the tg laterally with temperatures of either 0° or 25° Celsius.

Superfusion and infusion of gas mixtures: The different gas mixtures were mixed before application in a gas syringe and each type of experiment was repeated 5 to 8 times.

Terminology

A confusing multitude of synonyms exist for the ovipositor valves and linked structures (see Scudder [21]). For reasons of comprehensibility we use "gonapophysis" as a descriptive term to refer to all three valves forming the ovipositor shaft in *Acheta* without implying homology. The numbering of some muscles was taken from the descriptions for *Gryllus assimilis* by Snodgrass [19]. The nerve roots of the terminal ganglion were named according to the abdominal segment that they supply, e.g. 8d supplying the dorsal region of the 8th segment and 8v for the ventral region.

Additional file

Additional file 1: Video. The clip is divided into three sections. (1) Bases of the gonapophyses 8 are shown in ventral view displaying the left-right alternating pumping of the ovipositor pulsatile organs 8 (subgenital plate removed). The compression of one bulge supports the dilation of the other bulge as it fills up with hemolymph drawn from the abdominal body cavity. Note the alternating lateral shift of the flat cuticular area between the two bulges which narrows the hemolymph connection to the abdominal hemocoel and which probably prevents backflow. (2a) Countercurrent flow of hemocytes in the efferent ga8m (flow direction: left) and the afferent ga8l (flow direction: right), ventral view. The flow pulses alternate between the right (upper one in the video) and left gonapophysis 8 (2b) Collapsing tracheae in the proximal gonapophysis 8, ventral view. (3) Dorsal view of the muscles of the ovipositor pulsatile organ 9, which are active in the alternating pumping movements visible below the terminal ganglion (overexposed; cercal nerves cut). The whitish processes at the end of ovipositor pulsatile organ 9 are the median gonapophyses 9.

Abbreviations

Appended numbers 8 and 9 refer to the concerned abdominal segment; cm: Compressor muscle; cns: Central nervous system; cpg: Central pattern generator; cs: Coxosternite; ga: Gonapophysis; gc: Genital chamber; opo: Ovipositor pulsatile organ; opo-in: Ovipositor pulsatile organ interneuron; tg: Terminal ganglion.

Competing interests

The authors declare that they have no competing interests.

Authors' contributions

RH and GP discovered the opos in *Acheta* independently and later combined their efforts. The morphological investigations were carried out by AB, RH, and GP, the neuroanatomical and physiological analysis by RH and MF. The text was written by RH, GP and AB. All authors read and approved the final version of the manuscript.

Acknowledgements

The authors thank Julia Bauder for her careful technical assistance and Christina Heindl for taking photographs. The microCT scan was performed at the Department of Theoretical Biology of the University Vienna and we acknowledge Brian Metscher for his help. Many thanks also to John Plant for improving the English. The study was financially supported by the Austrian science fund FWF project 23251-B17.

Author details

[1]Department of Neurobiology, JFB-Institute for Zoology, University of Göttingen, Berliner Straße 28, 37073 Göttingen, Germany. [2]Department of Integrative Zoology, University of Vienna, Althanstraße 14, 1090 Vienna, Austria.

References

1. Pass G: Accessory pulsatile organs. In *Microscopic Anatomy of Invertebrates, Volume 11B*. Edited by Harrison F, Locke M. New York: Wiley-Liss; 1998:621–640.
2. Pass G: Accessory pulsatile organs: evolutionary innovations in insects. *Annu Rev Entomol* 2000, 45:495–518.
3. Pass G, Gereben-Krenn B-A, Merl M, Plant J, Szucsich NU, Tögel M: Phylogenetic relationships of the orders of Hexapoda: contributions from the circulatory organs for a morphological data matrix. *Arthr Syst Phyl* 2006, 64:165–203.
4. Murray JA: Morphology of the cercus in *Blattella germanica* (Blattaria: Pseudomopinae). *Ann Entomol Soc Am* 1967, 60:10–16.
5. Wasserthal LT: The open hemolymph system of Holometabola and its relation to the tracheal space. In *Microscopic Anatomy of Invertebrates, Volume 11B*. Edited by Harrison F, Locke M. New York: Wiley-Liss; 1998:583–620.
6. Hustert R: Accessory hemolymph pump in the middle legs of locusts. *Int J Insect Morphol & Embryol* 1999, 28:91–96.
7. Miller TA: Structure and physiology of the circulatory system. In *Comprehensive insect physiology, biochemistry and pharmacology, Volume 3*. Edited by Kerkut GA, Gilbert LI. Oxford: Pergamon Press; 1985:289–353.
8. Miller TA: Control of circulation in insects. *Gen Pharmacol* 1997, 29:23–38.
9. Hertel W, Pass G: An evolutionary treatment of the morphology and physiology of circulatory organs in insects. *Comp Biochem and Physiol A* 2002, 133:555–575.
10. Miller T, Pass G: Circulatory system. In *Encyclopedia of insects*. 2nd edition. Edited by Resh VH, Cardé RT. Burlington: Elsevier; 2009:169–173.
11. Pass G: Gross and fine structure of the antennal circulatory organ in cockroaches (Blattodea, Insecta). *J Morphol* 1985, 185:255–268.
12. Hertel W, Pass G, Penzlin H: Electrophysiological investigation of the antennal heart of *Periplaneta americana* and its reactions to proctolin. *J Insect Physiol* 1985, 31:563–572.
13. Pass G, Agricola H, Birkenbeil H, Penzlin H: Morphology of neurones associated with the antennal heart of *Periplaneta americana* (Blattodea, Insecta). *Cell Tissue Res* 1988, 253:319–326.
14. Pass G, Sperk G, Agricola H, Baumann E, Penzlin H: Octopamine in a neurohaemal area within the antennal heart of the American cockroach. *J Exp Biol* 1988, 135:495–498.
15. Hertel W, Richter M: Contributions to physiology of the antenna-heart in *Periplaneta americana* (L.) (Blattodea: Blattidae). *J Insect Physiol* 1997, 43:1015–1021.
16. Hertel W, Neupert S, Eckert M: Proctolin in the antennal circulatory system of lower Neoptera: a comparative pharmacological and immunohistochemical study. *Physiol Entomol* 2012, 37:160–170.
17. Frisch M, Hustert R: A central pattern generator in the terminal ganglion of a cricket. In *Gene - Brain – Behaviour*. Edited by Elsner N, Heisenberg M. Stuttgart: Thieme Verlag; 1993:80.
18. Hustert R, Mashaly AM: Spontaneous behavioral rhythms in the isolated CNS of insects – Presenting new model systems. *J Physiol-Paris* 2013, 107:147–151.
19. Snodgrass RE: Morphology of the insect abdomen. II. The genital ducts and the ovipositor. *Smith Misc Coll* 1933, 89:1–147.
20. Snodgrass RE: *Principles of Insect Morphology*. New York: McGraw-Hill; 1935.
21. Scudder GGE: The comparative morphology of the insect ovipositor. *Trans Roy Entomol Soc London* 1961, 113:25–40.
22. Klass KD, Ulbricht J: The female genitalic region and gonoducts of Embioptera (Insecta), with general discussions on female genitalia in insects. *Org Div & Evol* 2009, 9:115–154.
23. Sakai M, Kumashiro M: Copulation in the cricket is performed by chain reaction. *Zool Sci* 2004, 21:705–718.
24. Hustert R: Rhythmogenesis in a segmental neural network for respiration in an insect. In *Rhythmogenesis in neurons and networks*. Edited by Elsner N, Richter D. Stuttgart: Thieme Verlag; 1992:62.
25. Pass G: The "cercus heart" in stoneflies - a new type of accessory circulatory organ in insects. *Naturwiss* 1987, 74:440–441.
26. Pass G: Functional morphology and evolutionary aspects of unusual antennal circulatory organs in *Labidura riparia* Pallas (Labiduridae), *Forficula auricularia* L. and *Chelidurella acanthopygia* Géné (Forficulidae) (Dermaptera: Insecta). *Int J Insect Morph Embryol* 1988, 17:103–112.
27. Pass G: Antennal circulatory organs in Onychophora, Myriapoda and Hexapoda: functional morphology and evolution. *Zoomorphology* 1991, 110:145–164.
28. Gereben-Krenn BA, Pass G: Circulatory organs of abdominal appendages in primitive insects (Hexapoda: Archaeognatha, Zygentoma and Ephemeroptera). *Acta Zool-Stockholm* 2000, 81:285–292.
29. Debasieux P: Organes pulsatiles des tibias de Notonectes. *Ann Soc Sci Brux B* 1936, 56:77–87.

30. Hantschk AM: **Functional morphology of accessory circulatory organs in the legs of Hemiptera.** *Int J Insect Morphol & Embryol* 1991, **6**:259–273.

31. Wasserthal LT: **Oscillating haemolymph 'circulation' and discontinuous tracheal ventilation in the giant silk moth Attacus atlas L.** *J Comp Physiol* 1980, **145**:1–15.

32. Sugawara T, Loher W: **Oviposition behavior of the cricket *Teleogryllus commodus*: Observation of external and internal events.** *J Insect Physiol* 1986, **32**:179–188.

33. Steffens GR, Kutsch W: **Homonomies within the ventral muscle system and the associated motoneurons in the locust, *Schistocerca gregaria* (Insecta, Caelifera).** *Zoomorpology* 1995, **115**:133–155.

34. Otto D, Janiszewsky J: **Interneurones originating in the suboesophageal ganglion that control ventilation in two cricket species: Effects of the interneurones (SD-AE neurones) on the motor output.** *J Insect Physiol* 1989, **35**:483–491.

35. Ogawa H, Kagaya K, Saito M, Yamaguchi T: **Neural mechanism for generating and switching motor patterns of rhythmic movements of ovipositor valves in the cricket.** *J Insect Physiol* 2011, **57**:326–338.

36. Bustami HP, Hustert R: **Typical ventilatory pattern of the intact locust is produced by the isolated CNS.** *J Insect Physiol* 2000, **46**:1285–1293.

37. Schoofs A, Niederegger S, Spieß R: **From behavior to fictive feeding: Anatomy, innervation and activation pattern of pharyngeal muscles of *Calliphora vicina* 3rd instar larvae.** *J Insect Physiol* 2009, **55**:218–230.

38. Ryckebusch S, Laurent G: **Rhythmic patterns evoked in locust leg motor neurons by the muscarinic agonist pilocarpine.** *J Neurophysiol* 1993, **69**:1583–1595.

39. Johnston RM, Levine RB: Crawling motor patterns in the isolated larval nerve cord of Manduca sexta induced by pilocarpine. In: Neural Systems and Behaviour. Proceedings of the Fourth International Congress of Neuroethology, Cambridge UK, 3.-8.Sept: **Edited by Burrows M, Matheson T, Newland PL, Schuppe H.** *Stuttgart: Thieme Verlag* 1995, **1995**:470.

40. Wilson DM: **The central nervous control of flight in a locust.** *J Exp Biol* 1961, **38**:471–490.

41. Rast GF, Bräunig P: **Pilocarpine-induced motor rhythms in the isolated locust suboesophageal ganglion.** *J Exp Biol* 1997, **200**:2197–2207.

42. Bowdan E, Wyse GA: **Temporally patterned activity recorded from mandibular nerves of the isolated subesophageal ganglion of *Manduca*.** *J Insect Physiol* 2000, **46**:709–719.

43. Swale LS, Lane NJ: **Embryonic development of glial cells and their junctions in the locust central nervous system.** *J Neurosci* 1985, **5**:117–127.

44. Longley A, Edwards JS: **Tracheation of abdominal ganglia and cerci in the house cricket *Acheta domesticus* (Orthoptera, Gryllidae).** *J Morphol* 1997, **159**:233–243.

45. Clark MA, Eaton DC: **Effect of CO_2 on neurons of the house cricket, *Acheta domestica*.** *J Neurobiol* 1983, **14**:237–250.

46. Markus B: **Untersuchungen zur Sensomotorik des Ovipositors bei Grillen.** *Diplomarbeit*, Universität Köln 1985. As cited. In *Cricket behaviour and neurobiology*. Edited by Huber F, Moore TE, Loher W. Ithaca and London: Comstock/Cornell University Press; 1989:74.

47. Wirkner S, Tögel M, Pass G: **The arthropod circulatory system.** In *Arthropod Biology and Evolution, Molecules, Development, Morphology*. Edited by Minelli A, Boxshall G, Fusco G. Berlin, Heidelberg: Springer; 2013:343–391.

48. Crampton GC: **The terminal abdominal structures of female insects compared throughout the orders from the standpoint of phylogeny.** *J New York Entomol Soc* 1929, **37**:453–496.

49. Rousset A: **Squelette et musculature des regions génitals et postgénitales de la femelle de *Thermobia domestica* (Packard). Comparaison avec la region génital de *Nicoletia* sp. (Insecta: Apterygota: Lepismatida).** *Int J Insect Morph & Embryol* 1973, **2**:25–85.

50. Klass KD, Matushkina NA: **The exoskeleton of the female genitalic region in *Petrobiellus takunagae* (Insecta: Archaeognatha): Insect-wide terminology, homologies, and functional interpretations.** *Arthropod Struct Dev* 2012, **41**:575–591.

51. Eibl E: **Morphology of the sense organs in the proximal part of the tibiae of *Gryllus campestris* L. and *Gryllus bimaculatus* de Geer (Insecta, Ensifera).** *Zoomorphology* 1978, **89**:185–205.

Permissions

All chapters in this book were first published in FZ, by BioMed Central; hereby published with permission under the Creative Commons Attribution License or equivalent. Every chapter published in this book has been scrutinized by our experts. Their significance has been extensively debated. The topics covered herein carry significant findings which will fuel the growth of the discipline. They may even be implemented as practical applications or may be referred to as a beginning point for another development.

The contributors of this book come from diverse backgrounds, making this book a truly international effort. This book will bring forth new frontiers with its revolutionizing research information and detailed analysis of the nascent developments around the world.

We would like to thank all the contributing authors for lending their expertise to make the book truly unique. They have played a crucial role in the development of this book. Without their invaluable contributions this book wouldn't have been possible. They have made vital efforts to compile up to date information on the varied aspects of this subject to make this book a valuable addition to the collection of many professionals and students.

This book was conceptualized with the vision of imparting up-to-date information and advanced data in this field. To ensure the same, a matchless editorial board was set up. Every individual on the board went through rigorous rounds of assessment to prove their worth. After which they invested a large part of their time researching and compiling the most relevant data for our readers.

The editorial board has been involved in producing this book since its inception. They have spent rigorous hours researching and exploring the diverse topics which have resulted in the successful publishing of this book. They have passed on their knowledge of decades through this book. To expedite this challenging task, the publisher supported the team at every step. A small team of assistant editors was also appointed to further simplify the editing procedure and attain best results for the readers.

Apart from the editorial board, the designing team has also invested a significant amount of their time in understanding the subject and creating the most relevant covers. They scrutinized every image to scout for the most suitable representation of the subject and create an appropriate cover for the book.

The publishing team has been an ardent support to the editorial, designing and production team. Their endless efforts to recruit the best for this project, has resulted in the accomplishment of this book. They are a veteran in the field of academics and their pool of knowledge is as vast as their experience in printing. Their expertise and guidance has proved useful at every step. Their uncompromising quality standards have made this book an exceptional effort. Their encouragement from time to time has been an inspiration for everyone.

The publisher and the editorial board hope that this book will prove to be a valuable piece of knowledge for researchers, students, practitioners and scholars across the globe.

List of Contributors

Shantala Arundathi Hari Dass
School of Biological Sciences, Nanyang Technological University, 60 Nanyang Drive, Nanyang 37551, Republic of Singapore

Ajai Vyas
School of Biological Sciences, Nanyang Technological University, 60 Nanyang Drive, Nanyang 37551, Republic of Singapore

Aina Børve
Sars International Centre for Marine Molecular Biology, University of Bergen, Thormøhlensgate 55, 5008 Bergen, Norway

Andreas Hejnol
Sars International Centre for Marine Molecular Biology, University of Bergen, Thormøhlensgate 55, 5008 Bergen, Norway

Marian Y Hu
Institute of Cellular and Organismic Biology, Academia Sinica, Taipei City, Taiwan

Ying-Jey Guh
Institute of Cellular and Organismic Biology, Academia Sinica, Taipei City, Taiwan

Meike Stumpp
Institute of Cellular and Organismic Biology, Academia Sinica, Taipei City, Taiwan

Jay-Ron Lee
Institute of Cellular and Organismic Biology, Academia Sinica, Taipei City, Taiwan

Ruo-Dong Chen
Institute of Cellular and Organismic Biology, Academia Sinica, Taipei City, Taiwan

Po-Hsuan Sung
Department of Life Science, National Taiwan Normal University, Taipei City, Taiwan

Yu-Chi Chen
Department of Life Science, National Taiwan Normal University, Taipei City, Taiwan

Pung-Pung Hwang
Institute of Cellular and Organismic Biology, Academia Sinica, Taipei City, Taiwan

Yung-Che Tseng
Department of Life Science, National Taiwan Normal University, Taipei City, Taiwan

Alejandro Cabezas-Cruz
Center for Infection and Immunity of Lille (CIIL), INSERM U1019 – CNRS UMR 8204, Université Lille Nord de France, Institut Pasteur de Lille, Lille, France
SaBio. Instituto de Investigación de Recursos Cinegéticos, IREC-CSIC-UCLM-JCCM, Ciudad Real 13005, Spain

James J Valdés
Institute of Parasitology, Biology Centre of the Academy of Sciences of the Czech Republic, České Budějovice 37005, Czech Republic

Christine Morrow
Queen's University Belfast, Marine Laboratory, Portaferry BT22 1PF, Northern Ireland, UK

Paco Cárdenas
Department of Organismal Biology, Division of Systematic Biology, Evolutionary Biology Centre, Uppsala University, Norbyvägen 18D, 752 36 Uppsala, Sweden
Department of Medicinal Chemistry, Division of Pharmacognosy, BioMedical Centre, Husargatan 3, Uppsala University, 751 23 Uppsala, Sweden

Anna V Weber
Department of Integrative Zoology, University of Vienna, Althanstraße 14, 1090 Vienna, Austria

Andreas Wanninger
Department of Integrative Zoology, University of Vienna, Althanstraße 14, 1090 Vienna, Austria

Thomas F Schwaha
Department of Integrative Zoology, University of Vienna, Althanstraße 14, 1090 Vienna, Austria

Shuichi Shigeno
Department for Marine Biodiversity Research, Japan Agency for Marine-Earth Science and Technology, 2-15 Natsushima-cho, Yokosuka 237-0061, Kanagawa, Japan

Atsushi Ogura
Nagahama Institute of Bio-Science and Technology, Institute of Bio-Science and Technology, 1266 Tamura-Cho, Nagahama 526-0829, Shiga, Japan

Tsukasa Mori
Nihon University, 1866 Kameino, Fujisawa 252-0880, Kanagawa, Japan

Haruhiko Toyohara
Division of Applied Biosciences, Kyoto University, Graduate School of Agriculture, Laboratory of Marine Biological Function, Kitashirakawa Oiwake-cho, Sakyo-ku, Kyoto 606-8602, Japan

Takao Yoshida
Department for Marine Biodiversity Research, Japan Agency for Marine-Earth Science and Technology, 2-15 Natsushima-cho, Yokosuka 237-0061, Kanagawa, Japan

Shinji Tsuchida
Department for Marine Biodiversity Research, Japan Agency for Marine-Earth Science and Technology, 2-15 Natsushima-cho, Yokosuka 237-0061, Kanagawa, Japan

Katsunori Fujikura
Department for Marine Biodiversity Research, Japan Agency for Marine-Earth Science and Technology, 2-15 Natsushima-cho, Yokosuka 237-0061, Kanagawa, Japan

Ronald Seidel
Max Planck Institute of Colloids and Interfaces, Potsdam-Golm Science Park, Am Mühlenberg 1 OT Golm, 14476 Potsdam, Germany
Museum für Naturkunde, Leibniz Institute for Evolution and Biodiversity Science, Invalidenstrasse 43, 10115 Berlin, Germany

Carsten Lüter
Museum für Naturkunde, Leibniz Institute for Evolution and Biodiversity Science, Invalidenstrasse 43, 10115 Berlin, Germany

Eduardo E Zattara
Department of Biology, University of Maryland, College Park, MD 20740, USA
Department of Biology, Indiana University, 915 E. Third Street, Myers Hall 150, Bloomington, IN 47405-7107, USA

Alexandra E Bely
Department of Biology, University of Maryland, College Park, MD 20740, USA

Elizabeth A Williams
Max Planck Institute for Developmental Biology, Spemannstrasse 35, Tübingen 72076, Germany

Markus Conzelmann
Max Planck Institute for Developmental Biology, Spemannstrasse 35, Tübingen 72076, Germany

Gáspár Jékely
Max Planck Institute for Developmental Biology, Spemannstrasse 35, Tübingen 72076, Germany

Yung-Che Tseng
Department of Life Science, National Taiwan Normal University, Taipei City, Taiwan

Sian-Tai Liu
Department of Life Science, National Taiwan Normal University, Taipei City, Taiwan

Marian Y Hu
Institute of Cellular and Organismic Biology, Academia Sinica, Nankang, Taipei City, Taiwan

Ruo-Dong Chen
Institute of Cellular and Organismic Biology, Academia Sinica, Nankang, Taipei City, Taiwan

Jay-Ron Lee
Institute of Cellular and Organismic Biology, Academia Sinica, Nankang, Taipei City, Taiwan

Pung-Pung Hwang
Institute of Cellular and Organismic Biology, Academia Sinica, Nankang, Taipei City, Taiwan

Reinhold Hustert
Department of Neurobiology, JFB-Institute for Zoology, University of Göttingen, Berliner Straße 28, 37073 Göttingen, Germany

Matthias Frisch
Department of Neurobiology, JFB-Institute for Zoology, University of Göttingen, Berliner Straße 28, 37073 Göttingen, Germany

Alexander Böhm
Department of Integrative Zoology, University of Vienna, Althanstraße 14, 1090 Vienna, Austria

Günther Pass
Department of Integrative Zoology, University of Vienna, Althanstraße 14, 1090 Vienna, Austria

www.ingramcontent.com/pod-product-compliance
Lightning Source LLC
Chambersburg PA
CBHW050443200326

41458CB00014B/5047